Representation Learning for Natural Language Processing

Zhiyuan Liu • Yankai Lin • Maosong Sun
Editors

Representation Learning for Natural Language Processing

Second Edition

 Springer

Editors
Zhiyuan Liu
Department of Computer Science
and Technology
Tsinghua University
Beijing, China

Yankai Lin
Gaoling School of Artificial Intelligence
Renmin University of China
Beijing, China

Maosong Sun
Department of Computer Science
and Technology
Tsinghua University
Beijing, China

This work was supported by Tsinghua University

ISBN 978-981-99-1602-3 ISBN 978-981-99-1600-9 (eBook)
https://doi.org/10.1007/978-981-99-1600-9

This Springer imprint is published by the registered company Springer Nature Singapore Pte Ltd.
The registered company address is: 152 Beach Road, #21-01/04 Gateway East, Singapore 189721, Singapore

Paper in this product is recyclable.

Preface

In conventional natural language processing (NLP) systems, language items such as words and phrases are handled as distinct symbols. Many classical methods, such as n-gram and bag-of-words models, were proposed and have been widely used until now. All these methods take words as the minimum units for semantic representation, either used to estimate the conditional probabilities of the next word given previous words (e.g., n-gram) or used to represent semantic meanings of text (e.g., bag-of-words models). Even when people find it necessary to model word meanings, they either manually build linguistic knowledge bases such as WordNet or use context words to represent word meaning (i.e., distributional representation). All these semantic representation methods are still based on symbols!

With the development of NLP techniques for years, it has been realized that symbolic representation has caused many issues in NLP. First, symbolic representation always suffers from the data sparsity problem. Take statistical NLP methods, such as n-gram with large-scale corpora, for example. Due to the intrinsic power-law distribution of words, the performance will decay dramatically for those few-shot words, even after many smoothing methods have been developed to calibrate the estimated probabilities.

There are also multiple-grained items in natural languages ranging from words, phrases, and sentences to documents. It is impossible to find a unified symbol set to represent the semantic meanings for all of them. Moreover, many NLP tasks require semantic relatedness between language items at different levels. For example, we have to measure semantic relatedness between words/phrases and documents in Information Retrieval. Due to the absence of a unified scheme for semantic representation, distinct approaches were proposed for different tasks in NLP, which sometimes makes NLP seem not a compatible community. More than that, a deep understanding of natural language requires rich human knowledge, and it is non-trivial for symbolic representation in NLP to identify and incorporate sophisticated external knowledge.

As an alternative approach to symbolic representation, distributed representation was initially proposed by Geoffrey E. Hinton in a technique report in 1984. The report was then included in the well-known two-volume book *Parallel Distributed*

Processing (PDP) that introduced neural networks to model human cognition and intelligence. According to this report, distributed representation is inspired by the neural computation scheme of humans and other animals, and the essential idea is as follows:

> Each entity is represented by a pattern of activity distributed over many computing elements, and each computing element is involved in representing many different entities.

It means that a language item will be represented by multiple neurons, and each neuron is involved in representing many items. It also indicates the meaning of *distributed* in distributed representation. In contrast to distributed representation, another view assumes one neuron corresponds to a specific object. That is, there may be an individual neuron only activated by one specific object, such as someone's grandmother, known as the grandmother-cell hypothesis or local representation. We can see the direct connection between the grandmother-cell hypothesis and symbolic representation.

About 20 years after distributed representation was initiated, Yoshua Bengio presented the neural probabilistic language model for natural languages in 2003, where words are represented as low-dimensional and real-valued vectors based on the idea of distributed representation. However, it was until 2013 that a simpler and more efficient framework *word2vec* was proposed to learn word distributed representations from large-scale corpora. We come to the popularity of distributed representation and neural network techniques in NLP. The recent 10 years witnessed significant improvement in almost all NLP tasks with the support of distributed representation, especially with the development of the neural architecture Transformer in 2017 and pre-trained language models after 2018.

This book aims to overview recent advances in distributed representation learning for NLP. It covers how representation learning takes part in various topics related to NLP, paying particular attention to why representation learning can improve NLP. As an exciting and active research area, we also focus on what remaining challenges are still not well addressed by distributed representation.

Book Organization

This book is organized into 14 chapters with 4 parts. The first part depicts key components in NLP and how representation learning works for them. This part includes (1) the basics of representation learning and why it is vital for NLP (Chap. 1); (2) a comprehensive review of representation learning techniques on multiple-grained entries in NLP, including word representation (Chap. 2), phrase representation as known as compositional semantics (Chap. 3), and sentence and document representation (Chap. 4); (3) the most advanced techniques in recent years, pre-trained models (Chap. 5).

The second part presents representation learning for those topics closely related to NLP. This part includes (1) graph representation handling structured information around natural languages such as relations of either sentences or documents (Chap. 6); (2) cross-modal representation connecting natural languages to other modalities such as visual data (Chap. 7); (3) robust representation talking about the vulnerability of distributed representation to backdoor attack, adversarial attack, and out-of-distribution data (Chap. 8).

A deep understanding of natural languages requires the support of rich human languages. The third part is about knowledge representation and the framework of knowledge-guided NLP. This part includes (1) a general introduction to knowledge representation with entity-based world knowledge as the example (Chap. 9); (2) linguistic and commonsense knowledge representation with the form of sememes, which are defined as the minimum semantic units in human languages (Chap. 10); (3) domain knowledge representation, taking legal (Chap. 11) and biomedical (Chap. 12) domains as examples.

In the fourth part, we take an open-resource platform as an example to introduce big model systems for representation learning, including pre-training, fine-tuning, model compression, and inference (Chap. 13). Finally, we outlook the future research directions of representation learning for NLP by summarizing ten key problems of pre-trained models (Chap. 14).

Although the book is about representation learning for NLP, those theories and algorithms can also be applied in other related domains, such as machine learning, social network analysis, semantic Web, information retrieval, data mining, and computational biology.

Book Cover

We designed the book cover to correspond to three revolutionized stages of cognition and representation in human history. It is an oracle bone divided into three parts.

The left part shows oracle scripts, the earliest known form of Chinese writing characters used on oracle bones in the late 1200 BC. It represents the emergence of human languages, especially writing systems. We regard this as the first representation revolution of human beings about the world, i.e., **representation symbolization**. It makes information and knowledge transmitted from person to person and from generation to generation.

The upper right part shows the digitalized representation of information and signals. Since the invention of electronic computers in the 1940s, big data and knowledge can be efficiently represented and processed in computer programs. We regard this as the second representation revolution of human beings about the world, i.e., **representation digitalization**. It enables large-scale information and knowledge to be stored, accumulated, searched, and utilized with computer systems.

The bottom right part shows the distributed representation in artificial neural networks initiated in the 1980s. As the representation basis of deep learning, it has extensively revolutionized many fields in artificial intelligence, including NLP, CV, and speech recognition, since the 2010s. We regard this as the third representation revolution of human beings about the world, i.e., **representation intellectualization**. It enables sophisticated knowledge to be effectively acquired, represented, and utilized in AI systems. This book focuses on the theory, methods, and applications of distributed representation learning in natural language processing.

Note for the Second Edition

Our team has studied representation learning in NLP since 2014, starting with word representation, graph representation, and knowledge representation. As shown throughout the 2010s, distributed representation and deep learning have brought a significant paradigm shift to NLP. This book aims to summarize and showcase critical advances in representation learning in NLP from our point of view, also consisting of related works from our team in those directions. The book's first edition was published in 2020, and its preparation can be traced back to 2016; it is a team work with many efforts of the professors, graduate students, and research assistants, as listed in the Acknowledgments of the 2020 edition.

We have witnessed the great success of Transformer and pre-trained models in recent years, which further enhances our knowledge and understanding of representation learning. We regard pre-training-fine-tuning as another paradigm shift to NLP after deep learning. Hence, we prepare the second edition to reflect these critical changes and advances. There are the following critical modifications as compared to the first edition:

1. **Book Organization.** From our experience organizing the 2020 edition, we realize limited persons cannot master all detailed advances of each direction of representation learning in NLP. In this edition, we invite young researchers and senior graduate students to work with us together on writing or updating specific chapters. All of them have rich research experiences on the topics of their participating chapters. We work together and frequently discuss to ensure all chapters are consistent with each other, following the same goal as elaborated in Chap. 1. We also work together to summarize the key problems of pre-trained models as an outlook to representation learning, as shown in Chap. 14.

   We also optimize writing guidelines and improve writing styles in many aspects. For example, in the 2020 edition, we sometimes tended to emphasize those related works from our team, which make the structure of some parts not so fluent and coherent. In this edition, we comprehensively revise these aspects with one goal, i.e., to thoroughly introduce the key advances of representation learning and help readers grasp the development trends from the past and the present to the future.

2. **Supplements and Updates.** We supplement five new chapters for pre-trained models and other emerging topics. For the recent advances in pre-trained models, we introduce the general knowledge about pre-trained models in Chap. 5 and take

an open-source platform OpenBMB as an example to introduce the programming details of training, tuning, compressing, and inference of big models in Chap. 13. We also add three new chapters on new topics of representation learning, including robustness (Chap. 8) and new knowledge types of the legal domain (Chap. 11) and the biomedical domain (Chap. 12).

Pre-training techniques have a revolutionary impact on almost all areas. With the new perspective, we also have a deeper understanding of representation learning. Hence we update all remaining chapters by either re-organizing the chapter structure (Chaps. 2, 3, 4 and 9) or providing recent advances (Chaps. 6, 7 and 10). Moreover, we merge document representation into sentence representation to better demonstrate the advances in neural language models. Based on the updates of all chapters, we improve the general introduction to representation learning and NLP in Chap. 1, such as providing clues about the intellectual origins of distributed representation.

3. **Corrections.** The first edition of the book was prepared for more than four years from 2016 to 2020, which could have not been published without the contributions of so many students and contributors in our team as indicated in the Acknowledgements. Meanwhile, even after several rounds of revision and proofreading before publication, we still find some inconsistent expressions and equations when integrating chapters, unintentional overlaps with others' articles introduced from the initial draft prepared by a research assistant, and other mistakes in the published version. We sincerely apologize for these mistakes. In this new edition, we have updated all contents and corrected all issues we found. In the future, we will also try our best to make these cases never happen again by writing, proofreading, and applying duplicate detection more carefully to each material released by our team.

Prerequisites

This book is designed for advanced undergraduate and graduate students, post-doctoral fellows, researchers, lecturers, industrial engineers, and anyone interested in representation learning, NLP, knowledge engineering, and AI. This is not a textbook, and we don't introduce basic knowledge. We expect the readers to have prior knowledge of Probability, Linear Algebra, and Machine Learning.

We recommend that readers interested in NLP read the first part (Chaps. 1–5) in sequence. The remaining parts can be read according to readers' interests. Many areas of representation learning are still evolving quickly. Hence, we list some books, reviews, and resources at the end of each chapter for readers to learn more about the topics.

Contact Information

In this book, we try our best to summarize the advances of representation learning in NLP. When preparing the 2020 edition and this edition, we grow to acknowledge that we cannot be experts in each aspect of this area. We may unintentionally conduct wrong summarizations, omit critical works, make incorrect predictions, or introduce other flaws. We are always expecting feedback, corrections, and suggestions on the book from readers, and improving the book and our research accordingly. The messages may be sent to liuzy@tsinghua.edu.cn or raised as issues on the following GitHub repository,

https://github.com/thunlp/Book_RL4NLP

We will keep updating and supplementing the book on the GitHub repository, where readers can find up-to-date news, errata, and updates about the book.

Beijing, China   Zhiyuan Liu
January, 2023   Yankai Lin
                                                                    Maosong Sun

Acknowledgments

Acknowledgments for the Second Edition

In the second edition, we list the contributors at the end of each chapter respectively. Here we express our thanks to those who have offered help and support to the whole book. We thank Chaojun Xiao for unifying the styles of figures, thank Shengding Hu for making the notation table, and unifying the notations and organizing references across chapters, and thank Zhenning Dai for making the acronym table. We thank Ms. Ruiqi Shao for designing the new book cover.

We also thank the Springer senior editor Dr. Celine Lanlan Chang, for making the second edition of the book happen and for enlightening suggestions on improvements to the first edition.

Acknowledgments for the First Edition

The authors are very grateful to the contributions of our students and research collaborators who have prepared initial drafts of some chapters or have given us comments, suggestions, and corrections. We list main contributors for preparing initial drafts of each chapter as follows:

- Chapter 1: Tianyu Gao, Zhiyuan Liu.
- Chapter 2: Lei Xu, Yankai Lin.
- Chapter 3: Yankai Lin, Yang Liu.
- Chapter 4: Yankai Lin, Zhengyan Zhang, Cunchao Tu, Hongyin Luo.
- Chapter 5: Yankai Lin, Zhenghao Liu, Haozhe Ji.
- Chapter 6: Fanchao Qi, Chenghao Yang.
- Chapter 7: Ruobing Xie, Xu Han.
- Chapter 8: Cheng Yang, Jie Zhou, Zhengyan Zhang.
- Chapter 9: Ji Xin, Yuan Yao, Deming Ye, Hao Zhu.

- Chapter 10: Xu Han, Zhengyan Zhang, Cheng Yang.
- Chapter 11: Cheng Yang, Zhiyuan Liu.

For the whole book, we thank Chaojun Xiao and Zhengyan Zhang for drawing model figures, thank Chaojun Xiao for unifying the styles of figures and tables in the book, thank Shengding Hu for making the notation table and unifying the notations across chapters, thank Jingcheng Yuzhi and Chaojun Xiao for organizing the format of reference, thank Jingcheng Yuzhi, Jiaju Du, Haozhe Ji, Sicong Ouyang, and Ayana for the first-round proofreading, and thank Weize Chen, Ganqu Cui, Bowen Dong, Tianyu Gao, Xu Han, Zhenghao Liu, Fanchao Qi, Guangxuan Xiao, Cheng Yang, Yuan Yao, Shi Yu, Yuan Zang, Zhengyan Zhang, Haoxi Zhong, and Jie Zhou for the second-round proofreading. We also thank Cuncun Zhao for designing the book cover.

In this book, there is a specific chapter talking about sememe knowledge representation. Many works in this chapter are carried out by our research group. These works have received great encouragement from the inventor of HowNet, Mr. Zhendong Dong, who died at 82 on February 28, 2019. HowNet is the great linguistic and commonsense knowledge base composed by Mr. Dong for about 30 years. At the end of his life, he and his son Mr. Qiang Dong decided to collaborate with us and released the open-source version of HowNet, OpenHowNet. As a pioneer of machine translation in China, Mr. Zhendong Dong devoted his whole life to natural language processing. He will be missed by all of us forever.

We thank our colleagues and friends, Yang Liu and Juanzi Li at Tsinghua University, and Peng Li at Tencent Wechat, who offered close and frequent discussions which substantially improve this book. We also want to express our special thanks to Prof. Bo Zhang. His insights to deep learning and representation learning, and sincere encouragements to our research of representation learning on NLP, have greatly stimulated us to move forward with more confidence and passion.

We proposed the plan of this book in 2015 after discussing it with the Springer senior editor Dr. Celine Lanlan Chang. As the first of the time of preparing a technique book, we were not expecting it took so long to finish this book. We thank Celine for providing insightful comments and incredible patience to the preparation of this book. We are also grateful to Springer's Assistant Editor Jane Li, for offering invaluable help during manuscript preparation.

Finally, we give our appreciations to our organizations, Department of Computer Science and Technology at Tsinghua University, Institute for Artificial Intelligence at Tsinghua University, Beijing Academy of Artificial Intelligence (BAAI), Chinese Information Processing Society of China, and Tencent Wechat, who have provided outstanding environment, supports, and facilities for preparing this book.

Contents

Contributors

Ganqu Cui Department of Computer Science and Technology, Tsinghua University, Beijing, China

Ning Ding Department of Computer Science and Technology, Tsinghua University, Beijing, China

Xu Han Department of Computer Science and Technology, Tsinghua University, Beijing, China

Shengding Hu Department of Computer Science and Technology, Tsinghua University, Beijing, China

Yankai Lin Gaoling School of Artificial Intelligence, Renmin University of China, Beijing, China

Zhiyuan Liu Department of Computer Science and Technology, Tsinghua University, Beijing, China

Yujia Qin Department of Computer Science and Technology, Tsinghua University, Beijing, China

Maosong Sun Department of Computer Science and Technology, Tsinghua University, Beijing, China

Chaojun Xiao Department of Computer Science and Technology, Tsinghua University, Beijing, China

Cheng Yang School of Computer Science, Beijing University of Posts and Telecommunications, Beijing, China

Yuan Yao Department of Computer Science and Technology, Tsinghua University, Beijing, China

Zheni Zeng Department of Computer Science and Technology, Tsinghua University, Beijing, China

Guoyang Zeng ModelBest Inc. and OpenBMB, Beijing, China

Zhengyan Zhang Department of Computer Science and Technology, Tsinghua University, Beijing, China

Acronyms

AI	Artificial Intelligence
API	Application Programming Interface
BERT	Bidirectional Encoder Representations from Transformers
BiLSTM	Bidirectional Long Short-Term Memory
BIO	The Begin-Inside-Outside tagging format for sequence tagging
BOW	Bag-of-Words
CBOW	Continuous Bag-of-Words
CNN	Convolutional Neural Network
COT	Chain-of-Thought
CPM	Chinese Pre-trained Model introduced by BAAI and Tsinghua
CPU	Central Processing Unit
CRF	Conditional Random Field
CV	Computer Vision
DNA	Deoxyribonucleic Acid
EM	Expectation-Maximization Algorithm
FNN	Feed-Forward Neural Network
GCN	Graph Convolutional Network
GPT	Generative Pre-trained Model
GPU	Graph Processing Unit
GRU	Gated Recurrent Unit
HCI	Human-Computer Interaction
HMM	Hidden Markov Model
IDF	Inverse Document Frequency
IR	Information Extraction
KB	Knowledge Base
KG	Knowledge Graph
KRL	Knowledge Representation Learning
LDA	Latent Dirichlet Allocation
LM	Language Model
LSTM	Long Short-Term Memory
MLM	Mask Language Modeling

MLP	Multi-layer Perceptron
MoE	Mixture-of-Experts
MT	Machine Translation
NER	Named Entity Recognition
NLI	Natural Language Inference
NLP	Natural Language Processing
NPLM	Neural Probabilistic Language Model
ODE	Ordinary Differential Equations
OFAI	Old-Fashioned AI
PDP	Parallel Distributed Processing
PLM	Pre-trained Language Model
PMI	Pointwise Mutual Information
POS	Part-of-Speech
PTM	Pre-trained Model
QA	Question Answering
RE	Relation Extraction
ReLU	Rectified Linear Unit activation function
RNA	Ribonucleic Acid
RNN	Recurrent Neural Network
SVD	Singular Value Decomposition
SVM	Support Vector Machine
TF-IDF	Term Frequency-Inverse Document Frequency
VQA	Visual Question Answering

Symbols and Notations

$\odot$	element-wise multiplication
$\otimes$	Kronecker product of two matrices
$*$	convolution operator
$\wedge$	logical and
$\leftarrow$	update the value of left-hand side value with right-hand side value
$\propto$	proportional to
∇	gradient of a function
$\frac{\partial f}{\partial x}$	partial derivative of a function
$\top$	transpose
$\lVert \cdot \rVert_p$	p norm of a vector
$\lvert \cdot \rvert$	size of a set
argmax	find the arguments that give the maximum value from a target function
argmin	find the arguments that give the minimum value from a target function
lim	limitation
$\mathcal{L}(\cdot)$	loss function
$O(\cdot)$	complexity of an algogrithm
$s(\cdot)$	similarity, especially cosine similarity
$\exp(\cdot)$	exponential function
$\mathrm{ATT}(\cdot)$	self-attention layer
$\mathrm{concat}(\cdot; \ldots; \cdot)$	concatenate a list of vectors/tensors
$\mathrm{LN}(\cdot)$	layer norm layer in transformers
$\mathrm{FFN}(\cdot)$	feed-forward network in transformers
$\mathrm{MLP}(\cdot)$	multilayer perceptron
$\mathrm{Softmax}(\cdot)$	softmax function
$\mathrm{Sigmoid}(\cdot)$	sigmoid function
$\tanh(\cdot)$	hyperbolic tangent function
$\mathcal{N}(x; \mu, \Sigma)$	normal distribution
$\mathbf{h}$	hidden representation

$\mathbf{W};\mathbf{M}$	weight matrix
Θ	model parameter
P	probability
$\mathcal{K}$	external knowledge
$\mathcal{R}$	relationship rule
$\mathbb{R}$	set of real numbers
$\mathbb{Z}_+$	set of positive integers
[CLS]	a special token to aggregate semantics in PTMs
[MASK]	a special token for the masked tokens in PTMs

Chapter 1
Representation Learning and NLP

Zhiyuan Liu and Maosong Sun

Abstract Natural language processing (NLP) aims to build linguistic-specific programs for machines to understand and use human languages. Conventional NLP methods heavily rely on feature engineering to constitute semantic representations of text, requiring careful design and considerable expertise. Meanwhile, representation learning aims to automatically build informative representations of raw data for further application and achieves significant success in recent years. This chapter presents a brief introduction to representation learning, including its motivation, history, intellectual origins, and recent advances in both machine learning and NLP.

1.1 Motivation

Machine learning addresses the problem of automatically learning computer programs from data. A typical machine learning system consists of three components [13]:

$$\text{Machine Learning} = \text{Representation} + \text{Objective} + \text{Optimization}.$$

We first transform helpful information from raw data into internal representations such as feature vectors to build an effective machine learning system. Afterward, by designing appropriate objective functions, we can employ optimization algorithms to find the optimal parameter settings for the system.

Data representation methods determine what and how valuable information can be extracted from raw data for further classification or prediction. If more information is transformed from raw data to feature representations, the performance of classification or prediction will be better. Hence, data representation is a crucial component of supporting effective machine learning.

Z. Liu (✉) · M. Sun
Department of Computer Science and Technology, Tsinghua University, Beijing, China
e-mail: liuzy@tsinghua.edu.cn; sms@tsinghua.edu.cn

Z. Liu et al. (eds.), *Representation Learning for Natural Language Processing*,
https://doi.org/10.1007/978-981-99-1600-9_1

Conventional machine learning systems adopt careful feature engineering as preprocessing to build feature representations from raw data. Feature engineering needs careful design and considerable expertise. A specific task usually requires customized algorithms for feature engineering, which makes the process labor-intensive, time-consuming, and inflexible.

Representation learning aims to learn informative representations of objects from raw data automatically. The learned representations can be further fed as input to machine learning systems for prediction or classification. This way, machine learning algorithms will be more flexible and desirable while handling large-scale and noisy unstructured data, such as speech, images, videos, time series, and texts.

Deep learning [22] is a typical approach for representation learning, which has recently achieved great success in speech recognition, computer vision (CV), and natural language processing (NLP). Deep learning has two distinguishing features:

Distributed Representation Deep learning algorithms represent each object with a low-dimensional and real-valued dense vector. The representation form is usually named as *distributed representation* or *embedding*. Compared to conventional symbolic representation, distributed representation is more compact and smooth by mapping data in the low-dimensional and continuous space, as shown in Fig. 1.1. Hence, it is more robust to address the sparsity issue that is ubiquitous and inevitable due to the power-law distribution in large-scale data.

Deep Architecture Deep learning algorithms usually learn a *deep hierarchical architecture* to represent objects, known as multilayer neural networks. Deep architecture may capture informative features and complicated patterns of objects from raw data. Take the sentence "you are a night owl." For example, as illustrated

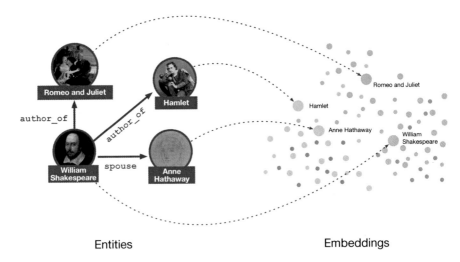

Entities   Embeddings

Fig. 1.1 Distributed representation of words and entities in human languages. (The images are obtained from wikimedia.org.)

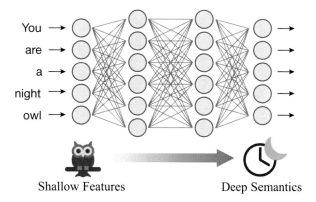

Fig. 1.2 Deep architecture enables representation learning to capture informative features and complicated patterns of human languages. Icons in this figure are bought or freely downloaded from IconFinder (https://www.iconfinder.com/)

in Fig. 1.2, the deep architecture of neural networks will be able to understand the deep semantics of the sentence, indicating a person stays up late using a metaphor, beyond the surface and shallow meanings. Hence, it is regarded as an important reason for the great success of deep learning for speech recognition, CV, and NLP.

The success of deep learning happens in speech recognition and CV first in around the 2010s, and in the following years, NLP also achieves significant improvements by following the deep learning approach. At the beginning of the revolution, deep learning for NLP significantly reduced feature engineering in NLP. In recent years, with the development of pre-trained language model techniques [17] in deep learning, the performance of almost all NLP tasks has achieved consistent and groundbreaking improvements. Hence, a growing number of researchers have devoted to developing effective deep learning methods for NLP.As per standard style, a footnote is not in the figure caption. So Footnote 1 has been moved to the corresponding citation, and the remaining footnotes are renumbered accordingly. Please check if okay.

In this chapter, we will first discuss why representation learning is essential for NLP and briefly review the development history and intellectual origins of representation learning for NLP. After that, we will introduce typical approaches of contemporary representation learning and summarize existing and potential applications of representation learning. Finally, we will introduce the general organization of this book.

1.2 Why Representation Learning Is Important for NLP

NLP aims to build linguistic-specific programs for machines to understand and use languages. Natural language texts are typically unstructured data with multiple granularities used in multiple domains. A deep understanding of natural languages also requires considerable human knowledge. There are multiple NLP tasks addressing different goals of NLP. These characteristics make NLP challenging to achieve satisfactory performance.

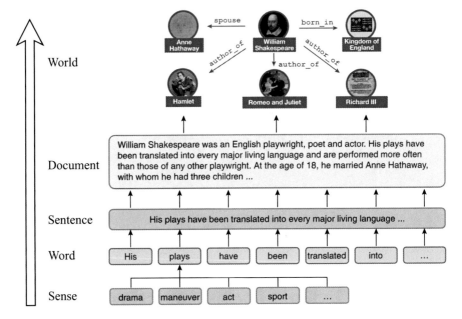

Fig. 1.3 There are multiple-grained language items, including characters, senses, words, phrases, sentences, paragraphs, documents, World Wide Web, as well as external world knowledge, with complicated compositional semantics. This figure shows some of them composed with each other. (The images are obtained from wikimedia.org.)

1.2.1 Multiple Granularities

NLP is concerned about multiple levels of language items, including but not limited to characters, senses, words, phrases, sentences, paragraphs, and documents. As shown in Fig. 1.3, each word may be composed of multiple senses, a sentence is composed of multiple words, and a document is composed of multiple sentences. The World Wide Web is even composed of billions of documents linked to each other, which is not shown in the figure. Moreover, human languages connect with the physical world, which may be described as world knowledge.

These language items are composed of each other following complicated patterns of compositional semantics. NLP is about understanding and processing these language items, and the key challenge is to model the complicated composition patterns. Representation learning is able to represent the semantic meanings of all language items in a unified semantic space. This significantly contributes to model complex semantic relations among these language items.

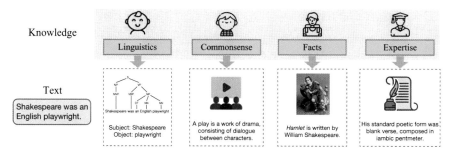

Fig. 1.4 Deep understanding of natural languages requires the support of multiple external knowledge such as linguistic knowledge, commonsense knowledge, world knowledge, and domain knowledge. Icons in the figure are bought or freely downloaded from IconFinder

1.2.2 Multiple Knowledge

A deep understanding of natural languages requires external human knowledge such as linguistic, commonsense, world, cognitive, and domain knowledge. The types of knowledge will be introduced in detail in Chap. 9. People and machines with different knowledge will have different-level understandings of the text.

As shown in Fig. 1.4, let's take the sentence "Shakespeare was an English playwright," for example. With the support of linguistic knowledge, we can capture the subject, and the object from the sentence by parsing the syntactic structure. With the commonsense knowledge of *A play is a work of drama, consisting of dialogue between characters*, we know most of Shakespeare's plays consist of character dialogues. If some persons also have some factual knowledge about Shakespeare, such as *Hamlet is written by William Shakespeare*, we can infer that *Hamlet* is an English play. A person with expert knowledge of literature may further think about the poetic form of Shakespeare.

Knowledge should be provided as much as possible to make machines more intelligent. For this goal, people have built many knowledge bases of multiple types and organized them in different structured forms. However, it is difficult for symbolic text and knowledge to work together due to their diverse representation forms, which are usually remedied by additional engineering efforts such as entity linking and suffered from error propagation. Representation learning, in contrast, can easily incorporate multiple types of structured knowledge into NLP systems by encoding both sequential text and structured knowledge into unified embedding forms.

1.2.3 Multiple Tasks

Many NLP tasks have been proposed and studied based on the same input to meet the needs of different scenarios, aspects, and levels. Take the sentence in Fig. 1.5 for example. We can perform multiple tasks on the same sentence as follows:

- **Part-of-speech (POS) tagging** aims to classify each word in a text into corresponding part-of-speech types, such as nouns, verbs, adjectives, adverbs, and prepositions, based on its context. In this figure, we show the annotated part-of-speech tags (e.g., NNP, VBD) following *Penn Part of Speech Tags* [34].
- **Dependency parsing** is a language grammar to build syntactic relations between language items in a sentence. Here we show the binary dependencies of language items and the dependency types. It can identify complicated syntactic relations inside a sentence, which is important in statistical NLP.
- **Named entity recognition** aims to find named entities mentioned in the text with pre-defined classes such as person names, organizations, locations, and time expressions.
- **Entity linking** further links named entity mentions to corresponding entities in external knowledge graphs by resolving those entities with the same names. The task is important for grounding human language understanding with the real world.
- **Relation extraction** aims to find relations between two entities expressed by the sentence. Relation extraction is a core task of information extraction to acquire structured knowledge from unstructured text and complete large-scale knowledge graphs.
- **Question answering** is to read a text and find answers for a given question. The task is important for the service of user information acquisition beyond search engines.
- **Machine translation** automatically translates the sentence from one language to another language. Machine translation is a long-standing NLP task to break the language barrier among people all over the world.

Here we only show several NLP tasks, and there are many more tasks concerning different goals and specific languages. For example, since there are no natural space marks between words in the text of Chinese and Japanese, automatic word segmentation has been proposed for these languages.

It is evident that all NLP tasks rely on accurate understanding and representation of given text input. In this case, building a unified and learnable representation of an input for multiple tasks will be more efficient and robust: on the one hand, a better and unified text representation will help to promote all NLP tasks, and on the other hand, taking advantage of more learning signals from multitask learning may contribute to building better semantic representations of natural languages. Hence, representation learning can benefit from multitask learning and further promote the performance of multiple tasks.

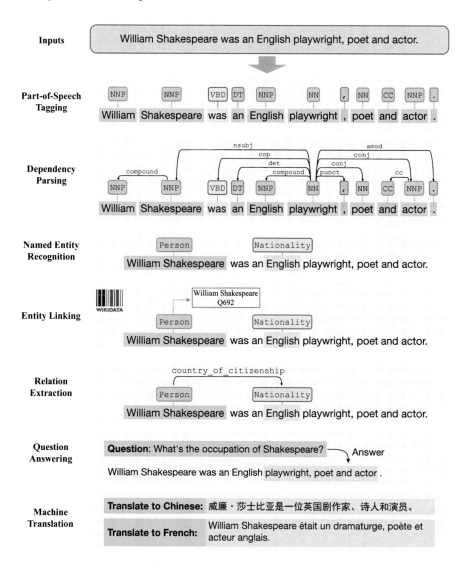

Fig. 1.5 There are various NLP tasks given the same sentence input, such as part-of-speech tagging, dependency parsing, named entity recognition, entity linking, relation extraction, question answering, and machine translation

1.2.4 Multiple Domains

Natural language texts may be generated from multiple domains, including but not limited to news articles, scientific articles, literary works, and online user-generated content such as product reviews. Moreover, we can also regard texts in different languages as multiple domains. Conventional NLP systems must design specific feature extraction algorithms for each domain according to its characteristics. In contrast, representation learning can take advantages of large-scale domain-specific data and can also transfer representation knowledge across multiple domains, especially from a much larger general domain to those specific domains.

1.3 Development of Representation Learning for NLP

We give a brief introduction to the development history of representation learning for NLP, from which we can see the paradigm shift of representation from symbolic representation to distributed representation, accompanied by the paradigm shift of machine learning from statistical learning to deep learning and further to pre-trained models. The development timeline is also shown in Fig. 1.6.

1.3.1 Symbolic Representation and Statistical Learning

Words would be a good start for studying representation schemes in NLP, because words are the minimum units in natural languages. The easiest way to represent a word in a computer-readable way (e.g., using a vector) is **one-hot vector**, which has the dimension of the vocabulary size and assigns 1 to the corresponding index of the word to be represented and 0 to others. It is apparent that one-hot vectors hardly contain semantic information about words other than distinguishing them from each other.

The idea of one-hot word representation can be further used for document representation, i.e., **bag-of-words (BOW) models** [18]. BOW models regard a document as a bag of its words, neglecting the orders of these words in this document. BOW represents a document as a vocabulary-size vector, with each word in the document corresponding to a nonzero dimension and other words to a zero dimension. The entry value of a word can be used to indicate the importance score of this word in the document, e.g., the number of its occurrences. BOW can be regarded as a combination of one-hot representations of all words in the document. BOW models are straightforward and work great in applications like spam filtering, text classification and clustering, and information retrieval. For example, in information retrieval, we build BOW vectors of a query and a document and compute the cosine distances as the semantic similarity for document ranking. Those documents that also attach importance to the important words in the query will be ranked higher. It proves that the distributions of words can serve as a good representation of documents.

One of the earliest ideas of word representation learning can date back to **n-gram models** [35]. It is easy to understand: when we want to predict the next word in a sentence, we usually look back at some previous words (and in the case of n-gram, they are the previous $n - 1$ words). And if going through a large-scale corpus, we can count and estimate a reasonable probability of a word under the

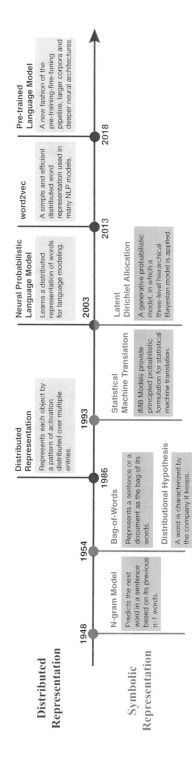

Fig. 1.6 The timeline for the development of representation learning in NLP. With the growing computing power and large-scale text data, distributed representation trained with neural networks from large corpora has become the mainstream

condition of all combinations of $n - 1$ previous words. These probabilities can predict word sequences and form vector representations for word meanings because similar words usually share similar probabilistic distributions of previous words.

The idea of n-gram models is coherent with the idea of **distributional hypothesis**: language items sharing similar distributions of context have similar meanings [18]. In another phrase, "a word is characterized by the company it keeps" [14]. The distributional hypothesis has been a fundamental idea of many NLP models, from n-gram in statistical NLP to word2vec and BERT in neural NLP.

In the above cases of one-hot word representation, BOW document representation, and n-gram models, each entry in the representation explicitly matches one language item (e.g., word scores in BOW models). This one-to-one correspondence between representation entries and language items is called **local representation** or **symbolic representation**.

The idea of symbolic representation is natural and straightforward, and most NLP algorithms in the stage of statistical learning in the 1980s–2000s are based on symbolic representation. Here we give another two iconic examples, IBM Model and latent Dirichlet allocation. IBM model [7] is a classical word alignment algorithm in statistical machine translation. It automatically builds lexical translation probabilities between the words of two languages from their parallel sentences, where words are regarded as symbolic items without considering their internal semantic representation. Latent Dirichlet allocation (LDA) [4] is a classical word and document representation algorithm. LDA builds latent topics to represent words and documents. These learned topics are typically interpretable, even capable of being labeled with symbolic names [29]. By regarding each latent topic as a meaningful symbol, we can also regard LDA as an example of symbolic representation, especially when learning with Gibbs Sampling [16] by iteratively assigning latent topics for each word in documents.

1.3.2 Distributed Representation and Deep Learning

Distributed representation, on the other hand, represents an object by a pattern of activation distributed over multiple entries, i.e., a low-dimensional and real-valued dense vector, and each computing entry can be involved in representing multiple objects [27]. Distributed representation has been proved to be more efficient because it usually has low dimensions. It also prevents from the sparsity issue that is inevitable for the symbolic representation due to the power-law distribution in large-scale data. Beneficial hidden properties can be learned from large-scale data and emerge in distributed representation.

Word embeddings can also learn complicated word relations automatically from large-scale corpora. As revealed by word2vec, we can identify the analogical properties of words such as

$$\mathbf{w}(\text{king}) - \mathbf{w}(\text{man}) \approx \mathbf{w}(\text{queen}) - \mathbf{w}(\text{woman}), \tag{1.1}$$

or

$$\mathbf{w}(\text{king}) - \mathbf{w}(\text{man}) + \mathbf{w}(\text{woman}) \approx \mathbf{w}(\text{queen}). \qquad (1.2)$$

It indicates the embeddings of both king and queen accurately encode similar semantic meanings with each other except for gender. The example shows the powerful capabilities of word embeddings for semantic representations.

The idea of distributed representation was initially inspired by the neural computing scheme of humans and other animals [20]. Brains can use various activation patterns of neurons to represent different objects. In distributed representation, the values of an entry in the low-dimensional vector can be regarded as the activation state of the specific neuron. It is named *distributed* because an object is represented as the activation pattern distributed over multiple neurons, and the activation state of one specific neuron does not mean anything.

With the great success of deep learning, distributed representation has become the most commonly used approach for representation learning. One of the pioneering practices of distributed representation in NLP is **neural probabilistic language model (NPLM)** [3]. A language model predicts the conditional probability of the next word given those previous words in a sentence. n-gram models can be regarded as simple language models based on symbolic representation. NPLM assigns a low-dimensional vector for each word (i.e., **word embedding**) and then uses a neural network to predict the next word based on distributed representations of previous words (i.e., context embedding, a combination of embeddings of previous words). By going through the training corpora, NPLM successfully learns word embeddings as model parameters to optimize the conditional probability of the next word or the joint probability of a sentence. Although it is hard to tell what each entry of a word embedding means, the vectors indeed encode semantic meanings about the words, verified by the performance of NPLM.

Inspired by NPLM, many methods have been proposed to learn word embeddings as model parameters optimized with language modeling objective, such as **word2vec** [30], **GloVe** [31], and **fastText** [6]. Although different in model and algorithm details, these methods are all very efficient for learning from large-scale corpora and have been widely used as word embeddings in many NLP models. Word embeddings can map discrete words into low-dimensional vectors as informative features in the NLP pipeline and help to shine a light on neural networks in computing languages. It makes representation learning a critical part of NLP.

1.3.3 Going Deeper and Larger with Pre-training on Big Data

The research on representation learning in NLP further takes a great leap by **ELMo** [32] and **BERT** [11]. These models apply larger corpora, more parameters, and more computing resources to build deeper and larger models. Moreover, they consider the complicated context of the text to learn richer knowledge of human languages. Instead of mapping a word to a fixed vector, ELMo and BERT use multilayer neural networks to build dynamic contextualized representations of each word based on its specific context in text, which is especially useful for those words with multiple meanings. Moreover, BERT starts a new fashion (although not originated from it) of the **pre-training-fine-tuning** pipeline. As shown in Fig. 1.7, previous word embeddings learned from large corpora were adopted as initialization of input representations of neural networks for downstream tasks; starting from BERT, it becomes a common practice to take the whole neural network structure such as BERT and all parameters pre-trained on large text corpora to downstream tasks, with those parameters further fine-tuned on supervised data of downstream tasks.

The models like BERT are pre-trained through language modeling objectives on large corpora, thus named as **pre-trained language models (PLM)**. PLMs take advantage of large-scale text corpora and have achieved state-of-the-art on almost all NLP benchmarks. Hence, although not a big theoretical breakthrough, PLMs have attracted wide attention in the NLP and machine learning community. Of course, PLMs reveal many distinct characteristics compared to conventional deep learning methods, such as parameter-efficient tuning capabilities [12] and in-context few-shot learning capabilities [8]. Some experiments of knowledge probing demonstrate that PLMs implicitly encode a variety of linguistic and world knowledge and patterns inside the multilayer neural network parameters [19, 24]. All these notable performances and interesting analyses suggest that there are a lot of open problems to explore in PLMs, as the future of representation learning for NLP.

In summary, representation learning for NLP has evolved from symbolic to distributed representation following the distributional hypothesis. Starting from word2vec, word embeddings learned from large corpora have shown outstanding performance in many NLP tasks. Recent PLMs learn complicated contextualized representations from large-scale text corpora and start the new paradigm of the pre-training-fine-tuning pipeline. Representation learning has revolutionized NLP in the past decades. What will be the next big breakthrough in representation learning for NLP? We hope this book can give some inspiration by introducing the evolutionary paths and most recent advances of representation learning for NLP.

1.4   Intellectual Origins of Distributed Representation

For this vital revolution in artificial intelligence (AI), one may be interested in the intellectual origins of the essential idea of distributed representation. To our knowledge, the exact term "distributed representation" was first proposed in parallel distributed processing (PDP) [27]. Still, the idea of distributed representation may have its prototypes in different areas. Here we try to find some clues between distributed representation and related areas, including neuroscience, AI, machine learning, and linguistics.

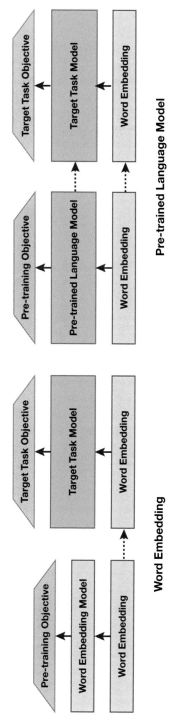

Fig. 1.7 This figure shows how word embeddings and pre-trained language models work in NLP pipelines. They both learn distributed representations for language items (e.g., words) through pre-training objectives and transfer them to target tasks. Furthermore, pre-trained language models can also transfer model parameters

1.4.1 Representation Debates in Cognitive Neuroscience

The most direct intellectual origin of distributed representation is **cognitive neuroscience**. A central topic in cognitive neuroscience is how information and knowledge are represented in human brains, and a long-standing debate is whether representation is *localized* or *distributed*. In the history of cognitive neuroscience, researchers used the terms *local* and *distributed* in different ways. For example, they may be used to describe the views that particular knowledge is either stored in specific brain regions or spread across the entire cortex [15].

Here we focus on the most well-known version from the classical book of parallel distributed processing (PDP) [27]. In this book, they raise two opposite representation schemes where the definition of local representation is

> Given a network of simple computing elements and some entities to be represented, the most straightforward scheme is to use one computing element for each Entity. This is called a *local* representation. (from [27])

and the definition of distributed representation is

> Each Entity is represented by a pattern of activity distributed over many computing elements, and each computing element is involved in representing many different entities. (from [27])

We can build a metaphor between one-hot representation in NLP and local representation in neuroscience, by regarding each entry in the vocabulary-size vector as a neuron in human brains, with the value 1 indicating active and the value 0 inactive. The view of local representation was also referred to as the famous *grandmother cell hypothesis*, which assumes a hypothetical neuron can encode and respond to a specific and complicated entity such as someone's grandmother [2]. From the view of one-hot representation, there will be an entry corresponding to someone's grandmother.

Local representation is straightforward because it is a simple mirror of the knowledge structure, with each object and concept corresponding to distinct neurons. Based on local representation, high-level knowledge can be organized into symbolic systems, such as concept hierarchy, propositional networks, semantic networks, schemas, frames, and scripts [1].

The above metaphor also works on distributed representation between NLP and neuroscience by regarding each vector entry as a neuron. The entry values indicate the status of neurons, active or inactive. The distributed representation scheme is not so straightforward, but it is already widely accepted that visual signals may activate millions of neurons throughout many regions of the visual cortex.

Although it is debatable whether individual neurons encode high-level concepts or objects, distributed representation seems to be a general solution for information processing at different levels, ranging from visual stimulus to high-level concepts. As discussed in PDP [27], distributed representation has better representation capacity, automatic generalization to novel entities, and learnability to changing environments. All these characteristics are valuable to our modern world with rich data for machine learning and have been verified in the recent revolution of deep learning.

1.4.2 Knowledge Representation in AI

An essential branch of philosophy is the theory of knowledge, also known as epistemology. Epistemology studies the nature, origin, and organization of human knowledge and concerns the problems like where knowledge is from and how knowledge is organized.

In philosophy, for the problem of **how knowledge is organized**, philosophers have developed many tools and methods, typically in the *symbolic* form, to describe human knowledge, and some of them, such as formal logic, have played an essential role in computer science. For the problem of **where knowledge is from**, there are two basic views: *rationalism* regards reason as the chief source of human knowledge and regards intellectual and deductive as truth criteria; *empiricism* generally appreciates the role of sensory experience in the generation of knowledge.

By building a metaphor between AI and humans, knowledge representation can be regarded as the epistemology for machines in AI. AI is also concerned about the above two problems, and many works have been done following the two lines.

For the problem of **how knowledge is organized** in AI, we can conclude two main approaches, *symbolism* and *connectionism*.

Symbolism aims to develop formal symbolic systems to organize knowledge of machine intelligence. Most pioneering works in AI follow the approach of symbolism, ranging from general problem solvers by Newell and Simon in 1959 to expert systems and knowledge bases by Ed Feigenbaum in the 1980s. Hence, some news articles on AI may imply symbolism as an obsolete approach, named old-fashioned AI (OFAI). With the rise of the Internet, WWW, and big data, remarkable works such as semantic Web by Tim Berners-Lee in the 2000s and recent large-scale Knowledge Graphs by Google in the 2010s can also be regarded as following the symbolism approach for knowledge representation.

Connectionism is the approach inspired by cognitive neuroscience, rooted in the works such as the perceptron by Frank Rosenblatt in the 1950s and parallel distributed processing (PDP) [27] in the 1980s. We usually regard deep learning in the 2010s as the great success of this approach, which has been almost 40 years since distributed representation was first proposed in the 1980s.

For the problem of **where knowledge is from** in AI, the approaches can be divided into *rationalism* and *empiricism*. The rationalism approach indicates that knowledge, including facts and rules, is directly designed or collected by human experts, i.e., the creators of AI agents; AI agents can complete tasks based on the given knowledge. The expert systems in the 1980s typically follow this approach. It is obvious that the manually organized knowledge is not flexible and dynamic, cannot generalize well to novel cases, and cannot evolve as the environment changes. With the development of the Internet and big data, the empiricism approach becomes feasible to learn from large-scale data, including unlabeled general data and labeled task-specific data. Statistical learning starting from the end of the 1980s and deep learning starting from the 2010s both follow the empiricism approach.

Epistemology of philosophy only developed formal and symbolic tools extensively. But for computational epistemology in AI, the approach of knowledge source and the approach of knowledge form have been studied in mixed ways in different periods of AI history: in the preliminary stage of the 1950s–1980s, most works followed a mix of symbolism and rationalism, also named as old-fashioned AI (OFAI); in statistical learning 1980s–2000s, the mainstream is a mix of symbolism and empiricism; and in deep learning from 2010s, it becomes the mix of connectionism and empiricism.

Note that, although *rationalism* and *empiricism* represent two distinct approaches for where knowledge is from, it does not mean they don't or cannot work together. On the one hand, machine learning models typically learn from data following the empiricism approach. On the other hand, the design of model architectures and algorithms involves the wisdom of human experts following the rationalism approach. It is the same to *symbolism* and *connectionism*. McCulloch and Pitts designed a computational scheme of symbolic logics using elementary units of neural systems in 1943 [28]; in the era of deep learning, neural-symbolic networks are also an active research area for reasoning and planning with neural networks [38].

It seems not feasible to explicitly mix rationalism and connectionism. However, as Noam Chomsky and many researchers indicated, human brains are not blank slates [9]. We can also regard the architectural design of neural networks as a kind of prior knowledge following the theory of rationalism. For humans and AI, we should not set up barriers between symbolism and connectionism or between rationalism and empiricism. All of them may play some roles in human and machine intelligence. *It doesn't matter whether a cat is black or white, as long as it catches mice.*

As we will show in this book, deep learning with distributed representation can manipulate symbols as well as other discrete information, such as instructions, operations, and codes (Chap. 5). We can also modify the architecture of neural networks given prior knowledge to fit downstream tasks better (Chap. 9). Hence, we believe distributed representation is a good foundation to take advantage of all promising approaches of knowledge sources and forms, with many open and interesting problems deserving further exploration.

1.4.3   Feature Engineering in Machine Learning

Feature engineering is a critical step in the pipeline of statistical learning, aiming to build feature vectors of instances for machine learning. It can be regarded as the representation learning of instances in the era of statistical learning. Feature engineering provides another intellectual origin of distributed representation, i.e., dimensionality reduction of raw data by mapping from a high-dimensional space into a low-dimensional space, usually with the term "embedding."

Feature engineering can be divided into feature selection and feature extraction. Feature selection techniques select the most informative features and remove redundant and irrelevant ones from large amounts of candidates to represent instances such as words and documents. This approach is expected to improve representation efficiency and remedy the curse of dimensionality. Many methods of feature selection and term weighting have been explored on specific NLP tasks such as text classification [36]. Since candidates for feature selection are usually symbols such as words, phrases, and n-grams, the selected feature vocabulary is also a symbolic representation.

Feature extraction aims to build a novel feature space from raw data, with each dimension of the feature space either interpretable or not. Latent topic models and dimensionality reduction can be regarded as representative approaches for feature extraction. Latent topic models represent each document and word as a distribution over latent topics, which can be regarded as interpretable space. Examples are probabilistic latent semantic analysis (pLSA) [21] and latent Dirichlet allocation (LDA) [4]. Dimensionality reduction methods learn to map objects into a low-dimensional and uninterpretable space. Examples are principal component analysis (PCA) and matrix factorization methods like singular value decomposition (SVD).

Note that the term **embedding** in machine learning refers to either the projection process (such as the algorithm locally linear embedding [33]) or the corresponding low-dimensional representation of objects. We can see that, without the metaphor between human brains and AI in deep learning and distributed representation, the idea of representing objects in a low-dimensional space has already been widely used in statistical learning. The representation scheme is the same between low-dimensional embedding and distributed representation in deep learning. Some neural networks, such as Autoencoder, used to be regarded as a dimensionality reduction method of data.

Hence, in recent years of deep learning and AI, the term *distributed representation* and *embedding* are used mutually to refer to each other. The difference is that the model architecture of most dimensionality reduction methods in statistical learning is usually shallow and straightforward and the algorithm is also restricted to specific data forms such as matrix decomposition. Those tasks that can easily organize data as matrix benefit much from these methods, such as recommender systems focusing on user-item interactions [23]. In contrast, the model architecture of deep learning is typically deep with multiple neural layers, capable of modeling complicated interactions and capturing sophisticated semantic compositions ubiquitous in human languages.

1.4.4 Linguistics

Human languages are regarded as the epitome of human intelligence, and linguistics aims to study the nature of human languages. Since human languages are regarded as one of the most complicated symbolic systems, linguistics typically follows the symbolism approach.

An influential theory of linguistics is *structuralism* derived from the founder of modern linguistics, Ferdinand de Saussure. Saussure proposed the following perspectives [10]: (1) a *symbol* (or sign) in human languages is composed of the *signified* (i.e., a concept in mind) and the *signifier* (i.e., a word token, or its sound or image). (2) For a symbol, "the bond between the signified and the signifier is arbitrary" [10]. For example, there is no *intrinsic* relationship between the concept of "sister" and the sound of the word "sister"; for another example, the words in different languages may refer to the same concept. (3) Hence, a symbol can only get its meaning from its relationship with other symbols. For example, the meaning of the word "parent" is related to the meaning of the corresponding word "child." In summary, the structuralism theory regards human languages as a symbolic system where each item is defined by its relationship to other items in the system [26].

The idea of distributed representation coincides in spirit with structuralism. By distributed representation learning, we can see that all language items we are interested in are projected into a unified low-dimensional semantic space. As demonstrated in Fig. 1.1, the geometric distance between two language items in the semantic space indicates their semantic relatedness; the semantic meaning of an item corresponds to its geometric relationships with other items, such as above-mentioned $\mathbf{w}(\text{queen}) \approx \mathbf{w}(\text{king}) - \mathbf{w}(\text{man}) + \mathbf{w}(\text{woman})$. In other words, the relative closeness with other items rather than its absolute position in the semantic space reveals an item's meaning.

Later, the structuralism theory evolved into a more computational version, *distributionalism*, arguing that the meanings of linguistic items are defined by their distribution in text corpora. The distributionalism is further developed into the *distributional hypothesis* formalized by American linguist Zellig Harris, arguing that language items sharing similar distributions of context have similar meanings [18]. The distributional hypothesis provides a computational way of following the empiricism approach to learning semantic representations of text from large-scale corpora, which is essential to distributed representation learning.

1.5 Representation Learning Approaches in NLP

In the history of AI, researchers have developed various effective and efficient approaches to learning semantic representations for NLP. Here we list some typical approaches.

1.5.1 Feature Engineering

As introduced above, semantic representations for NLP in the early stage often come from statistics instead of learning with optimization. Feature engineering is a typical approach to representation learning in statistical learning and can be divided into feature selection and feature extraction.

During the era of statistical learning, feature selection techniques have been extensively explored in NLP, focusing on selecting the most informative symbolic features because of the symbolic nature of human languages. For feature engineering of NLP, researchers should take care of issues such as feature set construction, feature weighting, and smoothing.

For statistical learning of various NLP tasks, we should determine what features should be considered. All syntactic and semantic features of language items, such as words and their part-of-speech (POS) tags, n-grams, word and entity types, semantic roles, and parse trees, may be helpful in specific NLP tasks. These linguistic features may be extracted by specific NLP systems or provided by given tasks. Even for the features of language items, how to select those most informative ones to form the feature set is also an important issue [36].

After the feature set is determined, measuring the feature weight for a specific instance is also essential. For example, in n-gram or bag-of-words models, entries in the representation are usually frequencies, occurrence numbers, or other weight scores of the corresponding language items counted in a given text or large-scale corpora. These feature scores indicate essential semantic characteristics of the given instance.

Moreover, the dimension of the feature space in statistical NLP is usually substantial, and the feature vector of a word or document correspondingly exhibits the sparsity issue, i.e., the curse of dimensionality in the context of NLP. To address the sparsity issue, besides dimensionality reduction techniques, researchers also developed smoothing techniques of semantic representation [37] by taking advantage of more context of the given document, such as its related documents.

In summary, in a long period before the era of distributed representation, researchers devoted lots of effort to manually designing, selecting, and weighing useful linguistic features and incorporating them as inputs of NLP models. The feature engineering pipeline heavily relies on human experts of specific tasks and domains, is thus time-consuming and labor-intensive, and cannot generalize well across objects, tasks, and domains.

1.5.2 Supervised Representation Learning

Distributed representations can emerge from the optimization of neural networks under supervised learning. In hidden layers of neural networks, the different activation patterns of neurons represent different objects. With a training objective (usually a loss function for the target task) and supervised signals (usually the gold-standard labels for training instances of the target task), the networks can learn to find better parameters for representing language items via optimization such as gradient descent. With proper training, the hidden states will become informative and generalized as good semantic representations of natural languages.

For example, to train a neural network for sentiment classification, the loss function is usually formalized as the cross entropy of model predictions considering gold-standard sentiment labels as supervision. By optimizing the objective with many supervised training instances, in company with the training loss getting smaller and the classification performance getting better, the model is expected to build better sentence representations as classification features.

1.5.3 Self-supervised Representation Learning

In many cases, we do not have human-labeled data for supervised learning. We need to find "labels" intrinsically from large-scale unlabeled data to acquire the training objective necessary for neural networks. The approach can be regarded as a mixed way between supervised and unsupervised learning, called self-supervised learning.

Language modeling is a typical self-supervised objective because it does not require human annotations. For the learning objective of predicting the next word given previous context words, we can effortlessly obtain the gold standard of the next words from large-scale corpora.

Another example of self-supervised representation learning is Autoencoder. An Autoencoder has a reduction (encoding) phase and a reconstruction (decoding) phase. The model will encode an object into a low-dimensional representation in the reduction phase and reconstruct the object from the intermediate representation in the reconstruction phase. Here the training objective is the reconstruction loss, taking the original data as the gold standard. During the training process, meaningful information will be encoded and kept in latent representations, and noisy or useless signals will be discarded.

The most advanced **pre-trained language models** combines the advantages of both self-supervised learning and supervised learning. In the **pre-training-fine-tuning** pipeline, pre-training can be regarded as self-supervised learning from large-scale unlabeled corpora, and fine-tuning is supervised learning with labeled task-specific data. Self-supervised learning has dramatically succeeded in NLP because the plain text contains abundant knowledge and patterns about languages. Self-supervised learning can effectively learn from almost infinite large-scale text corpora. Nowadays, it is still one of the most exciting research areas of representation learning for NLP, and a growing number of researchers are devoting their efforts to learning better pre-trained language models.

Besides, many other machine learning approaches have also been explored in representation learning for NLP, such as adversarial training, contrastive learning, few-shot learning, meta-learning, continual learning, and reinforcement learning. It is still an active research topic on developing effective and efficient representation learning methods for NLP from large-scale and complicated corpora and computing power.

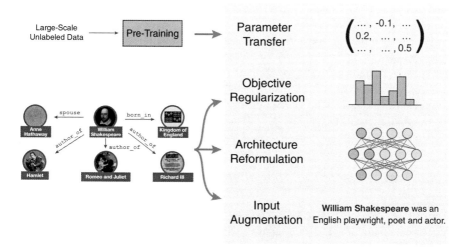

Fig. 1.8 The approaches of applying representation learning in NLP systems, including input augmentation, architecture reformulation, objective regularization, and parameter transfer. (The images are obtained from wikimedia.org.)

1.6   How to Apply Representation Learning to NLP

We summarize four typical approaches to applying representation learning of multiple objects to promote NLP systems, including input augmentation, architecture reformulation, objective regularization, and parameter transfer.

As shown in Fig. 1.8, a typical scenario is, if we have some structured knowledge, representation learning can help to incorporate the knowledge into various components of an NLP system, such as input, architecture, and objective. It also works with unstructured knowledge, such as word representations from large-scale text corpora. Moreover, the pre-training-fine-tuning pipeline in pre-trained language models offers parameter transfer to an NLP system.

1.6.1   Input Augmentation

The basic idea of input augmentation is to learn semantic representations of objects in advance. Then, object representations can be augmented as some parts of the input in downstream models. For example, word embeddings can be learned with language modeling from large-scale corpora and then used as input initialization for downstream NLP models. During the learning process of downstream models, we can either keep these word embeddings fixed and only tune other model parameters or tune all parameters considering word embeddings. There is no answer to which strategy is better. In practice, it should be determined by empirical comparison, influenced by many factors, such as the amount of supervised downstream data and the complexity of downstream tasks.

We can also introduce external knowledge related to input to augment inputs of downstream models. In this book, we will introduce world knowledge (Chap. 9), linguistic and commonsense knowledge (Chap. 10), and domain knowledge (Chaps. 11 and 12), whose representations can be learned based on either knowledge graphs or symbolic rules and then integrated into specific NLP systems as input augmentation for improving performance.

1.6.2 Architecture Reformulation

We can use objects (such as entity knowledge) and their distributed representations to restructure the architecture of neural networks for downstream tasks. For example, as we introduce in Chap. 10, we take sememe as the minimum indivisible unit of semantic meanings in human languages [5] and build linguistic and commonsense knowledge graphs with sememe-sense-word hierarchy. With the help of sememe knowledge, we can reformulate the next-word prediction task in neural language modeling into a pipeline of first predicting sememes of the next word, then predicting related senses, and finally predicting the next word. In this way, we make neural language models more interpretable and robust.

1.6.3 Objective Regularization

We can also apply object representations to regularize downstream model learning. As mentioned above, there are usually multiple language items in NLP tasks. Since all these items are mapped into a unified semantic space using representation learning, we can formalize various learning objectives to regularize model learning. For example, suppose we train neural language models from large-scale corpora and learn entity representations from a world knowledge graph. We can add a new learning objective of entity linking as regularization by minimizing the loss of predicting a mentioned entity in a sentence to the corresponding entity in the knowledge graph. With the help of more informative signals for learning, NLP models are expected to achieve better performance.

1.6.4 Parameter Transfer

The semantic composition and representation capabilities of language items such as sentences and documents lie in the weights within neural networks. We can directly transfer these pre-trained model parameters to downstream tasks in an end-to-end fashion. We have mentioned the approach of parameter transfer in the pre-training-fine-tuning pipeline of pre-trained language models. Most NLP tasks are at the levels of sentences and documents: the tasks like sentiment classification, natural language inference, machine translation, and relation extraction require sentence representation; the tasks like question answering and information retrieval require document representation. All these tasks can benefit from the capabilities of sentence or document representations from pre-trained language models on large-scale corpora. Moreover, many representation learning methods have been designed specifically for sentences and documents and benefit these NLP tasks, which will be introduced in the corresponding chapters of this book.

1.7 Advantages of Distributed Representation Learning

From the above brief introduction, we can summarize the following advantages of distributed representation learning for NLP.

Unified Representation Space As shown in Fig. 1.9, distributed representation can provide a unified representation scheme and space for natural languages. The unified scheme and space can facilitate knowledge transfer across multiple language items, multiple human knowledge, multiple NLP tasks, and multiple application domains, as discussed in Sect. 1.2, and significantly improve the effectiveness and robustness of NLP performance.

Learnable Representation The embeddings in distributed representation can be learned as a part of model parameters in supervised or self-supervised ways. It is the reason for the name "representation learning." Unlike previous feature-engineered

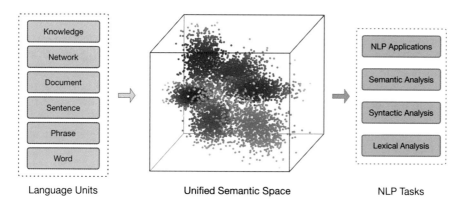

Language Units   Unified Semantic Space   NLP Tasks

Fig. 1.9 Distributed representation can provide unified semantic space for multiple language items and for multiple NLP tasks

representation methods, this enables distributed representation adaptable to NLP tasks by learning from task-specific data.

End-to-End Learning Feature engineering in the symbolic representation scheme usually consists of multiple learning components, such as feature selection and term weighting. These components are conducted step-by-step in a pipeline, which cannot be well optimized according to the ultimate goal of the given task. In contrast, distributed representation learning supports end-to-end learning via back-propagation across neural network hierarchies.

1.8 The Organization of This Book

The book focuses on the distributed representation scheme (i.e., embedding) in NLP and talks about recent advances in representation learning methods for (1) multiple language items including words, sentences, and documents; (2) closely related topics including graphs, cross-modality, and robustness; and (3) external knowledge including world knowledge, linguistic and commonsense knowledge, and domain knowledge.

We start the book from word representation. By giving a thorough introduction to word representation in Chap. 2, readers are expected to learn basic ideas of representation learning for NLP. After that, we introduce the techniques of sentence and document representation learning in Chap. 4, focusing on compositionally acquiring semantic representations of a higher-level language item from its components. We further introduce the most advanced techniques, pre-trained language models, in Chap. 5. After going through these chapters, readers will establish essential knowledge about deep learning techniques in NLP and realize the key to the deep learning revolution in NLP is distributed representation learning.

There are three essential and closely related topics for representation learning in NLP. First, the graph is also a natural way to represent objects and their relationships. In Chap. 6, we introduce representation learning techniques for modeling nodes, edges, and graphs; how graph representation learning can help NLP. Second, another important topic related to NLP is cross-modal representation learning. It studies how to build unified semantic representations across distinct modalities, such as texts, audios, images, and videos. In Chap. 7, we focus on the interaction between vision and text to introduce techniques and advances in cross-modal representation learning. Third, the robustness of semantic representations is critical for NLP applications, which will be introduced in Chap. 8.

In this book, we also argue that a deep understanding of natural languages requires the support of multiple human knowledge. Representation learning can incorporate external knowledge for NLP, known as knowledge-guided NLP, as shown in Fig. 1.10. Here, we introduce three typical forms of knowledge representation closely related to NLP: entity-based world knowledge, sememe-based linguistic and commonsense knowledge, and legal and biomedical domain knowledge.

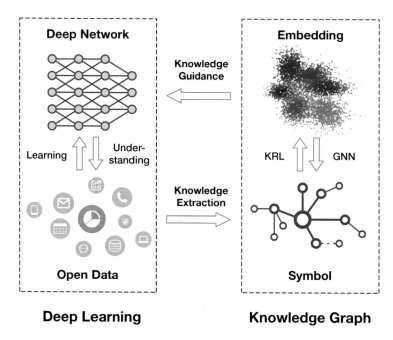

Fig. 1.10 The framework of knowledge representation learning (KRL), knowledge acquisition, and knowledge-guided NLP

In Chap. 9, we give a general introduction to knowledge representation learning (KRL) and take entity-based world knowledge as an example to introduce KRL methods and knowledge-guided NLP. World knowledge representation typically encodes world facts from knowledge graphs with entities and their relations into continuous semantic space. With world KRL, we can make NLP models knowledgeable of more information about those entities in text, such as rich attributes or relations with other entities.

Sememe representation encodes linguistic and commonsense knowledge of natural languages, where sememe is defined as the minimum indivisible unit of semantic meanings in human languages [5]. As shown in Chap. 10, with the help of sememe representation learning, we can get more interpretable and robust NLP models.

There is also rich and complicated domain knowledge along with large amounts of domain-specific texts. Domain knowledge is important for the accurate understanding of domain texts. In Chaps. 11 and 12, we take the legal and biomedical domains as examples to introduce how to represent domain knowledge of distinct forms and facilitate domain-specific NLP systems.

At the end of the book, we share some views about challenging topics in representation learning for NLP. We hope the outlook can inspire more readers to play a part in building more powerful representation learning for NLP and AI.

Acknowledgments We thank Tianyu Gao for preparing some initial draft materials for the first edition and thank Chaojun Xiao for drawing figures.

This is the introductory chapter of the second edition of the book *Representation Learning for Natural Language Processing*, with its first edition published in 2020 [25]. As compared to the first edition of this chapter, the main changes include the following: (1) we added the part Intellectual Origins of Distributed Representation, and (2) we comprehensively supplemented and updated the information, discussions, examples, and figures in other existing sections.

References

1. John Robert Anderson. *Cognitive psychology and its implications, seventh edition.* Worth Publishers, 2010.
2. H Barlow. Grandmother cells, symmetry, and invariance: how the term arose and what the facts suggest. *The cognitive neurosciences*, pages 309–320, 2009.
3. Yoshua Bengio, Réjean Ducharme, Pascal Vincent, and Christian Jauvin. A neural probabilistic language model. *Journal of Machine Learning Research*, 3, 2003.
4. David M Blei, Andrew Y Ng, and Michael I Jordan. Latent dirichlet allocation. *Journal of Machine Learning Research*, 3:993–1022, 2003.
5. Leonard Bloomfield. A set of postulates for the science of language. *Language*, 2(3):153–164, 1926.
6. Piotr Bojanowski, Edouard Grave, Armand Joulin, and Tomas Mikolov. Enriching word vectors with subword information. *Transactions of the Association for Computational Linguistics*, 2017.
7. Peter F. Brown, Vincent J. Della Pietra, Stephen A. Della Pietra, and Robert L. Mercer. The mathematics of statistical machine translation: Parameter estimation. *Computational Linguistics*, 19(2):263–311, 1993.
8. Tom Brown, Benjamin Mann, Nick Ryder, Melanie Subbiah, Jared D Kaplan, Prafulla Dhariwal, Arvind Neelakantan, Pranav Shyam, Girish Sastry, Amanda Askell, et al. Language models are few-shot learners. In *Proceedings of NeurIPS*, 2020.
9. Noam Chomsky et al. *Language and mind.* Cambridge University Press, 2006.
10. Ferdinand De Saussure. *Course in general linguistics.* Columbia University Press, 2011.
11. Jacob Devlin, Ming-Wei Chang, Kenton Lee, and Kristina Toutanova. BERT: pre-training of deep bidirectional transformers for language understanding. In *Proceedings of NAACL-HLT*, 2019.
12. Ning Ding, Yujia Qin, Guang Yang, Fuchao Wei, Zonghan Yang, Yusheng Su, Shengding Hu, Yulin Chen, Chi-Min Chan, Weize Chen, et al. Delta tuning: A comprehensive study of parameter efficient methods for pre-trained language models. *arXiv preprint arXiv:2203.06904*, 2022.
13. Pedro Domingos. A few useful things to know about machine learning. *Communications of the ACM*, 55(10):78–87, 2012.
14. John R Firth. A synopsis of linguistic theory, 1930-1955. *Studies in Linguistic Analysis*, 1957.
15. Tim van Gelder. *The MIT Encyclopedia of the cognitive sciences (MITECS)*, chapter Distributed vs. local representation, pages 236–238. MIT Press, 2001.
16. Thomas L Griffiths and Mark Steyvers. Finding scientific topics. *Proceedings of the National academy of Sciences*, 101:5228–5235, 2004.
17. Xu Han, Zhengyan Zhang, Ning Ding, Yuxian Gu, Xiao Liu, Yuqi Huo, Jiezhong Qiu, Yuan Yao, Ao Zhang, Liang Zhang, et al. Pre-trained models: Past, present and future. *AI Open*, 2021.
18. Zellig S Harris. Distributional structure. *Word*, 10(2-3):146–162, 1954.
19. John Hewitt and Christopher D. Manning. A structural probe for finding syntax in word representations. In *Proceedings of NAACL-HLT*, 2019.

20. Geoffrey E Hinton, James L McClelland, and David E Rumelhart. *Parallel distributed processing*, chapter Distributed representations, pages 77 – 109. MIT Press, 1986.
21. Thomas Hofmann. Probabilistic latent semantic indexing. In *Proceedings of SIGIR*, 1999.
22. Goodfellow Ian, Yoshua Bengio, and Aaron Courville. Deep learning. Book in preparation for MIT Press, 2016.
23. Yehuda Koren, Robert Bell, and Chris Volinsky. Matrix factorization techniques for recommender systems. *Computer*, 42(8), 2009.
24. Nelson F. Liu, Matt Gardner, Yonatan Belinkov, Matthew E. Peters, and Noah A. Smith. Linguistic knowledge and transferability of contextual representations. In *Proceedings of NAACL-HLT*, 2019.
25. Zhiyuan Liu, Yankai Lin, and Maosong Sun. *Representation Learning for Natural Language Processing*. Springer, 2020.
26. Peter Hugoe Matthews. *The concise Oxford dictionary of linguistics, third edition*, chapter Structural linguistics. Oxford University Press, 2014.
27. James L McClelland, David E Rumelhart, PDP Research Group, et al. *Parallel distributed processing*. MIT Press, 1986.
28. Warren S McCulloch and Walter Pitts. A logical calculus of the ideas immanent in nervous activity. *The bulletin of mathematical biophysics*, 5(4):115–133, 1943.
29. Qiaozhu Mei, Xuehua Shen, and ChengXiang Zhai. Automatic labeling of multinomial topic models. In *Proceedings of KDD*, pages 490–499, 2007.
30. T Mikolov and J Dean. Distributed representations of words and phrases and their compositionality. In *Proceedings of NeurIPS*, 2013.
31. Jeffrey Pennington, Richard Socher, and Christopher Manning. GloVe: Global vectors for word representation. In *Proceedings of EMNLP*, 2014.
32. Matthew Peters, Mark Neumann, Mohit Iyyer, Matt Gardner, Christopher Clark, Kenton Lee, and Luke Zettlemoyer. Deep contextualized word representations. In *Proceedings of NAACL-HLT*, 2018.
33. Sam T Roweis and Lawrence K Saul. Nonlinear dimensionality reduction by locally linear embedding. *Science*, 290(5500):2323–2326, 2000.
34. Beatrice Santorini. Part-of-speech tagging guidelines for the penn treebank project. 1990.
35. Claude E Shannon. A mathematical theory of communication. *Bell system technical journal*, 27(3):379–423, 1948.
36. Yiming Yang and Jan O Pedersen. A comparative study on feature selection in text categorization. In *Proceedings of ICML*, pages 412–420, 1997.
37. Chengxiang Zhai and John Lafferty. A study of smoothing methods for language models applied to information retrieval. *ACM Transactions on Information Systems (TOIS)*, 22(2):179–214, 2004.
38. Jing Zhang, Bo Chen, Lingxi Zhang, Xirui Ke, and Haipeng Ding. Neural, symbolic and neural-symbolic reasoning on knowledge graphs. *AI Open*, 2:14–35, 2021.

Chapter 2
Word Representation Learning

Shengding Hu, Zhiyuan Liu, Yankai Lin, and Maosong Sun

Abstract Words are the building blocks of phrases, sentences, and documents. Word representation is thus critical for natural language processing (NLP). In this chapter, we introduce the approaches for word representation learning to show the paradigm shift from symbolic representation to distributed representation. We also describe the valuable efforts in making word representations more informative and interpretable. Finally, we present applications of word representation learning to NLP and interdisciplinary fields, including psychology, social sciences, history, and linguistics.

2.1 Introduction

The nineteenth-century philosopher Wilhelm von Humboldt described language as *the infinite use of finite means*, which is frequently quoted by many linguists such as Noam Chomsky, the father of modern linguistics. Apparently, the vocabulary in human language is a finite set of words that can be regarded as a kind of *finite means*. Words can be *infinitely used* as building blocks of phrases, sentences, and documents. As human beings start learning languages from words, machines need to understand each word first so as to master the sophisticated meanings of human languages. Hence, effective word representations are essential for natural language processing (NLP), and it is also a good start for introducing representation learning in NLP.

We can consider word representations as the knowledge of the semantic meanings of words. As discussed in Chap. 1, we can investigate word representations

S. Hu · Z. Liu (✉) · M. Sun
Department of Computer Science and Technology, Tsinghua University, Beijing, China
e-mail: shengdinghu@gmail.com; liuzy@tsinghua.edu.cn; sms@tsinghua.edu.cn

Y. Lin
Gaoling School of Artificial Intelligence, Renmin University of China, Beijing, China
e-mail: yankailin@ruc.edu.cn

© The Author(s) 2023
Z. Liu et al. (eds.), *Representation Learning for Natural Language Processing*,
https://doi.org/10.1007/978-981-99-1600-9_2

from two aspects, how knowledge is organized and where knowledge is from, i.e., the **form** and **source** of word representations.

The form of word representation can be divided into the **symbolic representation** (Sect. 2.2) and the **distributed representation** (Sect. 2.3), which respectively correspond to *symbolism* and *connectionism* mentioned in Chap. 1. Both forms represent words into vectors to facilitate computer processing. The essential difference between these two approaches lies in the meaning of each dimension. In symbolic word representation, each dimension has clear meanings, corresponding to concrete concepts such as words and topics. The symbolic representation form is straightforward to human understanding and has been adopted by linguists and old-fashioned AI (OFAI). However, it's not optimal for computers due to high dimensionality and sparsity issues: computers need large storage for these high-dimensional representations, and computation is less meaningful because most entries of the representations are zeros. Fortunately, the distributed word representation overcomes these problems by representing words as *low-dimensional* and *real-valued* dense vectors. In distributed word representation, each dimension in isolation is meaningless because semantics is distributed over all dimensions of the vector. Distributed representations can be obtained by factorizing the matrices of symbolic representations or learned by gradient descent optimization from data. In addition to overcoming the aforementioned problems of symbolic representation, it handles emerging words easily and accurately.

The effectiveness of word representation is also determined by the source of word semantics. A word in most alphabetic languages, such as English, is usually a sequence of characters. The internal structure usually reflects its speech or sound but helps little in understanding word semantics, except for some informative prefixes and suffixes. By taking human languages as a typical and complicated symbolic system as *structuralism* suggests (Chap. 1), words obtain their semantics from their relationship to *other words*. Given a word, we can find its hypernyms, synonyms, hyponyms, and antonyms from a human-organized **linguistic knowledge base (KB)** like WordNet [52] to represent word semantics. By extending *structuralism* to the *distributional hypothesis*, i.e., *you shall know a word by the company it keeps* [24], we can build word representations from their rich context in **large-scale text corpora**. Since most linguistic knowledge graphs are usually annotated by linguists, they are convenient to be used by humans but difficult to comprehensively and immediately reflect the dynamics of human languages. Meanwhile, word representations obtained from large-scale text corpora can capture up-to-date semantics of words in the real world with few subjective biases.

We can summarize existing methods of word representation as a mix of the above two perspectives. In the era of statistical NLP, word representation follows the symbolic form, obtained either from a linguistic knowledge graph (Fig. 2.1a) or from large-scale text corpora (Fig. 2.1b), which will be introduced in Sect. 2.2.

In the era of deep learning, distributed word representation follows the spirits of *connectionism* and *empiricism*. It learns powerful low-dimensional word vectors from large-scale text corpora and achieves ground-breaking performance on numerous tasks (Fig. 2.1c). In Sect. 2.3, we will present representative works

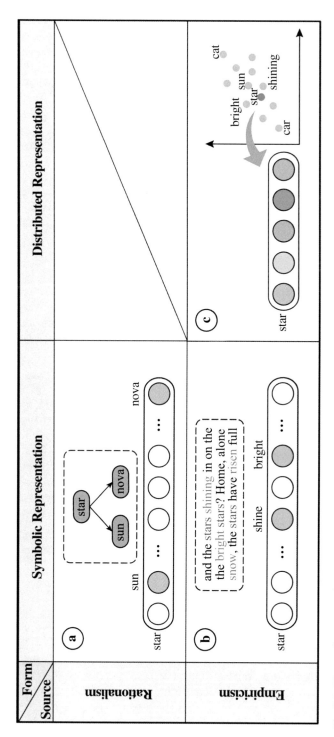

Fig. 2.1 The word representations can be divided according to their form of representation and source of the semantics: (**a**) shows the symbolic representations that use the knowledge base as the source, which is adopted by conventional linguistics; (**b**) shows the symbolic representations that adopt the distributional hypothesis as the foundation of the semantic source; (**c**) shows the distributed representation learned from large-scale corpora based on the distributional hypothesis, which is the mainstream of nowadays word representation learning

of distributed word representation such as word2vec [48] and GloVe [57]. These methods typically assign a fixed vector for each word and learn from text corpora. To address those words with multiple meanings under different contexts, researchers further propose contextualized word representation to capture sophisticated word semantics dynamically. The idea also inspires subsequent pre-trained models, which will be introduced in Chap. 5.

Many efforts have been devoted to constructing more *informative* word representations by encoding more information, such as multilingual data, internal character information, morphology information, syntax information, document-level information, and linguistic knowledge, as introduced in Sect. 2.4. Moreover, it would be a bonus if some degree of *interpretability* is added to word representation, and we will also briefly describe improvements in interpretable word representation.

Word representation learning has been widely used in many applications in NLP and other areas. In NLP, word representations can be applied to word-level tasks such as word similarity and analogy and simple downstream tasks such as sentiment analysis. We note that, with the advancement of deep learning and pre-trained models, word representations are less used in isolation in NLP but more as building blocks of neural language models, as shown in Chaps. 3, 4, and 5. Meanwhile, word representations play indispensable roles in interdisciplinary fields such as computational social sciences for studying social bias and historical change.

2.2 Symbolic Word Representation

Since the ancient days of knotted strings, human ancestors have used symbols to record and share information. As time progressed, isolated symbols gradually merged to form a symbol system. This system is human language. In fact, human language is probably the most complex and systematic symbol system that humans have ever built. In human language, each word is a discrete symbol that contains a wealth of semantic meaning. Therefore, ancient linguists also regard each word as a discrete symbol.

This common practice can also apply to NLP in modern computer science. In this section, we introduce three traditional symbolic approaches to word representations, i.e., one-hot word representation, linguistic KB-based word representation, and corpus-based word representation.

2.2.1 One-Hot Word Representation

One-hot representation is the simplest way for symbol-based word representation, which can be formalized as follows. Given a finite set of word vocabulary $V = \{w^{(1)}, w^{(2)}, \ldots, w^{(|V|)}\}$, where $|V|$ is the vocabulary size, one-hot representation represents an i-th word $w^{(i)}$ with a $|V|$-dimensional vector $\mathbf{w}^{(i)}$, where only the i-th dimension has a value 1 while all other dimensions are 0. That is, each dimension $\mathbf{w}_j^{(i)}$ is defined as:

$$
\mathbf{w}_j^{(i)} = \begin{cases} 1, & \text{if } j = i, \\ 0, & \text{otherwise.} \end{cases} \tag{2.1}
$$

In essence, the one-hot word representation maps each word to an index of the vocabulary. However, it can only distinguish between different words and does not contain any syntactic or semantic information. For any two words, their one-hot vectors are orthogonal to each other. That is, the cosine similarity between *cat* and *dog* is the same as the similarity between *cat* and *sun*, which are both zeros.

Although we do not have much to talk about one-hot word representation itself, it is the foundation of bag-of-words models for document representations, which are widely used in information retrieval and text classification. Readers can refer to document representation learning methods in Chap. 4.

As mentioned, there is no internal semantic structure in the one-hot representation. To incorporate semantics in the representation, we will present two methods with different sources of semantics: linguistic KB and natural corpus.

2.2.2 Linguistic KB-based Word Representation

As we introduced in Chap. 1, *rationalism* regards the introspective reasoning process as the source of knowledge. Therefore, the researchers construct a complex word-to-word network by reflecting on the relationship between words. For example, human linguists manually annotate the synonyms and hypernyms[1] of each word. In the well-known linguistic knowledge base WordNet [52], the hypernyms and hyponyms of *dog* are annotated as Fig. 2.2. To represent a word, we can use the vector forms just like one-hot representation as follows:

$$
\mathbf{w}_j^{(i)} = \begin{cases} 1 & \text{if } (w^{(i)}, \texttt{has\_hypernym}, w^{(j)}), \\ -1 & \text{if } (w^{(i)}, \texttt{has\_hyponym}, w^{(j)}), \\ 0 & \text{otherwise.} \end{cases} \tag{2.2}
$$

But it is clear that this representation has limited expressive power, where the similarity of two words without common hypernyms and hyponyms is 0. It would be better to directly adopt the original graph form, where the similarity between the two words can be derived using metrics on the graph. For synonym networks, we can

---

[1] Hypernyms are words whose meaning includes a group of other words, which are instances of the former. Word u is a hyponym of word v if and only if word v is a hypernym of word u.

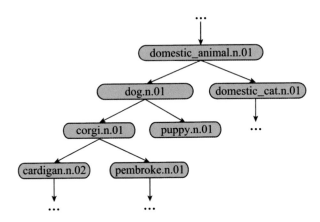

Fig. 2.2 Hypernyms and hyponyms of *dog* in WordNet [52]. The *dog.n.01* denotes the first synset of *dog* used as a noun

calculate the distance between two words on the network as their semantic similarity (i.e., the shortest path length between the two words). Hierarchical information can be utilized to better measure the similarity for hypernym-hyponym networks. For example, the information content (IC) approach [61] is proposed to calculate the similarity based on the assumption that the lower the frequency of the closest hypernym of two words is, the closer the two words are.

Formally, we define the similarity s as follows:

$$s(w_1, w_2) = \max_{w \in C(w_1, w_2)}[-\log P(w)], \qquad (2.3)$$

where $C(w_1, w_2)$ is the common hypernym set of w_1 and w_2 and $P(w)$ is the probability of word w's appearance in the corpus.[2] Intuitively, $P(w)$ is the generality of the word w. It indicates that if all common hypernyms of w_1 and w_2 are very general, then $s(w_1, w_2)$ will be very small. But if some hypernyms of w_1 and w_2 are specific, $s(w_1, w_2)$ will have a higher score, which indicates that these two words are closely related to each other. A vivid example is shown in Fig. 2.3.

2.2.3 Corpus-based Word Representation

The process of constructing a linguistic KB is labor-intensive. In contrast, it is much easier to collect a corpus. This motivation is also supported by *empiricism*, which emphasizes knowledge from naturally produced data.

---

[2] Estimated by dividing the frequency of a word by the total number of words in the corpus.

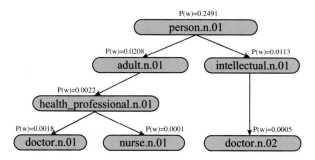

Fig. 2.3 *doctor.n.01* and *nurse.n.01* share a rare ancestor *health_professional.n.01*, their similarity is large. But the closest common ancestor of *doctor.n.02* and *nurse.n.01* is *people.n.01*, which is common. Therefore the similarity between them is small

The correctness of automatically derived representations from a corpus relies on the linguistic hypothesis behind them. We start with the **bag-of-words hypothesis**. To illustrate this hypothesis, we temporarily shift our attention to document representation. This hypothesis states that we can ignore the order of words in a document and simply treat the document as a bag (i.e., a multiset[3]) of words. Then the frequencies of the words in the bag can reflect the content of the document [66]. In this way, a document is represented by a *row vector* in which each element indicates the presence or frequency of a word in the document. For example, the value of the entry corresponding to word *cat* being 3 means that *cat* occurs three times in the document, and an entry corresponding to a word being 0 means that the word is not in the document [67]. In this way, we have automatically constructed a representation of the document.

How does this inspire us to construct the word representations of greater interest in this chapter? In fact, as we stack the row vectors of each document to form a document (row)-word (column) matrix, we can shift our attention from rows to columns [17]. Each column now represents the occurrence of a word in a stack of documents. Intuitively, if the words *rat* and *cat* tend to occur in the same documents, their statistics in the columns will be similar.

In the above approach, a document can be considered as the *context* of a word. Actually, more flexibility can be added in defining the context of a word to obtain other kinds of representations. For example, we can define a fixed-size window centered on a word and use the words inside the window as the context of that word. This corresponds to the well-known **distributional hypothesis** that the meaning of a word is described by its companions [24]. Then we count the words that appear in a word's neighborhood and use a dictionary as a word representation, where each key is a context word whose value is the frequency of the occurrence of that context word within a certain distance.

---

[3] A multiset is a set where duplicated elements are allowed, e.g., (1,2,2,3) and (2,2,3,1) are the same multiset.

To further extend the context of a word, several works propose to include dependency links [56] or links induced by argument positions [21]. Interested readers can refer to a summary of various contexts used for corpus-based distributional representations [65].

In summary, in symbolic representations, each entry of the representation has a clear and interpretable meaning. The clear interpretable meaning can correspond to a specific word, synset, or term, and that is why we call it "symbolic representation."

2.3 Distributed Word Representation

Although simple and interpretable, symbolic representations are not the best choice for computation. For example, the very sparse nature of the symbolic representation makes it difficult to compute word-to-word similarities. Methods like information content [61] cannot naturally generalize to other symbolic representations.

The difficulty of symbolic representation is solved by the distributed representation.[4] Distributed representation represents a subject (here is a word) as a fixed-length real-valued vector, where no clear meaning is assigned to every single dimension of the vector. More specifically, semantics is scattered over all (or a large portion) of the dimensions of the representation, and one dimension contributes to the semantics of all (or a large proportion) of the words.

We must emphasize that the "*distributed representation*" is completely different from and orthogonal to the "*distributional representation*" (induced by "*distributional hypothesis*"). Distributed representation describes the form of a representation, while distributional hypothesis (representation) describes the source of semantics.

2.3.1 Preliminary: Interpreting the Representation

Although each dimension is uninterpretable in distributed representation, we still want ways to interpret the meaning conveyed by the representation approximately. We introduce two basic computational methods to understand distributed word representation: similarity and dimension reduction.

Suppose the representations of two words are $\mathbf{u} = [u_1, \ldots, u_d]$ and $\mathbf{v} = [v_1, \ldots, v_d]$,[5] we can calculate the similarity or perform dimension reduction as follows.

---

[4] Models of distributed representations are also called vector space models (VSMs).

[5] In the following sections, we use the row vector for distributed word representation.

Similarity The Euclidean distance is the $L2$-norm of the difference vector of $\mathbf{u}$ and $\mathbf{v}$.

$$d_{\text{Euclidean}}(\mathbf{u}, \mathbf{v}) = \|\mathbf{u} - \mathbf{v}\|_2 = \sqrt{\sum_{i=1}^{d} |u_i - v_i|^2}. \tag{2.4}$$

Then the Euclidean similarity can be defined as the inverse of distance, i.e.,

$$s_{\text{Euclidean}}(\mathbf{u}, \mathbf{v}) = 1 \Big/ \sqrt{\sum_{i=1}^{d} |u_i - v_i|^2}. \tag{2.5}$$

Cosine similarity is also common. It measures the similarity by the angle between the two vectors:

$$s_{\text{cosine}}(\mathbf{u}, \mathbf{v}) = \frac{\mathbf{u} \cdot \mathbf{v}}{\|\mathbf{u}\|_2 \|\mathbf{v}\|_2}. \tag{2.6}$$

Dimension Reduction Distributed representations, though being lower dimensional than symbolic representations, still exist in manifolds higher than three dimensions. To visualize them, we need to reduce the dimension of the vector to 2 or 3. Many methods have been proposed for this purpose. We will briefly introduce principal component analysis (PCA).

PCA transforms the vectors into a set of new coordinates using an orthogonal linear transformation. In the new coordinate system, an axis is pointed in the direction which explains the data's most variance while being orthogonal to all other axes. Under this construction, the later constructed axes explain less variance and therefore are less important to fit the data. Then we can use only the first two to three axes as the principal components and omit the later axes. A case of PCA on two-dimensional data is in Fig. 2.4. Formally, denoting the new axes by a set of unit row vectors $\{\mathbf{d}_j | j = 1, \ldots, k\}$, where k is the number of unit row vectors. An original vector $\mathbf{u}$ of the sample u can be represented in the new coordinates by

$$\mathbf{u} = \sum_{j=1}^{k} a_j \mathbf{d}_j \approx \sum_{j=1}^{r} a_j \mathbf{d}_j, \tag{2.7}$$

where a_j is the weight of the vector $\mathbf{d}_j$ for representing $\mathbf{u}$, and $\mathbf{a} = [a_1, \ldots, a_r]$ forms the new vector representation of u. In practice, we only set $r = 2$ or 3 for visualization.

The set of new coordinates $\{\mathbf{d}_j | j = 1, \ldots, k\}$ can be computed by *eigen-decomposition* of the covariance matrix or using *singular value decomposition (SVD)*. Then we introduce SVD-based PCA and present its resemblance to latent

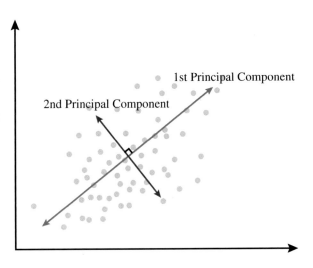

Fig. 2.4 The PCA specifies new axis directions (called principal components), and the axes are ordered from largest to smallest variance in their directions so that keeping the coordinates on the first few axes retains most of the distribution information

semantic analysis (LSA) in the next subsection. In SVD, a real-valued matrix can be decomposed into

$$\mathbf{U} = \mathbf{W} \Sigma \mathbf{D}, \tag{2.8}$$

such that Σ is a diagonal matrix of positive real numbers, i.e., the singular values, and $\mathbf{W}$ and $\mathbf{D}$ are singular matrices formed by orthogonal vectors. For a data sample (e.g., i-th row in $\mathbf{U}$):

$$\mathbf{U}_{i,:} = \sum_{j=1}^{k} \sigma_j \mathbf{W}_{i,j} \mathbf{D}_{j,:} . \tag{2.9}$$

Thus in Eq. (2.7), $a_j = \sigma_j \mathbf{W}_{i,j}$ and $\mathbf{d}_j = \mathbf{D}_{j,:}$.

Although widely adopted in high-dimensional data visualization, PCA is unable to visualize representations that form nonlinear manifolds. Other dimensionality reduction methods, such as t-SNE [73], can solve this problem.

2.3.2 Matrix Factorization-based Word Representation

Distributed representations can be transformed from symbolic representations by matrix factorization or neural networks. In this subsection, we introduce the matrix factorization-based methods. We introduce latent semantic analysis (LSA), its probabilistic version PLSA, and latent Dirichlet allocation (LDA) as the representative approaches. Readers who are only interested in neural networks can jump to the next section to continue reading.

Latent Semantic Analysis (LSA) LSA [17] utilizes singular value decomposition (SVD) to perform the transformation from matrices of symbolic representations. Suppose we have a word-document matrix $\mathbf{M} \in \mathbb{R}^{n \times d}$, where n is the number of words and d is the number of documents. By linear algebra, it can be uniquely[6] decomposed into the multiplication of three matrices $\mathbf{W} \in \mathbb{R}^{n \times n}$, $\Sigma \in \mathbb{R}^{n \times d}$, and $\mathbf{D} \in \mathbb{R}^{d \times d}$:

$$\mathbf{M} = \mathbf{W}\Sigma\mathbf{D}^\top, \tag{2.10}$$

such that Σ is a diagonal matrix of positive real numbers, i.e., the singular values, and columns of $\mathbf{W}, \mathbf{D}$ are left-singular vectors and right-singular vectors, respectively.[7]

Now let's try to interpret the two orthogonal matrices. The i-th row of the matrix $\mathbf{M}$ that represents the i-th word's symbolic representation (denoted by $\mathbf{M}_{i,:}$) is decomposed into:

$$\mathbf{M}_{i,:} = \mathbf{W}_{i,:}\Sigma\mathbf{D}^\top. \tag{2.11}$$

From Eq. (2.11), we can see only the i-th row of $\mathbf{W}$ contributes to the i-th word's symbolic representation. More importantly, since $\mathbf{D}$ is an orthogonal matrix, the similarity between words w_i and w_j is given by:

$$\mathbf{M}_{i,:}\mathbf{M}_{j,:}^\top = \mathbf{W}_{i,:}\Sigma^2\mathbf{W}_{j,:}^\top. \tag{2.12}$$

Thus we can take $\mathbf{W}_{i,:}\Sigma$ as the distributed representation for word w_i. Note that taking $\mathbf{W}_{i,:}\Sigma$ or $\mathbf{W}_{i,:}$ as the distributed representation is both ok because $\mathbf{W}_{i,:}\Sigma$ is $\mathbf{W}_{i,:}$ stretched along each axis j with ratio $\Sigma_{j,j}$, and the relative positions of points in the two spaces are similar.

Suppose we arrange the eigenvalues in descending order. In that case, the largest K singular values (and their singular vectors) contribute the most to matrix $\mathbf{M}$. Thus we only use them to approximate $\mathbf{M}$. Now Eq. (2.11) becomes:

$$\mathbf{M}_{i,:} \approx \mathbf{W}_{i,:K} \odot [\sigma_1, \sigma_2, \ldots, \sigma_K]\mathbf{D}_{:,:K}^\top, \tag{2.13}$$

where $\odot$ is the element-wise multiplication and σ_i is the i-th diagonal element of Σ.

To sum up, in LSA, we perform SVD to the counting statistics to get the distributed representation $\mathbf{W}_{i,:K}$. Usually, with a much smaller K, the approximation can be sufficiently good, which means the semantics in the high-dimensional

---

[6] Up to permutations of rows, columns, and signs.

[7] In fact, the mathematical backbone of LSA is the same as PCA. We repeat it here for the convenience of those readers who skipped the previous section.

symbolic representation are now compressed and distributed into a much lower-dimensional real-valued vector. LSA has been widely used to improve the recall of query-based document ranking in information retrieval since it supports ambiguous semantic matching.

One challenge of LSA comes from the computational cost. A full SVD on an $n \times d$ matrix requires $\mathcal{O}(\min\{n^2 d, nd^2\})$ time, and the parallelization of SVD is not trivial. A solution is random indexing [36, 64] that overcomes the computational difficulties of SVD-based LSA and avoids expensive preprocessing of a huge word-document matrix. In random indexing, each document is assigned a randomly-generated high-dimensional sparse ternary vector (named as *index vector*). Note that random vectors in high-dimensional space should be (nearly) orthogonal, analogous to the orthogonal matrix $\mathbf{D}$ in SVD-based LSA. For each word in a document, we add the document's index vector to the word's vector. After passing the whole text corpora, we can get accumulated word vectors. Random indexing is simple to parallelize and implement, and its performance is comparable to the SVD-based LSA [64].

Probabilistic LSA (PLSA) LSA further evolves into PLSA [34]. To understand PLSA based on LSA, we can treat the $\mathbf{W}_{i,:K}$ as a distribution over latent factors $\{z_k | k = 1, \ldots, K\}$, where

$$\mathbf{W}_{i,k} = P(w_i | z_k), \quad k = 1, \ldots, K. \tag{2.14}$$

We can understand these factors as "topics" as we will assign meanings to them later. Similarly, $\mathbf{D}_{i,:K}$ can also be regarded as a distribution, where

$$\mathbf{D}_{j,k} = P(d_j | z_k), \quad k = 1, \ldots, K. \tag{2.15}$$

And the $\{\sigma_k | k = 1, \ldots, K\}$ are the prior probabilities of factors z_k, i.e., $P(z_k) = \sigma_k$. Thus, the word-document matrix becomes a joint probability of word w_j and document d_j:

$$\mathbf{M}_{i,j} = P(w_i, d_j) = \sum_{k=1}^{K} P(w_i | z_k) P(z_k) P(d_j | z_k). \tag{2.16}$$

With the help of Bayes' theorem and Eq. (2.16), we can compute the conditional probability of w_i given a document d_j is

$$P(w_i | d_j) = \frac{P(w_i, d_j)}{P(d_j)} = \sum_{k=1}^{K} P(w_i | z_k) \frac{P(d_j | z_k) P(z_k)}{P(d_j)} = \sum_{k=1}^{K} P(w_i | z_k) P(z_k | d_j). \tag{2.17}$$

Now we can see a generative process is defined from Eq. (2.17). To generate a word in the document d_j, we first sample a latent factor z_k from $P(z_k | d_j)$ and

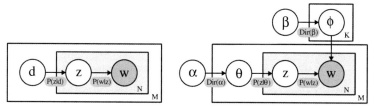

Fig. 2.5 The generative process of words in documents in the PLSA model (left) and LDA model (right). N is the number of words in a document, M is the number of documents, and K is the number of topics. w is the only observed variable, and we need to estimate the other white variables based on the observation. The figure is redrawn according to the Wikipedia entry "Latent Dirichlet allocation"

then sample a word w_i from the conditional probability $P(w_i|z_k)$. The process is represented in Fig. 2.5.

Note that to make Eq. (2.14)~Eq. (2.17) rigorous, the elements of $\mathbf{W}, \Sigma, \mathbf{D}$ have to be nonnegative. To do optimization under such probabilistic constraints, a different loss from SVD is used [34]:

$$\mathcal{L} = -\sum_j \sum_i \mathbf{M}_{i,j} \log P(w_i, d_j).   \qquad (2.18)$$

The following works prove that optimizing the above objective is the same as nonnegative matrix factorization [20, 38]. We omit the mathematical details here.

Latent Dirichlet Allocation (LDA)   PLSA is further developed into latent Dirichlet allocation (LDA) [10], a popular topic model that is widely used in document retrieval. LDA adds hierarchical Bayesian priors to the generative process defined by Eq. (2.14). The generative process for words in document j becomes:

1. Choose $\boldsymbol{\theta}_j \in \mathbb{R}^K \sim \text{Dir}(\boldsymbol{\alpha})$, where $\text{Dir}(\boldsymbol{\alpha})$ is a Dirichlet distribution (typically each dimension of $\boldsymbol{\alpha} < 1$), where K is the number of topics. This is the probability distribution of topics in the document d_j.
2. Choose $\boldsymbol{\phi}_z \in \mathbb{R}^{|V|} \sim \text{Dir}(\boldsymbol{\beta})$ for each topic z, where $|V|$ is the size of the vocabulary. Typically each dimension of $\boldsymbol{\beta}$ is less than 1. This is the probability distribution of words produced by topic z.
3. For each word w_i in the document d_j:

   a. Choose a topic $z_{i,j} \sim \text{Multinomial}(\boldsymbol{\theta}_j)$.
   b. Choose a word from $w_{i,j} = P(w_j|d_j) \sim \text{Multinomial}(\boldsymbol{\phi}_{z_{i,j}})$.

The generative process in LDA is in Fig. 2.5 (right). We will not dive into the mathematical details of LDA.

We would like to emphasize two points about LDA: (1) the hyper-parameter $\boldsymbol{\alpha}, \boldsymbol{\beta}$ in Dirichlet prior is typically set to be less than 1, resulting in a "sparse prior", i.e., most dimensions of the sampled $\boldsymbol{\theta}$ and $\boldsymbol{\phi}$ are close to zero, and the mass of the distribution will be concentrated in a few values. This is consistent with our common sense that a document will always have only a few topics and that a topic will only produce a small number of words. Moreover, the total number of topics K is pre-defined as a relatively small integer. The sparsity and interpretability make LDA essentially a kind of *symbolic representation*, and LDA can be seen as *a bridge between distributed representations and symbolic representations*. (2) Although PLSA and LDA are more often used in document retrieval, the distribution of a word over different topics (latent factors) $P(w|z_i)$ can be used as an effective word representation, i.e., $\mathbf{w} = [P(w|z_1), \ldots, P(w|z_K)]$.

However, the information source, i.e., counting matrix $\mathbf{M}$, of matrix factorization-based methods is still based on the bag-of-words hypothesis. These methods lose the word order information in the documents, so their expressiveness capability remains limited. Therefore, these classical methods are less often used when neural network-based methods that can model word order information emerge.

2.3.3   Word2vec and GloVe

The neural networks, revived in the 2010s, are similar to the neurons of the human brain, where the neurons inside a neural network perform distributed computation. One neuron is responsible for the computation of multiple pieces of information, and one input activates multiple neurons at the same time. This property coincides with distributed representation. Hence, distributed representation plays a dominant role in the era of neural networks. Moreover, neural models are optimized on large-scale data. The data dependency makes the distributional hypothesis particularly important in optimizing such distributed representations. In the following section, we first present **word2vec** [48], a milestone work of distributional distributed word representation using neural approaches. After that, we introduce **GloVe** [57] that improves word2vec with a global word-occurrence matrix.

Word2vec Word2vec adopts the distributional hypothesis but does not take a count-based approach. It directly uses gradient descent to optimize the representation of a word toward its neighbors' representations. Word2vec has two specifications, namely, **continuous bag-of-words (CBOW)** and **skip-gram**. The difference is that CBOW predicts a center word based on multiple context words, while skip-gram predicts multiple context words based on the center word.

CBOW predicts the center word given a window of context. Figure 2.6 shows the idea of CBOW with a window of $five$ words.

Formally, CBOW predicts w_i according to its contexts as:

$$P(w_i|w_{i-l}, \ldots, w_{i-1}, w_{i+1}, \ldots, w_{i+l}) = \text{Softmax}\left(\mathbf{W}\left(\sum_{i-l \leq j \leq i+l, j \neq i} \mathbf{w}_j^\top\right)\right),$$

$$(2.19)$$

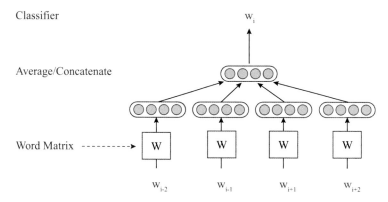

Fig. 2.6 The architecture of the CBOW model. (The figure is redrawn according to Fig. 1 from Mikolov et al. [49])

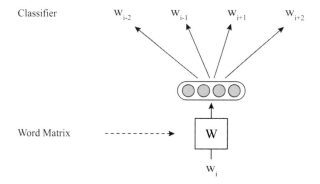

Fig. 2.7 The architecture of the skip-gram model. (The figure is redrawn according to Fig. 1 from Mikolov et al. [49])

where $P(w_i|w_{i-l}, \ldots, w_{i-1}, w_{i+1}, \ldots, w_{i+l})$ is the probability of word w_i given its contexts, $2l + 1$ is the size of training contexts, $\mathbf{w}_j$ is the word vector of word w_j, $\mathbf{W}$ is the weight matrix in $\mathbb{R}^{|V| \times m}$, V indicates the vocabulary, and m is the dimension of the word vector.

The CBOW model is optimized by minimizing the sum of the negative log probabilities:

$$\mathcal{L} = -\sum_i \log P(w_i|w_{i-l}, \ldots, w_{i-1}, w_{i+1}, \ldots, w_{i+l}). \qquad (2.20)$$

Here, the window size l is a hyper-parameter to be tuned. A larger window size may lead to higher accuracy as well as longer training time.

Contrary to CBOW, *skip-gram* predicts the context given the center word. Figure 2.7 shows the model.

Formally, given a word w_i, skip-gram predicts its context as:

$$P(w_j|w_i) = \text{Softmax}(\mathbf{W}\mathbf{w}_i^\top), \quad \text{where} \quad i - l \leq j \leq i + l, j \neq i, \tag{2.21}$$

where $P(w_j|w_i)$ is the probability of context word w_j given w_i and $\mathbf{W}$ is the weight matrix. The loss function is similar to CBOW but needs to sum over multiple context words:

$$\mathcal{L} = -\sum_i \sum_{i-l \leq j \leq i+l, j \neq i} P(w_j|w_i). \tag{2.22}$$

In the early stages of the deep learning renaissance, computational resources are still limited, and it is time-consuming to optimize the above objectives directly. The most time-consuming part is the softmax layer since the softmax layer uses the scores of predicting all words in the vocabulary V in the denominator:

$$P(w_j|w_i) = \text{Softmax}(\mathbf{W}\mathbf{w}_i^\top) = \frac{\exp(\mathbf{W}_{j,:}\mathbf{w}_i^\top)}{\sum_{j, w_j \in V} \exp(\mathbf{W}_{j,:}\mathbf{w}_i^\top)}. \tag{2.23}$$

An intuitive idea to improve efficiency is obtaining a reasonable but faster approximation of the softmax score. Here, we present two typical approximation methods, including **hierarchical softmax** and **negative sampling**. We explain these two methods using CBOW as an example.

The idea of hierarchical softmax is to build hierarchical classes for all words and to estimate the probability of a word by estimating the conditional probability of its corresponding hierarchical classes. Figure 2.8 gives an example. Each internal node of the tree indicates a hierarchical class and has a feature vector, while each leaf node of the tree indicates a word. The conditional probabilities, e.g., p_0 and p_1 in Fig. 2.8, of two child nodes are computed by the feature vector of each node and the context vector. For example,

$$p_0 = \frac{\exp(\mathbf{w}_0 \mathbf{w}_c^\top)}{\exp(\mathbf{w}_0 \mathbf{w}_c^\top) + \exp(\mathbf{w}_1 \mathbf{w}_c^\top)}, \tag{2.24}$$

$$p_1 = 1 - p_0, \tag{2.25}$$

where $\mathbf{w}_c$ is the context vector, $\mathbf{w}_0$ and $\mathbf{w}_1$ are the feature vectors.

Then, the probability of a word can be obtained by multiplying the probabilities of all nodes on the path from the root node to the corresponding leaf node. For example, the probability of the word *the* is $p_0 \times p_{01}$, while the probability of *cat* is $p_0 \times p_{00} \times p_{001}$.

The tree of hierarchical classes is generated according to the word frequencies, which is called the Huffman tree. Through this approximation, the computational complexity of the probability of each word is $\mathcal{O}(\log|V|)$.

Fig. 2.8 An illustration of
hierarchical softmax

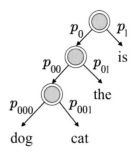

Negative sampling is a more straightforward technique. It directly samples k words as negative samples according to the word frequency. Then, it computes a softmax over the $k + 1$ words (1 for the positive sample, i.e., the target word) to approximate the conditional probability of the target word.

GloVe   The word2vec and matrix factorization-based methods have complementary advantages and disadvantages. In terms of learning efficiency and scalability, word2vec is superior because word2vec uses an online learning (or batch learning paradigm in deep learning) approach and is able to learn over large corpora. However, considering the preciseness of distribution modeling, the matrix factorization-based methods can exploit global co-occurrence information by building a global co-occurrence matrix. In comparison, word2vec is a local window-based method that cannot see the frequency of word pairs in a global corpus in a single optimization step. Therefore, GloVe [57] is proposed to combine the advantages of word2vec and matrix factorization-based methods.

To learn from global count statistics, GloVe firstly builds a co-occurrence matrix $\mathbf{M}$ over the entire corpus but does not directly factorize it. Instead, it takes each entry in the co-occurrence matrix $\mathbf{M}_{ij}$ and optimizes the following target:

$$\mathcal{L} = \sum_{i,j=1}^{|V|} f(\mathbf{M}_{ij})d(\mathbf{w}_i, \mathbf{w}_j, \mathbf{M}_{ij}). \tag{2.26}$$

The $d(\mathbf{w}_i, \mathbf{w}_j, \mathbf{M}_{ij})$ is a metric that compares the distributed representations of word w_i and w_j with the ground-truth statistics $\mathbf{M}_{ij}$. $f(\mathbf{M}_{ij})$ is a weight term measuring the importance of the word pair w_i, w_j. Specifically, GloVe adopts the following form as the metric d:

$$d(\mathbf{w}_i, \mathbf{w}_j, \mathbf{M}_{ij}) = (\mathbf{w}_i \mathbf{w}_j^\top + b_i + b_j - \log \mathbf{M}_{ij})^2, \tag{2.27}$$

where b_i and b_j are bias terms for word w_i and w_j. Interested readers can read the original paper [57] for the derivation details.

For the weight term $f(\mathbf{M}_{ij})$, most previous approaches set the weight of all word pairs to 1. However, common word pairs may have too weak semantics, so we should lower their weights and increase the weights of rare word pairs slightly. Thus, GloVe observes that it should satisfy three constraints:

- $f(0) = 0$ and $\lim_{x \to 0} f(x) \log^2 x$ is finite.
- A nondecreasing function.
- Truncated for large values of x to avoid overfitting to common words (stop words).

A possible choice is the following, where α is taken as $\frac{3}{4}$ in the original GloVe paper:

$$f(x) = \begin{cases} (x/x_{\max})^\alpha, & \text{if } x < x_{\max}, \\ 1, & \text{otherwise.} \end{cases} \tag{2.28}$$

In summary, GloVe uses weighted squared loss to optimize the representation of words based on the elements in the global co-occurrence matrix. Compared with word2vec, it captures global statistics. Compared with matrix factorization, it (1) reasonably reduces the weights of the most frequent words at the level of matrix entries, (2) reduces the noise caused by non-discriminative word pairs by implicitly optimizing the ratio of co-occurrence frequencies, and (3) enables fitting on large corpus by iterative optimization. Since the number of nonzero elements of the co-occurrence matrix is much smaller than $|V|^2$, the efficiency of GloVe is ensured in practice.

Word2vec as Implicit Matrix Factorization It seems so far that the neural network-based methods like word2vec and matrix factorization-based methods are two distinct paradigms for deriving distributed representation. But in fact, they have close theoretical connections. Omer et al. [41] prove that word2vec is factorizing pointwise mutual information matrix (PMI), where

$$\text{PMI}_{i,j} = \log \frac{P(w_i, w_j)}{P(w_i) P(w_j)}. \tag{2.29}$$

Omer et al. [41] then compare the performance of factoring the PMI matrix using SVD and skip-gram with the negative sampling (SGNS) model. SVD achieves a significantly better objective value when the embedding size is smaller than 500 dimensions and the number of negative samples is 1. With more negative samples and higher embedding dimensions, SGNS gets a better objective value. For downstream tasks, under several conditions, SVD achieves slightly better performance on word analogy and word similarity. In contrast, skip-gram with negative sampling achieves better performance by 2% on syntactical analogy.

2.3.4   Contextualized Word Representation

In natural language, the semantic meaning of an individual word can usually be different with respect to its context in a sentence. For example, in the two sentences: "*willows lined the bank of the stream.*", "*a bank account.*",[8] although the word *bank* is always the same, their meanings are different. This phenomenon is prevalent in any language. However, most of the traditional word embeddings (CBOW, skip-gram, GloVe, etc.) cannot well understand the different nuances of the meanings of words in the different surrounding texts. The reason is that these models only learn a unique and specific representation for each word. Therefore these models cannot capture how the meanings of words change based on their surrounding contexts.

Matthew et al. [58] propose ELMo to address this issue, whose word representation is a function of the whole input. More specifically, rather than having a look-up table of word embedding matrix, ELMo converts words into low-dimensional vectors on-the-fly by feeding the word and its context into a deep neural network. ELMo utilizes a bidirectional language model to conduct word representation. Formally, given a sequence of N words $(w_1, \ldots, w_N)$, a forward language model (LM)[9] models the probability of the sequence by predicting the probability of each word w_k according to the historical context:

$$P(w_1, w_2, \ldots, w_N) = \prod_{k=1}^{N} P(w_k \mid w_1, \ldots, w_{k-1}). \tag{2.30}$$

The forward LM in ELMo is a multilayer long short-term memory (LSTM) network [33], which is a kind of widely used neural network for modeling sequential data, and the j-th layer of the LSTM-based forward LM will generate the context-dependent word representation $\overrightarrow{\mathbf{h}}_{k,j}^{\text{LM}}$ for the word w_k. The backward LM is similar to the forward LM. The only difference is that it reverses the input word sequence to $(w_N, w_{N-1}, \ldots, w_1)$ and predicts each word according to the future context:

$$P(w_1, w_2, \ldots, w_N) = \prod_{k=1}^{N} P(w_k \mid w_{k+1}, \ldots, w_N). \tag{2.31}$$

Similar to the forward LM, the j-th backward LM layer generates the representations $\overleftarrow{\mathbf{h}}_{k,j}^{\text{LM}}$ for the word w_k.

When used in a downstream task, ELMo combines all layer representations of the bidirectional LM into a single vector as the *contextualized word representation*. The way to do the combination is flexible. For example, the final representation can be the weighting of all bidirectional LM layer's hidden representation, and the

---

[8] Examples are taken from the Oxford Dictionary of English [69].

[9] The details of the language model are in Chap. 5.

weights are task-specific:

$$\mathbf{w}_k = \alpha^{\text{task}} \sum_{j=0}^{L} s_j^{\text{task}} \mathbf{h}_{k,j}^{\text{LM}}, \tag{2.32}$$

where $\mathbf{s}^{\text{task}} = [s_1^{\text{task}}, \ldots, s_L^{\text{task}}]$ are softmax-normalized weights and α^{task} allows the task to scale the whole representation, and $\mathbf{h}_{k,j}^{\text{LM}} = \text{concat}(\overrightarrow{\mathbf{h}}_{k,j}^{\text{LM}}; \overleftarrow{\mathbf{h}}_{k,j}^{\text{LM}})$.

Due to the superiority of contextualized representations, a group of works led by ELMo [58], BERT [19], and GPT [59] has started to emerge since 2017, eventually leading to a unified paradigm of pre-training-fine-tuning across NLP. Please refer to Chap. 5 for further reading.

2.4 Advanced Topics

In the previous section, we introduced the basic models of word representation learning. These studies promoted more work on pursuing better word representations. In this section, we introduce the improvement from different aspects. Before we dive into the specific methods, let's first discuss the essential features of a **good** word representation.

Informative Word Representation A key point where representation learning differs from traditional prediction tasks is that when we construct representations, we do not know what information is needed for downstream tasks. Therefore, we should compress as much information as possible into the representation to facilitate various downstream tasks. From the development of one-hot representations to distributional and contextualized representations, the information in the representations is indeed increasing. And we still expect to incorporate more information into the representations.

Interpretable Word Representation For distributed representations, a single dimension is not responsible for explaining the factors of semantic change, and the semantics is entangled in multiple dimensions. As a result, distributed representations are difficult to interpret. As Bengio et al. [9] pointed out, a good distributed representation should *"disentangle the factors of variation."* There is always a desire for an interpretable distributed representation. Although PLSA and LDA have already increased interpretability, we would like to see more developments in this direction.

In this section, we will introduce the efforts that enhance the distributed word representations in terms of the above criteria.

2.4.1 Informative Word Representation

To make the representations informative, we can learn word representations from universal training data, including multilingual corpus. Another key direction for being informative is incorporating as much additional information into the representation as possible. From small to large information granularity, we can utilize character, morphological, syntactic, document, and knowledge base information. We will describe the related work in detail.

Multilingual Word Representation There are thousands of languages in the world. Making the vector space applicable for multiple languages not only improves the performance of word representation in low-resource languages but also can absorb information from the corpora of multiple languages. The bilingual word embedding model [78] proposes to make use of the word alignment pairs available in machine translation. It maps the embeddings of the source language to the embeddings of the target language and vice versa.

Specifically, a set of source words' representations that are trained on monolingual source language corpus is used to initialize the words in the target language:

$$\mathbf{w}_{\text{t-init}} = \sum_{s=1}^{S} \frac{N_{ts} + 1}{N_t + S} \mathbf{w}_s, \tag{2.33}$$

where $\mathbf{w}_s$ is the trained embeddings of the source word and $\mathbf{w}_{\text{t-init}}$ is the initial embedding of the target word, respectively. N_{ts} is the number of times that the target word t is aligned with the source word s. N_t is the total times of word t in the target corpus. The add-on terms $+1$ and $+S$ are the Laplace smoothing. S is the number of source words. Intuitively, the initialization of target word embedding is the weighted average of the aligned words in the source corpus, which ensure the two sets of embeddings are in the same space initially.

Then the source and target representation is optimized on their unlabeled corpora with target $\mathcal{L}_s$ and $\mathcal{L}_t$, respectively. To improve the alignment during training, alignment matrices $\mathbf{N}_{t \to s}$ and $\mathbf{N}_{s \to t}$ are used, where each element $\mathbf{N}_{ij}$ denotes the count of a word w_i is aligned with source word w_j normalized across all source words. Then a translation equivalence objective is used:

$$\begin{aligned} \mathcal{L}_{s \to t} &= \|\mathbf{w}_t - \mathbf{N}_{t \to s}\mathbf{w}_s\|_2^2, \\ \mathcal{L}_{t \to s} &= \|\mathbf{w}_s - \mathbf{N}_{s \to t}\mathbf{w}_t\|_2^2. \end{aligned} \tag{2.34}$$

Thus the unified objective becomes $\mathcal{L}_s + \lambda_1 \mathcal{L}_{t \to s} \mathcal{L}_t + \lambda_2 \mathcal{L}_{s \to t}$ for source words and target words, respectively, where λ_1 and λ_2 are the coefficients to weight the different sub-objectives.

However, this model performs poorly when the seed lexicon is small. Some works introduce virtual alignment between languages to tackle this limitation. Let's take Zhang et al. [77] as an example. In addition to monolingual word embedding learning and bilingual word embedding alignment based on seed lexicon, this work proposes an integer latent variable vector $\mathbf{m} \in \mathbb{N}^{V^T}$ (V^T is the size of target vocabulary, and $\mathbb{N}$ is the set of natural numbers) representing which source word is linked by a target word w_t. So $\mathbf{m}_t \in \{0, 1, \ldots, V^S\}$. $\mathbf{m}$ is randomly initialized and then optimized through an expectation-maximization algorithm [18] together with the word representations. In the E-step, the algorithm fixes the current word representation and finds the best matching $\mathbf{m}$ that can align the source and target representations. And in the M-step, it treats the mapping as fixed and known, just like Zou et al. [78], and optimizes the source and target word representations.

Character-Enhanced Word Representation Many languages, such as Chinese and Japanese, have thousands of characters compared to other languages containing only dozens of characters. And the words in Chinese and Japanese are composed of several characters. Characters in these languages have richer semantic information. Hence, the meaning of a word can be learned not only from its context but also from the composition of characters. This intuitive idea drives Chen et al. [14] to propose a joint learning model for character and word embeddings (CWE). In CWE, a word representation $\mathbf{w}$ is a composition of the original word embedding $\mathbf{w}_0$ trained on corpus and its character embeddings $\mathbf{c}_i$. Formally,

$$\mathbf{w} = \mathbf{w}_0 + \frac{1}{|w|} \sum_i \mathbf{c}_i, \qquad (2.35)$$

where $|w|$ is the number of characters in the word. Note that this model can be integrated with various models such as skip-gram, CBOW, and GloVe.

Further, position-based and cluster-based methods are proposed to address the issue that characters are highly ambiguous. In the position-based approach, each character is assigned three vectors that appear in *begin*, *middle*, and *end* of a word, respectively. Since the meaning of a character varies when it appears in the different positions of a word, this method can significantly resolve the ambiguity problem. However, characters that appear in the same position may also have different meanings. In the cluster-based method, a character is assigned K different vectors for its different meanings, in which a word's context is used to determine which vector to be used for the characters in this word.

Introducing character embeddings can significantly improve the representation of low-frequency words. Besides, this method can deal with new words while other methods fail. Experiments show that the joint learning method can perform better on both word similarity and analogy tasks.

Morphology-Enhanced Word Representation Many languages, such as English, have rich morphological information and plenty of rare words. However, most word representation models ignore the rich morphology information. This is a limitation because a word's affixes can help infer a word's meaning. Moreover, in traditional models, word representation is independent of each other. So when facing rare words without enough context to learn the representations, the representations tend to be inaccurate.

Fortunately, in morphology-enhanced word representation, morphologies' representation can enrich word embeddings and are shared among words to assist the representation of rare words. Piotr et al. [11] propose to represent a word as a bag of morphology n-grams. This model substitutes word vectors in skip-gram with the sum of morphology n-gram vectors. When creating the dictionary of morphology n-grams, they select all morphology n-grams with a length greater or equal to 3 and smaller or equal to 6. To distinguish prefixes and suffixes from other affixes, they also add special characters to indicate the beginning and the end of a word. This model is efficient and straightforward, which achieves good performance on word similarity and word analogy tasks, especially when the training set is small. Ling et al. [44] further use a bidirectional LSTM to generate word representation by composing morphologies. This model significantly reduces the number of parameters since only the morphology representations and the weights of LSTM need to be stored.

Syntax-Enhanced Word Representation Continuous word embeddings should combine the semantic and syntactic information of words. However, existing word representation models depend solely on linear contexts and have more semantic information than syntactic information. To inject the embeddings with more syntactic information, the dependency-based word embedding [40] uses the dependency-based context. Dependency-based representation contains less topical information than the original skip-gram representation and shows more functional similarity. It considers the dependency parsing tree's information when learning word representations. The contexts of a target word w are the modifiers m_i of this word, i.e., $(m_1, r_1), \ldots, (m_k, r_k)$, where r_i is the type of dependency relation between the target node and the modifier. During training, the model optimizes the probability of dependency-based contexts rather than neighboring contexts. This model gains some improvements on word similarity benchmarks compared with skip-gram. Experiments also show that words with syntactic similarity are more similar in the vector space.

Document-Enhanced Word Representation Word embedding methods like skip-gram simply consider the context information within a window to learn word representation. However, the information in the whole document, e.g., the topics of the document, could also help word representation learning. Topical word embedding (TWE) [45] introduces topic information generated by latent Dirichlet allocation (LDA) to help distinguish different meanings of a word. The model is defined to minimize the following objective:

$$\mathcal{L} = -\frac{1}{N} \sum_{i=1}^{N} \sum_{i-l \leq j \leq i+l, j \neq i} \left(\log P(\mathbf{w}_j | \mathbf{w}_i) + \log P(\mathbf{w}_j | \mathbf{z}_i) \right), \quad (2.36)$$

where $\mathbf{w}_i$ is the word embedding and $\mathbf{z}_i$ is the topic embedding of w_i. Each word w_i is assigned a unique topic, and each topic has a topic embedding. The topical word embedding model shows the advantage of contextual word similarity and document classification tasks.

TopicVec [42] further improves the TWE model. TWE simply combines the LDA with word embeddings and lacks statistical foundations. Moreover, the LDA topic model needs numerous documents to learn semantically coherent topics. TopicVec encodes words and topics in the same semantic space. It can learn coherent topics when only one document is presented.

Knowledge-Enhanced Word Representation People have also annotated many knowledge bases that can be used in word representation learning as additional information. Yu et al. [76] introduce relational objectives into the CBOW model. With the objective, the embeddings can predict their contexts and words with relations. The objective is to minimize the sum of the negative log probability of all relations as:

$$\mathcal{L} = -\frac{1}{N} \sum_{i=1}^{N} \sum_{w \in R_{w_i}} \log P(w|w_i), \tag{2.37}$$

where R_{w_i} indicates a set of words that have a relation with w_i. The external information helps train a better word representation, showing significant improvements in word similarity benchmarks.

Moreover, retrofitting [22] introduces a post-processing step that can introduce knowledge bases into word representation learning. It is more modular than other approaches which consider knowledge base during training. Let the word embeddings learned by existing word representation approaches be $\hat{\mathbf{W}}$. Retrofitting attempts to find a knowledgeable embedding space $\mathbf{W}$, which is close to $\hat{\mathbf{W}}$ but considers the relations in the knowledge base. It optimizes $\mathbf{W}$ toward $\hat{\mathbf{W}}$ and simultaneously shrinks the distance between knowledgeable representations of words w_i, w_j with relations. Formally,

$$\mathcal{L} = \sum_{i} \left(\alpha_i \|\mathbf{w}_i - \hat{\mathbf{w}}_i\|_2 + \sum_{(w_i, w_j) \in R} \beta_{ij} \|\mathbf{w}_i - \mathbf{w}_j\|_2 \right), \tag{2.38}$$

where α and β are hyper-parameters indicating the strength of the associations, and R is a set of relations in the knowledge base. With knowledge bases such as the paraphrase database [28], WordNet [52], and FrameNet [3], this model can achieve consistent improvement on word similarity tasks. But it may also reduce the performance of the analogy of syntactic relations if it emphasizes semantic knowledge. Since it is a post-processing approach, it is compatible with various distributed representation models.

In addition to the aforementioned synonym-based knowledge bases, there are also sememe-based knowledge bases, in which the sememe is defined as the minimum semantic unit of word meanings. Due to the importance of sememe in computational linguistics, we introduce it in detail in Chap. 10.

2.4.2 Interpretable Word Representation

Although distributed word representation achieves ground-breaking performance on numerous tasks, it is less interpretable than conventional symbolic word representations. It would be a bonus if the distributed representations also enjoy some degree of interpretability. We can improve the interpretability from three directions. The first is to increase the interpretability of the vector representation among its neighbors. Since a word has multiple meanings, especially those polysemy words, the vectors of different meanings should locate in different neighborhoods. Therefore, we introduce work on disambiguated word representations. Another direction is to increase the interpretability of each dimension of the representation. A group of nonnegative and sparse word representations is shown to be well interpretable in each dimension. The third direction is to increase the interpretability of the embedding space by introducing more spatial properties in addition to the translational semantics in word2vec. In this section, we illustrate related work in these three directions.

Disambiguated Word Representation Using only one single vector to represent a word is problematic due to the ambiguity of words. A single vector that implies multiple meanings is naturally difficult to interpret, and distinguishing different meanings can lead to a more accurate representation.

In the multi-prototype vector space model, Banerjee et al. [5] use clustering algorithms to cluster different word meanings. Formally, it assigns a different word representation $\mathbf{w}_i(w_1)$ to the same word w_1 in each different cluster i. When the multi-prototype embedding is used, the similarity between two words w_1, w_2 is computed by comparing each pair of prototypes, i.e.,

$$\mathrm{AvgSim}(w_1, w_2) = \frac{1}{K^2} \sum_{i=1}^{K} \sum_{j=1}^{K} s(\mathbf{w}_i(w_1), \mathbf{w}_j(w_2)),$$

$$\mathrm{MaxSim}(w_1, w_2) = \max_{1 \le i, j \le K} s(\mathbf{w}_i(w_1), \mathbf{w}_j(w_2)),$$

(2.39)

where K is a hyper-parameter indicating the number of clusters and $s(\cdot)$ is a similarity function of two vectors, such as cosine similarity. When contexts are available, the similarity can be computed more precisely as

$$\mathrm{AvgSimC}(w_1, w_2) = \frac{1}{K^2} \sum_{i=1}^{K} \sum_{j=1}^{K} s_{c,w_1,i} s_{c,w_2,j} s(\mathbf{w}_i(w_1), \mathbf{w}_j(w_2)),$$

$$\mathrm{MaxSimC}(w_1, w_2) = s(\hat{\mathbf{w}}(w_1), \hat{\mathbf{w}}(w_2)),$$

(2.40)

where $s_{c,w_1,i} = s(\mathbf{w}(c), \mathbf{w}_i(w_1))$ is the likelihood of context c belonging to cluster i, $\mathbf{w}(c)$ is the context representation, and $\hat{\mathbf{w}}(w_1) = \mathbf{w}_{\arg\max_{1 \le i \le K} s_{c,w_1,i}}(w_1)$ is the maximum likelihood cluster for w_1 in context c. With multi-prototype

Fig. 2.9 The framework of Chen et al. [13]. We use the center word to predict both the context word and the context word's sense. The figure is redrawn according to Fig. 1 from Chen et al. [13]

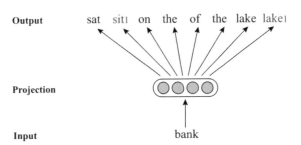

embeddings, the accuracy of the word similarity task is significantly improved, but the performance is still sensitive to the number of clusters.

The multi-prototype embedding method can effectively cluster different meanings of a word via its contexts. However, the clustering is offline, and the number of clusters is fixed and needs to be pre-defined. It is difficult for a model to select an appropriate amount of meanings for different words; to adapt to new senses, new words, or new data; and to align the senses with prototypes. A unified model is proposed for both word representation and word sense disambiguation [13]. It uses available knowledge bases such as WordNet [52] to provide the list of possible senses of a word, perform the disambiguation based on the original word vectors, and update the word vectors and sense vectors. More specifically, as shown in Fig. 2.9, it first initalizes the word vectors to be the skip-gram vectors learned from large-scale corpora. And then, it aggregates the words in the definition of the sense (provided by knowledge bases) to form the sense initialization, where only the words in the definition whose similarity with the original word is larger than a threshold are considered. After initialization, it uses the sense vector to update the context vectors. For example, in the sentence "*He sat on the bank of the lake*," the "*bank*" which means "*the land alongside the lake*" is closer to the context vectors formed by words "*sat, bank, lake*," and then the representation of $bank_1$ is utilized to update the context vectors. The process is repeated for all words with multiple senses. After the disambiguation, a joint objective is used to optimize the senses and word vectors together

$$\mathcal{L} = -\sum_{i} \sum_{i-l \leq j \leq i+l, j \neq i} P(w_j | w_i) P(M(w_j) | w_i)), \qquad (2.41)$$

where $M(w_j)$ is the disambiguated sense of word w_j and $2l + 1$ is the window size. With the joint learning framework, not only performance on word representation learning tasks are enhanced, but the representations of concrete senses are more interpretable than the representations of polysemy words.

Nonnegative and Sparse Word Representation Another aspect of interpretability comes from the interpretability of each dimension of distributed word representations. Murphy et al. [53] introduce nonnegative and sparse embeddings (NNSE), where each dimension indicates a unique concept. This method factorizes the corpus statistics matrix $\mathbf{M} \in \mathbb{R}^{|V| \times |D|}$ into a word embedding matrix $\mathbf{W} \in \mathbb{R}^{|V| \times m}$ and a document statistics matrix $\mathbf{D} \in \mathbb{R}^{m \times |D|}$, where $|V|$, $|D|$ and m are the vocabulary size, the number of documents, and the dimension of the distributed representation, respectively. Its training objective is

$$\arg \min_{\mathbf{W,D}} \frac{1}{2} \sum_{i=1}^{|V|} \|\mathbf{M}_{i,:} - \mathbf{W}_{i,:}\mathbf{D}\|_2 + \lambda \|\mathbf{W}_{i,:}\|_1,$$

$$\text{s.t.} \ \mathbf{D}_{i,:}\mathbf{D}_{i,:}^{\top} \leq 1, \ \text{for } 1 \leq i \leq m,$$

$$\mathbf{W}_{i,j} \geq 0, \ \text{for } 1 \leq i \leq |V|, 1 \leq j \leq m. \tag{2.42}$$

The sparsity is ensured by $\lambda \|\mathbf{W}_{i,:}\|_1$, and non-negativity is guaranteed by $\mathbf{W}_{i,j} \geq 0$. By iteratively optimizing $\mathbf{W}$ and $\mathbf{D}$ via gradient descent, this model can learn nonnegative and sparse embeddings for words. Since embeddings are nonnegative, words with the highest scores on each dimension show high similarity to more words. Therefore, they can be regarded as superordinate concepts of more specific words. Again, since embeddings are sparse and only a few words correspond to each dimension, each dimension can be interpreted as the concept (word) with the highest value in that dimension.

A word intrusion task is designed to assess the interpretability of the word representation. For each dimension, we pick the N words with the largest value for that dimension as the positive words. Then we select the noisy words with the value of that dimension in the small half. Finally, we let human annotators pick out these noise words. The performance of the human annotators in all dimensions is the interpretability score of the model.

Fyshe et al. [27] further improve NNSE by enforcing the compositionality of the interpretable dimensions. For a phrase p composed of words w_i and w_j, the following constraint can be applied:

$$\mathbf{W}_{p,:} = f(\mathbf{W}_{i,:}, \mathbf{W}_{j,:}). \tag{2.43}$$

Therefore, the objective becomes

$$\arg \min_{\mathbf{W,D}} \frac{1}{2} \sum_{i=1}^{|V|} \|\mathbf{M}_{i,:} - \mathbf{W}_{i,:}\mathbf{D}\|_2 + \lambda_1 \|\mathbf{W}_{i,:}\|_1$$

$$+ \lambda_2 \sum_{\text{phrase } p=(w_i, w_j)} (\mathbf{W}_{p,:} - f(\mathbf{W}_{i,:}, \mathbf{W}_{j,:}))^2, \tag{2.44}$$

where λ_1 and λ_2 are the coefficients to weigh different sub-objectives.

The f has many choices. The authors define it to be a weighted addition between $\mathbf{W}_{i,:}$ and $\mathbf{W}_{j,:}$, i.e.,

$$\mathbf{W}_{p,:} = \alpha \mathbf{W}_{i,:} + \beta \mathbf{W}_{j,:}. \tag{2.45}$$

The resulting word representations are more interpretable since the multiple dimensions can form compositional meanings.

The above method applies to the matrix factorization paradigm, which encounters difficulty when the corpus and global co-occurrence is large. Can we apply the same nonnegative regularization for neural word representations such as word2vec? Luo et al. [47] present a nonnegative skip-gram model OIWE (online interpretable word embeddings), which adds constraints to the gradient descent process. Specifically, the update rule of the parameter is

$$\mathbf{w}_k \leftarrow \max(0, \mathbf{w}_k + \gamma \nabla \mathcal{L}(\mathbf{w}_k)), \tag{2.46}$$

where $\mathbf{w}$ is the word representation that needs to be updated, k is its k-th dimension, and γ is the learning rate.

However, directly using this update rule leads to unstable optimization because the update deviates from the true update too much. What we need is to make fewer dimensions of $\mathbf{w}_k + \gamma \nabla f(\mathbf{w}_k)$ less than 0. To achieve this goal, we use a dynamic learning rate. The learning rate is chosen to make the following violation ratio small:

$$R(\gamma) = \frac{|\{k | \mathbf{w}_k > 0, \mathbf{w}_k + \gamma \nabla \mathcal{L}(\mathbf{w}_k) < 0\}|}{K}, \tag{2.47}$$

where K is the number of dimensions.

The resulting word representation exceeds NNSE in both word similarity and word intrusion detection tasks.

Non-Euclidean Word Representation Interpretability also comes from an embedding space with comprehensible spatial properties. For example, the translation property of word2vec makes the difference between *male* and *female* interpretable (i.e., relation *gender*). Therefore, we are looking for more interpretable spatial properties. We introduce two special embedding spaces, i.e., Gaussian distribution space and hyperbolic space. Both of them enjoy hierarchical spatial properties that are understandable by humans. Vilnis et al. [74] propose to encode words into Gaussian distribution $N(x; \mu, \Sigma)$, where the mean μ of the Gaussian distribution is similar to traditional word embedding and the variance Σ becomes the uncertainty of the word meaning. The similarity between two representations can be defined either using asymmetric similarity (e.g., the KL divergence) or symmetric similarity (e.g., the continuous inner product between two Gaussian distributions):

$$\int_{x \in \mathbb{R}^n} N(x; \mu_i, \Sigma_i) N(x; \mu_j, \Sigma_j) dx = N(0; \mu_i - \mu_j, \Sigma_i + \Sigma_j), \tag{2.48}$$

where n is the dimension of vectors.

Note that the focus of Vilnis et al. [74] is on the uncertainty estimation of word meanings, which increases the interpretability of word meanings in terms of uncertainty estimation. But on the other hand, it is very easy to define entailment relations between two Gaussian embeddings of different sizes (variances), thus natural to encode the hierarchy into the representation, which increases the interpretability of the embedding in terms of ontology. This line of work further develops into representations based on the Gaussian mixture model [2] and elliptical word embedding [54].

Another line of work focuses on hyperbolic embeddings. Hyperbolic spaces $\mathbb{H}^n$ are spaces with constant negative curvature. The volume of the circle in hyperbolic space grows exponentially with radius. This property makes it suitable for encoding the tree structure, where the number of nodes grows exponentially with depth. Hence, it is a suitable space for encoding hierarchical structures. For example, Nickel et al. [55] use a special hyperbolic space, namely, Poincaré Ball, as the embedding space. They propose to encode word relations such as those explicitly given by the hierarchy in WordNet using supervised learning. A subsequent work [72] successfully applies Poincaré embeddings in a completely unsupervised manner. Specifically, they propose Poincaré GloVe, a modified target of GloVe, for encoding hyperbolic geometry. Considering the GloVe target in Eq. (2.26), it can be generalized into more general metrics as follows:

$$
\begin{aligned}
d(\mathbf{w}_i, \mathbf{w}_j, \mathbf{M}_{ij}) &= (\mathbf{w}_i \mathbf{w}_j^\top + b_i + b_j - \log \mathbf{M}_{ij})^2 \\
&= (-\frac{1}{2} \|\mathbf{w}_i - \mathbf{w}_j\|_2^2 + b'_i + b'_j - \log \mathbf{M}_{ij})^2 \qquad (2.49) \\
&= (-\frac{1}{2} h_{\text{Euclidean}}(\|\mathbf{w}_i - \mathbf{w}_j\|_2) + b'_i + b'_j - \log \mathbf{M}_{ij})^2,
\end{aligned}
$$

where $\mathbf{M}_{ij}$ is the global co-occurrence matrix and $h_{\text{Euclidean}} = (\cdot)^2$ is the metric for accessing the similarity of the two embeddings. Now it can be substituted with $h_{\text{hyperbolic}} = \cosh^2(\cdot)$. The embedding is optimized using Riemannian optimization [8]. The use of this word vector for inference (e.g., word analogy tasks) requires using the corresponding hyperbolic space operators, which we omit the details.

In summary, encoding more information and improving interpretability have been pursued by researchers. With such efforts, word representations have become the basis of modern NLP and are widely used in many practical tasks.

2.5   Applications

Word representation, as a milestone breakthrough in NLP, has not only spawned subsequent work in NLP itself but has also been widely applied in other disciplines, catalyzing many highly influential interdisciplinary works. Therefore, in this section, we first introduce the applications of NLP itself, and then we introduce interdisciplinary works such as in psychology, history, and social science.

2.5.1 NLP

In the early stages of the introduction of neural networks into NLP, research on the application of word representations was very vigorous. For example, word representations are helpful in word-level tasks such as word similarity, word analogy, and ontology construction. They can also be applied to simple higher-level downstream tasks such as sentiment analysis. The performance of word representations on these tasks can measure the quality of word representations, so they can also be considered as *evaluation tasks of word representations*. Next, we introduce word similarity, word analogy, ontology construction, and sentence-level tasks.

Word Similarity and Relatedness Word similarity and relatedness both measure how close a word is to another word. Similarity means that the two words express similar meanings. And relatedness refers to a close linguistic relationship between the two words. Words that are not semantically similar could still be related in many ways, such as meronymy (*car* and *wheel*) or antonymy (*hot* and *cold*).

To measure word similarity and relatedness, researchers collect a set of word pairs and compute the correlation between human judgment and predictions made by word representations. So far, many datasets have been collected and made public. Some datasets focus on word similarity, such as RG-65 [62] and SimLex-999 [32]. Other datasets concern word relatedness, such as MTurk [60]. WordSim-353 [23] is a very popular dataset for word representation evaluation, but its annotation guideline does not differentiate similarity and relatedness. Agirre et al. [1] conduct another round of annotation based on WordSim-353 and generate two subsets, one for similarity and the other for relatedness. We summarize some information about these datasets in Table 2.1.

After collecting the datasets, quantitative evaluations of the word representations for the datasets are needed. Researchers usually select cosine similarity as the metric. After that, Spearman's correlation coefficient ρ is then used to evaluate

Table 2.1 Datasets for evaluating word similarity/relatedness

Dataset	Similarity type
RG-65 [62]	Word similarity
WordSim-353 [23]	Word similarity and relatedness
WordSim-353 REL [1]	Word relatedness
WordSim-353 SIM [1]	Word similarity
MTurk-287 [60]	Word relatedness
SimLex-999 [32]	Word similarity

the coherence between human annotators and the word representation. A higher Spearman's correlation coefficient indicates they are more similar.

Given the datasets and evaluation metrics, different kinds of word representations can be compared. Agirre et al. [1] address different word representations with different advantages. They point out that linguistic KB-based methods perform better on similarity than on relatedness, while distributed word representation shows similar performance on both. With the development of distributed representations, a study [68] in 2015 compares a series of word representations on a wide variety of datasets and gives the conclusion that distributed representations achieve state of the art in both similarity and relatedness.

Besides evaluation with deliberately collected datasets, the word similarity measurement can come in an alternative format, the TOEFL synonyms test. In this test, a cue word is given, and the test is required to choose one from four words that are the synonym of the cue word. The exciting part of this task is that the performance of a system could be compared with human beings. Landauer et al. [39] evaluate the system with the TOEFL synonyms test to address the knowledge inquiring and representing of LSA. The reported score is 64.4%, which is very close to the average rating of the human test-takers. On this test set with 80 queries, Sahlgren et al. [63] report a score of 72.0%. Freitag et al. [25] extend the original dataset with the help of WordNet and generate a new dataset (named as WordNet-based synonymy test) containing thousands of queries.

Word Analogy Besides word similarity, the word analogy task is an interesting task to infer words and serves as an alternative way to measure the quality of word representations. This task gives three words w_1, w_2, and w_3, and then it requires the system to predict a word w_4 such that the relation between w_1 and w_2 is the same as that between w_3 and w_4. This task has been used since the proposal of word2vec [49, 51] to exploit the structural relations among words. Here, word relations can be divided into two categories, including semantic relations and syntactic relations. Word analogy quickly becomes a standard evaluation metric once the dataset is released. Unlike the TOEFL synonyms test, the prediction is chosen from the whole vocabulary instead of the provided options. This test favors distributed word representations because it emphasizes the structure of word space. The comparison between different models on the word analogy task measured by accuracy could be found in [7, 68, 70, 75].

Ontology Construction Another usage of word representation is to construct the ontology knowledge bases. Section 2.4.1 talked about injecting knowledge base information into word representations. But conversely, learned word embeddings also help build the knowledge base. Since word representation is better at common words than rare words, word representations are more suitable for building ontology graphs[10] than building factual knowledge graphs.

---

[10] Ontology graph connects different abstract concepts in a graph according to their semantic relationships.

In ontologies, perhaps the most important relation is the is_a relation. Traditional word2vec models are good at expressing analogous relations, such as *man-woman* ≈ *king-queen* but not good at hierarchical relations, such as *mammal-cat* ≉ *celestial body-sun*. To model such relationships, Fu et al. [26] propose to use a linear projection rather than a simple embedding offset to represent the relationship. The model optimizes the projection as

$$\mathbf{W}^* = \arg\min_{\mathbf{W}} \sum_{i,j} \|\mathbf{w}_i \mathbf{W} - \mathbf{w}_j\|_2, \qquad (2.50)$$

where $\mathbf{w}_i$ and $\mathbf{w}_j$ are hypernym and hyponym embeddings and $\mathbf{W}$ is the transformation matrix.

The non-Euclidean word representations introduced in Sect. 2.4.2 also help build the ontology network. Another knowledge base that word embedding can help is the sememe knowledge introduced in Chap. 10.

Sentence-Level Tasks Besides word-level tasks, word representations can also be used alone in some simple sentence-level tasks. However, word representations trained under purely co-occurrence objectives may not be optimal for a given task, and we can include task-relevant objectives in training. Take sentiment analysis as an example. Most word representation methods capture syntactic and semantic information while ignoring the sentiment of the text. This is questionable because words with similar syntactic polarity but opposite sentiment polarity obtain close word vectors. Tang et al. [71] propose to learn sentiment-specific word embeddings (SSWE). An intuitive idea is to jointly optimize the sentiment classification model using word embeddings as its feature, and SSWE minimizes the cross entropy loss to achieve this goal. To better combine the unsupervised word embedding method and the supervised discriminative model, they further use the words in a window rather than a whole sentence to classify sentiment polarity. To get massive training data, they use distant supervision to generate sentiment labels for a document. On sentiment classification tasks, sentiment embeddings outperform other strong baselines, including SVM [15] and other word embedding methods. SSWE also shows strong polarity consistency, where the closest words of a word are more likely to have the same sentiment polarity compared with existing word representation models. This sentiment-specific word embedding method provides us with a general way to learn task-specific word embeddings, which is to design a joint loss function and generate massive labeled data automatically.

Interestingly, as subsequent research in NLP progressed, including the development of sentence representations (Chap. 4) and the introduction of pre-trained models (Chap. 5), simple word vectors gradually ceased to be used alone. We point out the following reasons for this:

- High-level (e.g., sentence level) semantic units require combinations between words, and simple arithmetic operations between word representations are not sufficient to model high-level semantic models.

- Most word representation models do not consider word order and cannot model utterance probabilities, much less generate language.

We recommend that readers continue reading subsequent chapters to become familiar with more advanced methods.

2.5.2 Cognitive Psychology

In cognitive psychology, a famous behavioral test examines the correlation in the subconscious mind, named the implicit association test (IAT) [30]. This test is widely used to detect biases, such as gender biases, religion biases, occupation biases, etc.

IAT is based on a hypothesis. The hypothesis says that people's reaction time decreases when faced with similar concepts and increases when faced with conflicting concepts. For example, given *target words* (*woman, man*) and *attribute words* (*beautiful, strong*), we want to test a subject's perspective: which attribute is associated more closely with each target. If a person believes that *woman* and *beautiful* are close and *man* and *strong* are close, then, when faced with category *A* (*woman, beautiful*) and category *B* (*man, strong*), he/she will quickly categorize the word *charming* into the former group. Whereas if he/she is faced with two categories *A* (*woman, strong*) and *B* (*man, beautiful*), then faced with the word *charming*, he/she will hesitate between the two pairs of words, thus increasing the reaction time substantially, although the correct answer is clear that *charming* should be grouped into *B* since it's an attribute word, and thus should be classified according to *beautiful* or *strong*. A part of an IAT is shown in Fig. 2.10.

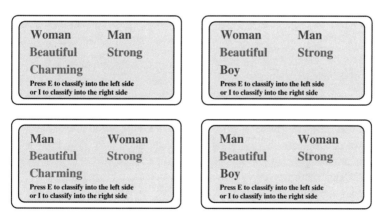

Fig. 2.10 Parts of the IAT. The box in the second row is considered to be harder than the first row for people who have an implicit association between *women* and *beautiful* (*man* and *strong*)

IAT can detect implicit thoughts and biases in the human mind. Considering that bias is so prevalent in people's perceptions, it will likely be reflected in the texts written by humans. A pioneering article [12] in Science magazine proposes WEAT (word-embedding association test) to detect bias in texts. Given two sets X, Y as the target words and A, B as two sets of attribute words, WEAT defines a difference between the target sets to the two sets of attribute words as follows:

$$s(X, Y; A, B) = \sum_{x \in X} s(x; A, B) - \sum_{y \in Y} s(y; A, B), \qquad (2.51)$$

and

$$s(x; A, B) = \text{mean}_{a \in A} \cos(\mathbf{x}, \mathbf{a}) - \text{mean}_{a \in B} \cos(\mathbf{x}, \mathbf{b}), \qquad (2.52)$$

where $\mathbf{x}, \mathbf{y}, \mathbf{a}, \mathbf{b}$ are the vector representation of word x, y, a, b. $\cos(\mathbf{x}, \mathbf{a})$ that can be treated as the response time in the IAT. And mean$(\cdot)$ is the average function. Thus, $s(x; A, B)$ measures the closeness of x to two sets of attributes. And $s(X, Y; A, B)$ measures the bias difference between X and Y to A and B. If X, Y are not biased differently for A and B, then they should not exceed at least the bias difference of (X_i, Y_i), where $X \cap Y$ is randomly divided into two sets X_i, Y_i of equal size. Hence, we test whether the following metric is small:

$$P[s(X_i, Y_i; A, B) > s(X, Y; A, B)]. \qquad (2.53)$$

After some statistical derivations, we are able to calculate the above probabilities. In their experiments, WEAT is capable of capturing the occupational gender bias, where the occupational gender association calculated from the word representation is highly correlated with the publicly available proportion of female workers in each industry. For the name gender association, WEAT can find a similar pattern.

2.5.3 History and Social Science

It is possible to detect human thoughts without conducting live experiments using tests built on texts. This makes it very helpful to study the thoughts of ancient people. That is, we can explore the thoughts of the ancients through the texts they wrote, and this is precisely the important role of word representations in history and social sciences. In this section, we talk about how to use word representations to study the changes in history and society across time.

In order to track the chronological changes in word meanings, we first need to have a corpus of different chronologies. Google NGram Book Corpus [43] is a relatively early chronological corpus. This dataset counts the words/phrases' frequency used during the last five centuries and includes 6% of all published books, which is a very large dataset. Another COHA dataset [16] has 400 million words, documenting the development of American English between 1810 and 2009, contains a wide variety of genres, and is relatively balanced across genres.

The work on tracking word sense changes [31, 37] divides these datasets into bins of equal size by time. They then train word representation models on the text within each period. For example, in the work [31], the SVD decomposition of the PPMI (positive pointwise mutual information) matrix and SGNS (skip-gram with negative sampling) are used as two base models. As mentioned earlier, the dimensions are not aligned for different groups of distributed representations, even if they are derived from the same counting matrix. Therefore, the authors propose to optimize a mapping matrix that maps the word vectors of the previous period to the word vector space of the latter period.

Aligning the word vectors after training them usually does not yield satisfying results because simple transformations cannot always align the two vector spaces, and complex transformations carry the risk of overfitting. **Time-sensitive word representation** is developed to address these issues. Bamler et al. [4] propose a dynamic skip-gram model which connects several Bayesian skip-gram models [6] using Kalman filters [35]. In this model, the embeddings of words in different periods could affect each other. For example, a word that appears in the 1990s document can affect the embeddings of that word in the 1980s and 2000s. Moreover, this model puts all the embeddings into the same semantic space, significantly improving against other methods and making word embeddings in different periods comparable. Experimental results show that the cosine distance between two words changes much more smoothly in this model than in those that simply divide the corpus into bins.

We can arrive at interesting *societal observations* using word representations from different periods. Hamilton et al. [31] perform two analyses, the first of which computes the time series formed by the cosine value of a word pair over time. They use Spearman correlation coefficients of the time series against time to estimate whether this change is an upward or downward trend and how significant the trend is. For the second analysis, the authors track the degree of change in the word vector of the same word over time to see the semantic drifts of a word across periods. They have come to some interesting conclusions. (1) Some established word sense shifts can be confirmed from the corpus. For example, the shift of *gay* change from *happy* to *homosexual* is observed from the word representation. (2) The authors also find the ten words that changed the most from 1900 to 1990. (3) Combining some experimental observations, the authors found two statistical rules of semantic variation. The first is that common words change their meanings more slowly, and rare words change their meanings more quickly. The second is that words with multiple meanings change their meanings more quickly.

The follow-up work [29] is based on the same diachronic word vectors as Hamilton et al. [31] but makes some observations with more depth in social science. Specifically, it compares trends in gender and race stereotypes over 100 years. To get the stereotype information from word representations, this article computes the difference in the association scores of an attribute word (e.g., *intelligent*) to two groups of words (e.g., *woman, female* versus *man, male*). The association score can be calculated by either cosine similarity or Euclidean distance. This is similar to the WEAT mentioned in Sect. 2.5.2. The work further compares the association score with publicly available data about the gender per occupation statistics over the year. They find the two trends match almost exactly. When studying the association of adjectives to genders, a clear phase shift is found. The similarity of adjectives' association scores to genders is similar within the 1910s~1960s and the 1960s~1990s, respectively, but differs substantially between the two time periods. The phase shift in the 1960s corresponds to the US women's movement in history.

2.6 Summary and Further Readings

In this chapter, we focus on words, which are the basic semantic units. We introduce the representative methods in symbolic representation and distributed representation. Some well-known models are introduced, such as one-hot representation, LSA, PLSA, LDA, word2vec, GloVe, and ELMo. We also present the methods to make the representations more informative and interpretable. At last, the applications of word representations are introduced, where interdisciplinary applications are emphasized. For further readings, firstly, we encourage the readers to read Chaps. 3 and 4 for higher-level representations as they are more widely used in practical tasks. We also encourage the readers to review historical research, such as the review paper on representation learning by Yoshua Bengio [9], and the review paper on pre-trained distributed word representations by Tomas Mikolov, the author of word2vec [50].

Acknowledgments Zhiyuan Liu, Yankai Lin, and Maosong Sun designed the overall architecture of this chapter; Shengding Hu drafted this chapter. Zhiyuan Liu and Yankai Lin proofread and revised this chapter.

We also thank Ning Ding, Yujia Qin, Si Sun, Yusheng Su, Zhitong Wang, Xingyu Shen, Zheni Zeng, and Ganqu Cui for proofreading the chapter and Lei Xu for preparing the initial draft materials for the first edition.

This is the word representation learning chapter about of the second edition of the book *Representation Learning for Natural Language Processing*, with its first edition published in 2020 [46]. Compared to the first edition of this chapter, the main changes include the following: (1) we rewrote the sections before Sect. 2.4 by systematically restructuring the works, (2) we restructured and summarized the advanced topics into two directions and polish the writing of advanced topics, and (3) we added a new section to introduce word representation's applications.

References

1. Eneko Agirre, Enrique Alfonseca, Keith Hall, Jana Kravalova, Marius Paşca, and Aitor Soroa. A study on similarity and relatedness using distributional and WordNet-based approaches. In *Proceedings of NAACL-HLT*, 2009.
2. Ben Athiwaratkun and Andrew Wilson. Multimodal word distributions. In *Proceedings of ACL*, 2017.

3. Collin F Baker, Charles J Fillmore, and John B Lowe. The berkeley framenet project. In *Proceedings of ACL*, 1998.
4. Robert Bamler and Stephan Mandt. Dynamic word embeddings via skip-gram filtering. *arXiv preprint arXiv:1702.08359*, 2017.
5. Arindam Banerjee, Inderjit S Dhillon, Joydeep Ghosh, and Suvrit Sra. Clustering on the unit hypersphere using von Mises-Fisher distributions. *Journal of Machine Learning Research*, 2005.
6. Oren Barkan. Bayesian neural word embedding. In *Proceedings of AAAI*, 2017.
7. Marco Baroni, Georgiana Dinu, and Germán Kruszewski. Don't count, predict! a systematic comparison of context-counting vs. context-predicting semantic vectors. In *Proceedings of ACL*, 2014.
8. Gary Bécigneul and Octavian-Eugen Ganea. Riemannian adaptive optimization methods. In *Proceedings of ICLR*, 2019.
9. Yoshua Bengio, Aaron Courville, and Pascal Vincent. Representation learning: A review and new perspectives. *IEEE Transactions on Pattern Analysis and Machine Intelligence*, 2013.
10. David M. Blei, Andrew Y. Ng, and Michael I. Jordan. Latent Dirichlet Allocation. In *Proceedings of NeurIPS*, 2001.
11. Piotr Bojanowski, Edouard Grave, Armand Joulin, and Tomas Mikolov. Enriching word vectors with subword information. *Transactions of the Association for Computational Linguistics*, 2017.
12. Aylin Caliskan, Joanna J Bryson, and Arvind Narayanan. Semantics derived automatically from language corpora contain human-like biases. *Science*, 2017.
13. Xinxiong Chen, Zhiyuan Liu, and Maosong Sun. A unified model for word sense representation and disambiguation. In *Proceedings of EMNLP*, 2014.
14. Xinxiong Chen, Lei Xu, Zhiyuan Liu, Maosong Sun, and Huanbo Luan. Joint learning of character and word embeddings. In *Proceedings of IJCAI*, 2015.
15. Corinna Cortes and Vladimir Vapnik. Support-vector networks. *Machine Learning*, 1995.
16. Mark Davies. The corpus of historical American english: 400 million words, (1810-2009), 2010.
17. Scott Deerwester, Susan T Dumais, George W Furnas, Thomas K Landauer, and Richard Harshman. Indexing by latent semantic analysis. *Journal of the American Society for Information Science*, 1990.
18. Arthur P Dempster, Nan M Laird, and Donald B Rubin. Maximum likelihood from incomplete data via the EM algorithm. *Journal of the Royal Statistical Society: Series B (Methodological)*, 1977.
19. Jacob Devlin, Ming-Wei Chang, Kenton Lee, and Kristina Toutanova. BERT: pre-training of deep bidirectional transformers for language understanding. In *Proceedings of NAACL-HLT*, 2019.
20. Chris Ding, Tao Li, and Wei Peng. On the equivalence between non-negative matrix factorization and probabilistic latent semantic indexing. *Computational Statistics & Data Analysis*, 2008.
21. Katrin Erk and Sebastian Padó. A structured vector space model for word meaning in context. In *Proceedings of EMNLP*, 2008.
22. Manaal Faruqui, Jesse Dodge, Sujay Kumar Jauhar, Chris Dyer, Eduard Hovy, and Noah A Smith. Retrofitting word vectors to semantic lexicons. In *Proceedings of NAACL-HLT*, 2015.
23. Lev Finkelstein, Evgeniy Gabrilovich, Yossi Matias, Ehud Rivlin, Zach Solan, Gadi Wolfman, and Eytan Ruppin. Placing search in context: The concept revisited. In *Proceedings of WWW*, 2001.
24. John R Firth. A synopsis of linguistic theory, 1930-1955. *Studies in Linguistic Analysis*, 1957.
25. Dayne Freitag, Matthias Blume, John Byrnes, Edmond Chow, Sadik Kapadia, Richard Rohwer, and Zhiqiang Wang. New experiments in distributional representations of synonymy. In *Proceedings of CoNLL*, 2005.
26. Ruiji Fu, Jiang Guo, Bing Qin, Wanxiang Che, Haifeng Wang, and Ting Liu. Learning semantic hierarchies via word embeddings. In *Proceedings of ACL*, 2014.

27. Alona Fyshe, Leila Wehbe, Partha Pratim Talukdar, Brian Murphy, and Tom M Mitchell. A compositional and interpretable semantic space. In *Proceedings of NAACL-HLT*, 2015.
28. Juri Ganitkevitch, Benjamin Van Durme, and Chris Callison-Burch. PPDB: The paraphrase database. In *Proceedings of NAACL-HLT*, 2013.
29. Nikhil Garg, Londa Schiebinger, Dan Jurafsky, and James Zou. Word embeddings quantify 100 years of gender and ethnic stereotypes. *Proceedings of the National Academy of Sciences*, 2018.
30. Anthony G Greenwald and Shelly D Farnham. Using the implicit association test to measure self-esteem and self-concept. *Journal of Personality and Social Psychology*, 2000.
31. William L. Hamilton, Jure Leskovec, and Dan Jurafsky. Diachronic word embeddings reveal statistical laws of semantic change. In *Proceedings of ACL*, 2016.
32. Felix Hill, Roi Reichart, and Anna Korhonen. Simlex-999: Evaluating semantic models with (genuine) similarity estimation. *Computational Linguistics*, 2015.
33. Sepp Hochreiter and Jürgen Schmidhuber. Long short-term memory. *Neural Computation*, 1997.
34. Thomas Hofmann. Probabilistic latent semantic indexing. In *Proceedings of SIGIR*, 1999.
35. Rudolph Emil Kalman et al. A new approach to linear filtering and prediction problems. *Journal of Basic Engineering*, 1960.
36. Pentti Kanerva, Jan Kristofersson, and Anders Holst. Random indexing of text samples for latent semantic analysis. In *Proceedings of CogSci*, 2000.
37. Yoon Kim, Yi-I Chiu, Kentaro Hanaki, Darshan Hegde, and Slav Petrov. Temporal analysis of language through neural language models. In *Proceedings of ACL Workshop*, 2014.
38. Da Kuang, Jaegul Choo, and Haesun Park. Nonnegative matrix factorization for interactive topic modeling and document clustering. *Partitional Clustering Algorithms*, page 215, 2014.
39. Thomas K Landauer and Susan T Dumais. A solution to Plato's problem: The latent semantic analysis theory of acquisition, induction, and representation of knowledge. *Psychological Review*, 1997.
40. Omer Levy and Yoav Goldberg. Dependency-based word embeddings. In *Proceedings of ACL*, 2014.
41. Omer Levy and Yoav Goldberg. Neural word embedding as implicit matrix factorization. In *Proceedings of NeurIPS*, 2014.
42. Shaohua Li, Tat-Seng Chua, Jun Zhu, and Chunyan Miao. Generative topic embedding: a continuous representation of documents. In *Proceedings of ACL*, 2016.
43. Yuri Lin, Jean-Baptiste Michel, Erez Aiden Lieberman, Jon Orwant, Will Brockman, and Slav Petrov. Syntactic annotations for the Google Books NGram corpus. In *Proceedings of the ACL*, 2012.
44. Wang Ling, Chris Dyer, Alan W Black, Isabel Trancoso, Ramón Fermandez, Silvio Amir, Luis Marujo, and Tiago Luís. Finding function in form: Compositional character models for open vocabulary word representation. In *Proceedings of EMNLP*, 2015.
45. Yang Liu, Zhiyuan Liu, Tat-Seng Chua, and Maosong Sun. Topical word embeddings. In *Proceedings of AAAI*, 2015.
46. Zhiyuan Liu, Yankai Lin, and Maosong Sun. *Representation Learning for Natural Language Processing*. Springer, 2020.
47. Hongyin Luo, Zhiyuan Liu, Huanbo Luan, and Maosong Sun. Online learning of interpretable word embeddings. In *Proceedings of EMNLP*, 2015.
48. T Mikolov and J Dean. Distributed representations of words and phrases and their compositionality. In *Proceedings of NeurIPS*, 2013.
49. Tomas Mikolov, Kai Chen, Greg Corrado, and Jeffrey Dean. Efficient estimation of word representations in vector space. In *Proceedings of ICLR*, 2013.
50. Tomas Mikolov, Edouard Grave, Piotr Bojanowski, Christian Puhrsch, and Armand Joulin. Advances in pre-training distributed word representations. In *Proceedings of LREC*, 2018.
51. Tomas Mikolov, Wen-tau Yih, and Geoffrey Zweig. Linguistic regularities in continuous space word representations. In *Proceedings of NAACL-HLT*, 2013.
52. George A Miller. WordNet: a lexical database for English. *Communications of the ACM*, 1995.

53. Brian Murphy, Partha Talukdar, and Tom Mitchell. Learning effective and interpretable semantic models using non-negative sparse embedding. In *Proceedings of COLING*, 2012.
54. Boris Muzellec and Marco Cuturi. Generalizing point embeddings using the Wasserstein space of elliptical distributions. In *Proceedings of NeurIPS*, 2018.
55. Maximillian Nickel and Douwe Kiela. Poincare embeddings for learning hierarchical representations. In *Proceedings of NeurIPS*, 2017.
56. Sebastian Padó and Mirella Lapata. Dependency-based construction of semantic space models. *Computational Linguistics*, 2007.
57. Jeffrey Pennington, Richard Socher, and Christopher Manning. GloVe: Global vectors for word representation. In *Proceedings of EMNLP*, 2014.
58. Matthew Peters, Mark Neumann, Mohit Iyyer, Matt Gardner, Christopher Clark, Kenton Lee, and Luke Zettlemoyer. Deep contextualized word representations. In *Proceedings of NAACL-HLT*, 2018.
59. Alec Radford, Karthik Narasimhan, Tim Salimans, and Ilya Sutskever. Improving language understanding with unsupervised learning. 2018.
60. Kira Radinsky, Eugene Agichtein, Evgeniy Gabrilovich, and Shaul Markovitch. A word at a time: computing word relatedness using temporal semantic analysis. In *Proceedings of WWW*, 2011.
61. Philip Resnik. Semantic similarity in a taxonomy: An information-based measure and its application to problems of ambiguity in natural language. *Journal of artificial intelligence research*, 1999.
62. Herbert Rubenstein and John B Goodenough. Contextual correlates of synonymy. *Communications of the ACM*, 1965.
63. Magnus Sahlgren. Vector-based semantic analysis: Representing word meanings based on random labels. In *Proceedings of SKAC Workshop*, 2001.
64. Magnus Sahlgren. An introduction to random indexing. In *Proceedings of TKE*, 2005.
65. Magnus Sahlgren. *The Word-Space Model: Using Distributional Analysis to represent syntagmatic and paradigmatic relations between words in high-dimensional vector spaces.* PhD thesis, Institutionen för lingvistik, 2006.
66. Gerard Salton. *The SMART retrieval system—experiments in automatic document processing.* Prentice-Hall, Inc., 1971.
67. Gerard Salton, Anita Wong, and Chung-Shu Yang. A vector space model for automatic indexing. *Communications of the ACM*, 1975.
68. Tobias Schnabel, Igor Labutov, David Mimno, and Thorsten Joachims. Evaluation methods for unsupervised word embeddings. In *Proceedings of EMNLP*, 2015.
69. Angus Stevenson. *Oxford dictionary of English.* 2010.
70. Fei Sun, Jiafeng Guo, Yanyan Lan, Jun Xu, and Xueqi Cheng. Learning word representations by jointly modeling syntagmatic and paradigmatic relations. In *Proceedings of ACL*, 2015.
71. Duyu Tang, Furu Wei, Nan Yang, Ming Zhou, Ting Liu, and Bing Qin. Learning sentiment-specific word embedding for Twitter sentiment classification. In *Proceedings of ACL*, 2014.
72. Alexandru Tifrea, Gary Bécigneul, and Octavian-Eugen Ganea. Poincaré glove: Hyperbolic word embeddings. *arXiv preprint arXiv:1810.06546*, 2018.
73. Laurens Van der Maaten and Geoffrey Hinton. Visualizing data using t-SNE. *Journal of Machine Learning Research*, 2008.
74. Luke Vilnis and Andrew McCallum. Word representations via Gaussian embedding. In *Proceedings of ICLR*, 2015.
75. Dani Yogatama, Manaal Faruqui, Chris Dyer, and Noah A Smith. Learning word representations with hierarchical sparse coding. In *Proceedings of ICML*, 2015.
76. Mo Yu and Mark Dredze. Improving lexical embeddings with semantic knowledge. In *Proceedings of ACL*, 2014.
77. Meng Zhang, Haoruo Peng, Yang Liu, Huan-Bo Luan, and Maosong Sun. Bilingual lexicon induction from non-parallel data with minimal supervision. In *Proceedings of AAAI*, 2017.
78. Will Y Zou, Richard Socher, Daniel M Cer, and Christopher D Manning. Bilingual word embeddings for phrase-based machine translation. In *Proceedings of EMNLP*, 2013.

Chapter 3
Representation Learning for Compositional Semantics

Ning Ding, Yankai Lin, Zhiyuan Liu, and Maosong Sun

Abstract Many important applications in NLP fields rely on understanding complex language units composed of words, such as phrases, sentences, and documents. A key problem is semantic composition, i.e., how to represent the semantic meanings of complex language units by composing the semantic meanings of those words in them. Semantic composition is a much more complicated process than a simple sum of the parts, and compositional semantics remains a core task in computational linguistics and NLP. In this chapter, we first introduce representative approaches for binary semantic composition, including additive models and multiplicative models, to demonstrate the nature of natural languages. After that, we present typical modeling methods for N-ary semantic composition, including sequential order, recursive order, and convolutional order methods, which follows a similar track for sentence and document representation learning and will be introduced in detail in the next chapter.

3.1 Introduction

Following the distributional hypothesis, one could project the semantic meaning of a word into a low-dimensional real-valued vector according to its context information. Here comes a further problem: how to compress a higher semantic unit, such as a phrase, into a vector or other kinds of mathematical representations like a matrix or a tensor? In this chapter, we introduce the representation learning approach to model semantic composition functions from the linguistic perspective.

Compositionality enables natural languages to construct complicated semantic meanings from the combinations of basic semantic elements with particular rules.

N. Ding · Z. Liu (✉) · M. Sun
Department of Computer Science and Technology, Tsinghua University, Beijing, China
e-mail: dingn18@mails.tsinghua.edu.cn; liuzy@tsinghua.edu.cn; sms@tsinghua.edu.cn

Y. Lin
Gaoling School of Artificial Intelligence, Renmin University of China, Beijing, China
e-mail: yankailin@ruc.edu.cn

© The Author(s) 2023
Z. Liu et al. (eds.), *Representation Learning for Natural Language Processing*,
https://doi.org/10.1007/978-981-99-1600-9_3

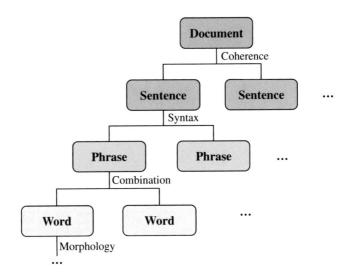

Fig. 3.1 Higher-level linguistic units are composed of basic linguistic units guided by certain rules

The basic principle is the semantic meaning of a whole is a function of the semantic meanings of its several parts. Therefore, the semantic meanings of complex structures will depend on how their semantic elements combine. There are many previous studies dedicated to the representation learning of compositional semantics. Among them, the article *composition in distributional models of semantics* [7] proposes a comprehensive framework of compositional semantics and becomes a rather representative summarization of this line of work. In this chapter, we adopt the framework to introduce compositional semantics along with our understanding and discussion (Fig. 3.1).

Here, we consider two basic semantic units and use **u** and **v** to denote them, respectively. And the most intuitive way to define the joint representation could be formulated by directly building a mapping function:

$$\mathbf{p} = f(\mathbf{u}, \mathbf{v}), \tag{3.1}$$

where **p** corresponds to the representation of the joint semantic unit $(\mathbf{u}, \mathbf{v})$. Generally, **u** and **v** could denote words, phrases, sentences, paragraphs, or even higher-level semantic units.

However, given the representations of two semantic constituents, it is not enough to derive their joint embedding without syntactic information. For instance, although the phrase *machine learning* and *learning machine* have the same vocabulary, they contain different meanings: *machine learning* refers to a research field in artificial intelligence, while *learning machine* means some specific learning algorithms. That is to say, the way how the units are mixed is also an essential part of the semantic composition. This phenomenon stresses the importance of syntactic and order information in a compositional unit. Hence, an improved version of the framework is to take the role of syntactic and order information into consideration [9]. Specifically, in terms of formulation, we can introduce an $\mathcal{R}$ to represent the relationship between units.

The complex semantics of a combined unit in the real world can also be influenced by human background knowledge. In other words, some sentences are difficult to understand by merely paying attention to the constituent units and the syntax. For example, the sentence *Tom and Jerry is one of the most popular comedies in that style.* needs two main backgrounds: firstly, *Tom and Jerry* is a special noun phrase or knowledge entity that indicates a cartoon comedy, rather than two ordinary people. The other prior knowledge should be *that style*, which needs further explanation in the previous sentences. Hence, a full understanding of compositional semantics needs to take external knowledge into account. We consummate the formulation by adding another item $\mathcal{K}$ to denote such background knowledge. The complete composition function in Eq. (3.1) is redefined to combine the syntactic relationship rule $\mathcal{R}$ between the semantic units $\mathbf{u}$ and $\mathbf{v}$ and human background knowledge $\mathcal{K}$:

$$\mathbf{p} = f(\mathbf{u}, \mathbf{v}, \mathcal{R}, \mathcal{K}). \tag{3.2}$$

From the perspective of computational linguistics, the meaning of a word is not isolated from the context. That is, the semantics of the combined unit are constituted from its components, and the meanings of the components are meanwhile derived from the combined unit [3]. This also echoes the concept of structuralism stated in Chap. 1. Compositionality is also a matter of degree instead of an either-or issue [8]. We could divide linguistic structures into several groups according to the degree of compositionality. For example, the fully compositional structure means that the combined high-level semantics is completely composed of the independent semantics of basic units (e.g., white fur). In partly compositional expressions, basic units still have separate meanings, but when combined together, they derive additional semantics (e.g., take care). In non-compositional idioms or multi-word expressions, the combined semantics have little relationship with the semantics of basic units (e.g., at large).

From the above equations formulating composition function, it could be concluded that composition could be viewed as more than a specific binary operation. The syntactic information could help to indicate a particular approach, while background knowledge helps to explain some obscure words or specific context-dependent entities such as pronouns. To this end, we could realize that the complexity of language comes from the nearly infinite combination of finite elements. This chapter can be seen as the transition from word representation to sentence and document representation, aiming to introduce the basic concepts and methods for dealing with compositional semantics from a linguistic point of view. We will first explain basic binary composition functions in Sect. 3.2, including the additive model and multiplicative model and then give a brief introduction to modeling methods for more complex N-ary composition in Sect. 3.3, including sequential order, recursive order, and convolutional order modeling. We will introduce the specific methods of learning sentence and document representation in modern NLP in detail in the next chapter.

3.2 Binary Composition

The goal of compositional semantics is to construct vector representations for higher-level linguistic units with basic units via binary composition. Without loss of generality, we assume that each constituent of a phrase (or even higher-level linguistic units) is embedded into a computable vector which will be further used to generate a representation vector for the phrase.

In this section, we focus on binary composition, where two objects will be involved in each operation. We now consider phrases consisting of two components: a head and a modifier or complement. If we cannot model the binary composition (or phrase representation), it is almost impossible to build more complex compositional representations for higher-level linguistic units. Even in today's age of neural networks, the concept of binary composition is still important. For example, in the transformer architecture, the calculation of the attention of two units could be regarded as a type of binary operation.

Given a phrase with two constituent words *machine learning*, as well as the representations $\mathbf{u}$ and $\mathbf{v}$ representing the words *machine* and *learning*, respectively, our primary goal is to construct a representation vector $\mathbf{p}$ of the phrase according to the representations of the words. With a simple semantic space where each vector is represented by five integers, we let the hypothetical vectors for *machine* and *learning* be $[0, 3, 1, 5, 2]$ and $[1, 4, 2, 2, 0]$, respectively. And if we simply use the add operator to represent the phrase *machine learning*, it becomes $[0, 3, 1, 5, 2] + [1, 4, 2, 2, 0] = [1, 7, 3, 7, 2]$. The key to this problem is designing a primitive composition function as a binary operator. Based on this function, one could apply it to a word sequence recursively to derive composition for longer text.

Modeling the binary composition function is a well-studied but still challenging problem. There are mainly two perspectives on this question, including the additive model and the multiplicative model according to the basic operators. We will introduce the basic concepts and computation principles of these two approaches in this section.

3.2.1 Additive Model

The additive model, as the name implies, is a modeling method with addition as the basic operation. Recall that in the introductory part, we have derived a formulation that may contain complex relationships between units and external background knowledge, which would make our discussion exceedingly broad. In this section, to narrow the space of our considered function and establish fundamental understandings of compositional semantics, we start by simplifying the formula to $\mathbf{p} = f(\mathbf{u}, \mathbf{v})$ and omitting the relationship and background items. Naturally, if we aim to perform addition correctly, $\mathbf{p}$, $\mathbf{u}$, and $\mathbf{v}$ should lie in the same semantic space.

One of the simplest ways is to directly use the sum to represent the joint representation:

$$\mathbf{p} = \mathbf{u} + \mathbf{v}. \tag{3.3}$$

As computed in the foregoing part, the sum of the two vectors representing *machine* and *learning* would be $\mathbf{w}(machine) + \mathbf{w}(learning) = [1, 7, 3, 7, 2]$. It assumes that the composition of different constituents is a symmetric function where $\mathbf{p} = \mathbf{u} + \mathbf{v} = \mathbf{v} + \mathbf{u}$. That is, it does not consider the order of constituents. Although having lots of drawbacks such as a lack of the ability to model word orders and the absence of background syntactic or knowledge information, this approach still provides a relatively strong baseline [6].

To overcome the word order issue, one easy variant is applying a weighted sum instead of uniform weights. This is, the composition has the following form:

$$\mathbf{p} = \alpha \mathbf{u} + \beta \mathbf{v}, \tag{3.4}$$

where α and β correspond to different weights for two vectors. Under this setting, two sequences (u, v) and (v, u) have different representations when $\alpha \neq \beta$, which is consistent with real language phenomena. For example, *machine learning* and *learning machine* have different meanings and require different representations. To this end, we assign different importance scores to different components. For instance, we set α to 0.3 and β to 0.7, the $0.3 \times \mathbf{w}(machine) = [0, 0.9, 0.3, 1.5, 0.6]$ and $0.7 \times \mathbf{w}(learning) = [0.7, 2.8, 1.4, 1.4, 0]$, and *machine learning* is represented by their addition $0.3 \times \mathbf{w}(machine) + 0.7 \times \mathbf{w}(learning) = [0.7, 3.7, 1.7, 2.9, 0.6]$.

Now, we attempt to incorporate prior knowledge and syntax information into the additive model in a straightforward way. To achieve that, one could combine K nearest neighborhood semantics into composition, deriving:

$$\mathbf{p} = \mathbf{u} + \sum_{i=1}^{L} \mathbf{m}_i + \mathbf{v} + \sum_{i=1}^{K} \mathbf{n}_i, \tag{3.5}$$

where $\mathbf{m}_1, \mathbf{m}_2, \ldots, \mathbf{m}_L$ denote semantic neighbors (i.e., synonyms) of $\mathbf{u}$, and $\mathbf{n}_1, \mathbf{n}_2, \ldots, \mathbf{n}_K$ denote semantic neighbors of $\mathbf{v}$. To this end, this method could ensemble such synonyms of the component as a smoothing factor into the composition function, which reduces the variance of language. For example, if in the composition of *machine* and *learning*, the chosen neighbors are *computer* and *optimizing* with $\mathbf{w}(computer) = [1, 0, 0, 0, 1]$ and $\mathbf{w}(optimizing) = [1, 5, 3, 2, 1]$, respectively. This leads to the situation that the representation of *machine learning* becomes $\mathbf{w}(machine) + \mathbf{w}(computer) + \mathbf{w}(learning) + \mathbf{w}(optimizing) =$

[3, 12, 6, 9, 4]. Although it is a simple strategy, the use of synonyms to improve the robustness of language models is still a very effective practice in modern NLP.

When it comes to the measurement of similarity between representations, the cosine function is a natural approach in the semantic space. We will take a closer look to understand the additive model by computing the cosine similarity between $\mathbf{p} = \mathbf{u} + \mathbf{v}$ (we go back to the naive additive model for computation simplicity) and an arbitrary word $\mathbf{w}$. The cosine similarity, denoted as $s(\cdot)$ could be derived as:

$$s(\mathbf{p}, \mathbf{w}) = \frac{\mathbf{p} \cdot \mathbf{w}}{\|\mathbf{p}\| \cdot \|\mathbf{w}\|} = \frac{(\mathbf{u} + \mathbf{v})\mathbf{w}}{\|\mathbf{u} + \mathbf{v}\| \|\mathbf{w}\|} \tag{3.6}$$

$$= \frac{\|\mathbf{u}\|}{\|\mathbf{u} + \mathbf{v}\|} s(\mathbf{u}, \mathbf{w}) + \frac{\|\mathbf{v}\|}{\|\mathbf{u} + \mathbf{v}\|} s(\mathbf{v}, \mathbf{w}). \tag{3.7}$$

From the derivation ahead, it could be concluded that this composition function composes of both magnitude and directions of two component vectors. And the composition similarity of two linguistic units could be viewed as a linear combination of the similarity of two components. In other words, if one vector dominates the magnitude, it will also dominate the similarity. For example, if $\|\mathbf{u}\| = 10^3$ and $\|\mathbf{v}\| = 10^{-3}$, the similarity between $\mathbf{p}$ and $\mathbf{w}$ will be mostly determined by the semantics of $\mathbf{u}$. This could happen if $\mathbf{u}$ is an entity with a strong specific meaning like *Europe* while $\mathbf{v}$ is an empty word like *there*. Further disassembly, in terms of the norm of compositional semantics, we have:

$$\|\mathbf{p}\| = \|\mathbf{u} + \mathbf{v}\| \le \|\mathbf{u}\| + \|\mathbf{v}\|. \tag{3.8}$$

This lemma suggests that the semantic unit with a deeper-rooted parsing tree could determine the joint representation when combined with a shallow unit. That is, the closer the unit to the final semantic combined unit, the more likely it is to exert a greater influence on the overall semantics.

3.2.2 Multiplicative Model

Though the additive model achieves considerable success in semantic composition, the simplification may also restrict it from performing more complex interactions. Different from the additive model that regards composition as a simple linear transformation, the three-order multiplicative model aims to make higher-order interactions by using multiplication as the basic operator. Among all models from this perspective, the most intuitive approach tried to apply the pairwise product as a composition function approximation. In this method, the composition function is shown as the following:

$$\mathbf{p} = \mathbf{u} \odot \mathbf{v}, \tag{3.9}$$

where, $\mathbf{p}_i = \mathbf{u}_i \cdot \mathbf{v}_i$, which implies each dimension of the output only depends on the corresponding dimension of two input vectors. However, similar to the simplest additive model, this model is also suffering from the lack of the ability to model word order and the absence of background syntactic or knowledge information.

In the additive model, we have $\mathbf{p} = \alpha \mathbf{u} + \beta \mathbf{v}$ to alleviate the word order issue by assigning different weights to different items. Here, α and β are two scalars, which can also be naturally changed to two matrices. The composition function could be represented as:

$$\mathbf{p} = \mathbf{W}_\alpha \cdot \mathbf{u} + \mathbf{W}_\beta \cdot \mathbf{v}, \tag{3.10}$$

where $\mathbf{W}_\alpha$ and $\mathbf{W}_\beta$ are weight matrices that indicate the importance of components $\mathbf{u}$ and $\mathbf{v}$ to the combined unit $\mathbf{p}$. With this expression, the composition could be more expressive and flexible, although much harder to train.

By generalizing the multiplicative model ahead, another approach is to utilize tensors as a multiplicative descriptor, and the composition function could be viewed as:

$$\mathbf{p} = \vec{\mathbf{W}} \cdot \mathbf{uv}, \tag{3.11}$$

where $\vec{\mathbf{W}}$ denotes a three-order tensor, i.e., the formula above could be written as $\mathbf{p}_k = \sum_{i,j} \mathbf{W}_{ijk} \cdot \mathbf{u}_i \cdot \mathbf{v}_j$. Hence, this model makes sure that each element of $\mathbf{p}$ could be influenced by all elements of both $\mathbf{u}$ and $\mathbf{v}$, with a relationship of linear combination by assigning each (i, j) a unique weight.

Starting from this simple but general baseline, some researchers proposed to make the function not symmetric to consider word order in the sequence, paying more attention to the first element. The composition function could be:

$$\mathbf{p} = \vec{\mathbf{W}} \cdot \mathbf{uuv}, \tag{3.12}$$

where $\vec{\mathbf{W}}$ denotes a four-order tensor. This method could be understood as replacing the linear transformation of $\mathbf{u}$ and $\mathbf{v}$ to a quadratic in $\mathbf{u}$ asymmetrically. So this is a variant of the tensor multiplicative compositional model.

Different from expanding a simple multiplicative model to complex ones, other kinds of approaches are proposed to reduce the parameter space. With the reduction of parameter size, people could make compositions much more efficient rather than having an $O(n^3)$ time complexity in the tensor-based model. Thus, some compression techniques could be applied to the original tensor model. One representative instance is the circular convolution model, which could be shown as:

$$\mathbf{p} = \mathbf{u} \circledast \mathbf{v}, \tag{3.13}$$

where $\circledast$ represents the circular convolution operation with the following definition:

$$\mathbf{p}_i = \sum_j \mathbf{u}_j \cdot \mathbf{v}_{i-j}. \tag{3.14}$$

If we assign each pair with unique weights, the composition function will be:

$$\mathbf{p}_i = \sum_j \mathbf{W}_{ij} \cdot \mathbf{u}_j \cdot \mathbf{v}_{i-j}. \tag{3.15}$$

The circular convolution model could be viewed as a special instance of a tensor-based composition model. If we write the circular convolution in the tensor form, we have $\mathbf{W}_{ijk} = 0$, where $k \neq i + j$. Thus, the parameter number could be reduced from n^3 to n^2, while maintaining the interactions between each pair of dimensions in the input vectors.

Both in the additive and multiplicative models, the basic condition is all components lie in the same semantic space as the output. Nevertheless, different modeling types of words in different semantic spaces could bring us different perspectives. For instance, given $(\mathbf{u}, \mathbf{v})$, the multiplicative model could be reformulated as:

$$\mathbf{p} = \mathbf{W} \cdot (\mathbf{u} \cdot \mathbf{v}) = \mathbf{U} \cdot \mathbf{v}. \tag{3.16}$$

This implies that each left unit could be treated as an operation on the representation of the right one. In other words, each remaining unit could be formulated as a transformation matrix, while the right one should be represented as a semantic vector. This argument could be meaningful, especially for some kinds of phrase compositions. Baroni et al. [2] argue that for *adj-noun* phrases, the joint semantic information could be viewed as the conjunction of the semantic meanings of two components. Given a phrase *red car*, its semantic meaning is the conjunction of all red things and all different kinds of cars. Thus, *red* could be formulated as an operator on the vector of *car*, deriving the new semantic vector, which expressed the meaning of *red car*. These observations lead to another genre of semantic compositional modeling: semantic matrix-composition space.

3.3 *N*-ary Composition

In real-world NLP tasks, the input is typically a sequence of multiple words or tokens rather than just a pair of words. Therefore, besides designing a suitable binary compositional operator, the order to apply binary operations is also important. In this section, we will introduce mainstream strategies in *N*-ary composition by taking language modeling as an example. To illustrate the language modeling task more clearly, the composition problem to model a sentence or even a document could be formulated as follows. Given a sentence/document consisting of a word sequence $\{w_1, w_2, ..., w_N\}$, we aim to design the following functions to obtain the joint semantic representation of the whole sentence/document:

1. A semantic representation method like semantic vector space or compositional matrix space.
2. A binary compositional operation function $f(\mathbf{u}, \mathbf{v})$ like we introduced in the previous sections. Here the input $\mathbf{u}$ and $\mathbf{v}$ denote the representations of two constitute semantic units, while the output is also the representation in the same space.
3. An order to apply the binary function in step 2. To describe in detail, we could use a bracket to identify the order to apply the composition function. For instance, we could use $((w_1, w_2), w_3)$ to represent the sequential order from beginning to end.

Methods to model sentence semantics and tackle the above problems could be classified by word-level order: sequential order and convolution order. These composition methods can be particularly implemented by neural networks with corresponding structures. We will introduce the fundamental concepts of the modeling and leave the specific neural network methods to the next chapter.

Sequential Order To design orders to apply binary compositional functions, the most intuitive method is utilizing sequentiality. Namely, the sequence order should be $s_n = (s_{n-1}, w_n)$, where s_{n-1} is the order of the first $n - 1$ words. In this case, the most suitable neural network is the recurrent neural network (RNN). An RNN applies the composition function sequentially and derives the representations of hidden semantic units. Based on these hidden semantic units, we could use them on some specific NLP tasks like sentiment analysis or text classification. Also, note that basic RNNs only utilize the sequential information from head to tail of a sentence/document. To improve the representation ability, RNNs could be enhanced by considering sequential and reverse-sequential information. In RNNs, each hidden state is controlled by the previous hidden state and the input embeddings at the current timestep, thereby forming the composition function of the sequential order.

Convolutional Order In addition to the sequential and recursive order from linguistic intuition, we can also model high-level semantics from the convolutional order. Naturally, this is implemented by a convolutional neural network (CNN), which extracts local features by a convolution layer and then integrates local features via pooling operations to produce sentence-level representations. The starting point of such methods is also to model local features for basic units and then synthesize the universal representation of the entire input. The difference from the previous approaches is that it does not follow the sequence order or syntactic structure but lets the convolutional layer complete this combination automatically.

For the sake of simplicity, this chapter ignores the relationship $\mathcal{R}$ and external knowledge $\mathcal{K}$ of compositional semantics when introducing them. And these two items are challenging to be heuristically defined and applied in traditional computational linguistics. However, the modern NLP, typically based on deep neural networks, brings a twist to the situation. The tremendous capacity enables neural networks to model almost arbitrarily complex semantic structures in an implicit way, which could be regarded as modeling the $\mathcal{R}$ item (will be introduced in Chap. 4). And advances in knowledge representation learning and knowledge-guided NLP could be naturally seen as a process to model the $\mathcal{K}$ item (will be introduced in Chap. 9).

3.4 Summary and Further Readings

In this chapter, we first introduce the semantic space for compositional semantics. Afterward, we take phrase representation as an example to introduce representative models for binary semantic composition, including additive models and multiplicative models. Finally, we introduce typical methods for N-ary semantic composition. We use fundamental principles and concepts to illustrate the core idea of compositional semantics: to build complex semantics with the combinations of basic components. For further understanding of compositional semantics, readers can refer to some recommended surveys and books that comprehensively introduce the area. For example, the framework applied in this chapter is from the inspiring article of Pelletier et al. [10].

For better modeling compositional semantics, some directions require further efforts in the future. For example, neurobiology-inspired compositional semantics is a promising research topic that explores the neurobiological insights of compositional semantics [11]. The analysis of how language builds meaning and lays out directions in neurobiological research may bring some instructive reference for modeling compositional semantics in representation learning. It is valuable to design novel compositional forms inspired by recent neurobiological advances. There are also studies that attempt to consider discrete symbols in deep neural networks [1, 4], triggering new research issues on the combination of neural models and symbolic models.

Generally speaking, modeling complex semantics distributed in sentences and even documents could be extremely difficult. It may be difficult for us to complete the modeling through heuristic methods. At this point, the powerful fitting and generalization capability of the neural networks are needed to play a role. In the next chapter, we will introduce concepts, methodologies, and applications of sentence and document representation and particularly put focus on the neural network approaches.

Acknowledgments Zhiyuan Liu, Yankai Lin, and Maosong Sun designed the overall architecture of this chapter; Ning Ding and Yankai Lin drafted the chapter. Zhiyuan Liu and Yankai Lin proofread and revised this chapter.

We thank Yang Liu for providing some initial materials in the first edition. We thank Ganqu Cui, Yuan Yao, Shi Yu, Yulin Chen, Xingtai Lv, and Suyuan Zhao for proofreading this chapter.

This is the representation learning for compositional semantics chapter of the second edition of the book *Representation Learning for Natural Language Processing*, with its first edition published in 2020 [5]. As compared to the first edition of this chapter, main changes include the following: (1) we reorganized the structure and moved the detailed introduction of neural networks (e.g., RNNs and CNNs) to the next chapter; (2) we polished and rewrote the content of the introduction and binary composition; and (3) we removed the semantic space section of the first edition to make this chapter more focused on the basic concepts of semantic composition modeling.

References

1. Jacob Andreas, Marcus Rohrbach, Trevor Darrell, and Dan Klein. Neural module networks. In *Proceedings of CVPR*, 2016.
2. Marco Baroni and Roberto Zamparelli. Nouns are vectors, adjectives are matrices: Representing adjective-noun constructions in semantic space. In *Proceedings of EMNLP*, 2010.
3. Gottlob Frege. Die grundlagen derarithmetik. *Eine logisch mathematische Untersuchung u'ber den Begrijfder Zahl. Breslau: Koebner*, 1884.
4. Chen Liang, Jonathan Berant, Quoc Le, Kenneth Forbus, and Ni Lao. Neural symbolic machines: Learning semantic parsers on freebase with weak supervision. In *Proceedings of ACL*, 2017.
5. Zhiyuan Liu, Yankai Lin, and Maosong Sun. *Representation Learning for Natural Language Processing*. Springer, 2020.
6. Jeff Mitchell and Mirella Lapata. Vector-based models of semantic composition. In *Proceedings of ACL*, 2008.
7. Jeff Mitchell and Mirella Lapata. Composition in distributional models of semantics. *Cognitive science*, 34(8):1388–1429, 2010.
8. Geoffrey Nunberg, Ivan A Sag, and Thomas Wasow. Idioms. *Language*, pages 491–538, 1994.
9. Barbara Partee. Lexical semantics and compositionality. *An Invitation to Cognitive Science: Language*, 1:311–360, 1995.
10. Francis Jeffry Pelletier. The principle of semantic compositionality. *Topoi*, 13(1):11–24, 1994.
11. Liina Pylkkänen. The neural basis of combinatory syntax and semantics. *Science*, 366(6461):62–66, 2019.

Chapter 4
Sentence and Document Representation Learning

Ning Ding, Yankai Lin, Zhiyuan Liu, and Maosong Sun

Abstract Sentence and document are high-level linguistic units of natural languages. Representation learning of sentences and documents remains a core and challenging task because many important applications of natural language processing (NLP) lie in understanding sentences and documents. This chapter first introduces symbolic methods to sentence and document representation learning. Then we extensively introduce neural network-based methods for the far-reaching language modeling task, including feed-forward neural networks, convolutional neural networks, recurrent neural networks, and Transformers. Regarding the characteristics of a document consisting of multiple sentences, we particularly introduce memory-based and hierarchical approaches to document representation learning. Finally, we present representative applications of sentence and document representation, including text classification, sequence labeling, reading comprehension, question answering, information retrieval, and sequence-to-sequence generation.

4.1 Introduction

A natural language sentence is a linguistic unit that conveys complete semantic information, which is composed of words and phrases guided by grammatical rules. Although all the elements in a sentence come from a finite set, they could constitute almost infinite semantics with complex sequential and hierarchical structures. Transforming sentence-level information into computable numerical representations is an intriguing and meaningful research issue for broad tasks of natural language processing.

N. Ding · Z. Liu (✉) · M. Sun
Department of Computer Science and Technology, Tsinghua University, Beijing, China
e-mail: dingn18@mails.tsinghua.edu.cn; liuzy@tsinghua.edu.cn; sms@tsinghua.edu.cn

Y. Lin
Gaoling School of Artificial Intelligence, Renmin University of China, Beijing, China
e-mail: yankailin@ruc.edu.cn

In the early stage before deep learning, symbolic strategies are widely adopted to represent sentences. Following the bag-of-words assumption, sentences could be represented as one-hot or term frequency-inverse document frequency (TF-IDF) vectors. However, such methods would bring the computational efficiency problem since the dimension of such representation vectors is usually up to thousands or millions. And these methods also neglect the syntactic structure of a sentence, which is the core of constituent words to express different semantics. By contrast, the n-gram probabilistic language model that assigns probabilities to sequences of words could consider the context while modeling sentences. Despite the simpleness of the probabilistic language model, it inspires the subsequent state-of-the-art neural language models that are based on deep neural networks, such as convolutional neural networks and recurrent neural networks, etc. Compared with conventional symbolic sentence representations, deep neural networks can capture the internal structures of sentences, e.g., sequential and dependency information, through convolutional, recurrent, or self-attention operations, yielding significant success in sentence modeling and NLP tasks.

Documents, usually regarded as the highest-level linguistic unit of natural language, are constituted when there are enough sentences, and they are organized in a particularly logical way. With the rapid development of the Internet, how to effectively retrieve and mine the vast new-coming information from massive amounts of online text becomes a crucial problem for natural language processing. Therefore, document representation plays a vital role in a series of real-world applications and becomes an intriguing research problem. In principle, the aforementioned symbolic or neural-based methods for sentence representation learning could also be applied to documents. But it is also easy to see that the coherence between sentences provides space for more complex combinations to form document-level semantics, thereby producing new challenges. Common approaches to tackle document representation include memory-based and hierarchical methods.

In this chapter, we first introduce symbolic sentence representation learning methods in Sect. 4.2, including the bag-of-words model and probabilistic language models. Then we detail the techniques of neural language models in Sect. 4.3, including feed-forward neural networks, recurrent neural networks, convolutional neural networks, and Transformers. Memory-based and hierarchical methods to model document-level information are elaborated in Sect. 4.4. Finally, we comprehensively introduce representative applications of sentence and document representation in Sect. 4.5, including text classification, sequence labeling, reading comprehension, question answering, information retrieval, recommendation, etc.

4.2 Symbolic Sentence Representation

When words and phrases form sentences, they obtain complete semantics. Similar to word representations in Chap. 2, sentences can also be represented symbolically. But with a slight difference, the sentence is not the smallest unit in this recipe.

A symbol-based sentence representation is composed of multiple symbolic word representations. In this section, we introduce the bag-of-words model and the probabilistic language model for symbolic sentence representation learning.

4.2.1   Bag-of-Words Model

As introduced in Chap. 2, one-hot representation is the most straightforward symbolic method for words and phrases. This approach represents each word with a fixed-length binary vector. For a vocabulary $V = \{w_1, w_2, \ldots, w_{|V|}\}$, the one-hot representation of word w is $\mathbf{w} = [0, \ldots, 0, 1, 0, \ldots, 0]$. Based on the one-hot word representation and the vocabulary, it can be extended to represent a sentence $s = \{w_1, w_2, \ldots, w_N\}$ based on the bag-of-words hypothesis. Bag-of-words model represents sentences as a multiset of its words while ignoring the order and other grammatical rules:

$$\mathbf{s} = \sum_{i=1}^{N} \mathbf{w}_i, \tag{4.1}$$

where N indicates the length of the sentence s. The sentence representation $\mathbf{s}$ is the sum of the one-hot representations of N words within the sentence, i.e., each element in $\mathbf{s}$ represents the term frequency (TF) of the corresponding word. In practice, to prevent it from being biased toward longer texts, it is usually normalized according to the number of words in the whole text.

However, TF alone cannot properly represent a sentence or document since not all the words are equally important. For example, the function words such as *a*, *an*, and *the* usually appear in almost all sentences and reserve little semantics that could represent the sentence or document. Therefore, the inverse document frequency (IDF) is developed to measure the prior importance of w_i in V as follows:

$$\text{idf}_{w_i} = \log \frac{|D|}{\text{df}_{w_i}}, \tag{4.2}$$

where $|D|$ is the number of all sentences or documents in the corpus D and df_{w_i} represents the document frequency (DF) of w_i, which is the number of documents that w_i appears. With the importance of each word, the sentences are represented more precisely as follows:

$$\hat{\mathbf{s}} = \mathbf{s} \odot \text{idf}, \tag{4.3}$$

where $\odot$ is the element-wise product.

Here, $\hat{\mathbf{s}}$ is the TF-IDF representation of the sentence s, and it could be naturally applied to both the sentence and document levels. The insight behind it is that the more frequently a word appears and the less it appears in other texts, the more it represents the uniqueness of the current text and thus will be assigned more weight. TF-IDF is one of the most popular methods in information retrieval and recommender system [76, 81].

4.2.2 Probabilistic Language Model

One-hot sentence representation identifies important terms to construct the representation and neglects the structural information in a sentence. In this section, we introduce the probabilistic language model, a symbolic sentence representation approach that takes context into account.

A standard probabilistic language model defines the probability of a sentence $s = \{w_1, w_2, \ldots, w_N\}$ by the chain rule of probability:

$$P(s) = P(w_1)P(w_2|w_1)P(w_3|w_1, w_2) \ldots P(w_N|w_1, \ldots, w_{N-1}) \quad (4.4)$$

$$= \prod_{i=1}^{N} P(w_i|w_1, \ldots, w_{i-1}). \quad (4.5)$$

The probability of each word is determined by all the preceding words. And the conditional probabilities of all the words jointly compute the probability of the sentence. However, the model indicated in the Eq. (4.5) is not practicable due to its enormous parameter space for long texts. This is where the n-gram model comes to play, whose core idea is not to use all previous words but $n - 1$ words to predict the current word. We show an example of the n-gram model in Fig. 4.1.

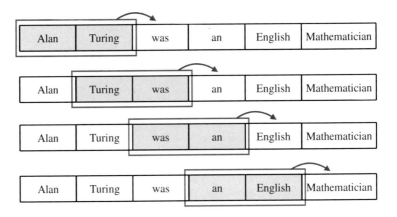

Fig. 4.1 An example of the n-gram language model, where $n = 3$

In practice, we set such $n - 1$-sized context windows in the probabilistic language model, assuming that the probability of word w_i only depends on $\{w_{i-n+1} \cdots w_{i-1}\}$. More specifically, an n-gram language model predicts word w_i in the sentence s based on its previous $n - 1$ words:

$$P(w_i|w_1, \ldots, w_{i-1}) \approx P(w_i|w_{i-n+1}, \ldots, w_{i-1}). \qquad (4.6)$$

After simplifying the language model problem, how to estimate the conditional probability is crucial. In practice, a common approach is maximum likelihood estimation (MLE), which is generally in the following form:

$$P(w_i|w_{i-n+1}, \ldots, w_{i-1}) = \frac{P(w_{i-n+1}, \ldots, w_i)}{P(w_{i-n+1}, \ldots, w_{i-1})}. \qquad (4.7)$$

In this equation, the denominator and the numerator can be estimated by counting the frequencies in the corpus. To avoid the probability of some n-gram sequences from being zero, researchers also adopt several types of smoothing approaches, which assign some of the total probability mass to unseen words or n-grams, such as "add-one" smoothing, Good-Turing discounting [31], or back-off models [45].

n-gram model is a typical probabilistic language model for predicting the next word in an n-gram sequence, which follows the Markov assumption that the probability of the target word only relies on the previous $n - 1$ words. The idea is employed by most current sentence modeling methods, where the n-gram language model serves as an approximation of the true language model. This hypothesis is crucial because it substantially simplifies the problem of learning the parameters of language models from data. Recent works on word representation learning [1, 69, 72] are mainly based on the n-gram language model.

The introductory part of this chapter states that the semantic information of a sentence not only exists in constituent words but is also closely related to its flexible syntactic structure. Obviously, despite its simplicity, symbolic approaches treat constituent words as independent symbols and are not capable of representing rich semantic information. Symbolic methods for sentence representation learning have been extensively introduced by many classical textbooks [42]. In this chapter, we mainly focus on sentence representations based on neural networks, which is a common practice in modern NLP.

4.3   Neural Language Models

Although the aforementioned symbolic methods are cornerstones to represent sentences in inchoate NLP, they still face challenges in modeling rich semantic information and universal information distributed in flexible structures of sentences.

To this end, a set of more powerful modeling tools, neural networks, are developed for language modeling. Different from symbolic methods, neural language models use continuous representations to represent all words, which enjoy better generalization and modeling capability for longer texts.

A neural network could also be viewed as an estimator of the language model function, and the architecture could be flexible in this setting. Similar to n-gram probabilistic language models, neural language models are constructed and trained to model a probability distribution of a target word conditioned on previous words:

$$P(s) = \prod_{i=1}^{N} P(w_i | w_{i-n+1}, \dots, w_{i-1}), \qquad (4.8)$$

where the conditional probability of the selecting word w_i can be calculated by multiple kinds of neural networks and the common choices include the feed-forward neural network, recurrent neural network, convolutional neural network, etc. The training of neural language models is achieved by optimizing the cross-entropy loss function:

$$\mathcal{L} = -\sum_{i=1}^{N} \log P(w_i | w_{i-n+1}, \dots, w_{i-1}). \qquad (4.9)$$

The parameters of the language model will be iteratively optimized during training and result in a language model that could predict the next word based on the context. In the following sections, we will detail these neural language models.

4.3.1 Feed-Forward Neural Network

Whether it is a probabilistic language model or a neural language model, the primary goal is to estimate the conditional probability $P(w_i | w_1, \dots, w_{i-1})$. And as stated, adopting the idea of n-gram to approximate the conditional probability is a common approach, where each word is determined by its $n - 1$ context words, i.e., $P(w_i | w_1, \dots, w_{i-1}) \approx P(w_i | w_{i-n+1}, \dots, w_{i-1})$. In this section, we first introduce language modeling with the feed-forward neural network (FFN).

The architecture of the FFN language model is proposed by Bengio et al. [1] (illustrated in Fig. 4.2). Although more sophisticated neural architectures could be applied to the problem, the FFN language model first elaborates on the methodology of neural-based language modeling. To evaluate the conditional probability of the word w_i, it first projects its $n - 1$ context-related words to their word vector representations $[\mathbf{w}_{i-n+1}, \dots, \mathbf{w}_{i-1}]$ and concatenate the representations

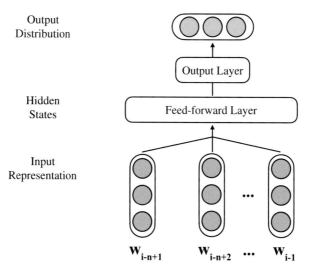

Fig. 4.2 The architecture of the feed-forward neural network

$\mathbf{x} = \text{concat}(\mathbf{w}_{i-n+1}; \ldots; \mathbf{w}_{i-1})$ to feed them into a FFN. The formulation can be generally written as follows:

$$\mathbf{h} = \mathbf{M} f(\mathbf{W}_1 \mathbf{x} + \mathbf{b}) + \mathbf{W}_2 \mathbf{x} + \mathbf{d}, \tag{4.10}$$

where $f(\cdot)$ is an activation function, $\mathbf{W}_1, \mathbf{W}_2$ are weighted matrices to transform word vectors into hidden representations, $\mathbf{M}$ is a weighted matrix for the connections between the hidden layer and the output layer, and $\mathbf{b}, \mathbf{d}$ are bias terms. And then, the conditional probability of the word w_i can be calculated by a Softmax function:

$$P(w_i | w_{i-n+1}, \ldots, w_{i-1}) = \text{Softmax}(\mathbf{h}). \tag{4.11}$$

4.3.2 Convolutional Neural Network

Convolutional neural networks (CNNs) use convolutional layers to conduct the basic operation. This type of neural network layer represents the context by extracting hierarchical information from it [23]. For the input words $\{w_1, \ldots, w_l\}$, we first obtain their word embeddings $[\mathbf{w}_1, \ldots, \mathbf{w}_N]$. Let d denote the dimension of the hidden states. The convolutional layer involves a sliding window with the size of k of the input vectors centered on each word vector using a kernel matrix $\mathbf{W}_c$. And the hidden representation could be calculated by

$$\mathbf{h} = f(\mathbf{X} * \mathbf{W}_c + \mathbf{b}), \tag{4.12}$$

Fig. 4.3 The architecture of
the convolutional neural
network

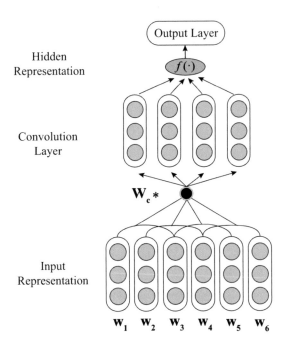

where $*$ is the convolution operation, $f(\cdot)$ is a nonlinear activation function (e.g., a sigmoid or tangent function), $\mathbf{X} \in \mathbb{R}^{l \times d}$ is the matrix of word embeddings, $\mathbf{W}_c \in \mathbb{R}^{k \times d \times d'}$ (d' is the kernel size), and $\mathbf{b} \in \mathbb{R}^{d'}$ are learned parameters. The sliding window prevents the model from seeing the subsequent words so that $\mathbf{h}$ does not learn information from future words. For each sliding step, the hidden state of the current word is computed based on the previous k words and then further fed to an output layer to calculate the probability of the present word. The architecture of a CNN is shown in Fig. 4.3. In practice, we can use distinct lengths of sliding windows to form multi-channel operations to learn local information with different scales.

4.3.3 Recurrent Neural Network

To address the lack of ability to model long-term dependency in the FFN language model, Mikolov et al. [70] propose a recurrent neural network (RNN) language model which applies an RNN in language modeling. RNNs are different from FFNs in a fundamental way in that they operate in an internal state space where representations can be sequentially processed. Therefore, the RNN language model can deal with those sentences of arbitrary length. At every time step, its input is the vector of its previous word instead of the concatenation of vectors of its $n - 1$

previous words. The information of all other previous words can be considered by its internal state.

Given the input word embeddings $\mathbf{x} = [\mathbf{w}_1, \mathbf{w}_2, \dots, \mathbf{w}_N]$, at timestep t, the current hidden state $\mathbf{h}_t$ is computed based on the current input $\mathbf{w}_t$ and the hidden state of the last timestep $\mathbf{h}_{t-1}$. Formally, the RNN language model can be defined as

$$\mathbf{h}_t = f(\mathbf{W} \text{ concat}(\mathbf{w}_t; \mathbf{h}_{t-1}) + \mathbf{b}), \tag{4.13}$$

$$\mathbf{y} = \text{Softmax}(\mathbf{Mh}_t + \mathbf{d}), \tag{4.14}$$

where $f(\cdot)$ is a nonlinear activation function, $\mathbf{y}$ represents a probability distribution over the given vocabulary, $\mathbf{W}$ and $\mathbf{M}$ are weighted matrices and $\mathbf{b}, \mathbf{d}$ are bias terms. As the increase of the length of the sequence, a common issue of the RNN language model is the vanishing gradients problem. The architecture of the RNN language model is shown in Fig. 4.4. Here, the RNN unit can also be implemented in other variants of recurrent neural networks, e.g., long short-term memory (LSTM) and gated recurrent unit (GRU).

LSTM Since the raw RNN only utilizes the simple tangent function, it is hard to obtain the long-term dependency of a long sentence/document. Hochreiter et al.

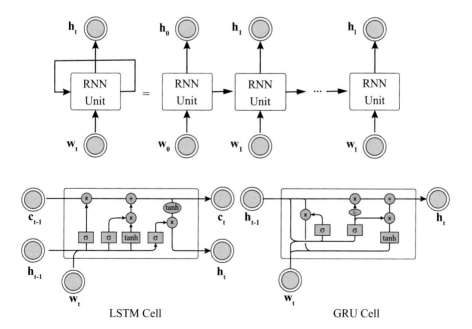

Fig. 4.4 The architecture of recurrent neural networks. The figure is re-drawn according to the blog for introducing LSTM models (https://colah.github.io/posts/2015-08-Understanding-LSTMs/)

[37] propose long short-term memory (LSTM) networks to strengthen the ability to model long-term semantic dependency in RNN.

LSTM introduces a cell state $\mathbf{c}_t$ to represent the current information at timestep t, which is computed from the cell state at the last timestep $\mathbf{c}_{t-1}$ and the candidate cell state of the current timestep $\tilde{\mathbf{c}}_t$. And the representation of the current timestep $\mathbf{h}_t$ is calculated based on $\mathbf{c}_t$. Formally,

$$\tilde{\mathbf{c}}_t = \tanh(\mathbf{W}_c \text{concat}(\mathbf{w}_t; \mathbf{h}_{t-1}) + \mathbf{b}_c), \tag{4.15}$$

$$\mathbf{c}_t = \mathbf{f}_t \odot \mathbf{c}_{t-1} + \mathbf{i}_t \odot \tilde{\mathbf{c}}_t, \tag{4.16}$$

$$\mathbf{h}_t = \mathbf{o}_t \odot \tanh(\mathbf{c}_t), \tag{4.17}$$

where $\odot$ is the element-wise multiplication operation, $\mathbf{W}_c$ and $\mathbf{b}_c$ are learnable parameters and $\mathbf{f}_t$, $\mathbf{i}_t$, and $\mathbf{o}_t$ are different gates introduced in LSTM to control the information flow. Specifically, $\mathbf{f}_t$ is the forgetting gate to determine how much information of the cell state at the last timestep $\mathbf{c}_{t-1}$ should be forgotten, $\mathbf{i}_t$ is the input gate to control how much information of the candidate cell state at the current timestep $\tilde{\mathbf{c}}_t$ should be reserved, and $\mathbf{o}_t$ is the output gate to control how much information of the current cell state $\mathbf{c}_t$ should be output to the representation $\mathbf{h}_t$. And all these gates are computed by the representation of the last timestep $\mathbf{h}_{t-1}$ and the current input $\mathbf{w}_t$. Formally, it could be written as

$$\mathbf{f}_t = \text{Sigmoid}(\mathbf{W}_f \text{concat}(\mathbf{w}_t; \mathbf{h}_{t-1}) + \mathbf{b}_f), \tag{4.18}$$

$$\mathbf{i}_t = \text{Sigmoid}(\mathbf{W}_i \text{concat}(\mathbf{w}_t; \mathbf{h}_{t-1}) + \mathbf{b}_i), \tag{4.19}$$

$$\mathbf{o}_t = \text{Sigmoid}(\mathbf{W}_o \text{concat}(\mathbf{w}_t; \mathbf{h}_{t-1}) + \mathbf{b}_o), \tag{4.20}$$

where $\mathbf{W}_f$, $\mathbf{W}_i$, $\mathbf{W}_o$ are weight matrices and $\mathbf{b}_f$, $\mathbf{b}_i$, $\mathbf{b}_o$ are bias terms in different gates. It is generally believed that LSTM could model longer text than the vanilla RNN model.

GRU To simplify LSTM and obtain more efficient algorithms, Chung et al. [19] propose to utilize a simple but comparable RNN architecture, named gated recurrent unit (GRU), which also utilizes the gating mechanism to handle information flow. But compared to several gates with different functionalities, GRU uses an update gate $\mathbf{z}_t$ to control the information flow. And a reset gate $\mathbf{r}_t$ is adopted to control how much information from the last step hidden state $\mathbf{h}_{t-1}$ would flow into the candidate hidden state of the current step $\tilde{\mathbf{h}}$. Formally, the computation flow of GRU is as follows:

$$\mathbf{z}_t = \text{Sigmoid}(\mathbf{W}_z \text{concat}(\mathbf{w}_t; \mathbf{h}_{t-1}) + \mathbf{b}_z), \tag{4.21}$$

$$\mathbf{r}_t = \text{Sigmoid}(\mathbf{W}_r \text{concat}(\mathbf{w}_t; \mathbf{h}_{t-1}) + \mathbf{b}_r), \tag{4.22}$$

$$\tilde{\mathbf{h}}_t = \tanh(\mathbf{W}_h \text{concat}(\mathbf{w}_t; \mathbf{r}_t \odot \mathbf{h}_{t-1}) + \mathbf{b}_h), \tag{4.23}$$

$$\mathbf{h}_t = (1 - \mathbf{z}_t) \odot \mathbf{h}_{t-1} + \mathbf{z}_t \odot \tilde{\mathbf{h}}_t, \tag{4.24}$$

where $\mathbf{W}_z$, $\mathbf{W}_r$, $\mathbf{W}_h$ are weight matrices and $\mathbf{b}_z$, $\mathbf{b}_r$, $\mathbf{b}_h$ are bias terms. The update gate z in GRU simultaneously manages the historical and current information. Moreover, the model also omits cell modules $\mathbf{c}$ in LSTM and directly uses hidden states $\mathbf{h}$ in the computation. GRU has fewer parameters, which brings higher efficiency and could be seen as a simplified version of LSTM.

Generally, compared to CNNs, RNNs are more suitable for the sequential characteristic of textual data. However, the nature of each step's hidden state is dependent on the previous step also makes RNNs difficult to perform parallel computation and thus slower in training.

4.3.4 Transformer

Since 2017, a more powerful neural architecture, the Transformer [96] model, which is equipped with a self-attention mechanism, has received extensive attention from the NLP community. Compared to RNNs, Transformers could handle sequential data in parallel instead of processing a word at a timestep. The Transformer model has become a mainstream choice of neural networks to model natural language and pre-trained language models based on deep Transformers have achieved state-of-the-art results on various NLP tasks. In this section, we introduce the mechanism of the Transformer model. We will use the next chapter to introduce the progress and research issues of representation learning brought by pre-trained models.

Structure A Transformer is a nonrecurrent encoder-decoder architecture with a series of attention-based blocks. For the encoder, there are multiple layers, and each layer is composed of a multi-head attention sublayer and a position-wise feed-forward sublayer. And there is a residual connection and layer normalization of each sublayer. The decoder also contains multiple layers, and each layer is slightly different from the encoder. First, sublayers of multi-head attention and feed-forward with identical structures with the encoder are adopted. And the input of the multi-head attention sublayer is from both the encoder and the previous sublayer, which is additionally developed. This sublayer is also a multi-head attention sublayer that performs self-attention over the outputs of the encoder. And the sublayer adopts a masking operation to prevent the decoder from seeing subsequent tokens. The architecture of the Transformer is shown in Fig. 4.5.

Attention There are several attention heads in the multi-head attention sublayer. A *head* represents a scaled dot-product attention structure, which takes the query

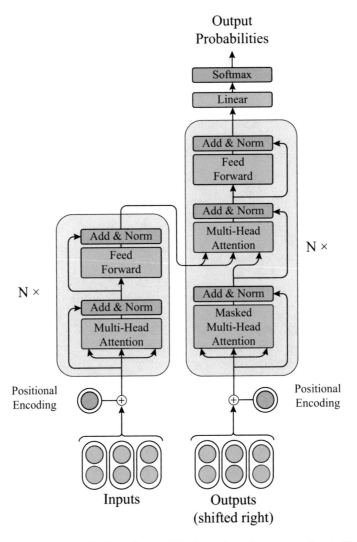

Fig. 4.5 The architecture of a Transformer. This figure is re-drawn according to Fig. 1 from Google's Transformer paper [96]

matrix $\mathbf{Q}$, the key matrix $\mathbf{K}$, and the value matrix $\mathbf{V}$ as the inputs, and the output is computed by

$$\text{ATT}(\mathbf{Q}, \mathbf{K}, \mathbf{V}) = \text{Softmax}\left(\frac{\mathbf{Q}\mathbf{K}^{\top}}{\sqrt{d_k}}\right)\mathbf{V}, \qquad (4.25)$$

where d_k is the dimension of the query matrix; note that in language models, $\mathbf{Q}$, $\mathbf{K}$, and $\mathbf{V}$ usually come from the same source, i.e., the input sequences. Specifically, they are obtained by the multiplication of the input embedding $\mathbf{H}$ and three weight matrices $\mathbf{W}^Q$, $\mathbf{W}^K$, and $\mathbf{W}^V$, respectively. The dimensions of query, key, and value vectors are d_k, d_k, and d_v, respectively. The computation in Eq. (4.25) is typically known as the self-attention mechanism.

The multi-head attention sublayer linearly projects the input hidden states $\mathbf{H}$ several times into the query matrix, the key matrix, and the value matrix for h heads. The multi-head attention sublayer could be formulated as follows:

$$\text{Multihead}(\mathbf{H}) = [\text{head}_1, \text{head}_2, \ldots, \text{head}_h]\mathbf{W}^O, \tag{4.26}$$

where $\text{head}_i = \text{ATT}(\mathbf{QW}_i^Q, \mathbf{KW}_i^K, \mathbf{VW}_i^V)$ and $\mathbf{W}_i^Q$, $\mathbf{W}_i^K$, and $\mathbf{W}_i^V$ are linear projections. $\mathbf{W}^O$ is also a linear projection for the output.

After operating self-attention, the output would be fed into a fully connected position-wise feed-forward sublayer, which contains two linear transformations with ReLU activation:

$$\text{FFN}(\mathbf{x}) = \mathbf{W}_2 \max(0, \mathbf{W}_1\mathbf{x} + \mathbf{b}_1) + \mathbf{b}_2. \tag{4.27}$$

Input Tokenization   Tokenization is a crucial step in NLP to process the raw input sequences. Generally, tokenization converts the input sequence into "tokens" and feeds them to subsequence processing modules. A simple approach is to directly regard a word as a token, whereas such a method cannot well handle unknown out-of-vocabulary (OOV) words and cannot grasp the correlations of similar words. For example, it is more intuitive to tokenize "apples" into "apple" and "s" than a separate token independent of "apple." In modern NLP, more mature methods like byte pair encoding (BPE) and wordpiece are extensively applied to Transformer-based models. Taking BPE as an example, it iteratively replaces two adjacent units with a new unit, which ensures that common words will remain as a whole and uncommon words are split into multiple subwords. Practically, BPE is applied to many pre-trained models such as RoBERTa [64] and GPT-2 [79], and wordpiece is used to pre-train BERT [24].

Positional Encoding   Positional encoding indicates the position of each token in an input sequence. The self-attention mechanism of Transformers does not involve positional information. Thus, the model needs to represent positional information of the input sequence additionally. Transformers do not use integers to represent positions because the value range varies with the input length. For example, positional values may become very large if the model process a long text, which will restrain the generalization over texts of different lengths.

Specifically, each position is encoded to a particular vector with the same dimension d of the hidden states to represent the positional information. For the k-th token, let $\mathbf{p}^k$ be the positional vector; the i-th element of the positional encoding $\mathbf{p}_i^k$ is calculated by

$$\mathbf{p}_i^k = \sin\left(\frac{k}{10{,}000^{\frac{2j}{d}}}\right), \text{ if } i = 2j, \tag{4.28}$$

$$\mathbf{p}_i^k = \cos\left(\frac{k}{10{,}000^{\frac{2j}{d}}}\right), \text{ if } i = 2j + 1. \tag{4.29}$$

In this way, for each positional encoding vector, the frequency would decrease along with the dimension. We can imagine that at the end of each vector, $k/10{,}000^{\frac{2j}{d}}$ is near to 0 since the denominator becomes very large, which makes $\sin(k/10{,}000^{\frac{2j}{d}})$ approximates 0 and $\cos(k/10{,}000^{\frac{2j}{d}})$ approximates 1. Assuming the state of alternating 0s and 1s is a kind of "stable point," for different positions k, the "speed" to reach such a stable point is also different. That is, the later the token is (larger k), the later the value $k/10{,}000^{\frac{2j}{d}}$ will be close to 0. Moreover, no matter the text lengths the model is currently processing, the encoding values are stable and range from -1 to 1. Alternatively, learnable positional embeddings could also be applied to Transformers and could consistently yield similar performance. Pre-trained language models like BERT [24] adopt learnable position embeddings rather than sinusoidal encoding.

Although the Transformer model was proposed to tackle machine translation, the powerful capability to model sequential data makes it the most popular backbone of NLP applications. For example, it has become the standard architecture for pre-trained language models, and GPT is a representative example of using a Transformer for the language modeling task. As stated, the overall objective is $\mathscr{L} = -\sum_{i=1}^{N} \log P(w_i | w_{i-n+1}, \ldots, w_{i-1})$. Here, we use the decoder of a Transformer to adopt the self-attention mechanism to the previous $n-1$ words of the current word, and the output will be further fed into the feed-forward sublayer. After multiple layers of propagation, the final probability distribution P is computed by a softmax function acting on the hidden representation. Compared to RNNs, Transformers could better model the long-term dependency, where all tokens will be equally considered and computed during the attention operation.

4.3.5 Enhancing Neural Language Models

The foregoing parts have described representative neural language models. Next, we introduce some techniques that can further improve the performance of such models, including word classification and the caching approach.

Word Classification   Researchers [9, 32] propose a class-based language model to adopt word classification to improve the performance and speed of the language model. In this class-based language model, all words are assigned to a unique class, and the conditional probability of a word given its context can be decomposed into the probability of the word's class given its previous words and the probability of the word given its class and history, which is formally defined as

$$P(w_i|w_{i-n+1}, \ldots, w_{i-1}) = \sum_{c(w_i) \in C} P(w_i|c(w_i)) P(c(w_i)|w_{i-n+1}, \ldots, w_{i-1}),$$

$$(4.30)$$

where C indicates the set of all classes and $c(w_i)$ indicates the class of word w_i.

Moreover, Morin et al. [73] propose a hierarchical neural network language model, which extends word classification to hierarchical binary clustering of words in the language model. Instead of simply assigning each word a unique class, it first builds a hierarchical binary tree of words according to the word similarity obtained from WordNet. Next, it assigns a unique bit vector $[c_1(w_i), c_2(w_i), \ldots, c_N(w_i)]$ for each word, which indicates the hierarchical classes of them. And then, the conditional probability of each word can be defined as

$$P(w_i|w_{i-n+1}, \ldots, w_{i-1})$$

$$= \prod_{j=1}^{N} P(c_j(w_i)|c_1(w_i), \ldots, c_{j-1}(w_i), w_{i-n+1}, \ldots, w_{i-1}). \quad (4.31)$$

The hierarchical neural network language model can achieve $\mathcal{O}(k/\log k)$ speed up as compared to a standard language model. However, the experimental results of [73] show that it performs worse than the standard language model. The reason is that the introduction of hierarchical architecture or word classes imposes a negative influence on word classification by neural network language models.

Caching   Caching is also one of the important extensions of neural language models. A type of cache-based language model assumes that each word in a recent context is more likely to appear again [90]. Hence, the conditional probability of a word can be calculated by the information from history and caching:

$$P(w_i|w_{i-n+1}, \ldots, w_{i-1}) = \lambda P_s(w_i|w_{i-n+1}, \ldots, w_{i-1})$$

$$+ (1 - \lambda) P_c(w_i|w_{i-n+1}, \ldots, w_{i-1}), \quad (4.32)$$

where $P_s(w_i|w_{i-n+1}, \ldots, w_{i-1})$ indicates the conditional probability generated by standard language models and $P_c(w_i|w_{i-n+1}, \ldots, w_{i-1})$ indicates the conditional probability retrieved from cache, and λ is a constant.

Another cache-based language model is also used to speed up the RNN language modeling [39]. The main idea of this approach is to store the outputs and states of language models for future predictions given the same contextual history.

Neural language models are among the most powerful techniques for sentence representations, which could comprehensively model the complex syntactical structures of long texts. Even for longer documents, modern approaches are based on neural networks. And in the next chapter, we will discuss how to model documents more effectively.

4.4 From Sentence to Document Representation

The aforementioned representation learning approaches could be applied to both sentence- and document-level texts since most existing works treat documents as "longer sentences" in practice. However, the interactions of multiple sentences in a document bring more complex semantics, thereby establishing new challenges. In this section, we introduce two types of document representation learning methods. Memory-based document representation treats the document as a whole to directly learn the representation, and hierarchical document representation performs the fusion of the information of different levels of linguistic units to obtain the final document representation.

4.4.1 Memory-Based Document Representation

A direct way to learn the document representation is to regard the document as a whole. We regard this type of method as the memory-based document representation whose intuition is to use inherent modules to remember the context with critical information of the target document.

Paragraph Vector Here, we extend the idea of word2vec to the document level, which is named paragraph vector (PV) [53]. Given a target word and the corresponding contexts from the document, the training objective of this strategy is to use the paragraph vector to predict the target word. More specifically, similar to word2vec, PV has two variants: distributed memory (denoted as PV-DM) and distributed bag-of-words (denoted as PV-DBOW).

As illustrated in Fig. 4.6, PV-DM adds an additional token in each document and uses the token representation to represent the document. By extending the idea of CBOW, PV-DM predicts the target word according to historical contexts and document representation in the training phase. There are multiple choices exploiting the document representation and word representations. For example, one can directly concatenate these representations or average them. It can be seen that the additional document representation here acts as a memory module that gradually

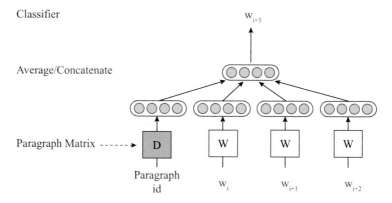

Fig. 4.6 The architecture of PV-DM model. This figure is re-drawn according to Fig. 2 from the paragraph vector paper [53]

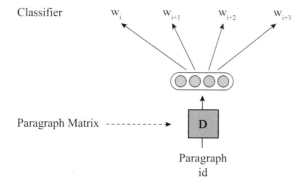

Fig. 4.7 The architecture of PV-DBOW model. This figure is re-drawn according to Fig. 3 from the paragraph vector paper [53]

captures the key semantics of the document as it participates in the training process of predicting words based on context. After training, the paragraph vectors can be regarded as the representations of the documents and be used as pre-trained document embeddings like pre-trained word embeddings.

Besides PV-DM, PV-DBOW extends the idea of skip-gram to learn document representation. As illustrated in Fig. 4.7, PV-DBOW ignores the context words in the input text and directly uses the document representation to predict the target word in a randomly sampled window. In the training phase, the model will randomly sample a window and then randomly sample a word to be the prediction target. Obviously, it is simpler in concept than PV-DM, and experiments have shown that this method is also effective for document representation.

However, when the document is too long, PV may not have enough capacity to remember enough information. These issues could be alleviated by modern deep neural networks. In the following parts, we introduce deep neural networks that yield superior performance in storing and representing historical information as memories.

Memory Networks In the era of deep learning, memory networks [104] have become one of the representative methods for learning document representation, which uses memory units to store and maintain long-term information. Compared to standard neural networks that utilize special aggregation operations to obtain the document representation, memory networks explicitly adopt memory neural modules to store information, which could prevent information forgetting. Given an input document with n sentences $d = \{s_1, s_2, \ldots, s_n\}$, for the i-th sentence s_i, the model firstly uses a feature extractor F to transform the input into a sentence-level representation $\mathbf{h}_i^s$:

$$\mathbf{h}_i^s = F(s_i). \tag{4.33}$$

A memory unit M is responsible for storing and updating memories according to the current inputs. In this case, the memories will be updated by certain operations. For the specific update mechanism of the memory module, there are many options to define M. Here we introduce one of the most straightforward methods by using slots. The basic idea of this approach is to store the representation of each input into a separate "slot":

$$\mathbf{m}_{H(s_i)} = F(s_i), \tag{4.34}$$

where $H(\cdot)$ is used to select a particular index of a slot for the input sentence s_i. In this case, M only updates one slot with index $H(s_i)$ given the input s_i and does not interfere with any other memories.

Given memories stored through the aforementioned process, we could find the k most relevant memories given a query q and generate a final output. An output module O is adapted to select supporting memories and generate the latent representation of the output of the current query q. We take $k = 1$ as an example, where the module selects one memory index:

$$o = O(q, \mathbf{m}) = \text{argmax}_j \, s_O(q, \mathbf{m}_j), \tag{4.35}$$

where $s_O(\cdot)$ is a score function to evaluate the relevance of the query and memories. Then we can use a decoder D to generate concrete tokens. In particular, if the final output y is a single word, given query q, memory $\mathbf{m}_o$ (o is the selected index of memory produced by O), and dictionary V, we use another score function $s_y(\cdot)$ that measures the candidate word and o to produce y:

$$y = \text{argmax}_{w \in V} \, s_y([q, \mathbf{m}_o], w). \tag{4.36}$$

The framework is illustrated in Fig. 4.8.

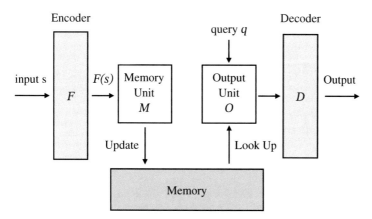

Fig. 4.8 The general architecture of memory networks

In this general framework, each module can be carefully designed to store various historical information. The framework is effective for document modeling and could be applied to many related tasks. For example, in reading comprehension question answering, we can store the representation of each sentence of the input passage as memory and then match the query against the memory to select the answer more accurately.

Variants of Memory Networks Subsequently, many memory network variants are designed from different perspectives. Generally, the improvement can be delivered from the training strategy and the memory form.

Training Strategy If the operation of each module is designed discretely, it is not easy to directly train the network via back-propagation.

The end-to-end memory network [91] presents a continuous version of this framework, which uses an RNN-based architecture (it can also be replaced with other neural backbones) to read the stored memories before outputting the results. Specifically, given a document $d = [s_1, s_2, \ldots, s_n]$, for a sentence s_i, an encoder F is adopted to obtain the representation $\mathbf{h}_i^s$, which is regarded as the raw memory for s_i. Given a query q, whose representation is $\mathbf{q}$, we need to extract relevant memories and produce the final output. As shown in Fig. 4.9, the model generates a memory vector $\mathbf{m}_i$ and an output vector $\mathbf{c}_i$ for each $\mathbf{h}_i^s$ with a trainable matrix $\mathbf{W}_m$ and another trainable matrix $\mathbf{W}_c$, respectively:

$$\mathbf{m}_i = \mathbf{W}_m \mathbf{h}_i^s, \quad \mathbf{c}_i = \mathbf{W}_c \mathbf{h}_i^s. \tag{4.37}$$

Memory vectors are used to compute matching scores p against the query q with a softmax function. Specifically for the i-th memory vector:

$$p_i = \text{Softmax}(\mathbf{m}_i^\top \mathbf{q}). \tag{4.38}$$

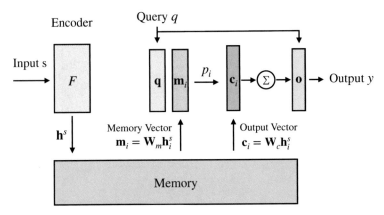

Fig. 4.9 The architecture of end-to-end memory network. This figure is re-drawn according to Fig. 1 from the end-to-end memory network paper [91]

The matching score p_i is then used as a weight of the output vector $\mathbf{c}_i$ to represent the relevance between q and $\mathbf{m}_i$. By conducting a weighted sum, we obtain a vector $\mathbf{o}$ that is responsible for the final output:

$$\mathbf{o} = \sum p_i \mathbf{c}_i. \tag{4.39}$$

Finally, the final output y given a query q could be derived from the query representation $\mathbf{q}$ and the vector $\mathbf{o}$:

$$y = \text{Softmax}(\mathbf{W}_o(\mathbf{q} + \mathbf{o})), \tag{4.40}$$

where $\mathbf{W}_o$ are trainable parameters. As we can see, in the training procedure, $\mathbf{W}_m$, $\mathbf{W}_c$, $\mathbf{W}_o$, and the encoder F will be optimized in an end-to-end manner by directly minimizing the cross-entropy loss between the prediction and the ground truth label.

Dynamic memory networks [48] present a similar methodology. After the model produces representations for all the input sentences and the current query, the query representation will trigger a retrieval procedure based on the attention mechanism. This procedure will iteratively read the stored memories and retrieve the relevant ones to produce the output. The transformation of memory networks into the end-to-end manner in terms of training strategies has further expanded its influence and inspired new research works [61, 62, 106, 108].

Memory Form In addition to the training strategy perspective, we can also improve memory networks from the perspective of the form of stored memories. It is easy to see that such a framework may be difficult to store vast amounts of information because it is hard to compute matching scores for large-scale memories. Hierarchical memory networks [10] give a solution that organizes memories in a hierarchical form. This method forms a group of memories, and then multiple

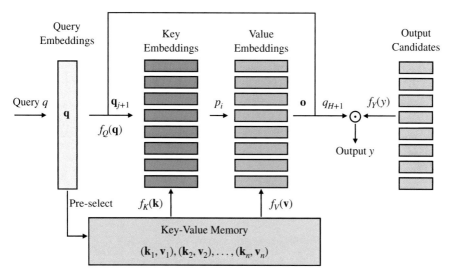

Fig. 4.10 The architecture of key-value memory network. This figure is re-drawn according to Fig. 1 from the key-value memory network paper [71]

groups can be reorganized into higher-level groups. It uses a maximum inner product search combined with the attention mechanism to efficiently retrieve desired memory. This approach effectively improves the efficiency of searching memory but may also risk losing some precision as the number of levels increases.

Key-value memory network (KV-MemNN) [71], as the name means, uses a key-value structure to store and organize memories. The design of such a structure is to boost the process of retrieving memories and could store information from different sources (e.g., text and knowledge graphs). The framework is illustrated in Fig. 4.10. Formally, the memories are pairs of key-values $(\mathbf{k}_i, \mathbf{v}_i)$. Suppose that we already have large-scale established key-value memories, given a query q and the corresponding representation $\mathbf{q}$; the model could use it to preselect a small group of memories by directly matching the words in the query and memories with the reverse index. After narrowing the search space, one can calculate the relevant score between the query and a key:

$$p_i = \text{Softmax}(f_Q(\mathbf{q}) \cdot f_K(\mathbf{k}_i)), \tag{4.41}$$

where $f_Q(\cdot)$ and $f_K(\cdot)$ are feature mapping functions. Similar to the end-to-end memory network, in the reading stage, the vector $\mathbf{o}$ responsible for the final output could be calculated by a weighted sum:

$$\mathbf{o} = \sum p_i f_V(\mathbf{v}_i), \tag{4.42}$$

where $f_V(\cdot)$ is a feature mapping function. It is noteworthy that the output vector leverages both the key and value representations. Another prominent feature of KV-MemNN is that the query can be iteratively updated during the training process. The intuition of this mechanism is that the retrieved memory key could be incorporated into the query and produce a new query to find more accurate memories. Formally, the updated query is

$$\mathbf{q}_{j+1} = \mathbf{W}_q(\mathbf{q}_j + \mathbf{o}), \qquad (4.43)$$

where $\mathbf{W}_q$ is a transformation matrix. Accordingly, the relevant score of the updated query and memories is $p_i = \text{Softmax}(\mathbf{q}_{j+1}^\top f_K(\mathbf{k}_i))$. The number of updates to the query H is a fixed value, which is treated as a hyperparameter. Thus, the final prediction of the model is

$$y = \text{argmax}_k \ \text{Softmax}(\mathbf{q}_{H+1}^\top f_Y(\mathbf{y}_k)), \qquad (4.44)$$

where $\mathbf{y}_k$ is an output candidate of all the possible outputs in a particular task, and $f_Y(\cdot)$ is a feature mapping function. At this time, key-value memories conduct interactions with the output candidates. In the training phase, all the aforementioned feature mapping functions and trainable parameters are optimized in an end-to-end manner. KV-MemNN could also be generalized to a variety of applications with different forms of knowledge by flexibly designing f_Q, f_K, and f_V. For example, for storing world knowledge in the form of a triplet, we can regard the head entity and the relation as the key and the tail entity as the value. For textual knowledge, we can encode sentences or words directly into both key and value in practice.

Although memory networks are proposed for better document modeling, it has profoundly influenced the academic community with this idea. We can use additional modules to store information explicitly and enhance the memory capacity of neural networks. To this day, this framework is still a common idea for modeling very long texts. There are three key points in designing such a network: representation learning of memory, the matching mechanism between memory and query, and how to perform memory retrieval based on the input efficiently.

4.4.2 Hierarchical Document Representation

As mentioned in the former sections, higher-level units in natural languages are often composed of lower-level units, and documents are composed of multiple sentences in a specific logical order. Therefore, an intuitive way to obtain sentence representations is to perform hierarchical modeling [55], where word representations are used to compose sentence representations, which in turn compose document representations. With the powerful representation capabilities of neural networks, we can explicitly develop this type of method. Here, we introduce several neural-based methods of learning document representation hierarchically.

Hierarchical Document Encoder The basic idea of the hierarchical document encoder is to use low-level representations to produce high-level representations. First, the word vectors obtained by pre-training with self-supervised methods can be directly used as the basic word representations. We can also optimize these word representations according to specific tasks. And there are various ways to get the sentence representation through the constituent word representation. For example, we can let it pass through a layer of multilayer perceptron (MLP) and then average over all the hidden states. Here we attempt to recurrently process the document and take LSTM as an example, and the input document is $d = \{s_1, \ldots, s_m\}$, where s_i is a sentence $s_i = \{w_1, \ldots, w_n\}$. We input a sentence s_j directly into LSTM (other neural networks like GRU and CNN can also be applied) and get the corresponding hidden states. In this way, according to the previous equations, the hidden state of each step is calculated from the hidden state of the previous step and the input of the current step:

$$\mathbf{h}_i^w = \text{LSTM}(\mathbf{w}_i, \mathbf{h}_{i-1}^w). \tag{4.45}$$

Thus, the hidden state of the last time step contains the semantic information of the whole sentence and can be used as a sentence representation:

$$\mathbf{s}_j = \mathbf{h}_n^w. \tag{4.46}$$

At this point, we get a representation of each sentence. Considering the sentence as a basic unit, we can build another LSTM on the sentence level to process the sentence representation sequentially. The hidden state at each sentence-level step is determined by the previous hidden state and the current sentence representation input, just like the word-level LSTM:

$$\mathbf{h}_j^s = \text{LSTM}(\mathbf{s}_j, \mathbf{h}_{j-1}^s). \tag{4.47}$$

Repeating the above operation, the hidden state of the last step of this LSTM contains all the information of the sentence representation. It thus can be regarded as a document representation:

$$\mathbf{d} = \mathbf{h}_m^s. \tag{4.48}$$

To this end, we introduce basic hierarchical modeling of document representation. When there is a supervised signal, we can use this document representation directly for neural network training with document-level classification. When there is no supervised signal, we can self-code the document representation, which can be decoded in the reverse order, i.e., first decode the document representation into a sentence representation and then generate words sequentially. The supervised and autoencoding frameworks are illustrated in Figs. 4.11 and 4.12, respectively.

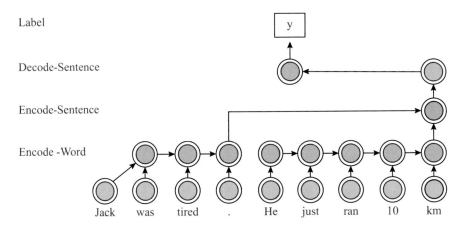

Fig. 4.11 The framework of hierarchical document representation for supervised learning. The figure is a modification of Fig. 2 of the hierarchical autoencoding paper [55]

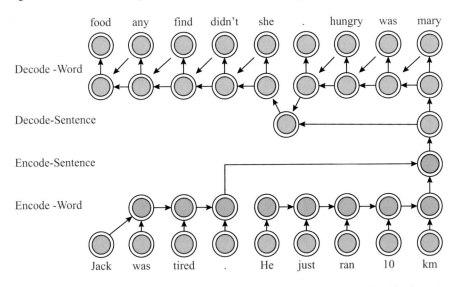

Fig. 4.12 The framework of hierarchical autoencoding of document representation. The figure is re-drawn according to Fig. 2 of the hierarchical autoencoding paper [55]

Hierarchical Attention Network Following the idea of hierarchical modeling, we can make various improvements to the model, such as replacing the LSTM with a more powerful neural network structure and adding attention mechanisms to enhance the transmission of long-dependency information. The hierarchical attention network (HAN [109]) is proposed to use attention mechanisms to capture the hierarchical correlations of documents. The key insight of this model is that while doing hierarchical modeling, different attention weights are assigned to components (words and sentences) using the attention mechanism to learn the

Fig. 4.13 The architecture of the hierarchical attention network. This figure is re-drawn according to Fig. 2 from the hierarchical attention network paper [109]

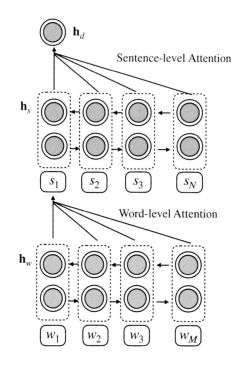

document's representation dynamically. The framework is illustrated in Fig. 4.13. It is worth noting that the intuition of hierarchical attention networks could be applied to various neural networks. We use GRU, another version of RNNs, as the backbone to introduce the approach.

The first step is also to model the basic linguistic units—words—using a bidirectional GRU to incorporate contextual information. The bidirectional hidden states for a word embedding $\mathbf{w}$ is computed by

$$\overrightarrow{\mathbf{h}}_w = \overrightarrow{\text{GRU}}(\mathbf{w}), \tag{4.49}$$

$$\overleftarrow{\mathbf{h}}_w = \overleftarrow{\text{GRU}}(\mathbf{w}). \tag{4.50}$$

By directly concatenating the two hidden states of both directions, we could obtain the final word representation:

$$\mathbf{h}_w = \text{concat}(\overrightarrow{\mathbf{h}}_w; \overleftarrow{\mathbf{h}}_w). \tag{4.51}$$

Then, following the spirit of hierarchical modeling, we need to construct sentence-level representations. Instead of directly feeding word-level representations to a higher-level neural network, an attention mechanism is adopted to automatically determine how important a word is to the sentence-level representation. First, we use a one-layer MLP to further extract the feature of one word:

$$\mathbf{u}_w = \tanh(\mathbf{W}\mathbf{h}_w + \mathbf{b}). \tag{4.52}$$

Then, the sentence representation is computed by

$$\mathbf{s} = \sum_i \alpha^i \mathbf{h}_w^i, \tag{4.53}$$

where α is an attention score and is computed by:

$$\alpha^i = \frac{\exp(\mathbf{u}_w^{i\top} \mathbf{u}_w)}{\sum_i \exp(\mathbf{u}_w^{i\top} \mathbf{u}_w)}. \tag{4.54}$$

Now we obtain the sentence representation $\mathbf{s}$. Logically, the foregoing procedures could be analogically applied to the sentence level and obtain the final document representation. We first still use a bidirectional GRU to capture the correlations between sentences:

$$\overrightarrow{\mathbf{h}}_s = \overrightarrow{\text{GRU}}(\mathbf{s}), \tag{4.55}$$

$$\overleftarrow{\mathbf{h}}_s = \overleftarrow{\text{GRU}}(\mathbf{s}). \tag{4.56}$$

Similarly, the hidden state of a sentence is the concatenation of the two directions of hidden states:

$$\mathbf{h}_s = \text{concat}(\overrightarrow{\mathbf{h}}_s; \overleftarrow{\mathbf{h}}_s). \tag{4.57}$$

Then exactly the same neural network and the attention mechanism are applied as follows:

$$\mathbf{u}_s = \tanh(\mathbf{W}\mathbf{h}_s + \mathbf{b}), \tag{4.58}$$

$$\alpha^i = \frac{\exp(\mathbf{u}_s^{i\top} \mathbf{u}_s)}{\sum_i \exp(\mathbf{u}_s^{i\top} \mathbf{u}_s)}, \tag{4.59}$$

$$\mathbf{h}_d = \sum_i \alpha^i \mathbf{h}_s^i. \tag{4.60}$$

Here, we again use the hierarchical spirit, equipped with the attention technique, to construct the document representation $\mathbf{h}_d$. This representation can be fed to an output layer for document-level classification, thereby training the model.

This section introduces two primary frameworks, memory-based and hierarchical approaches, to model documents. As opposed to directly treating documents as longer sentences and then directly applying neural language modeling, such methods more accurately grasp the characteristics of documents with complex structures.

4.5   Applications

Sentence and document representations play a crucial role in multifarious downstream tasks, many of which are the cornerstone tasks of modern information processing. In this section, we introduce typical applications of sentence and document representations in real-world scenarios, which could fall into three groups: classification, sequence labeling, and generation. For classification, we introduce text classification, information retrieval, reading comprehension, open-domain question answering, sequence labeling, its three representative applications, and sequence-to-sequence generation and its typical applications.

4.5.1   Text Classification

Text classification is a typical NLP application that covers many important real-world tasks, such as parsing and semantic analysis. Therefore, it has attracted the interest of many researchers. The conventional text classification models (e.g., the LDA [5] and tree kernel [78] models) focus on capturing more contextual information and correct word order by extracting more useful and distinct features but still expose a few issues (e.g., data sparseness) which has a significant impact on the classification accuracy. With the development of deep learning in the various fields of artificial intelligence, neural models have been introduced into the text classification field, given their abilities of text representation learning. This section will introduce the three typical text classification tasks, including topic classification, sentiment classification, and natural language inference (NLI).

Topic Classification   Topic classification aims to assign a sentence to an appropriate category (e.g., type of questions, type of news article), which is a fundamental task of the text classification application. Examples of topic classification are listed in Table 4.1.

Considering the effectiveness of the CNN-based models in capturing sentence semantics, many works use CNNs as representation encoders. The character-level CNN [110] is among the first few works to apply character-level information

Table 4.1   Some examples of topic classification

Sentence	Topic
One of the faculties of Stanford just won a Nobel Prize for her contributions to organic chemistry	Sci-Tech
After IPO, the company's share price has risen 147.4% in 2 weeks, and several media outlets are scrambling to cover the news	Business
The Golden State Warriors, led by Stephen Curry, won an NBA championship, and now they're eyeing contract extensions for their core players	Sports

modeling to topic classification. Increasing the depth of the CNNs [20] helps extract the hierarchical information from scattered characters to whole sentences. MG-CNN [111] captures multiple features from multiple sets of embeddings and concatenates them at the penultimate layer.

RNN-based models, which aim to capture the sequential information of sentences, are also widely used in sentence classification. Recurrent CNN [51] applies a recurrent structure to capture contextual information. Hierarchical attention networks [109] introduce word-level and sentence-level attention mechanisms into an RNN-based model as well as a hierarchical structure to capture the hierarchical information of the document for sentence classification. Combining an LSTM with a CNN [112] also shows better performance on text classification, as it captures both local and global features.

Sentiment Classification Sentiment classification is a particular task of the sentence classification application, whose objective is to determine the sentimental polarities of opinions a piece of text contains, e.g., favorable or unfavorable and positive or negative. This task appeals to the NLP community since it has many potential downstream applications, such as movie review suggestions. Examples of sentiment classification are illustrated in Table 4.2.

Similar to text classification, sentence representation based on neural models has also been widely explored for sentiment classification. Text-CNN [47] utilizes the CNNs trained on top of pre-trained word embeddings and achieves promising results on several sentiment classification datasets. The dynamic CNN model [44] can handle sentences of varying lengths and uses dynamic max-pooling over linear sequences, which could help the model capture both short-range and long-range semantic relations in sentences.

Xavier et al. [29] adopt a stacked denoising autoencoder in sentiment classification. Then, a series of studies based on recursive neural networks are presented to learn sentence representations for sentiment classification, including the recursive autoencoder (RAE) [88], matrix-vector recursive neural network (MV-RNN) [87], and recursive neural tensor network (RNTN) [89]. Besides, Johnson et al. [40] adopt a CNN to learn sentence-level representations and yield promising experimental results in sentiment classification.

The RNN models also benefit sentiment classification as they are able to capture sequential information. Studies [54, 93] investigate tree-structured LSTM

Table 4.2 Some examples of sentiment classification

Sentence	Sentiment
The plot and set design of this movie is breathtaking	Positive
He is immersed in sorrow	Negative
All the audience who saw the film stood up and clapped their hands, this is a masterpiece that deserves to be watched again and again	Positive
This book is written without any rules, and the author is very self-righteous	Negative

models on text classification. Hierarchical neural models are proposed to tackle the document-level sentiment classification problem [3, 94], which generate semantic representations at different levels within a document. Besides, an RNN-based multitask learning framework [63] learns across multiple sentence classification tasks and employs three different mechanisms of sharing information to model sentences with task-specific and shared layers. Moreover, the attention mechanism is also introduced into sentiment classification, which aims to determine the importance of each word contributing to the whole sentiment [109].

Natural Language Inference Natural language inference (NLI) is a classification task involving two sentences. Its objective is to determine whether the first sentence entails the second sentence or not. For example, *I was late for class on Monday* entails that *I had a class on Monday*. It could be viewed as a semantic matching problem of two sentences that requires a high-level understanding of sentence-level information. We provide more examples in Table 4.3 to help readers better understand the task. Same as other classification tasks, neural models can automatically learn the two-sentence representations, and a classifier is used for the detection of entailment. The RNN [7] is one of the baseline models for NLI tasks, which derives the representations for both sentences. Apart from using sentence representation directly, some also perform word-level matching to facilitate semantic learning [99]. Kim et al. [46] concatenate features from the attention mechanism with the original hidden states at each layer of RNNs and obtain better performance. Linguistic features like syntactic information [14] are also used to enhance LSTM representation. The recurrent entity network [35] is an entity-centered RNN, which contains several RNN cells, and each cell learns specific entity-related representations. It improves the memory capacity of the original RNN and achieves satisfactory results on NLI tasks.

Table 4.3 Some examples of natural language inference

Premise	Relation	Hypothesis
A cat jumped	Entailment	A cat moved
Some cats walked	Contradiction	No cats moved
Every cat jumped	Neutral	One cat ate
It is nice talking to you all righty	Neutral	I talk to you every day
Fun for adults and children	Contradiction	Fun only for children
Well it's been very interesting	Entailment	It has been very intriguing
You can access the database anytime you want	Entailment	The database is accessible to you
He smiled back at me	Neutral	He was so happy at that moment

4.5.2   Information Retrieval

In the Internet era, information retrieval becomes one of the most critical applications of sentence and document representations. Information retrieval aims to obtain relevant resources from a large-scale collection of information resources. As shown in Fig. 4.14, given the query "William Shakespeare" as input, the search engine (a typical information retrieval application) provides relevant webpages for users. Traditional information retrieval data consists of search queries and document collections D. And the ground truth is available through explicit human judgments or implicit user behavior data such as clickthrough rate.

For the given query q and document d, traditional information retrieval models estimate their relevance through lexical matches. Neural information retrieval models pay more attention to garnering the query and document relevance from semantic matches. Both lexical and semantic matches are essential for neural information retrieval. Thriving from neural network black magic, it helps information retrieval models catch more sophisticated matching features and have achieved the state of the art in the information retrieval task [22].

Neural ranking models typically fall into two groups: representation-based and interaction-based [34]. Studies in the early stage primarily focus on representation-based models. They learn informative representations and match them in the embedding space of queries and documents. On the other hand, interaction-based methods model the query-document matches from the interactions of their terms.

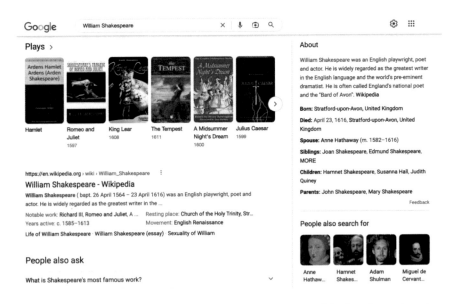

Fig. 4.14 An example of information retrieval. This is a screenshot of the Google search engine

4.5.3   Reading Comprehension

Reading comprehension is crucial to question-answering systems and therefore has been the focus of NLP research. The development of neural-based models has dramatically boosted the performance of reading comprehension. As shown in Fig. 4.15, machine reading comprehension aims to determine the answer given a question and a passage. The task could be viewed as a standard supervised learning task: with a set of training instances, our goal is to learn a mapping that takes the context (i.e., the passage) and related questions as inputs and outputs an answer. The input context can be either a single passage or multiple passages. Intuitively, the longer the provided context is, the more complex the task is. The evaluation metric is typically correlated with the answer type, which will be discussed in the following.

Generally, the current machine reading comprehension task could be divided into four groups according to the answer types [11], i.e., cloze style, multiple-choice, span prediction, and free-form answer.

Cloze Style The cloze style task such as CNN/DAILY MAIL [36] consists of fill-in-the-blank sentences where the question contains a placeholder to be filled in. The answer is either from a predefined candidate set or the vocabulary.

Multiple-Choice The multiple-choice task such as RACE [50] and MCTEST [83] aims to select the best answer from a set of answer choices. It is typical to use accuracy to measure the performance on these two tasks: the percentage of correctly

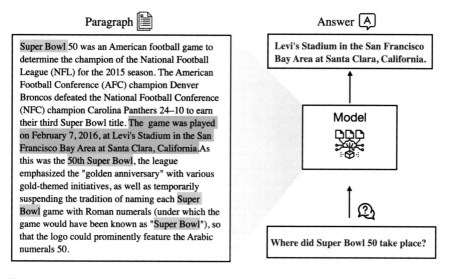

Fig. 4.15 An example of machine reading comprehension from SQuAD [80]

answered questions in the whole example set, since the question could be either correctly answered or not from the given hypothesized answer set.

Span Prediction The span prediction task such as SQuAD [80] is perhaps the most widely adopted task among all since it compromises flexibility and simplicity. It extracts a most likely text span from the passage as the answer to the question, which is usually modeled as predicting the start position and end position of the answer span. We typically use two evaluation metrics proposed by the SQuAD benchmark [80]. The exact match assigns a full score of 1.0 to the predicted answer span if it exactly equals the ground truth answer; otherwise, 0.0. F1-score measures the degree of overlap between prediction and truth by computing a harmonic mean of the precision and recall.

Free-Form Answer The free-form answer task such as MS MARCO [74] does not restrict the answer form or length and is also referred to as *generative question answering*. It is practical to model the task as a sequence generation problem, where the discrete token-level prediction was made. Currently, a consensus on the ideal evaluation metrics has not been achieved. It is common to adopt standard metrics in machine translation and summarization, including ROUGE [58] and BLEU [95].

Since the span prediction format is the most widely researched problem, the following part of this section will be mainly devoted to the mainstream methods in machine reading comprehension with span prediction. With neural networks, the machine reading comprehension system is commonly composed of three consecutive phases: the embedding phase, the reasoning phase, and the prediction phase. Like many other NLP tasks, the embedding phase often adopts pre-trained or contextual word embedding with RNNs, character embedding, or hybrid embeddings. The query and the context are separately encoded. The reasoning phase is responsible for joint learning based on the two representations and is the focus of most works. The prediction phase decides how the output is finally drawn. For extractive mode like span prediction, where a piece of text is extracted from the context, a standard operation is to predict the start position and the end position of the extracted part.

We will mainly introduce the different approaches in the reasoning phase. As shown in Fig. 4.16, while encoding the passage, the model retains the length of the sequence and encodes the question into a fixed-length hidden representation $\mathbf{q}$. The question's hidden vector is then used as a pointer to scan over the passage representation $\{\mathbf{p}_i\}_{i=1}^{n}$ and compute scores on every position in the passage. While maintaining this similar architecture, most machine reading comprehension models vary in the interaction methods between the passage and the question. In the following, we will introduce several classic reading comprehension architectures that follow this paradigm. Most of them merge the two lines of information from the query and the context with the attention mechanism. And they mainly differ in two aspects: the direction of attention and the dimension of attention. Direction refers to whether using only query-to-context attention (as shown in Fig. 4.16) or both directions. Dimension refers to whether attention is only calculated at the

Fig. 4.16 The architecture of classic machine reading comprehension models

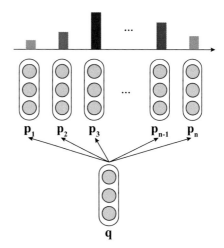

sentence representation level, which outputs a single-dimension vector, or at the word embedding level, where output is an embedding matrix.

Single Direction and Single Dimension The first attempt [36] to apply neural networks on machine reading comprehension constructs bidirectional LSTM reader models along with attention mechanisms. The work introduces two reader models, i.e., the attentive reader and the impatient reader. After encoding the passage and the query into hidden states using LSTMs, the attentive reader computes a scalar distribution over the passage tokens and uses it to calculate the weighted sum of the passage's hidden states. The impatient reader extends this idea further by repeatedly updating the weighted sum of passage hidden states after seeing each query token. Following Hermann et al. [36], Chen et al. [12] modify the method to compute attention and simplify the prediction layer in the attentive reader with a simple bilinear term.

Bidirectional Attention and Single Dimension The attention-over-attention reader [21] also computes both query-to-context and context-to-query attention but handles them differently. Instead of simply averaging the token-level query-to-context attention to obtain a final vector for prediction, attention-over-attention computes a weighted vector with a query word importance vector. The word importance vector is computed by averaging the context-to-query attention. This operation is considered to learn the contributions of individual question words explicitly.

Bidirectional Attention Flow and Multi-Dimension Instead of unifying the document and query representation to a single vector with query-to-context attention only, the BiDAF network [85] computes the attentive token representation of both query-to-context and context-to-query at each bidirectional long short-term memory (BiLSTM) layer to allow fine-grained information flow. It consists of the token embedding layer, the contextual embedding layer, the bidirectional attention flow layer, the LSTM modeling layer, and the Softmax output layer. At each layer, the input is the concatenation of the previous layer's hidden states, the query-to-context representation, and the context-to-query representation. The representation of multiple granularities and a bidirectional attention flow can fully capture the interaction between document and query for start and end position prediction.

The gated-attention reader [25] adopts the gated-attention module, where each token representation of the passage is scaled by the attended query vector after each BiGRU layer. This gated-attention mechanism allows the query to interact directly with the token embeddings of the passage at the semantic level. And such layer-wise interaction enables the model to learn conditional token representation given the question at different representation levels.

4.5.4   Open-Domain Question Answering

Open-domain QA (OpenQA) [33] aims to answer open-domain questions utilizing external resources such as collections of documents [98], webpages [15, 49], structured knowledge graphs [2, 6], or automatically extracted relational triples [28]. Recently, with the development of machine reading comprehension techniques [12, 25, 86, 102], researchers attempt to answer open-domain questions via performing reading comprehension on plain texts with neural-based models [13]. As illustrated in Fig. 4.17, a neural-based OpenQA system usually retrieves relevant articles or paragraphs of the question from a large-scale corpus (e.g., Wikipedia). It then generates answers from these texts by a reading comprehension model introduced in the last section. Open-domain question answering essentially combines two critical applications: information retrieval and reading comprehension.

The system [13], namely, DrQA, is composed of two modules: (1) one document retriever module to retrieve relevant articles or paragraphs and (2) one document reader to produce the final answers from the extracted articles.

Fig. 4.17 An example of open-domain question answering. This figure is re-drawn according to Fig. 1 in the DrQA paper [13]

The document retriever is used as a first quick skim to narrow the search space and focus on potentially relevant documents. The retriever builds TF-IDF weighted bag-of-words vectors for the documents and the questions and computes similarity scores for ranking. The retriever uses bigram counts with hash to further utilize local word order information while ensuring speed and memory efficiency. The document reader model takes in the top five Wikipedia articles yielded by the document retriever and extracts the final answer to the question. The document reader predicts an answer span with a confidence score for each article. The final prediction is made by maximizing the unnormalized exponential prediction scores across the documents.

Given each document, the document reader first builds a feature representation for each word in the document, which is often the concatenation of the following components: (1) Word embeddings: The pre-trained word embeddings like GloVe embeddings pre-trained on Wikipedia. (2) Manual features: The manual features combined with part-of-speech (POS) and named entity recognition tags and normalized term frequencies (TF). (3) Exact match: This feature indicates whether the word in the document can be precisely matched to one question word. (4) Aligned question embeddings: This feature aims to encode a soft alignment between words in the document and the question in the word embedding space.

Then the feature representation of the document is fed into a multilayer bidirectional LSTM (BiLSTM) to encode the contextual representation. For the question, the contextual representation is simply obtained by encoding the word embeddings using a multilayer BiLSTM. After that, the contextual representation is aggregated into a fixed-length vector using self-attention. In the answer prediction phase, the start and end probability distributions are calculated following the paradigm mentioned in the strategy in Sect. 4.5.3.

Despite its success, the DrQA system is prone to noise in retrieved texts which may hurt the performance of the system. Hence, several approaches [18, 100] are proposed to attempt to tackle the noise problem in DrQA by using two separate procedures for question answering: paragraph selection and answer extraction. However, they both only select the most relevant paragraph among all retrieved paragraphs to extract answers and may lose valuable information distributed in other paragraphs.

Wang et al. [101] adopt strength-based and coverage-based methods for re-ranking, aggregating the answers that existing methods retrieved from all the paragraphs. Nevertheless, the challenge of noisy data is still unsolved. To address this issue, a coarse-to-fine denoising OpenQA model [60] is developed to the first screen out relevant paragraphs and then retrieve correct answers.

4.5.5   Sequence Labeling

Sequence labeling is a classic application in natural language processing. In this paradigm, given an input sequence $\{w_1, \ldots, w_n\}$, we need to assign a label y_i

Fig. 4.18 An example of
sequence labeling

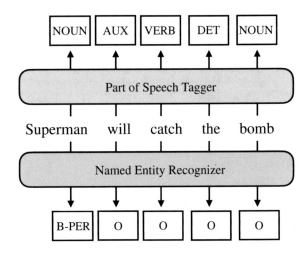

to each token w_i. Part-of-speech (POS) tagging and named entity recognition
(NER) are the two most representative sequence labeling tasks. Sequence labeling
requires the model to capture the correlations of words in the sequence accurately.
Hence, classic approaches use probabilistic graphical models (PGM) to represent
the dependency structure of different words. Modern methods use powerful deep
neural networks to produce richer representations and adopt conditional random
field (CRF) or direct token-level classification to conduct sequence labeling [38]. In
addition to these two tasks, word segmentation of languages without delimiters (e.g.,
Chinese) is typically treated as a sequence labeling task [26, 66, 107] (Fig. 4.18).

Part-of-Speech (POS) Tagging POS tagging aims to assign part-of-speech tags
to each word in a given piece of text, including nouns, verbs, adjectives, etc.
Some tags might be evident and static (e.g., proper nouns), while most words are
polysemy, and their part-of-speech attributes are context-dependent. For example,
the word "record" can be either a noun or a verb. Early on, Brill et al. [8]
propose rule-based methods that highly rely on expert knowledge and extraction
of rich linguistic features in syntax, morphology, and lexicon. Classical statistical
models like the hidden Markov model (HMM) [41] model the probability of tags
given words in a context-aware manner. Modern neural networks are based on
contextual representations of words and parameterize the predicted probability with
a conditional random field (CRF) layer and a simple MLP classifier head. CNNs
and RNNs are common backbones used for feature extraction [67, 77].

Named Entity Recognition (NER) In NER, we need to identify if a word in an
input sequence is a *named entity*, a term that could specifically indicate a real-
world object. Typical named entity types include *Person, Organization, Location*,
etc. A named entity could be one word or a phrase with multiple words. Hence,
in this task, a BIO label schema is universally adopted, where a word could be
classified at the beginning of an entity (B), inside an entity (I), and outside an

entity (O). Final entity prediction is extracted based on the word assigned tags, and evaluation is conducted at the entity level [27, 103]. Feature-based methods extract word-level and character-level features and adopt classic classification models for prediction. Bike et al. [4] and Mcnamee et al. [68] propose an HMM-based and support vector machine (SVM)-based NER system, respectively. Deep learning methods allow for richer feature representation. Apart from using pre-trained word embeddings like skip-gram, a series of works [16, 52, 56, 82] also learn character-level features and incorporate them with word representations for better performance. The bidirectional LSTM-CNN [16] encodes character-level features with a CNN and word-level features with a BiLSTM. The bidirectional LSTM-CRF model [38] also adds other features, including spelling features, context features, and gazetteer features, to enhance final representations in a BiLSTM-CRF model.

4.5.6   Sequence-to-Sequence Generation

Sequence-to-sequence generation refers to a group of tasks that require sequence generation based on an input sequence, including machine translation, text summarization, question generation, etc. A famous model structure for sequence-to-sequence problems is an encoder-decoder structure, where the model is composed of an encoder and a decoder. The encoder encodes the input source language $S = \{s_1, s_2, \ldots, s_n\}$ and passes the encoded representation to the decoder. The decoder decodes and outputs tokens in target language $T = \{t_1, t_2, \ldots, t_m\}$ based on encoder output. More specifically, output tokens are typically generated in an autoregressive manner, i.e., each t_i is generated depending on the previously generated tokens $\{t_1, t_2, \ldots, t_{i-1}\}$. Both structures are trained in an end-to-end fashion with parallel training data. Below is a formalized training objective for a sequence-to-sequence problem:

$$\arg\max \prod_{i=1}^{m} P(t_i | t_{j<i}, S) \tag{4.61}$$

Metrics First, it is essential to learn the commonly used metrics to evaluate a sequence-to-sequence system.

BLEU [75] is an adjusted precision calculation based on the count of n-grams. First, it extracts all n-grams in the output sequence. Then, it calculates the sum of occurrences of these n-grams in the reference sequence (i.e., the correct translation) against the total number of n-grams in the output sequence. For example, if the output is *the cat cat* and the reference is *the cat jumps*, all 2 grams in the output is *"the cat," cat cat* and the total number of their occurrence in the reference is 1 (*the cat*). So the score of 2-gram will be $p_2 = \frac{1}{2} = 0.5$. BLEU also takes a brevity penalty (BP) that penalizes the mismatch of output and reference length. Suppose we set a range for the number of grams involving the calculation as $[1, N]$, the

BLEU score is

$$\text{BLEU} = \text{BP} \cdot \exp\left(\sum_{i=1}^{N} w_i \log p_i\right), \tag{4.62}$$

$$\text{BP} = \begin{cases} 1 & c > r \\ e^{(1-r/c)} & c \leq r, \end{cases} \tag{4.63}$$

where w_i is a weight and can be set to $\frac{1}{N}$, c is the length of the output sequence, and r is the length of the reference sequence.

ROUGE [58] is a group of metrics often used in evaluating text summarization systems. ROUGE-N (most commonly ROUGE-1 and ROUGE-2) calculates the recall of n-grams. So take the example above; we can get ROUGE-1 $= 2/3$ and ROUGE-2 $= 1/2$. ROUGE-L concerns the ratio of the length of the longest common subsequence against the reference length. In the example above, ROUGE-L $= 2/3$.

Next, we introduce some representative models in machine translation and text summarization.

Machine Translation Machine translation aims to translate texts in one language into another language while retaining their semantic meanings. While traditional rule-based and statistical machine translation systems require abundant expert knowledge and often fail to capture meaning from context to handle polysemy, the development of deep neural networks has inspired neural machine translation systems and achieved competitive performance.

Kalchbrenner et al. [43] use a one-dimensional CNN as the encoder and a single-layer RNN as the decoder. Cho et al. [17] enhance the alignment scores calculation between phrases with an RNN encoder-decoder structure and improve on the traditional statistical machine translation system. Sutskever et al. [92] adopt a deep LSTM encoder-decoder.

GNMT [105] is the first NMT system put into production. It has an eight-layer LSTM encoder and 8-layer LSTM decoder, and the first layer of the encoder is bidirectional. The attention mechanism is also applied to the output of the encoder. In terms of decoding, it also adds coverage penalty and length normalization to encourage the generation of longer and high-quality sentences. And the Transformer [96], an encoder-decoder neural network, is proposed initially as a sequence-to-sequence model and used on the machine translation task. The model then achieves the new state-of-art performance on benchmark datasets compared to models based on LSTM.

Text Summarization Text summarization takes a long passage as its input and generates a relatively short one that summarizes the key points in the original passage. It is worth noting that typically sequence-to-sequence models can be simultaneously applied to machine translation and text summarization since the task is of the same form.

Pointer-generator network [84] is one of the most classical text summarization models that combine LSTM-attention-based encoder-decoder with pointer network [97]. The basic structure contains a single-layer bidirectional LSTM encoder with attention and a single-layer LSTM decoder. Apart from the standard encoder-decoder pipeline, it applies an extra pointer while decoding. The pointer depends on the encoder output, the current decoder hidden states, and decoder input and calculates a probability p_{gen} indicating how much we favor the decoder generated results. The final distribution from which the next token is drawn is a weighted sum of distribution given by the decoder and distribution given by attention weights of the encoder output, each weighted by p_{gen} and $1 - p_{gen}$. So the pointer serves as a mediator between generated tokens and copied tokens from the original input. It is especially beneficial for text summarization as copying original words from the input can help keep the semantics on the right track.

4.6   Summary and Further Readings

This chapter introduces basic concepts, methodologies, and applications of sentence and document representation learning, which encode sentences and documents into real-valued representation vectors. We first introduce the symbolic representation for sentences and probabilistic language models. Then we extensively introduce several neural language models, including adopting feed-forward neural networks, convolutional neural networks, recurrent neural networks, and Transformers for language models. We further introduce document representation learning methods, including memory-based and hierarchical approaches. Finally, we introduce several typical applications of sentence and document representation. Sentence and document representations provide an effective way of downstream tasks utilizing high-level semantic information and have significantly improved the performances of these tasks. For further understanding of sentence representation learning and its applications, there are also some recommended surveys and books that introduce neural network methods [30, 42], sentence representation methods [57], and Transformers [59].

More recently, pre-trained language models based on deep Transformers show state-of-the-art performance in this area. Meanwhile, it also spawns particular research issues of sentence and document representation learning. We will introduce and discuss this topic in the next chapter. In addition, the use of more efficient neural network architectures, the establishment of a more stable and universal representation of long text, and the development of a comprehensive evaluation approach are worthy research topics in this field.

Acknowledgments Zhiyuan Liu, Yankai Lin, and Maosong Sun designed the overall architecture of this chapter; Ning Ding and Yankai Lin drafted the chapter. Zhiyuan Liu and Yankai Lin proofread and revised this chapter.

We also thank Zhengyan Zhang, Cunchao Tu, Hongyin Luo, Zhenghao Liu, and Haozhe Ji for providing the initial materials for the first edition. And we thank Ganqu Cui, Yuan Yao, Shi Yu, Yulin Chen, Xingtai Lv, and Suyuan Zhao for proofreading this chapter.

This is the sentence and document representation learning chapter of the second edition of the book *Representation Learning for Natural Language Processing*, with its first edition published in 2020 [65]. As compared to the first edition of this chapter, the main changes include the following: (1) we merged the sentence representation and document representation to a new chapter by restructuring the organization, (2) we polished and rewrote the content of the methods and applications, and (3) we added new methods for document representation learning.

References

1. Yoshua Bengio, Réjean Ducharme, Pascal Vincent, and Christian Jauvin. A neural probabilistic language model. *Journal of Machine Learning Research*, 3, 2003.
2. Jonathan Berant, Andrew Chou, Roy Frostig, and Percy Liang. Semantic parsing on freebase from question-answer pairs. In *Proceedings of EMNLP*, 2013.
3. Parminder Bhatia, Yangfeng Ji, and Jacob Eisenstein. Better document-level sentiment analysis from rst discourse parsing. In *Proceedings of EMNLP*, 2015.
4. Daniel M. Bikel, Scott Miller, Richard Schwartz, and Ralph Weischedel. Nymble: a high-performance learning name-finder. In *Fifth Conference on Applied Natural Language Processing*, 1997.
5. David M Blei, Andrew Y Ng, and Michael I Jordan. Latent dirichlet allocation. *Journal of Machine Learning Research*, 3:993–1022, 2003.
6. Antoine Bordes, Nicolas Usunier, Sumit Chopra, and Jason Weston. Large-scale simple question answering with memory networks. *arXiv preprint arXiv:1506.02075*, 2015.
7. Samuel R. Bowman, Gabor Angeli, Christopher Potts, and Christopher D. Manning. A large annotated corpus for learning natural language inference. In *Proceedings of EMNLP*, 2015.
8. Eric Brill. A simple rule-based part of speech tagger. In *Proceedings of ANLP*, 1992.
9. Peter F Brown, Peter V Desouza, Robert L Mercer, Vincent J Della Pietra, and Jenifer C Lai. Class-based n-gram models of natural language. *Computational linguistics*, 18(4):467–479, 1992.
10. Sarath Chandar, Sungjin Ahn, Hugo Larochelle, Pascal Vincent, Gerald Tesauro, and Yoshua Bengio. Hierarchical memory networks. *arXiv preprint arXiv:1605.07427*, 2016.
11. Danqi Chen. *Neural Reading Comprehension and Beyond*. PhD thesis, Stanford University, 2018.
12. Danqi Chen, Jason Bolton, and Christopher D. Manning. A thorough examination of the cnn/daily mail reading comprehension task. In *Proceedings of ACL*, 2016.
13. Danqi Chen, Adam Fisch, Jason Weston, and Antoine Bordes. Reading wikipedia to answer open-domain questions. In *Proceedings of ACL*, 2017.
14. Qian Chen, Xiaodan Zhu, Zhen-Hua Ling, Si Wei, Hui Jiang, and Diana Inkpen. Enhanced LSTM for natural language inference. In *Proceedings of ACL*, 2017.
15. Tongfei Chen and Benjamin Van Durme. Discriminative information retrieval for question answering sentence selection. In *Proceedings of EACL*, 2017.
16. Jason PC Chiu and Eric Nichols. Named entity recognition with bidirectional lstm-cnns. *Transactions of the association for computational linguistics*, 4:357–370, 2016.

17. Kyunghyun Cho, Bart van Merriënboer, Caglar Gulcehre, Dzmitry Bahdanau, Fethi Bougares, Holger Schwenk, and Yoshua Bengio. Learning phrase representations using RNN encoder–decoder for statistical machine translation. In *Proceedings of EMNLP*, 2014.
18. Eunsol Choi, Daniel Hewlett, Jakob Uszkoreit, Illia Polosukhin, Alexandre Lacoste, and Jonathan Berant. Coarse-to-fine question answering for long documents. In *Proceedings of ACL*, 2017.
19. Junyoung Chung, Caglar Gulcehre, Kyunghyun Cho, and Yoshua Bengio. Gated feedback recurrent neural networks. In *Proceedings of ICML*, 2015.
20. Alexis Conneau, Holger Schwenk, Loïc Barrault, and Yann Lecun. Very deep convolutional networks for text classification. In *Proceedings of EACL*, 2017.
21. Yiming Cui, Zhipeng Chen, Si Wei, Shijin Wang, Ting Liu, and Guoping Hu. Attention-over-attention neural networks for reading comprehension. In *Proceedings of ACL*, 2017.
22. Zhuyun Dai, Chenyan Xiong, Jamie Callan, and Zhiyuan Liu. Convolutional neural networks for soft-matching n-grams in ad-hoc search. In *Proceedings of WSDM*, 2018.
23. Yann N Dauphin, Angela Fan, Michael Auli, and David Grangier. Language modeling with gated convolutional networks. In *Proceedings of ICML*, 2017.
24. Jacob Devlin, Ming-Wei Chang, Kenton Lee, and Kristina Toutanova. BERT: pre-training of deep bidirectional transformers for language understanding. In *Proceedings of NAACL-HLT*, 2019.
25. Bhuwan Dhingra, Hanxiao Liu, Zhilin Yang, William Cohen, and Ruslan Salakhutdinov. Gated-attention readers for text comprehension. In *Proceedings of ACL*, 2017.
26. Ning Ding, Dingkun Long, Guangwei Xu, Muhua Zhu, Pengjun Xie, Xiaobin Wang, and Hai-Tao Zheng. Coupling distant annotation and adversarial training for cross-domain chinese word segmentation. In *Proceedings ACL*, 2020.
27. Ning Ding, Guangwei Xu, Yulin Chen, Xiaobin Wang, Xu Han, Pengjun Xie, Hai-Tao Zheng, and Zhiyuan Liu. Few-nerd: A few-shot named entity recognition dataset. In *Proceedings of ACL*, 2021.
28. Anthony Fader, Luke Zettlemoyer, and Oren Etzioni. Open question answering over curated and extracted knowledge bases. In *Proceedings of KDD*, 2014.
29. Xavier Glorot, Antoine Bordes, and Yoshua Bengio. Domain adaptation for large-scale sentiment classification: A deep learning approach. In *Proceedings of ICML*, 2011.
30. Yoav Goldberg. Neural network methods for natural language processing. *Synthesis Lectures on Human Language Technologies*, 10(1):1–309, 2017.
31. Irving J Good. The population frequencies of species and the estimation of population parameters. *Biometrika*, 40(3–4):237–264, 1953.
32. Joshua Goodman. Classes for fast maximum entropy training. In *Proceedings of ASSP*, 2001.
33. Bert F Green Jr, Alice K Wolf, Carol Chomsky, and Kenneth Laughery. Baseball: an automatic question-answerer. In *Proceedings of IRE-AIEE-ACM*, 1961.
34. Jiafeng Guo, Yixing Fan, Qingyao Ai, and W.Bruce Croft. A deep relevance matching model for ad-hoc retrieval. In *Proceedings of CIKM*, 2016.
35. Mikael Henaff, Jason Weston, Arthur Szlam, Antoine Bordes, and Yann LeCun. Tracking the world state with recurrent entity networks. In *Proceedings of ICLR*, 2017.
36. Karl Moritz Hermann, Tomas Kocisky, Edward Grefenstette, Lasse Espeholt, Will Kay, Mustafa Suleyman, and Phil Blunsom. Teaching machines to read and comprehend. In *Proceedings of NeurIPS*, 2015.
37. Sepp Hochreiter and Jürgen Schmidhuber. Long short-term memory. *Neural Computation*, 1997.
38. Zhiheng Huang, Wei Xu, and Kai Yu. Bidirectional lstm-crf models for sequence tagging. *arXiv preprint arXiv:1508.01991*, 2015.
39. Zhiheng Huang, Geoffrey Zweig, and Benoit Dumoulin. Cache based recurrent neural network language model inference for first pass speech recognition. In *Proceedings of ICASSP*, 2014.
40. Rie Johnson and Tong Zhang. Effective use of word order for text categorization with convolutional neural networks. In *Proceedings of ACL-HLT*, 2015.

41. Nisheeth Joshi, Hemant Darbari, and Iti Mathur. Hmm based pos tagger for hindi. In *Proceeding of AISC*, 2013.
42. Dan Jurafsky and James H Martin. Speech and language processing. 3rd, 2022.
43. Nal Kalchbrenner and Phil Blunsom. Recurrent continuous translation models. In *Proceedings of EMNLP*, 2013.
44. Nal Kalchbrenner, Edward Grefenstette, and Phil Blunsom. A convolutional neural network for modelling sentences. In *Proceedings of ACL*, 2014.
45. Slava Katz. Estimation of probabilities from sparse data for the language model component of a speech recognizer. *IEEE Transactions on Acoustics, Speech, and Signal Processing*, 35(3):400–401, 1987.
46. Seonhoon Kim, Inho Kang, and Nojun Kwak. Semantic sentence matching with densely-connected recurrent and co-attentive information. In *Proceedings of AAAI*, 2019.
47. Yoon Kim. Convolutional neural networks for sentence classification. In *Proceedings of EMNLP*, 2014.
48. Ankit Kumar, Ozan Irsoy, Peter Ondruska, Mohit Iyyer, James Bradbury, Ishaan Gulrajani, Victor Zhong, Romain Paulus, and Richard Socher. Ask me anything: Dynamic memory networks for natural language processing. In *Proceedings of ICML*, 2016.
49. Cody Kwok, Oren Etzioni, and Daniel S Weld. Scaling question answering to the web. *TOIS*, pages 242–262, 2001.
50. Guokun Lai, Qizhe Xie, Hanxiao Liu, Yiming Yang, and Eduard Hovy. RACE: Large-scale reading comprehension dataset from examinations. *arXiv preprint arXiv:1704.04683*, 2017.
51. Siwei Lai, Liheng Xu, Kang Liu, and Jun Zhao. Recurrent convolutional neural networks for text classification. In *Proceedings of AAAI*, 2015.
52. Guillaume Lample, Miguel Ballesteros, Sandeep Subramanian, Kazuya Kawakami, and Chris Dyer. Neural architectures for named entity recognition. In *Proceedings of NAACL*, 2016.
53. Quoc V Le and Tomas Mikolov. Distributed representations of sentences and documents. In *Proceedings of ICML*, 2014.
54. Jiwei Li, Minh-Thang Luong, Dan Jurafsky, and Eduard Hovy. When are tree structures necessary for deep learning of representations? In *Proceedings of EMNLP*, 2015.
55. Jiwei Li, Thang Luong, and Dan Jurafsky. A hierarchical neural autoencoder for paragraphs and documents. In *Proceedings of ACL*, 2015.
56. Peng-Hsuan Li, Ruo-Ping Dong, Yu-Siang Wang, Ju-Chieh Chou, and Wei-Yun Ma. Leveraging linguistic structures for named entity recognition with bidirectional recursive neural networks. In *Proceedings of EMNLP*. Association for Computational Linguistics, 2017.
57. Ruiqi Li, Xiang Zhao, and Marie-Francine Moens. A brief overview of universal sentence representation methods: A linguistic view. *ACM Computing Surveys (CSUR)*, 55(3):1–42, 2022.
58. Chin-Yew Lin. ROUGE: A package for automatic evaluation of summaries. *Text Summarization Branches Out*, 2004.
59. Tianyang Lin, Yuxin Wang, Xiangyang Liu, and Xipeng Qiu. A survey of transformers. *arXiv preprint arXiv:2106.04554*, 2021.
60. Yankai Lin, Haozhe Ji, Zhiyuan Liu, and Maosong Sun. Denoising distantly supervised open-domain question answering. In *Proceedings of ACL*, 2018.
61. Fei Liu, Trevor Cohn, and Timothy Baldwin. Improving end-to-end memory networks with unified weight tying. In *Proceedings of ALTA*, 2017.
62. Fei Liu and Julien Perez. Gated end-to-end memory networks. In *Proceedings of EACL*, 2017.
63. Pengfei Liu, Xipeng Qiu, and Xuanjing Huang. Recurrent neural network for text classification with multi-task learning. In *Proceedings of IJCAI*, 2016.
64. Yinhan Liu, Myle Ott, Naman Goyal, Jingfei Du, Mandar Joshi, Danqi Chen, Omer Levy, Mike Lewis, Luke Zettlemoyer, and Veselin Stoyanov. RoBERTa: A robustly optimized BERT pretraining approach. *arXiv preprint arXiv:1907.11692*, 2019.
65. Zhiyuan Liu, Yankai Lin, and Maosong Sun. *Representation Learning for Natural Language Processing*. Springer, 2020.

66. Ji Ma, Kuzman Ganchev, and David Weiss. State-of-the-art chinese word segmentation with bi-lstms. In *Proceedings of EMNLP*, 2018.
67. Xuezhe Ma and Eduard Hovy. End-to-end sequence labeling via bi-directional LSTM-CNNs-CRF. In *Proceedings of ACL*, 2016.
68. Paul McNamee and James Mayfield. Entity extraction without language-specific resources. In *Proceedings of COLING*, 2002.
69. Tomas Mikolov, Kai Chen, Greg Corrado, and Jeffrey Dean. Efficient estimation of word representations in vector space. In *Proceedings of ICLR*, 2013.
70. Tomas Mikolov, Martin Karafiát, Lukas Burget, Jan Cernockỳ, and Sanjeev Khudanpur. Recurrent neural network based language model. In *Proceedings of InterSpeech*, 2010.
71. Alexander Miller, Adam Fisch, Jesse Dodge, Amir-Hossein Karimi, Antoine Bordes, and Jason Weston. Key-value memory networks for directly reading documents. In *Proceedings of EMNLP*, 2016.
72. Andriy Mnih and Yee Whye Teh. A fast and simple algorithm for training neural probabilistic language models. In *Proceedings of ICML*, 2012.
73. Frederic Morin and Yoshua Bengio. Hierarchical probabilistic neural network language model. In *Proceedings of AISTATS*, 2005.
74. Tri Nguyen, Mir Rosenberg, Xia Song, Jianfeng Gao, Saurabh Tiwary, Rangan Majumder, and Li Deng. MS MARCO: A human generated machine reading comprehension dataset. *arXiv preprint arXiv:1611.09268*, 2016.
75. Kishore Papineni, Salim Roukos, Todd Ward, and Wei-Jing Zhu. BLEU: A method for automatic evaluation of machine translation. In *Proceedings of ACL*, 2002.
76. Michael J Pazzani and Daniel Billsus. *Content-based recommendation systems*. Springer, 2007.
77. Barbara Plank, Anders Søgaard, and Yoav Goldberg. Multilingual part-of-speech tagging with bidirectional long short-term memory models and auxiliary loss. In *Proceedings of ACL*, 2016.
78. Matt Post and Shane Bergsma. Explicit and implicit syntactic features for text classification. In *Proceedings of ACL*, 2013.
79. Alec Radford, Jeffrey Wu, Rewon Child, David Luan, Dario Amodei, and Ilya Sutskever. Language models are unsupervised multitask learners. *OpenAI Blog*, 2019.
80. Pranav Rajpurkar, Jian Zhang, Konstantin Lopyrev, and Percy Liang. SQuAD: 100,000+ questions for machine comprehension of text. In *Proceedings of EMNLP*, 2016.
81. Juan Ramos. Using tf-idf to determine word relevance in document queries. In *Proceedings of ICML*, 2003.
82. Marek Rei, Gamal Crichton, and Sampo Pyysalo. Attending to characters in neural sequence labeling models. In *Proceedings of COLING*, 2016.
83. Matthew Richardson, Christopher JC Burges, and Erin Renshaw. MCTest: A challenge dataset for the open-domain machine comprehension of text. In *Proceedings of EMNLP*, 2013.
84. Abigail See, Peter J. Liu, and Christopher D. Manning. Get to the point: Summarization with pointer-generator networks. In *Proceedings of ACL*, 2017.
85. Minjoon Seo, Aniruddha Kembhavi, Ali Farhadi, and Hannaneh Hajishirzi. Bidirectional attention flow for machine comprehension. In *Proceedings of ICLR*, 2017.
86. Yelong Shen, Po-Sen Huang, Jianfeng Gao, and Weizhu Chen. ReasoNet: Learning to stop reading in machine comprehension. In *Proceedings of KDD*, 2017.
87. Richard Socher, Brody Huval, Christopher D Manning, and Andrew Y Ng. Semantic compositionality through recursive matrix-vector spaces. In *Proceedings of EMNLP*, 2012.
88. Richard Socher, Jeffrey Pennington, Eric H Huang, Andrew Y Ng, and Christopher D Manning. Semi-supervised recursive autoencoders for predicting sentiment distributions. In *Proceedings of EMNLP*, 2011.
89. Richard Socher, Alex Perelygin, Jean Wu, Jason Chuang, Christopher D. Manning, Andrew Ng, and Christopher Potts. Recursive deep models for semantic compositionality over a sentiment treebank. In *Proceedings of EMNLP*, 2013.

90. Daniel Soutner, Zdeněk Loose, Luděk Müller, and Aleš Pražák. Neural network language model with cache. In *Proceedings of ICTSD*, 2012.

91. Sainbayar Sukhbaatar, Jason Weston, Rob Fergus, et al. End-to-end memory networks. In *Proceedings of NeurIPS*, volume 28, 2015.

92. Ilya Sutskever, Oriol Vinyals, and Quoc V Le. Sequence to sequence learning with neural networks. In *Proceedings of NeurIPS*, 2014.

93. Kai Sheng Tai, Richard Socher, and Christopher D. Manning. Improved semantic representations from tree-structured long short-term memory networks. In *Proceedings of ACL*, 2015.

94. Duyu Tang, Bing Qin, and Ting Liu. Document modeling with gated recurrent neural network for sentiment classification. In *Proceedings of EMNLP*, 2015.

95. Ahmet Cüneyd Tantuğ, Kemal Oflazer, and Ilknur Durgar El-Kahlout. Bleu+: a tool for fine-grained bleu computation. In *Proceedings of LREC*, 2008.

96. Ashish Vaswani, Noam Shazeer, Niki Parmar, Llion Jones, Jakob Uszkoreit, Aidan N Gomez, and Lukasz Kaiser. Attention is all you need. In *Proceedings of NeurIPS*, 2017.

97. Oriol Vinyals, Meire Fortunato, and Navdeep Jaitly. Pointer networks. In *Proceedings of NeurIPS*, 2015.

98. Ellen M Voorhees et al. The trec-8 question answering track report. In *Proceedings of TREC*, 1999.

99. Shuohang Wang and Jing Jiang. Learning natural language inference with LSTM. In *Proceedings of NAACL*, 2016.

100. Shuohang Wang, Mo Yu, Xiaoxiao Guo, Zhiguo Wang, Tim Klinger, Wei Zhang, Shiyu Chang, Gerald Tesauro, Bowen Zhou, and Jing Jiang. R3: Reinforced ranker-reader for open-domain question answering. In *Proceedings of AAAI*, 2018.

101. Shuohang Wang, Mo Yu, Jing Jiang, Wei Zhang, Xiaoxiao Guo, Shiyu Chang, Zhiguo Wang, Tim Klinger, Gerald Tesauro, and Murray Campbell. Evidence aggregation for answer re-ranking in open-domain question answering. In *Proceedings of ICLR*, 2018.

102. Wenhui Wang, Nan Yang, Furu Wei, Baobao Chang, and Ming Zhou. Gated self-matching networks for reading comprehension and question answering. In *Proceedings of ACL*, 2017.

103. Ralph Weischedel, Martha Palmer, Mitchell Marcus, Eduard Hovy, Sameer Pradhan, Lance Ramshaw, Nianwen Xue, Ann Taylor, Jeff Kaufman, Michelle Franchini, et al. Ontonotes release 5.0 ldc2013t19. *Linguistic Data Consortium, Philadelphia, PA*, 23, 2013.

104. Jason Weston, Sumit Chopra, and Antoine Bordes. Memory networks. *arXiv preprint arXiv:1410.3916*, 2014.

105. Yonghui Wu, Mike Schuster, Zhifeng Chen, Quoc V Le, Mohammad Norouzi, Wolfgang Macherey, Maxim Krikun, Yuan Cao, Qin Gao, Klaus Macherey, et al. Google's neural machine translation system: Bridging the gap between human and machine translation. *arXiv preprint arXiv:1609.08144*, 2016.

106. Caiming Xiong, Stephen Merity, and Richard Socher. Dynamic memory networks for visual and textual question answering. In *Proceedings of ICML*, 2016.

107. Nianwen Xue. Chinese word segmentation as character tagging. In *Proceddings of IJCL*, 2003.

108. Tianyu Yang and Antoni B Chan. Learning dynamic memory networks for object tracking. In *Proceedings of ECCV*, 2018.

109. Zichao Yang, Diyi Yang, Chris Dyer, Xiaodong He, Alex Smola, and Eduard Hovy. Hierarchical attention networks for document classification. In *Proceedings of NAACL-HLT*, 2016.

110. Xiang Zhang, Junbo Zhao, and Yann LeCun. Character-level convolutional networks for text classification. In *Proceedings of NeurIPS*, 2015.

111. Ye Zhang, Stephen Roller, and Byron C Wallace. MGNC-CNN: A simple approach to exploiting multiple word embeddings for sentence classification. In *Proceedings of NAACL-HLT*, 2016.

112. Chunting Zhou, Chonglin Sun, Zhiyuan Liu, and F. Lau. A c-lstm neural network for text classification. *ArXiv, abs/1511.08630*, 2015.

Chapter 5
Pre-trained Models for Representation Learning

Yankai Lin, Ning Ding, Zhiyuan Liu, and Maosong Sun

Abstract Pre-training-fine-tuning has recently become a new paradigm in natural language processing, learning better representations of words, sentences, and documents in a self-supervised manner. Pre-trained models not only unify semantic representations of multiple tasks, multiple languages, and multiple modalities but also emerge high-level capabilities approaching human beings. In this chapter, we introduce pre-trained models for representation learning, from pre-training tasks to adaptation approaches for specific tasks. After that, we discuss several advanced topics toward better pre-trained representations, including better model architecture, multilingual, multi-task, efficient representations, and chain-of-thought reasoning.

5.1 Introduction

Representation learning is the critical component of machine learning systems, which aims to learn informative representations of objects from large-scale data. With the learned representations, machine learning systems thus can handle multiple tasks, languages, and modalities more flexibly and desirable. Representation learning for natural language processing (NLP) can be divided into three stages according to the learning paradigm: statistical learning, deep learning, and pre-trained models, with the paradigm shift of representation from symbolic representation to distributed representation.

Statistical learning started early in the 1940s [39, 93]. It requires domain experts to design task-specific rules according to their knowledge to transfer raw data into

Y. Lin
Gaoling School of Artificial Intelligence, Renmin University of China, Beijing, China
e-mail: yankailin@ruc.edu.cn

N. Ding · Z. Liu (✉) · M. Sun
Department of Computer Science and Technology, Tsinghua University, Beijing, China
e-mail: dingn18@mails.tsinghua.edu.cn; liuzy@tsinghua.edu.cn; sms@tsinghua.edu.cn

Z. Liu et al. (eds.), *Representation Learning for Natural Language Processing*,
https://doi.org/10.1007/978-981-99-1600-9_5

task-related representation task-by-task. This makes representation learning based on statistical learning fragmented in multiple granularities of text and multiple tasks. Later, distributed representation learning with deep learning techniques [45] was developed with larger datasets, more computing power, and advanced neural architectures. It utilizes deep neural networks such as convolutional neural networks (CNNs) and recurrent neural networks (RNNs) to extract task-related representations automatically. It makes an initial step toward unified representation learning: although deep learning still results in various model parameters for multiple tasks, the learned representations can be transferred to multiple NLP tasks, and the same neural network architecture can be applied to model the data of various NLP tasks.

Recently, pre-trained models (PTMs) for representation learning [7, 20], also known as foundation models [6], have become a new trend in NLP. As shown in Figs. 5.1 and 5.2, compared with conventional representation learning techniques, the pre-training-fine-tuning paradigm of PTMs enables them to learn unified representations for multiple tasks, languages, and modalities. Moreover, big PTMs have shown high-level capabilities like human beings. In the following, we introduce the new characteristics of PTMs in detail.

(1) **Pre-training-Fine-Tuning Paradigm.** Transfer learning [103] enables the knowledge (usually stored in model parameters) learned from one task/domain to be transferred to help the learning of other tasks/domains in the same model architecture. Inspired by the idea of transfer learning, PTMs learn general task-agnostic representations via self-supervised learning from large-scale unlabeled data and then adapt their model parameters to downstream tasks by task-specific fine-tuning. Different from conventional deep learning techniques that only learn from task-specific supervised data, the self-supervised pre-training objective enables PTMs to learn from larger unlabeled web-scale data without labeled task signals automatically. With the pre-training-fine-tuning pipeline,

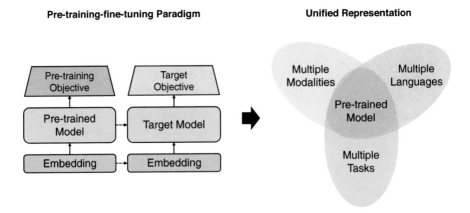

Fig. 5.1 Pre-training-fine-tuning paradigm and unified representation of pre-trained models

Fig. 5.2 The development trends of big PTMs. The size of the circles indicates the model scale. The figure is obtained from the official website of OpenBMB (https://openbmb.github.io/BMList/)

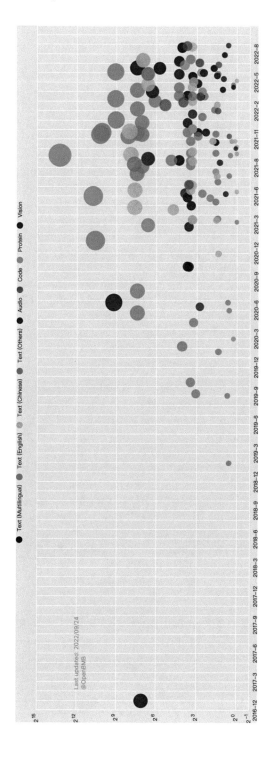

PTMs learn general knowledge in the self-supervised pre-training and then stimulate task-related knowledge to complete the downstream tasks through model adaptation.

(2) **Unified Representation.** The development of the Transformer architecture [104] unifies the encoders of multiple entries such as text, image, and video. Based on the Transformer architecture and the pre-training-fine-tuning paradigm, PTMs unify the paradigm for representation learning from three perspectives: task, language, and modality. A unified representation learned by pre-trained models can be adapted and utilized for multiple downstream tasks, multiple languages, and even multiple modalities.

(3) **Larger Models with Novel Capabilities.** With more training data and computation power available, constructing larger PTMs has become a new trend in representation learning research. We demonstrate the development trends of big PTMs in Fig. 5.2. As model sizes go larger, PTMs emerge with many fantastic abilities approaching human beings. For example, big PTMs can perform in-context learning [7], which learns the downstream tasks with a task instruction and some optional examples as the additional input text. Big PTMs can also perform chain-of-thought reasoning [112], which mimics the intuitive thought process from human-written reasoning chains as the additional input text, and behavior learning [72] which learns the human behavior such as operating search engine. It indicates that PTMs are quickly evolving into more intelligent agents than we have ever imagined.

In this section, we mainly introduce text-based PTMs since the fantastic Transformer-based PTMs for representation learning begin from NLP, leaving the introduction of the PTMs for graph in Chap. 6, multi-modality in Chap. 7, and knowledge in Chaps. 9, 10, 11, and 12. In the rest of this chapter, we first introduce pre-training tasks in Sect. 5.2, including word-level and sentence-level pre-training tasks. After that, we present how to adapt PTMs to downstream tasks, including full-parameter fine-tuning, delta tuning, and prompt learning in Sect. 5.3. Note that we only discuss the PTMs with pre-training-fine-tuning paradigm in this section, while the feature-based PTMs have been discussed in Chapter refchap:word. Finally, we overview four advanced topics, including better model architecture, multilingual learning, multi-task learning, efficient representations, and chain-of-thought reasoning in Sect. 5.4.

5.2 Pre-training Tasks

As introduced in Sect. 5.1, PTMs for representation learning typically consist of two phases: pre-training and adaptation (fine-tuning). During pre-training, PTMs learn the task-agnostic representations, which aim to capture the text's lexical, syntactic,

Table 5.1 A list of typical pre-trained models in NLP

Model	Architecture	Size
BERT [20]	Encoder	340 M
RoBERTa [68]	Encoder	340 M
SpanBERT [49]	Encoder	340 M
UniLM [23]	Encoder	340 M
ELECTRA [15]	Encoder	340 M
XLM [17]	Encoder	340 M
KnowBERT [76]	Encoder	340 M
K-BERT [66]	Encoder	340 M
ERNIE (Tsinghua) [132]	Encoder	110 M
ERNIE (Baidu) [100]	Encoder	340M
ELMo [75]	Decoder	–
GPT [86]	Decoder	340 M
XLNET [117]	Decoder	340 M
GPT-2 [87]	Decoder	1.5 B
CPM-1 [133]	Decoder	2.6 B
GPT-3 [7]	Decoder	175 B
GLM-130B [127]	Decoder	130 B
BART [59]	Encoder-decoder	340 M
T5 [88]	Encoder-decoder	11 B
CPM-2 [131]	Encoder-decoder	11 B
mT5 [115]	Encoder-decoder	13 B
OPT [129]	Encoder-decoder	175 B
Switch-Transformer [26]	Encoder-decoder	1.6 T

semantic, and discourse knowledge as well as the world and commonsense knowledge hiding in the text. The typical form of PTMs can be divided into three types: encoder-based PTMs, decoder-based PTMs, and encoder-decoder-based PTMs. We list typical PTMs in Table 5.1. Based on existing pre-training tasks designed for these PTMs, we conclude two major categories of existing pre-training tasks from them, including word-level and sentence-level pre-training tasks.

5.2.1   Word-Level Pre-training

Word-level pre-training tasks aim to learn contextualized word representations for PTMs from the large-scale unlabeled corpus. As discussed in Chap. 2, contextualized word representations can generate different representations according to different contexts, aiming to capture the lexical meaning of a word as well as the syntactic and semantic relations with its context words. Next, we present multiple widely used word-level pre-training objectives, including casual language

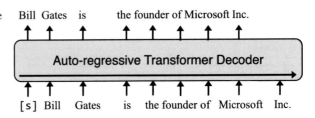

Fig. 5.3 The casual language model objective

modeling, masked language modeling (MLM), replaced language modeling (RLM), and denoising language modeling (DLM).

Casual Language Modeling (CLM) CLM is the most typical form of language modeling task. It is widely adopted as the pre-training objective of decoder-based PTMs such as GPT [86], which utilizes an auto-regressive Transformer decoder to model the probability of the input text. As shown in Fig. 5.3, CLM feeds the whole input text sequence into the Transformer decoder word-by-word auto-regressively and then asks the model to predict the next word at each position. Formally, given the input text $s = (w_1, w_2, \ldots, w_N)$ with N words, the pre-training objective of CLM is formulated as:

$$\mathscr{L}_{CLM} = -\sum_{i=1}^{N} \log P(w_i | w_0, w_1, \ldots, w_{i-1}), \qquad (5.1)$$

where w_0 is the start token [s] of the sentence and $P(w_i | w_0, w_1, \ldots, w_{i-1})$ is the probability of w_i modeled conditioned on the historical context generated by an auto-regressive Transformer decoder. In fact, CLM is widely used as the pre-training task when training big PTMs due to its effectiveness and efficiency.

Although CLM can learn the contextualized word representations simply and effectively, it can only encode the historical information in one direction in language understanding tasks. Hence, the downstream language understanding applications usually concatenate the word representations of left-to-right and right-to-left Transformer decoders learned by CLM, which can naturally combine the contextual information from both directions.

Masked Language Modeling (MLM) MLM is another widely used word-level pre-training objective for PTMs. MLM believes that the contextual information of a word is not a simple combination of the historical information captured by a left-to-right model and the future information captured by a right-to-left model. When generating a word, a deep bidirectional Transformer model should be utilized to consider both historical and future information. However, casual language modeling cannot be directly applied in pre-training the deep bidirectional Transformer model since it suffers from information leakage brought by the self-attention operation. Therefore, as shown in Fig. 5.4, MLM first masks out part of the words with [MASK] token in the input text and then asks the model to predict the masked words

Fig. 5.4 The masked
language model objective

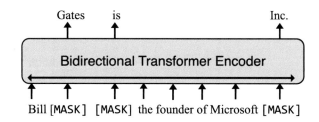

according to the remaining unmasked context. Formally, we denote the masked words as $s_{\text{mask}} = (w_{m_1}, \ldots, w_{m_M})$ where m_i is the index of the masked word and M is the number of the masked words and the masked sequence as $\bar{s}$. The pre-training objective of MLM is formulated as:

$$\mathscr{L}_{\text{MLM}} = - \sum_{w \in s_{\text{mask}}} \log P(w|\bar{s}). \tag{5.2}$$

MLM was first adopted by BERT [20] and later used by many other PTMs. To address the gap between pre-training and adaptation phases caused by the introduction of the [MASK] token, BERT further utilizes a masking strategy: for a randomly selected word to be masked, BERT replaces it with (1) the [MASK] token with an 80% probability, (2) the original word with a 10% probability, and (3) a random word with a 10% probability. A major limitation of word-level masking is that it may not sufficiently capture the linguistic and knowledge information at the span level, such as phrases and named entities. Span-level semantics are important for many downstream NLP tasks, such as named entity recognition and entity linking. Hence, SpanBERT [49] and ERNIE (Baidu) [100] further introduce a novel masking strategy for MLM: span-based masking. Span-based masking proposes to mask contiguous random spans instead of individual random words. The PTMs can better encode the span-level semantics in their learned representations by predicting the entire masked spans.

Although MLM can take advantage of the superior power of a bidirectional Transformer encoder, its pre-training objective can only cover part of the input text, e.g., BERT only masks 15% input words. The reason is that it must ensure that the contextual information in the remaining unmasked words is sufficient to recover the masked words to some extent. Hence, the training efficiency of MLM is lower than that of CLM, which predicts every input word. MLM requires more training steps for convergence.

Replaced Language Modeling (RLM) RLM is then proposed to improve the training efficiency of MLM, which is first adopted by ELECTRA [15]. RLM proposes to replace part of the words in random positions of the input text and then asks the model to predict which positions are replaced words. As shown in Fig. 5.5, RLM trains the PTMs in an adversarial manner. It uses a smaller bidirectional encoder as the generator, which generates replaced words that are

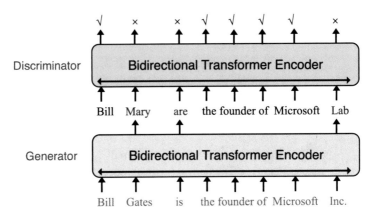

Fig. 5.5 The replaced language modeling objective

harder to be discriminated against. And then, it regards the PTMs as a discriminator to distinguish the replaced words from other unreplaced words. Hence, RLM can cover its pre-training objective in all input words. Let $\bar{s}$ denote the corrupted input text after random word replacement, and we can define the pre-training objective of RLM as:

$$\mathscr{L}_{\text{RLM}} = -\sum_{i=1}^{N} \log P(y_i|\bar{s}), \tag{5.3}$$

where y_i is the predicted label indicating whether the i-th word is replaced or not.

Denoising Language Modeling (DLM) DLM can cover nearly all the pre-training forms introduced above. It is widely used in encoder-decoder-based PTMs, which contain a bidirectional Transformer encoder and an auto-regressive Transformer decoder. With the encoder-decoder architecture, DLM allows more modifications to the input text sequence, which can help PTMs capture more lexical, syntactic, and semantic knowledge from the text. As shown in Fig. 5.6, DLM randomly modifies the input text in several strategies [59, 88, 97]:

- *Word Masking* is to mask parts of the words in the input text. This strategy is the corresponding form of the MLM pre-training task in DLM, ensuring that DLM can make the PTMs capture the information learned by MLM.
- *Text Infilling* is to mask a contiguous span of the input text with a single [MASK] token. It can be viewed as a harder version of word masking, where PTMs require learning to predict how many words the masked span originally has instead of only distinguishing whether the word is masked. It is similar to the span-level masking strategy of MLM.
- *Word Deletion* is to delete parts of the words from the input text randomly. It requires PTMs to decide which words have been deleted.

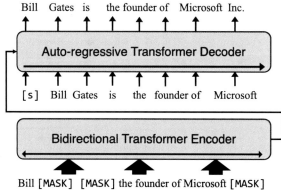

Fig. 5.6 The denoising language model objective

- *Word Permutation* is to shuffle all words from the input text in random order. It requires PTMs to understand the syntactic and semantic relations between words as well as the whole meaning of the sentence to recover the sentence order.

Besides the above word-level modifications, DLM also allows sentence-level modifications, such as *sentence permutation* and *document rotation*. The sentence- and document-level modifications help the word representation learned by PTMs capture high-level semantics, such as the discourse relations between different sentences. We will discuss the sentence-level pre-training tasks in the next subsection.

Let $\bar{s}$ denote the corrupted input text by applying several above input text modification strategies on s. The pre-training objective of DLM is then formulated as:

$$\mathscr{L}_{\text{DLM}} = -\sum_{i=1}^{N} \log P(w_i | \bar{s}, w_0, w_1, \ldots, w_{i-1}), \qquad (5.4)$$

where $P(w_i | \bar{s}, w_0, w_1, \ldots, w_{i-1})$ is the conditional probability of w_i, which is modeled with an encoder-decoder Transformer model.

5.2.2 Sentence-Level Pre-training

As discussed in Chap. 4, sentence representations are essential for many downstream NLP tasks such as information retrieval, question answering, machine translation, etc. Sentence-level pre-training aims to learn sentence representation

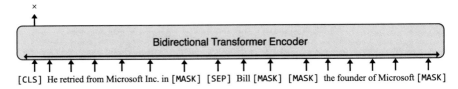

Fig. 5.7 The next sentence prediction objective

which can capture the global meanings of sentences as well as the relationships between sentences for PTMs. In this subsection, we introduce three typical sentence-level pre-training tasks for PTMs, including the next sentence prediction (NSP), sentence order prediction (SOP), and sentence contrastive learning (SCL) tasks.

Next Sentence Prediction (NSP) NSP is the first self-supervised pre-training objective to learn sentence representation for PTMs. As shown in Fig. 5.7, for the sentence s, NSP adds a [CLS] token in the front of the sentence and utilizes the representation of [CLS] as the sentence representation. After that, NSP adds another sentence s' at the end of s with a token [SEP] to indicate the sentence boundary. s' can be the next sentence of s in the document or a randomly selected sentence from the pre-training corpus. NSP aims to determine whether s' and s appear consecutively in the original text, which may help PTMs understand the relationship between sentences. Formally, NSP's pre-training objective can be formulated as:

$$\mathcal{L}_{\text{NSP}} = -\log P(y|s, s'), \tag{5.5}$$

where y is the predicted label indicating whether s' is the next sentence of s or not. In practice, BERT utilizes a uniform sampling strategy, i.e., choosing s' with (1) the original next sentence of s with a half chance and (2) the randomly selected sentence with a half chance.

NSP is adopted by BERT, which claims that NSP can help PTMs capture sentence-level semantics. However, RoBERTa [68] reimplements BERT and surprisingly finds that the performance of PTMs on most downstream NLP tasks is even better by removing the NSP objective and only pre-training on the MLM objective. ALBERT [55] further points out that the lack of task difficulty of NSP may be the key reason for its ineffectiveness. In fact, due to the big difference in topic distribution between the original next sentence and the randomly selected sentence, NSP usually suffers from a shortcut, i.e., it just requires PTMs to perform topic prediction. It is easy and already partly covered by the MLM objective.

Sentence Order Prediction (SOP) SOP is proposed to avoid the problem of NSP in modeling inter-sentence relationships. SOP also adds a `[CLS]` token in front of the sentence to obtain the sentence representation. After that, SOP randomly swaps the two consecutive sentences and asks the PTMs to predict the proper orders. In this way, the instances of SOP with correct or wrong sentence orders do not differ explicitly in topic distribution. Hence, SOP forces PTMs to distinguish discourse-level coherence relations between two input sentences rather than their topics. Formally, the objective of SOP can be formulated similarly to the NSP objective:

$$\mathscr{L}_{\text{SOP}} = -\log P(y|s, s'), \tag{5.6}$$

where y is the predicted label indicating whether s' and s are in order or not. The experimental results on several downstream tasks of ALBERT show that SOP can somewhat solve the problem of NSP, which may come from analyzing misaligned coherence cues.

Sentence Contrastive Learning (SCL) SimCSE [29] introduces sentence contrastive learning to pre-train the PTMs. Unlike NSP and SOP, which learn the sentence-level semantics by distinguishing the relations between different raw sentences, SimCSE simply predicts whether two input sentences are the same. The basic idea of SimCSE is that the representations of a sentence with different dropout masks should be closer than representation of other sentences. Formally, as shown in Fig. 5.8, let and $\mathbf{s}^{z'}$ denote the sentence representations of sentence s with dropout mask z and z', respectively. We can define the pre-training objective of SimSCE as:

$$\mathscr{L}_{\text{SimCSE}} = -\log \frac{\exp(\cos(\mathbf{s}^z, \mathbf{s}^{z'}))}{\sum_{i=1}^{M} \exp(\cos(\mathbf{s}^z, \mathbf{s}_i))}, \tag{5.7}$$

where $\mathbf{s}_i$ is the representation of the i-th negative sentence in the training batch, $\cos(\cdot)$ indicates the cosine similarity, and M is the batch size. In practice, the negative sentences are usually sampled from the same mini-batch for convenience. Although SimCSE is strikingly simple, it outperforms the NSP and SOP pre-training objectives by a large margin in a series of downstream NLP tasks [29]. Other concurrent works also adopt the idea of sentence contrastive learning for sentence-level pre-training, such as self-guidance contrastive learning [52], contrastive tension [46], and TSDAE [107].

Besides word-level and sentence-level pre-training tasks, knowledge-level pre-training tasks have also been widely explored to help PTMs better capture the world knowledge hiding behind the text. We will introduce how to pre-train PTMs at the knowledge level in Chap. 9.

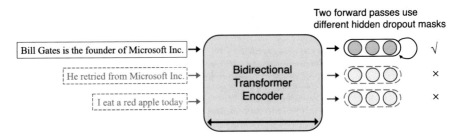

Fig. 5.8 The SimCSE objective. The figure is redrawn according to Fig. 1 from SimCSE paper [29]

5.3 Model Adaptation

Through self-supervised pre-training on the large-scale unlabeled corpus, PTMs have learned a strong ability to understand language and thus can generate task-agnostic informative representations. Then for downstream NLP tasks, it is natural to introduce task-specific objectives to adapt the PTMs, aiming to directionally stimulate the specific functionality of PTMs and obtain task-specific text representation. Now, the remaining question is how to adapt big PTMs to target downstream tasks effectively and efficiently. We introduce the model adaptation methods from full-parameter fine-tuning to optimization-efficient delta tuning and data-efficient prompt learning. In fact, between pre-training and model adaptation, some works also explore to pre-adapt the PTMs with multi-task learning or domain-specific learning. We remain the introduction of model preadaptation in Sect. 5.4.

5.3.1 Full-Parameter Fine-Tuning

Full-parameter fine-tuning is the most straightforward solution for adapting PTMs to downstream tasks. Full-parameter fine-tuning tunes all parameters of PTMs with the guidance of task-specific data, aiming to stimulate the task-specific abilities of PTMs. Given a PTM model $\Theta = \{\theta_1, \theta_2, \ldots, \theta_{|\Theta|}\}$ and the training data D of the downstream task, the goal of fine-tuning phase can be formulated as finding the parameter updates $\Delta\Theta$:

$$\Delta\Theta = \nabla f_\Theta(D), \tag{5.8}$$

where $f_\Theta(D)$ is the adaptation objective of the downstream task. That is, we can simply feed the task-specific inputs into PTMs and fine-tune all the parameters so that the parameters of PTMs for the downstream task can be obtained by $\Theta' = \Theta - \Delta\Theta$.

Now, the remaining problem is how to define the adaptation objective $f_\Theta(D)$. It can be divided into three categories according to the downstream task types: classification, sequence labeling, and generation.

Classification Classification is one of the typical forms of NLP tasks, such as topic classification, sentiment classification, natural language inference, etc. Formally, given the input sentence s and the output label y, the classification task models the conditional probability $P(y|s)$. As shown in Figs. 5.9, 5.10, and 5.11, a common solution to fine-tune PTMs is to add a task-specific classifier on the top of the sentence/document representation generated by PTMs, i.e., $P(y|s) = P(y|\mathbf{s})$. As for the sentence/document representation $\mathbf{s}$, we usually use (1) the representation of the [CLS] token for encoder-based PTMs, (2) the representation of the last word in the sentence for decoder-based PTMs, and (3) the representation of the start word in the Transformer-based decoder for encoder-decoder-based PTMs. Besides adding an external classifier, decoder-based and encoder-decoder-based PTMs also model

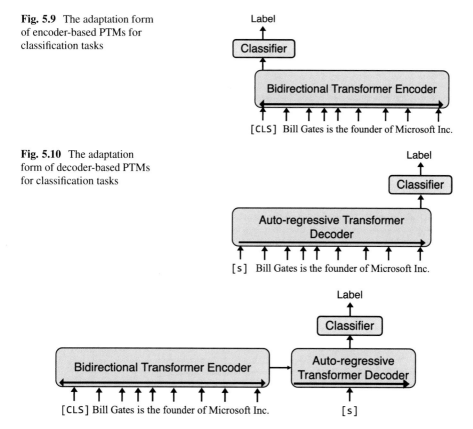

Fig. 5.9 The adaptation form of encoder-based PTMs for classification tasks

Fig. 5.10 The adaptation form of decoder-based PTMs for classification tasks

Fig. 5.11 The adaptation form of encoder-decoder-based PTMs for classification tasks

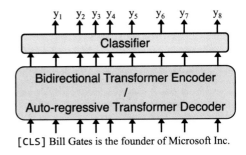

Fig. 5.12 The adaptation form of PTMs for sequence labeling tasks

the classification tasks as text generation, which directly generate the target labels in the decoder.

Sequence Labeling Sequence labeling is also a classical NLP task format, such as part-of-speech tagging, named entity recognition, etc. Formally, given the input sentence $s = (w_1, \ldots, w_N)$ and the corresponding output labels $y = (y_1, \ldots, y_N)$ for all words, the sequence labeling task models the conditional probabilities $P(y|s)$. It is usually modeled as the word-level classification form, in which the output labels of all words are conducted independently, i.e., $P(y|s) = \prod P(y_i|s)$. As shown in Fig. 5.12, we can add a task-specific classifier on top of the output representation $\mathbf{h}_i$ for the i-th word generated by either the bidirectional Transformer encoder (e.g., encoder-based PTMs and encoder-decoder-based PTMs) or the auto-regressive Transformer decoder (e.g., decoder-based PTMs), i.e., $P(y_i|s) = P(y_i|\mathbf{h}_i)$. Except for the basic word-level classification form, we can also regard the sequence labeling task as a generation task, i.e., directly generating the whole label sequence.

Generation As we have introduced in Chap. 4, many typical NLP tasks are in text generation form, such as machine translation, summarization, etc. Formally, given the source sentence s and the corresponding target sentence t, the generation task models the conditional probability $P(t|s)$ (for the language modeling task, we only model $P(t)$ without any condition). As shown in Fig. 5.13, for decoder-based PTMs, we can directly feed the input text into the auto-regressive Transformer decoder and ask it to generate the target sentence after the input text continually. As shown in Fig. 5.14, for encoder-decoder-based PTMs, we can feed the text into the bidirectional Transformer encoder and ask the auto-regressive Transformer decoder to generate the target sentence.

Fine-tuning the whole PTMs is simple and effective, showing superior performance in a wide range of downstream NLP tasks. However, performing full-parameter fine-tuning has two significant drawbacks. First, it is time- and resource-consuming, especially considering the growing model scale. Nowadays, researchers [7, 88] have revealed that the performance of PTMs can be continually improved as the PTMs get larger and the increasing scale has become an irreversible trend for developing PTMs. Full-parameter fine-tuning requires the PTMs to update all the model parameters during adaptation and storing the whole model for each

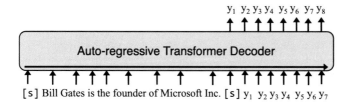

Fig. 5.13 The adaptation form of decoder-based PTMs for generation tasks

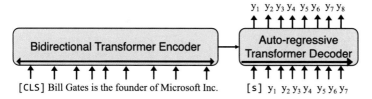

Fig. 5.14 The adaptation form of encoder-decoder-based PTMs for generation tasks

task. Second, it is hard to generalize from a few examples and thus still requires considerable training examples in the downstream tasks for model adaptation. In fact, when taking a closer look at model adaptations, we can find the gap between pre-training and full-parameter fine-tuning. Hence, this raises a new question: **how can we adapt PTMs more effectively?** Therefore, delta tuning and prompt learning target these two problems from **model optimization perspective** and **data utilization perspective**, respectively. We will introduce them as follows.

5.3.2 Delta Tuning

Delta tuning (a.k.a., parameter-efficient tuning) [22] proposes to only update part of the model parameters instead of full-parameter updating for adapting PTMs to downstream tasks, which improves the model adaptation from the optimization perspective. The basic assumption of delta tuning is that we can stimulate the necessary abilities for downstream tasks by only modifying a few model parameters. Formally, different from full-parameter fine-tuning that the number of updated parameters $|\Delta\Theta|$ is equal to the number of whole model parameters $|\Theta| (\Theta = \theta_1, \theta_2, \ldots, \theta_n)$, delta tuning only updates a small number of parameters while achieving the same adaptation objectives. From the perspective of representation learning, the general representations obtained by self-supervised pre-training can be adapted to task-specific representations with little cost.

We classify existing delta tuning methods into three main categories [22]: addition-based, specification-based, and reparameterization-based methods, as shown in Fig. 5.15. In this subsection, we detail these three types of delta tuning approaches.

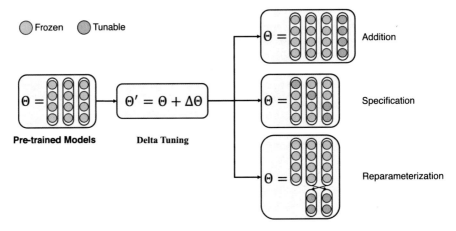

Fig. 5.15 The overall architecture of delta tuning. The figure is redrawn according to Fig. 4 from delta tuning paper [22]

Addition-Based Approach Addition-based approach keeps all the parameters in the original PTMs frozen and inserts new trainable neural modules or parameters (denoted as $\Delta\Theta = \Theta_{add} = \{\theta_{n+1}, \theta_{n+2}, \dots, \theta_{n+m}\}$ for tuning the downstream tasks). In practice, we have $m \ll n$ in the addition-based methods. In the following, we introduce two typical addition-based methods: adapter-based and prefix tuning.

Adapter-Based Methods Adapter-based methods insert tiny neural adapter modules into the middle of Transformer layers. It only tunes the parameters of the inserted adapter while keeping the PTMs frozen to adapt the model for downstream tasks. Vanilla adapter [42] first utilizes a two-layer feed-forward network as adapters and achieves comparable performance compared with full-parameter fine-tuning in a lot of downstream NLP tasks. As shown in Fig. 5.16, for an output hidden representation $\mathbf{h} \in \mathbb{R}^d$ of a PTM module, vanilla adapter first feeds $\mathbf{h}$ into a down-projection network which projects it into r-dimensional semantic space with a transform matrix $\mathbf{W}_{down} \in \mathbb{R}^{d \times r}(r < d)$ and then feeds the output into an up-projection network which projects it back to d-dimensional space with a transform matrix $\mathbf{W}_{up} \in \mathbb{R}^{r \times d}$. The process of vanilla adapter can be formulated as:

$$\mathbf{h} \leftarrow f(\mathbf{h}\mathbf{W}_{down})\mathbf{W}_{up} + \mathbf{h}, \tag{5.9}$$

where $\Delta\Theta = [\mathbf{W}_{down}, \mathbf{W}_{up}]$ are the tunable parameters (we highlight them by red color and underline) and $f(\cdot)$ is a nonlinear activation function.

In practice, the adapter modules are inserted in the middle of two Transformer blocks in the PTMs, and it can reduce the number of tunable parameters of PTMs to about 0.5–8%. Moreover, AdapterDrop [90] further proposes to dynamically remove adapter modules from lower Transformer layers to further reduce the computational cost for model inference.

Fig. 5.16 The illustration of adapter-based tuning methods

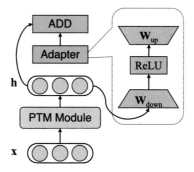

After the vanilla adapter, recent works continue to explore better forms of adapter modules. For example, Compacter [50] further reduces the number of tunable parameters of the adapter module with parameterized hypercomplex multiplication layer. Formally, it replaces the original projection matrix with the sum of the Kronecker products of two low-rank matrices:

$$\mathbf{W}_{up} = \sum_{i=1}^{l} \mathbf{A}_i \otimes \mathbf{B}_i, \tag{5.10}$$

where $\mathbf{A}_i \in \mathbb{R}^{l \times l}$ and $\mathbf{B}_i \in \mathbb{R}^{(d/l) \times (r/l)}$ and $\otimes$ indicates the Kronecker product operation. The formulation of $\mathbf{W}_{down}$ is similar. The experimental results of Compacter [50] show that it can effectively reduce the number of tunable parameters in the adapter modules into $\frac{1}{l}$ without hurting the model performance in the downstream tasks.

Although existing adapter-based methods can achieve the performance of nearly full-parameter fine-tuning with fewer modified parameters, it still requires back-propagation through the whole PTM. To address this issue, Ladder side tuning [101] further proposes to move the adapter modules out of the Transformer architecture of PTMs, bridging a ladder outside the backbone model. Hence, it can effectively save computation of backpropagation of the original PTMs while updating adapter modules and also save memory by shrinking the hidden size of representations.

Prefix Tuning Methods Prefix tuning [62] adds trainable prefix vectors to the hidden states at each layer instead of inserting adapter modules in the middle of the Transformer layers. Formally, as shown in Fig. 5.17, prefix tuning can be viewed as concatenating two prefix matrices $\mathbf{P}_K, \mathbf{P}_V \in \mathbb{R}^{l \times d}$ (l is the number of the inserted prefix vectors in the prefix matrix) to the input key hidden matrix K and value hidden matrix V of the multi-head attention layer, which is formulated as:

$$\mathbf{h}_i = \text{ATT}(\mathbf{x}\mathbf{W}_Q^i, \text{concat}(\underline{\mathbf{P}_K^i}; \mathbf{x}\mathbf{W}_K^i), \text{concat}(\underline{\mathbf{P}_V^i}; \mathbf{x}\mathbf{W}_V^i)), \tag{5.11}$$

Fig. 5.17 The illustration of
the prefix tuning method

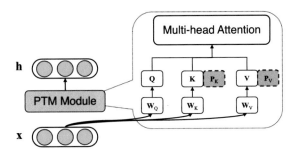

where $\mathbf{P}_K^i$ and $\mathbf{P}_V^i$ are the i-th sub-vectors of $\mathbf{P}_K$ and $\mathbf{P}_V$ for i-th attention head's calculation, ATT($\cdot$) indicates the self-attention function, and $\mathbf{x}$ is the input feature of Transformer blocks. For prefix tuning, we have $\Delta\Theta = \mathbf{P}_K \cup \mathbf{P}_V$. Empirically, directly optimizing $\mathbf{P}_K$ and $\mathbf{P}_V$ may be unstable and hurt the performance slightly, and thus prefix tuning proposes to reparametrize them with feed-forward neural networks:

$$\mathbf{P}_K = \text{MLP}(\mathbf{P}_K'),$$
$$\mathbf{P}_V = \text{MLP}(\mathbf{P}_V'), \tag{5.12}$$

and they only save $\mathbf{P}_K$ and $\mathbf{P}_V$ after training.

Prompt tuning [58] is a simplified form of prefix-tuning, which only adds prefix vectors (a.k.a., soft prompts) to the input layer instead of all layers. It shows that prompt tuning can achieve nearly the same performance as full-parameter fine-tuning when the model size increases. A significant limitation of prefix tuning approaches is that their extremely small parameter spaces make them challenging to optimize and thus require more training time to converge compared to full-parameter fine-tuning. This phenomenon is more severe in small-scale PTMs. Gu et al. [32] thus propose to pre-train the representations of soft prompt tokens in the pre-training stage. The experimental results demonstrate that pre-training soft prompts can effectively improve the performance of prompt learning in downstream tasks and even outperform full-parameter fine-tuning.

In summary, both prefix tuning and adapter-based methods insert new trainable parameters to learn the downstream tasks, and their major difference is the position of the inserted parameters.

Specification-Based Approach Specification-based approach proposes to specify part of the model parameters in the original PTMs to be tunable (denoted as $\Delta\Theta = \{\Delta\theta_{idx_1}, \Delta\theta_{idx_2}, \ldots, \Delta\theta_{idx_m}\}$ where $idx_i \in [1, n]$ is the index of tunable parameters) and also $m \ll n$.

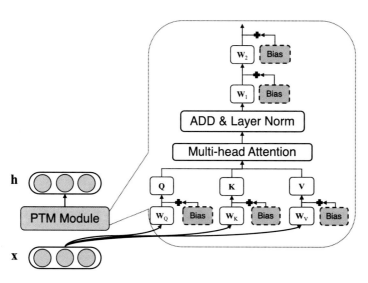

Fig. 5.18 The illustration of the BitFit method

BitFit [125] proposes to only optimize the bias terms inside the PTMs while freezing other parameters. Formally, as shown in Fig. 5.18, BitFit first specifies the multi-head attention layer in the Transformer block as:

$$\mathbf{h}_i = \text{ATT}(\mathbf{x}\mathbf{W}_Q^i + \underline{\mathbf{b}_Q^i}, \mathbf{x}\mathbf{W}_K^i + \underline{\mathbf{b}_K^i}, \mathbf{x}\mathbf{W}_V^i + \underline{\mathbf{b}_V^i}), \quad (5.13)$$

and then specifies the next feed-forward layer as:

$$\mathbf{h}' = \text{GeLU}(\mathbf{h}\mathbf{W}_1 + \underline{\mathbf{b}_1})\mathbf{W}_2 + \underline{\mathbf{b}_2}, \quad (5.14)$$

where $\text{GeLU}(\cdot)$ indicates the Gaussian error linear unit [41]. We do not show the layer-norm layers for convenience, but their bias terms are also tunable in BitFit. Experimental results in BitFit show that it can achieve over 95% performance as full-parameter fine-tuning on several benchmarks. They also find that different functionalities may be controlled by different parts of specified bias terms during model adaptation.

Besides BitFit which directly specifies the bias term to be tuned, diff pruning [34] proposes to learn to select part of the model parameters for model adaptation. The basic idea of diff pruning is to encourage the delta parameter $\Delta\Theta$ to be as sparse as possible. To this end, Diff pruning first decomposes $\Delta\Theta$ into a binary mask vector $\mathbf{z} \in \{0, 1\}^{|\Theta|}$ multiplied with a dense vector $\mathbf{w} \in \mathscr{R}^{|\Theta|}$:

$$\Delta\Theta = \mathbf{z} \odot \mathbf{w}, \quad (5.15)$$

and then it optimizes an expectation with respect to $\mathbf{z}$ under a Bernoulli distribution parameter α:

$$\min_{\alpha, \mathbf{w}} \mathscr{E}_{\mathbf{z} \sim p(\mathbf{z}; \alpha)}[\mathscr{L}(\Theta + \Delta\Theta) + \lambda \|\Delta\Theta\|_0], \qquad (5.16)$$

where $\mathscr{L}(\cdot)$ indicates the learning objective of the downstream task and the L_0-norm penalty is added to achieve the goal of sparsity. The idea of learning a binary mask vector for delta tuning is also proposed by Zhao et al. [135].

Reparameterization-Based Approach Reparameterization-based approach proposes to reparameterize part of existing parameters in PTMs to a parameter-efficient form by transformation. Let $\mathbf{P} = \{\mathbf{p}_1, \mathbf{p}_2, \dots, \mathbf{p}_m\}$ represent the set of parameter subsets to be reparameterized and $\Delta\Theta = \Theta + (\cup_{i=1}^m R(\mathbf{p}_i))$ where $R(\mathbf{p}_i)$ is used to reparameterize the parameter subset $\mathbf{p}_i$.

LoRA [43] decomposes the change of the original weight matrices in the multi-head attention modules into low-rank matrices. Its basic idea is inspired by Aghajanyan et al. [2] that the full-parameter fine-tuning phase of PTMs has a low intrinsic dimension. As shown in Fig. 5.19, LoRA utilizes four low-rank matrices to decomposite the changes of the transform matrices for key and value spaces, which can be formulated as:

$$\mathbf{h}_i = \text{ATT}(\mathbf{x}\mathbf{W}_Q^i, \mathbf{x}(\mathbf{W}_K^i + \underline{\mathbf{A}_K \mathbf{B}_K}), \mathbf{x}(\mathbf{W}_V^i + \underline{\mathbf{A}_V \mathbf{B}_V})), \qquad (5.17)$$

where $\mathbf{A}_K, \mathbf{A}_V \in \mathbb{R}^{d \times r}$ and $\mathbf{B}_K, \mathbf{B}_V \in \mathbb{R}^{r \times d}$. In the experiment on the GLUE benchmark, LoRA can nearly achieve comparable performance with full-parameter fine-tuning for the PTMs of various scales and architectures.

Understanding Delta Tuning from Ability Space Qin et al. [84] point out that for a particular delta tuning method, the adaptations of PTM for multiple downstream tasks can be reparameterized as optimizations in a unified low-dimension parameter space. Based on this work, Yi et al. [119] further find that the optimization of different delta tuning methods for adapting PTMs to downstream tasks can also be reparameterized into optimizations in a unified low-dimension parameter space.

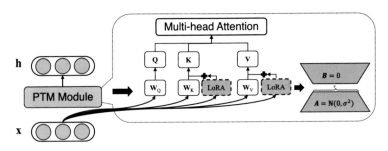

Fig. 5.19 The illustration of the LoRA method

This demonstrates the optimization space of PTMs' adaptation is intrinsically low-dimensional, which may explain why the adaptation of PTMs can be done with relatively small-scale downstream data. The intrinsic low-dimensional tuning parameter space may indicate parts of the parameters in the PTMs that are related to each other, which may be co-activated and controlled in a unified manner. This phenomenon is also observed by MoEfication [134].

5.3.3 Prompt Learning

Prompt learning [64] is proposed to overcome the limitation of full-parameter fine-tuning from the data utilization perspective. It reformulates the downstream tasks as the conditional language modeling form with a textual prompt as task instruction. This could effectively bridge the gap between model pre-training and adaptation for PTMs. Moreover, it incorporates the prior knowledge of domain experts into the model adaptation phase by elaborately designing the textual prompt, which can be viewed as feature engineering toward PTMs. Therefore, prompt learning can significantly reduce the requirements of extensive training data in the model adaptation phase while maintaining good performance.

Prompt learning is inspired by the in-context learning ability in GPT-3 [7]. In-context learning regards PTMs as a black box and utilizes the input to describe the downstream task with a task instruction and some optional examples to PTMs. It hopes PTMs to learn to proceed with the downstream task from the given descriptive context without updating the model parameters. Taking the English-to-Chinese translation task as an example, as shown in Fig. 5.20, in-context learning can be divided into two levels: (1) task instruction learning, which adds a task instruction (*Translate English to Chinese*) in front of the translated text sequence and requires the PTMs to perform zero-shot learning, and (2) example learning, which also adds some task examples besides the task instruction and requires the PTMs to perform few-shot learning based on the task-related context.

In-context learning provides a flexible way to utilize PTMs, with which we can describe many possible tasks, from text classification, named entity recognition,

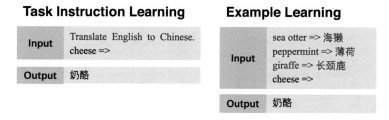

Fig. 5.20 The illustration of task instruction learning and example learning of in-context learning. The figure is redrawn according to Fig. 2.1 from OpenAI's GPT-3 paper [7]

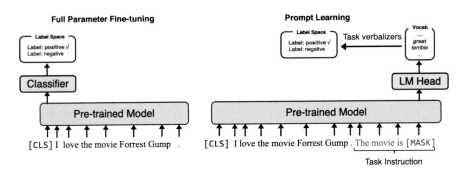

Fig. 5.21 An example of prompt learning

and question answering to machine translation. The experimental results in GPT-3 show that large PTM with in-context learning can even achieve better performance compared with the full-parameter fine-tuning in small PTMs.

From the fantastic results of in-context learning in GPT-3, researchers realize that we can stimulate PTMs' specific functionalities with textual prompts. After that, many researchers focus on exploring how to better stimulate PTMs with textual prompts, i.e., prompt learning. As shown in Fig. 5.21, prompt learning has two essential parts, including task instruction, which is a textual prompt (*The movie is [MASK]*) to stimulate the specific functionalities of PTMs for the downstream tasks, and task verbalizers, which maps the output words of the language modeling head to label space of the target task. Therefore, the research work on prompt learning focuses on how to design the optimal task instruction prompts and task verbalizers for downstream tasks.

Task Instruction Design Task instruction design aims to find the optimal task instruction prompts that can achieve the best performance in the downstream tasks. It can be divided into three categories, including manual, automatic, and knowledgeable methods.

Manual Design Early works [7, 19, 77, 120] usually design the task instruction prompts manually, based on the intuition of human experts. Although manual design methods are simple and effective, they still have two significant limitations: first, they require much time and expert experience. Second, the optimal task instruction prompts are highly related to specific PTMs and task datasets, and even experts may fail to find the optimal task instruction prompts.

Automatic Design Later, automatic design methods are proposed to learn or find the optimal task instruction prompts automatically. We categorize them into two typical types: (1) generate-then-rank. It first generates a candidate set of task instruction prompts by prompt mining [47], prompt paraphrasing [40, 122], or prompt generation [5, 28] and then ranks the best one according to the performance in the downstream tasks. (2) Gradient-based search. It searches over all words in the vocabulary to find short task instructions that can stimulate the specific

PTMs to generate the target output of the downstream tasks according to the gradients [95, 106].

Knowledgeable Design Knowledgeable design methods further incorporate external knowledge into the task instruction prompts. For example, Han et al. [38] propose prompt tuning with rules (PTR) to handle text classification tasks. It applies logic rules to guide the construction of task instruction prompts, encoding the prior knowledge of each class into prompt learning. Besides, Chen et al. [9] propose to insert the type markers in front of the head and tail entities to incorporate the entity type knowledge and insert a soft word with the average embeddings of the relation descriptions between the head and tail entities to incorporate the relation knowledge.

Task Verbalizer Design Task verbalizer design aims to find the optimal label word space of the verbalizer, i.e., the optimal words in the output vocabulary to map to the label words. Similar to task instruction design, it can also be divided into manual, automatic, and knowledgeable methods.

Manual Design Early manual task instruction designs usually accompany manual task verbalizer designs [19, 77, 120]. They ask the experienced experts to select the optimal words in the vocabulary as the task verbalizer, which is often based on specific downstream tasks such as sentiment classification, named entity recognition, etc. For example, as shown in Fig. 5.21, for sentiment analysis, it usually maps the probability of the word *great* into the probability of the *positive* sentiment and the probability of the word `terrible` to the `negative` sentiment.

Automatic Design After early manual designs, researchers have devoted much effort to automating the task verbalizer design. Its most typical form is to find a candidate word set by paraphrasing [47], searching [92], or generation [28, 121] that maps to task labels. After that, different from task instruction design, task verbalizer design usually selects the top-k candidate words/phrases as the verbalizer and sums up their probabilities as the label probabilities. The reason is that a task label may have multiple expressions in language. For example, we can describe that *The movie is great/interesting/fantastic/awesome*, and they are all mapped to `positive` sentiment.

Knowledgeable Approaches Knowledgeable task verbalizer design aims to utilize external knowledge information to help design or learn the label word space in the verbalizer. Hu et al. [44] first propose to utilize external knowledge bases to help to expand the verbalizer's label word space. Specially, for topic classification, they utilize the external topic-related vocabulary, and for sentiment classification, they use an external sentiment vocabulary to help expand the candidate words mapping to the label space of the verbalizer. Moreover, Cui et al. [18] extend the label word space from discrete words into soft embeddings and learn prototype vectors as verbalizers by self-supervised contrastive learning. Ding et al. [21] also learn to prototype vectors for entity typing tasks by self-supervised learning.

Connections Between Prompt Learning and Prompt Tuning Prompt learning directly utilizes textual prompts to stimulate the specific functionalities of PTMs for downstream tasks. However, the optimal textual prompt corresponds to many factors, such as the selection of PTMs, task data distribution, etc. The restricted discrete space of words limits the manual [7, 19, 77, 120], automatic [5, 28, 47, 95, 106], or even knowledgeable prompt learning [9, 38] to find optimal textual prompts. Stimulating PTMs' abilities with textual prompts still has a performance gap with full-parameter fine-tuning in many scenarios. Hence, prompt learning [18, 21] proposes to extend the space of textual prompts to a soft form, i.e., utilizing several additional tunable tokens instead of hard prompt tokens. This can be viewed as a kind of prompt tuning introduced in Sect. 5.3.2. In summary, while prompt tuning is a more parameter-efficient way compared to prompt learning, prompt learning utilizes the prior knowledge of human beings by designing explainable textual prompts and is a natural interface of PTMs which is more explainable for users.

5.4 Advanced Topics

In the previous section, we have introduced the basics of PTMs, including the pre-training and adaptation of text representations. In this section, we present several advanced topics of PTMs, including better model architecture, multilingual representation, multi-task representation, efficient representation, and chain-of-thought reasoning.

5.4.1 Better Model Architecture

Although Transformer-based PTMs have achieved promising results in a wide range of downstream tasks, we still have a question: is Transformer the optimal architecture for PTMs? In this subsection, we introduce the explorations in better model architecture, which can be categorized into three types:

Improving Model Capacity Recently, researchers have found that the strong ability of PTMs comes from their large-scale parameters, i.e., the bigger model leads to better performance. Therefore, researchers explore improving the Transformer architecture to increase the number of model parameters while keeping the same theoretical computation complexity.

Sparsity, which indicates that the model only activates a part of the parameters for a specific task, has been widely explored. In this way, model capacity can be significantly increased without proportionally increasing theoretical computation complexity. Sparsely gated mixture of experts layer (MoE) [94] is thus proposed to allow models to only activate a part of the parameters for each input sample. As shown in Fig. 5.22, the architecture of sparse-gated MoE consists of two parts: experts and a routing network. Each expert is usually a feed-forward neural network. The routing network is to determine which experts are activated when processing

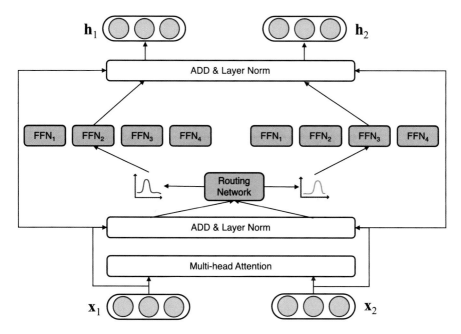

Fig. 5.22 The architecture of sparsely gated mixture of experts layer. The figure is redrawn according to Fig. 2 from the Switch Transformer paper [26].

each input sample. Compared with the vanilla Transformer, sparse-gated MoE only selects a part of experts for computation (the number of parameters is the same as the vanilla feed-forward layer), and thus does not increase the training and inference time.

However, the sparsely gated MoE still cannot be applied in real-world scenarios due to the training instability and the communication costs in GPU clusters. To address these issues, GShard [57], the first work to combine sparsely gated MoE with Transformer architecture, simplifies the routing strategy of sparsely gated MoE, which only assigns at most two experts for each instance and employs a capacity factor to balance the workload of each expert. Based on the improvement of GShard, Switch Transformer [26] extends sparsely gated MoE as the basic modeling block for PTMs, and only allows one expert for each input sample. GLaM [24] further improves the routing strategy by allowing PTMs to select two experts for each input sample, which provides more model capacity while restricting computation cost. The experimental results in Switch Transformer [26] and GLaM [24] show that PTMs with sparsely gated MoE can converge faster than that with vanilla Transformer architecture due to the significantly larger model capacity.

We believe sparse model architectures, which allow PTMs to stimulate a part of the neurons for each input sample, would be an essential feature of the next generation of PTMs' architecture. This corresponds to the phenomenon in neuroscience that each neuron tends to have fewer average connections to other neurons with the increasing number of neurons in a primate brain. In fact, Zhang et al. [134] also point out that the vanilla Transformer can be transformed into a sparse-gated MoE form by their proposed MoEfication strategy, i.e., the vanilla Transformer is a special case of sparse-gated MoE. It may demonstrate that sparsity is the intrinsic emergent characteristic of the neural network after pre-training, even without any constraint or pre-design.

Modeling Long-Term Dependency Besides the model capacity, another critical problem of the vanilla Transformer is that its self-attention mechanism's computational and memory footprints are quadratic with the length of the input sequence. Hence, a question is can we implement a quadratic Transformer so that the scale of computational and memory requirements are linear with the input sequence length? To this end, a natural solution is to approximate the original multi-head attention with faster attention mechanisms. We introduce several typical fast attention mechanisms widely used in PTMs.

Structured Sparse Attention Clark et al. [14] point out that the attention heads of PTMs exhibit specific patterns. For example, some tokens may attend to the [CLS] token, and some tokens may attend to the other tokens' specific positional offsets, etc. Motivated by this phenomenon, later works propose to replace the original full-connected multi-head attentions with several types of pre-defined structured sparse attentions, including (1) the sparse global attention with which the token is visible for all other tokens and typically employed in the [CLS] token, (2) the structured local attention which reduces the visible field for most other tokens with stride window form [4, 124] or blockwise form [12, 85], etc.

Low-Rank Approximation Since the attention heads of PTMs exhibit specific patterns, the learned attention matrices are low-rank. Hence, several recent works [13, 51, 108] propose to approximate the multi-head attention matrices with low-rank decomposition, reducing the multi-head attention to an operation which is linear with the length of the sentence.

Cluster-Based Sparse Attention Its basic idea is that tokens can only attend to similar tokens according to the routing mechanism of the multi-head attention layer. Hence, it learns to cluster tokens in the input text sequence according to their similarities and restricts that only the tokens in the same clusters are visible to each other in the attention layer. For example, Reformer [53] employs a locality-sensitive hashing strategy to cluster tokens for attention calculation, and routing Transformer [89] employs a ks-means algorithm to cluster tokens.

Retrieving External Information Researchers argue that it is a very unreasonable way for traditional PTMs to store all knowledge in model parameters due to their limited capacity compared with the endless knowledge. Therefore, REALM [36] proposes to teach PLMs to retrieve and use external knowledge during inference. REALM augments the BERT model with a latent knowledge retriever, allowing PTMs to retrieve relevant text information (i.e., documents) from a large-scale unlabeled corpus, such as Wikipedia. In the experiment, REALM demonstrates that it can achieve much better results compared to T5–11B, which has nearly 100 times parameters, verifying the effectiveness of retrieving external knowledge. Nevertheless, REALM is based on the BERT model, an encoder-based PTM, which is limited in classification tasks. To address this issue, RAG [60] further extends the idea of retrieval augmentation into the encoder-decoder-based PTM, allowing retrieval-based PTMs to handle text generation tasks.

5.4.2 Multilingual Representation

Big PTMs trained on the large-scale monolingual corpus, such as the English corpus, have shown superior performance in a wide range of NLP tasks. Nevertheless, there are thousands of languages in the world, and it is nearly impossible and unreasonable for us to train individual big PTMs for each language. The reason lies in two points: (1) there are many resource-scarce languages that we cannot easily collect a large amount of unlabeled text for pre-training; (2) there are many NLP tasks related to more than one language. In fact, semantics is independent of symbolic languages since people in the world can express the same meaning in different languages. Hence, training multilingual PTMs has recently attracted much attention from researchers. In this subsection, we introduce the explorations of learning multilingual PTMs in two main categories:

Nonparallel Pre-training Nonparallel pre-training is the initial attempt at learning multilingual PTMs, which directly pre-trains PTMs on nonparallel multilingual corpora with monolingual pre-training tasks. Its basic idea is that the lexical overlaps between languages can help to align the multilingual language representations of PTMs learned from corpora of multiple languages in the semantic space. It can be divided into three categories according to the model architecture: (1) encoder-based PTMs. Multilingual BERT (mBERT) [20] is the first multilingual PTM with nonparallel pre-training. It pre-trains with an MLM pre-training objective on multilingual Wikipedia corpora which have 104 languages but are nonparallel. (2) Decoder-based PTMs. Multilingual GPT (mGPT) [96] pre-trains with a CLM pre-training objective with Wikipedia and colossal clean crawled corpus, learning a multilingual PTM with 60 languages from 25 language families. (3) Encoder-decoder-based PTMs. mBART [67] and mT5 [115] extend the DLM pre-training objective to support multilingual pre-training. They simply add special language symbols to the end of the input text for the encoder and the start of the input text for the decoder of PTMs. Such special language symbols enable PTMs to realize the languages to be encoded and generated. With the development of multilingual PTMs with nonparallel pre-training, we still wonder how multilingual these PTMs can reach. Therefore, Pires et al. [79] take mBERT as an example for investigation and find that mBERT can achieve superior zero-shot performance in a wide range of cross-lingual NLP tasks, showing its ability in cross-lingual knowledge generalization. This verifies the reasonability of learning multilingual capabilities from the nonparallel multilingual corpora with the Transformer-based PTMs.

A major challenge of multilingual pre-training is how to alleviate the data unbalance problem between high-resource and low-resource languages. To address this issue, mBERT perform exponentially smoothed weighting of the data distribution of different languages during pre-training data construction. Furthermore, XLM-R [16] constructs a new nonparallel multilingual corpus named CC-100, which has 100 languages. Compared to the Wikipedia corpora used by mBERT, CC-100 has a larger scale, especially for those low-resource languages.

Although the monolingual pre-training objective can simply extend to train multilingual PTMs in nonparallel corpora, it cannot well utilize the language-alignment signals from parallel corpora. In fact, such language-alignment signals are essential for multilingual NLP tasks such as cross-lingual information retrieval and machine translation.

Parallel Pre-training Parallel pre-training is another typical approach for learning multilingual PTMs, which mainly focuses on designing multilingual pre-training tasks to better utilize the language-alignment signals from parallel corpora. This line of research work can be divided into three types according to the pre-training tasks: (1) cross-lingual masked language modeling. XLM [17] thus proposes the cross-lingual masked language modeling (CMLM) pre-training objective to better utilize the language-alignment signals from bilingual sentence pairs. Extending the MLM objective, CMLM concatenates two semantically matched sentences in two languages and asks PTMs to recover randomly masked tokens in the connected sentence. Compared to MLM, CMLM allows PTMs to recover the masked tokens not only from the monolingual context information but also from its aligned tokens in another language. (2) Cross-lingual denoising language modeling. XNLG [10] proposes cross-lingual denoising language modeling (CDLM). Different from DLM, CDLM assigns the inputs of the encoder and decoder of PTMs with text in different languages, similar to CMLM. (3) Cross-lingual contrastive learning. InfoXLM [11] further analyzes MLM and CMLM from the perspective of information theory and proposes a contrastive pre-training objective for learning multilingual PTMs based on the analysis. Based on InfoXLM, HICTL [113] further extends the idea of cross-lingual contrastive learning to help PTMs to learn with multilingual representations at the word level and sentence level.

Compared to nonparallel pre-training, performing parallel pre-training can learn semantic-aligned multilingual representations more effectively and thus achieves promising results in a series of cross-lingual NLP tasks. However, most existing parallel pre-training objective relies on a large number of parallel data at the sentence level and even word level, which is quite rare for many languages. To address this issue, ERNIE-M [74] proposes to expand the scale of parallel multilingual corpora using the back-translation technique as well as a back-translation masked language modeling (BTMLM) pre-training objective. Except for utilizing machine translation technique, ALM [116] proposes a code-switched pre-training objective, which directly replaces the tokens/spans in one language with the token/spans from either its semantic-aligned sentence in another language or bilingual lexicons and then performs CMLM on it.

5.4.3 Multi-Task Representation

Multi-task learning [8] has been widely explored in the representation learning of NLP. With the development of PTMs, pre-trained representations have become much more expressive, unifying text representations across a wide range of NLP tasks. Nevertheless, it still has no clear answer whether multi-task learning in downstream tasks can make the pre-trained representations more expressive. Therefore, researchers have devoted many efforts to exploring how multi-task learning of downstream tasks can promote the PTMs and stimulate the potential of pre-trained representations. We roughly divide the explorations into the following three directions:

Multi-Task Pre-training Multi-task pre-training unifies the learning paradigm of various kinds of NLP tasks during the pre-training stage for PTMs. The basic idea of multi-task pre-training is to introduce the learning signals of different NLP tasks into the pre-training phase. For example, T5 [88] unifies nearly all NLP tasks as text-to-text generation problems so that it can pre-train the encoder-decoder-based PTMs with all NLP task data besides self-supervised learning with unlabeled corpus. After that, Liu et al. [64] propose to unify the learning objective of all NLP tasks as prompt learning by inserting human-designed /automatically generated task prompts into the input text. This combines the idea of multi-task learning and prompt learning, which can further mitigate the gap between multi-task pre-training and task-specific model adaptation. Besides directly enhancing the PTMs by multi-task learning, some works also explore understanding the principle of task unification in the pre-training stage. Qin et al. [84] reveal that PTMs actually learn the capabilities to handle multiple tasks in the pre-training phase. Moreover, they find that there exists a unified low-dimensional task subspace to the task capabilities, and the task-specific model adaptation of big PTMs can be all reparameterized into optimizing the task vector in this space.

Multi-Task Preadaptation Multi-task preadaptation additionally adapts the big PTMs by adding intermediate auxiliary tasks between pre-training and model adaptation. The research of multi-task preadaptation can be roughly divided into three categories: (1) exploring the effectiveness of preadaptation. First, big PTMs could further learn more task capabilities that are not reflected in the self-supervised learning signals by incorporating the intermediate knowledge transfer from auxiliary tasks, such as text classification [78], named entity recognition [128], relation extraction [82], and question answering [30]. Second, preadaptation on domain-specific unlabeled data for downstream tasks could provide rich domain-specific knowledge for PTMs [33, 35, 56, 83]. (2) Understanding the working mechanism of preadaptation. Although simple and effective, the success of preadaptation is very sensitive to the selection of intermediate auxiliary tasks. Hence, recent works have focused on exploring the reason for this phenomenon. One on hand, Aghajanyan et al. [1] find that scaling the number of tasks as well as adopting task-heterogeneous batches and task-rebalancing loss scaling is important for multi-task preadaptation. On the other hand, Pruksachatkun et al. [81] explore the task capabilities big PTMs learn during the model preadaptation stage and find that the preadaptation tasks requiring high-level reasoning abilities lead to better downstream task performance. (3) Selecting intermediate auxiliary tasks for preadaptation. Researchers also explore how to efficiently select the optimal intermediate auxiliary tasks according to the knowledge transferability among different tasks, such as embedding-based methods [80], manually defined feature-based methods [63], task gradient-based methods [27], etc.

Multi-Task Model Adaptation Multi-task model adaptation aims to fine-tune PTMs so that their generated text representations can jointly solve multiple tasks. Researchers argue that PTMs have learned versatile knowledge during self-supervised pre-training in the large-scale unlabeled corpus, which may help a wide range of NLP tasks. Hence, big PTMs can be stimulated with multiple task signals to build a unified downstream task model that can handle a variety of downstream NLP tasks. However, in real-world applications, we usually suffer from the data imbalance problem, i.e., the data volume of different tasks varies a lot. Hence, simply performing typical multi-task learning for model adaptation will lead to underfitting in resource-rich tasks and over-fitting on resource-scarce tasks [3]. To address this problem, the basic idea is to learn task-specific model modules for the PTMs, which can be divided into three types: (1) task-specific layers which are added on top of the shared universal text representations of PTMs [65]; (2) task-specific controlling modules which generate weights of the existing layers such as the FFN layers of the Transformer [102]; and (3) delta tuning modules which reduce the number of newly added model parameters for multiple downstream tasks [70, 98].

5.4.4 Efficient Representation

Although the text representations learned by big PTMs have shown fantastic abilities in language understanding, it requires a large amount of inference time, making them impractical in real-world applications. Therefore, many recent works have explored how to generate efficient pre-trained representations, which can be mainly divided into three types, including model pruning, knowledge distillation, and parameter quantization.

Model Pruning Model pruning reduces the size of PTMs by omitting redundant model parameters of big PTMs. Model pruning has two main categories: (1) unstructured pruning, which directly prunes the model parameter at the neuron level. CompressingBERT [31] conducts a comprehensive analysis of the multi-head attention layers and feed-forward layers of Transformer blocks in PTMs and then prunes 30–40% of the weights in PTMs without loss of performance during the pre-training stage. This is because these weights do not encode any useful inductive bias for language understanding in the downstream tasks. Although unstructured pruning effectively makes PTMs more sparse, it cannot speed up the inference since the computation hardware cannot well deal with the pruned unstructured PTMs. (2) Structured pruning, which prunes the model parameter at the attention level or layer level. For layer-level pruning, Fan et al. [25] propose to randomly drop several layers so that they can dynamically pick up parts of the model layers during inference. Besides, DeeBERT [114] and CascadeBERT [61] learn to exit the inference in the shallow layer of PTMs in the downstream tasks. While these works all focus on the layer-wise early exiting for the classification tasks, TR-BERT [118] further extends the idea of early exiting into the inference of the sequence labeling tasks for PTMs. For attention-level pruning, researchers observe that there exist redundancy phenomena in attention heads, i.e., the same syntactic or semantic relations may be modeled by more than one attention head [71, 105], and thus they propose to remove the redundant attention heads. Compared to unstructured pruning, the PTMs after structured pruning is still structured and can be easily accelerated in typical computation hardware such as GPUs.

Knowledge Distillation Knowledge distillation learns a smaller student PTM to transfer the knowledge from a bigger teacher PTM, which aims to reduce both the inference time and memory cost while maintaining the performance of big PTMs. The main challenge of knowledge distillation is how to construct effective supervisions from the teacher PTMs, which can be divided into three types: from (1) the original output probabilities [91] of the self-supervised learning tasks or downstream tasks, (2) the hidden states in different layers [48, 99], and (3) the attention matrices [109]. Compared to directly training a smaller PTM, knowledge distillation can transfer the learned knowledge in larger PTMs, enhancing the representations generated by student PTMs.

Parameter Quantization Parameter quantization converts the precision of model parameters from a higher float point to the lower one. The original precision of PTMs is usually 32 bits, 16 bits, or mixed 32–16-bits. Q8BERT [123] first proposes to quantize the model parameters' coding of PTMs into 8-bit to speed up its inference speed. However, it is harder to reduce the parameter coding into extremely low-bit further (e.g., 1 or 2 bits) since the low fixed points have huge precision gaps with float points which may affect the output representations of PTMs. To address this issue, Q-BERT [110] further proposes to apply different levels of precisions for different kinds of modules in the PTMs according to their different precision requirements. Besides, TernaryBERT [130] proposes quantization-aware training for PTMs, which directly trains the quantized PTMs during the pre-training stage. However, extreme low-bit quantization is still limited in real-world applications since it relies on specially designed hardware implementation.

5.4.5 Chain-of-Thought Reasoning

Recent studies have revealed that even the extremely large-scale PTMs can still struggle with complex multi-step reasoning tasks, such as numerical reasoning and commonsense reasoning. Therefore, we have a question: do PTMs learn complex reasoning abilities in the pre-training stage? If yes, how can we stimulate the complex reasoning ability of PTMs?

To this end, chain-of-thought (COT) reasoning [112] is proposed to stimulate the complex reasoning ability of PTMs. The basic idea of COT reasoning is that a model-generated chain of thought can enable PTMs to mimic an intuitive thought process to perform reasoning. As shown in Fig. 5.23, COT reasoning adds a human-labeled explanation that describes the explicit intermediate reasoning path as the textual prompt for obtaining the final answer. COT reasoning hopes the PTMs can learn to decompose the complex reasoning task into multiple intermediate steps that are solved individually, and then PTMs can obtain the correct answer by reasoning over the generated path. In this way, PTMs can generate more interpretable solutions and improve the model performance in the samples requiring complex reasoning. Experimental results in the original paper [112] show that the complex reasoning ability emerges from PTMs when the model parameter grows up to about 100B with COT reasoning, and such big PTMs can achieve promising results on numerical reasoning and commonsense reasoning tasks. Later, Wang et al. [111] further propose an answer ensembling strategy to improve the reasoning accuracy for COT reasoning. They first sample a diverse set of reasoning paths with beam search and then perform reasoning over them. After that, they select the most consistent final answer from the generated answer set following these reasoning paths. Their experiments show that such a simple strategy can effectively improve the model performance without additional training for various PTMs with

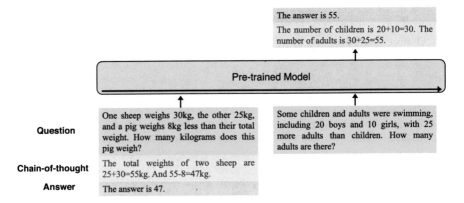

Fig. 5.23 An example of chain-of-thought reasoning. The text colored red is the added explanation. The figure is redrawn according to Fig. 1 from the chain-of-thought reasoning paper [112]

Fig. 5.24 An example of zero-shot chain-of-thought reasoning

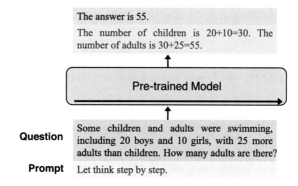

different scales. Although simple and effective, a major drawback of COT reasoning is that it requires expensive manually annotating explanations for different tasks and datasets. To address this problem, STaR [126] further proposes a bootstrapping approach to generate high-quality explanations for each example from a tiny seed training set and verifies its effectiveness in arithmetic, math word problems, and commonsense reasoning, especially in the few-shot settings.

Now, the remaining question is where does the complex reasoning ability of PTMs come from? One possibility is the intrinsic ability of PTMs that are learned in the self-supervised pre-training phase, while another possibility is learning from the provided reasoning explanations due to the few-shot learning ability of big PTMs. Recently, Kojima et al. [54] reveal that big PTMs can perform complex multi-step reasoning without any human-labeled explanation prompt. As shown in Fig. 5.24, they find that big PTMs can perform zero-shot COT reasoning by simply adding "Let's think step by step" before each answer as the textual prompt. They show that with such a simple prompt, the zero-shot performance of big PTMs achieves consistent improvements in diverse NLP tasks, including arithmetic, symbolic, and other reasoning tasks. This preliminarily demonstrates that the complex reasoning ability may be learned by PTMs during pre-training on the large-scale corpus.

Nevertheless, it is still unclear to us how PTMs learn such ability: do such reasoning text patterns exist in the training corpus and PTMs just learn a shortcut? Or have PTMs evolved into more intelligent agents we never imagined, i.e., the pre-trained representations of big PTMs may also exist other untapped and understudied fundamental "magic" abilities? This is still an open question. But we confirmedly believe that big PTMs are the foundation and future direction toward high-level cognitive intelligence.

For this question, an interesting recent finding is that big PTMs have the ability to behavior learning. For example, WebGPT [72] can learn how to operate an online search engine like Bing API[1] to answer open-domain questions. InstructGPT [73] can perform various types of tasks according to the corresponding task instructions

---

[1] https://www.microsoft.com/en-us/bing/apis/bing-web-search-api.

by learning from human feedback with reinforcement learning. Inspired by the idea of reinforcement learning from human feedback of InstructGPT, more recently, ChatGPT[2] have also demonstrated the fantastic dialogue ability of big PTMs, which learns from tens of thousands of human conversation behaviors. All these works make an initial exploration to more intelligent utilization of big PTMs: by learning from human behavior with reinforcement learning, we may mine the unexplored high-level cognitive intelligence hiding in big PTMs.

5.5 Summary and Further Readings

In this section, we review the current progress and the remaining challenges of pre-trained models for representation learning in NLP. First, we introduce the pre-training tasks, including word-level pre-training and sentence-level pre-training. After that, we turn to the model adaptation, from full-parameter fine-tuning to optimization-perspective delta tuning and data-perspective prompt learning. Finally, we discuss several advanced topics, such as better model architecture, multilingual learning, multi-task learning, efficient representations, and chain-of-thought reasoning.

For further understanding of pre-trained models for representation learning, you can find more related papers in our paper lists about pre-trained models,[3] delta tuning[4] and prompt learning.[5] On the survey of pre-trained models, Han et al. [37] give a comprehensive review of the history and recent breakthroughs of PTMs and also discuss its remaining open challenges. Ding et al. [22] give a detailed review of existing delta tuning methods. Bommasani et al. [6] systematically review the PTMs' developments from the capability, technical principle, application, and societal impact perspectives.

Acknowledgments Zhiyuan Liu, Yankai Lin, and Maosong Sun designed the overall architecture of this chapter; Yankai Lin and Ning Ding drafted the chapter. Zhiyuan Liu proofread and revised this chapter.

We also thank Xu Han, Zheni Zeng, Shengding Hu, Zhengyan Zhang, Yingfa Chen, and Zhiyuan Zeng for proofreading the chapter.

This is the newly complemented chapter about pre-trained models of the second edition of the book *Representation Learning for Natural Language Processing*. The first edition of the book was published in 2020 [69].

---

[2] https://openai.com/blog/chatgpt/.

[3] https://github.com/thunlp/PLMpapers.

[4] https://github.com/thunlp/DeltaPapers.

[5] https://github.com/thunlp/PromptPapers.

References

1. Armen Aghajanyan, Anchit Gupta, Akshat Shrivastava, Xilun Chen, Luke Zettlemoyer, and Sonal Gupta. Muppet: Massive multi-task representations with pre-finetuning. In *Proceedings of EMNLP*, 2021.
2. Armen Aghajanyan, Sonal Gupta, and Luke Zettlemoyer. Intrinsic dimensionality explains the effectiveness of language model fine-tuning. In *Proceedings of ACL-IJCNLP*, 2021.
3. Naveen Arivazhagan, Ankur Bapna, Orhan Firat, Dmitry Lepikhin, Melvin Johnson, Maxim Krikun, Mia Xu Chen, Yuan Cao, George Foster, Colin Cherry, et al. Massively multilingual neural machine translation in the wild: Findings and challenges. *arXiv preprint arXiv:1907.05019*, 2019.
4. Iz Beltagy, Matthew E Peters, and Arman Cohan. Longformer: The long-document transformer. *arXiv preprint arXiv:2004.05150*, 2020.
5. Eyal Ben-David, Nadav Oved, and Roi Reichart. Pada: A prompt-based autoregressive approach for adaptation to unseen domains. *arXiv preprint arXiv:2102.12206*, 2021.
6. Rishi Bommasani, Drew A Hudson, Ehsan Adeli, Russ Altman, Simran Arora, Sydney von Arx, Michael S Bernstein, Jeannette Bohg, Antoine Bosselut, Emma Brunskill, et al. On the opportunities and risks of foundation models. *arXiv preprint arXiv:2108.07258*, 2021.
7. Tom Brown, Benjamin Mann, Nick Ryder, Melanie Subbiah, Jared D Kaplan, Prafulla Dhariwal, Arvind Neelakantan, Pranav Shyam, Girish Sastry, Amanda Askell, et al. Language models are few-shot learners. In *Proceedings of NeurIPS*, 2020.
8. Rich Caruana. Multitask learning. *Machine learning*, 28(1):41–75, 1997.
9. Xiang Chen, Ningyu Zhang, Xin Xie, Shumin Deng, Yunzhi Yao, Chuanqi Tan, Fei Huang, Luo Si, and Huajun Chen. Knowprompt: Knowledge-aware prompt-tuning with synergistic optimization for relation extraction. In *Proceedings of WebConf*, 2022.
10. Zewen Chi, Li Dong, Furu Wei, Wenhui Wang, Xian-Ling Mao, and Heyan Huang. Crosslingual natural language generation via pre-training. In *Proceedings of AAAI*, 2020.
11. Zewen Chi, Li Dong, Furu Wei, Nan Yang, Saksham Singhal, Wenhui Wang, Xia Song, Xian-Ling Mao, He-Yan Huang, and Ming Zhou. Infoxlm: An information-theoretic framework for cross-lingual language model pre-training. In *Proceedings of NAACL-HLT*, 2021.
12. Rewon Child, Scott Gray, Alec Radford, and Ilya Sutskever. Generating long sequences with sparse transformers. *arXiv preprint arXiv:1904.10509*, 2019.
13. Krzysztof Choromanski, Valerii Likhosherstov, David Dohan, Xingyou Song, Andreea Gane, Tamas Sarlos, Peter Hawkins, Jared Davis, David Belanger, Lucy Colwell, et al. Masked language modeling for proteins via linearly scalable long-context transformers. *arXiv preprint arXiv:2006.03555*, 2020.
14. Kevin Clark, Urvashi Khandelwal, Omer Levy, and Christopher D. Manning. What does BERT look at? an analysis of bert's attention. In *Proceedings of ACL Workshop BlackboxNLP*, 2019.
15. Kevin Clark, Minh-Thang Luong, Quoc V Le, and Christopher D Manning. Electra: Pretraining text encoders as discriminators rather than generators. In *Proceedings of ICLR*, 2019.
16. Alexis Conneau, Kartikay Khandelwal, Naman Goyal, Vishrav Chaudhary, Guillaume Wenzek, Francisco Guzmán, Édouard Grave, Myle Ott, Luke Zettlemoyer, and Veselin Stoyanov. Unsupervised cross-lingual representation learning at scale. In *Proceedings of ACL*, pages 8440–8451, 2020.
17. Alexis Conneau and Guillaume Lample. Cross-lingual language model pretraining. In *Proceedings of NeurIPS*, 2019.
18. Ganqu Cui, Shengding Hu, Ning Ding, Longtao Huang, and Zhiyuan Liu. Prototypical verbalizer for prompt-based few-shot tuning. In *Proceedings of ACL*, 2022.
19. Leyang Cui, Yu Wu, Jian Liu, Sen Yang, and Yue Zhang. Template-based named entity recognition using bart. In *Findings of ACL*, 2021.
20. Jacob Devlin, Ming-Wei Chang, Kenton Lee, and Kristina Toutanova. BERT: pre-training of deep bidirectional transformers for language understanding. In *Proceedings of NAACL-HLT*, 2019.

21. Ning Ding, Yulin Chen, Xu Han, Guangwei Xu, Pengjun Xie, Hai-Tao Zheng, Zhiyuan Liu, Juanzi Li, and Hong-Gee Kim. Prompt-learning for fine-grained entity typing. *arXiv preprint arXiv:2108.10604*, 2021.

22. Ning Ding, Yujia Qin, Guang Yang, Fuchao Wei, Zonghan Yang, Yusheng Su, Shengding Hu, Yulin Chen, Chi-Min Chan, Weize Chen, et al. Delta tuning: A comprehensive study of parameter efficient methods for pre-trained language models. *arXiv preprint arXiv:2203.06904*, 2022.

23. Li Dong, Nan Yang, Wenhui Wang, Furu Wei, Xiaodong Liu, Yu Wang, Jianfeng Gao, Ming Zhou, and Hsiao-Wuen Hon. Unified language model pre-training for natural language understanding and generation. In *Proceedings of NeurIPS*, 2019.

24. Nan Du, Yanping Huang, Andrew M Dai, Simon Tong, Dmitry Lepikhin, Yuanzhong Xu, Maxim Krikun, Yanqi Zhou, Adams Wei Yu, Orhan Firat, et al. Glam: Efficient scaling of language models with mixture-of-experts. In *Proceedings of ICML*, 2022.

25. Angela Fan, Edouard Grave, and Armand Joulin. Reducing transformer depth on demand with structured dropout. In *Proceedings of ICLR*, 2019.

26. William Fedus, Barret Zoph, and Noam Shazeer. Switch transformers: Scaling to trillion parameter models with simple and efficient sparsity. *Journal of Machine Learning Research*, 23(120):1–39, 2022.

27. Chris Fifty, Ehsan Amid, Zhe Zhao, Tianhe Yu, Rohan Anil, and Chelsea Finn. Efficiently identifying task groupings for multi-task learning. In *Proceedings of NeurIPS*, 2021.

28. Tianyu Gao, Adam Fisch, and Danqi Chen. Making pre-trained language models better few-shot learners. In *Proceedings of ACL*, 2021.

29. Tianyu Gao, Xingcheng Yao, and Danqi Chen. Simcse: Simple contrastive learning of sentence embeddings. In *Proceedings of EMNLP*, 2021.

30. Michael Glass, Alfio Gliozzo, Rishav Chakravarti, Anthony Ferritto, Lin Pan, GP Shrivatsa Bhargav, Dinesh Garg, and Avirup Sil. Span selection pre-training for question answering. In *Proceedings of ACL*, 2020.

31. Mitchell Gordon, Kevin Duh, and Nicholas Andrews. Compressing bert: Studying the effects of weight pruning on transfer learning. In *Proceedings of RL4NLP*, 2020.

32. Yuxian Gu, Xu Han, Zhiyuan Liu, and Minlie Huang. Ppt: Pre-trained prompt tuning for few-shot learning. In *Proceedings of ACL*, 2022.

33. Yuxian Gu, Zhengyan Zhang, Xiaozhi Wang, Zhiyuan Liu, and Maosong Sun. Train no evil: Selective masking for task-guided pre-training. In *Proceedings of EMNLP*, 2020.

34. Demi Guo, Alexander M Rush, and Yoon Kim. Parameter-efficient transfer learning with diff pruning. In *Proceedings of ACL-IJCNLP*, 2021.

35. Suchin Gururangan, Ana Marasović, Swabha Swayamdipta, Kyle Lo, Iz Beltagy, Doug Downey, and Noah A Smith. Don't stop pretraining: Adapt language models to domains and tasks. In *Proceedings of ACL*, 2020.

36. Kelvin Guu, Kenton Lee, Zora Tung, Panupong Pasupat, and Mingwei Chang. Retrieval augmented language model pre-training. In *Proceedings of ICML*, 2020.

37. Xu Han, Zhengyan Zhang, Ning Ding, Yuxian Gu, Xiao Liu, Yuqi Huo, Jiezhong Qiu, Yuan Yao, Ao Zhang, Liang Zhang, et al. Pre-trained models: Past, present and future. *AI Open*, 2021.

38. Xu Han, Weilin Zhao, Ning Ding, Zhiyuan Liu, and Maosong Sun. PTR: Prompt tuning with rules for text classification. *arXiv preprint arXiv:2105.11259*, 2021.

39. Zellig S Harris. Distributional structure. *Word*, 10(2–3):146–162, 1954.

40. Adi Haviv, Jonathan Berant, and Amir Globerson. Bertese: Learning to speak to bert. In *Proceedings of EACL*, 2021.

41. Dan Hendrycks and Kevin Gimpel. Gaussian error linear units (gelus). *arXiv preprint arXiv:1606.08415*, 2016.

42. Neil Houlsby, Andrei Giurgiu, Stanislaw Jastrzebski, Bruna Morrone, Quentin De Laroussilhe, Andrea Gesmundo, Mona Attariyan, and Sylvain Gelly. Parameter-efficient transfer learning for nlp. In *Proceedings of ICML*, 2019.

43. Edward J Hu, Phillip Wallis, Zeyuan Allen-Zhu, Yuanzhi Li, Shean Wang, Lu Wang, Weizhu Chen, et al. LoRA: Low-rank adaptation of large language models. In *Proceedings of ICLR*, 2021.
44. Shengding Hu, Ning Ding, Huadong Wang, Zhiyuan Liu, Jingang Wang, Juanzi Li, Wei Wu, and Maosong Sun. Knowledgeable prompt-tuning: Incorporating knowledge into prompt verbalizer for text classification. In *Proceedings of ACL*, 2022.
45. Goodfellow Ian, Yoshua Bengio, and Aaron Courville. Deep learning. Book in preparation for MIT Press, 2016.
46. Sverker Janson, Evangelina Gogoulou, Erik Ylipää, Amaru Cuba Gyllensten, and Magnus Sahlgren. Semantic re-tuning with contrastive tension. In *Proceedings of ICLR*, 2021.
47. Zhengbao Jiang, Frank F Xu, Jun Araki, and Graham Neubig. How can we know what language models know? *Transactions of the Association for Computational Linguistics*, 8:423–438, 2020.
48. Xiaoqi Jiao, Yichun Yin, Lifeng Shang, Xin Jiang, Xiao Chen, Linlin Li, Fang Wang, and Qun Liu. Tinybert: Distilling bert for natural language understanding. In *Findings of EMNLP*, 2020.
49. Mandar Joshi, Danqi Chen, Yinhan Liu, Daniel S Weld, Luke Zettlemoyer, and Omer Levy. Spanbert: Improving pre-training by representing and predicting spans. *Transactions of the Association for Computational Linguistics*, 8:64–77, 2020.
50. Rabeeh Karimi Mahabadi, James Henderson, and Sebastian Ruder. Compacter: Efficient low-rank hypercomplex adapter layers. In *Proceedings of NeurIPS*, 2021.
51. Angelos Katharopoulos, Apoorv Vyas, Nikolaos Pappas, and François Fleuret. Transformers are rnns: Fast autoregressive transformers with linear attention. In *Proceedings of ICML*, 2020.
52. Taeuk Kim, Kang Min Yoo, and Sang-goo Lee. Self-guided contrastive learning for bert sentence representations. In *Proceedings of ACL-IJCNLP*, 2021.
53. Nikita Kitaev, Lukasz Kaiser, and Anselm Levskaya. Reformer: The efficient transformer. In *Proceedings of ICLR*, 2019.
54. Takeshi Kojima, Shixiang Shane Gu, Machel Reid, Yutaka Matsuo, and Yusuke Iwasawa. Large language models are zero-shot reasoners. *arXiv preprint arXiv:2205.11916*, 2022.
55. Zhenzhong Lan, Mingda Chen, Sebastian Goodman, Kevin Gimpel, Piyush Sharma, and Radu Soricut. Albert: A lite bert for self-supervised learning of language representations. In *Proceedings of ICLR*, 2019.
56. Jinhyuk Lee, Wonjin Yoon, Sungdong Kim, Donghyeon Kim, Sunkyu Kim, Chan Ho So, and Jaewoo Kang. BioBERT: a pre-trained biomedical language representation model for biomedical text mining. *Bioinformatics*, 36(4):1234–1240, 2020.
57. Dmitry Lepikhin, HyoukJoong Lee, Yuanzhong Xu, Dehao Chen, Orhan Firat, Yanping Huang, Maxim Krikun, Noam Shazeer, and Zhifeng Chen. Gshard: Scaling giant models with conditional computation and automatic sharding. In *Proceedings of ICLR*, 2020.
58. Brian Lester, Rami Al-Rfou, and Noah Constant. The power of scale for parameter-efficient prompt tuning. In *Proceedings of EMNLP*, 2021.
59. Mike Lewis, Yinhan Liu, Naman Goyal, Marjan Ghazvininejad, Abdelrahman Mohamed, Omer Levy, Veselin Stoyanov, and Luke Zettlemoyer. Bart: Denoising sequence-to-sequence pre-training for natural language generation, translation, and comprehension. In *Proceedings of ACL*, 2020.
60. Patrick Lewis, Ethan Perez, Aleksandra Piktus, Fabio Petroni, Vladimir Karpukhin, Naman Goyal, Heinrich Küttler, Mike Lewis, Wen-tau Yih, Tim Rocktäschel, et al. Retrieval-augmented generation for knowledge-intensive nlp tasks. In *Proceedings of NeurIPS*, 2020.
61. Lei Li, Yankai Lin, Deli Chen, Shuhuai Ren, Peng Li, Jie Zhou, and Xu Sun. Cascadebert: Accelerating inference of pre-trained language models via calibrated complete models cascade. In *Findings of EMNLP*, 2021.
62. Xiang Lisa Li and Percy Liang. Prefix-tuning: Optimizing continuous prompts for generation. In *Proceedings of ACL-IJCNLP*, 2021.

63. Yu-Hsiang Lin, Chian-Yu Chen, Jean Lee, Zirui Li, Yuyan Zhang, Mengzhou Xia, Shruti Rijhwani, Junxian He, Zhisong Zhang, Xuezhe Ma, et al. Choosing transfer languages for cross-lingual learning. In *Proceedings of ACL*, 2019.
64. Pengfei Liu, Weizhe Yuan, Jinlan Fu, Zhengbao Jiang, Hiroaki Hayashi, and Graham Neubig. Pre-train, prompt, and predict: A systematic survey of prompting methods in natural language processing. *arXiv preprint arXiv:2107.13586*, 2021.
65. Xiaodong Liu, Pengcheng He, Weizhu Chen, and Jianfeng Gao. Multi-task deep neural networks for natural language understanding. In *Proceedings of ACL*, 2019.
66. Ye Liu, Yao Wan, Lifang He, Hao Peng, and S Yu Philip. Kg-bart: Knowledge graph-augmented bart for generative commonsense reasoning. In *Proceedings of AAAI*, 2021.
67. Yinhan Liu, Jiatao Gu, Naman Goyal, Xian Li, Sergey Edunov, Marjan Ghazvininejad, Mike Lewis, and Luke Zettlemoyer. Multilingual denoising pre-training for neural machine translation. *Transactions of the Association for Computational Linguistics*, 8:726–742, 2020.
68. Yinhan Liu, Myle Ott, Naman Goyal, Jingfei Du, Mandar Joshi, Danqi Chen, Omer Levy, Mike Lewis, Luke Zettlemoyer, and Veselin Stoyanov. RoBERTa: A robustly optimized BERT pretraining approach. *arXiv preprint arXiv:1907.11692*, 2019.
69. Zhiyuan Liu, Yankai Lin, and Maosong Sun. *Representation Learning for Natural Language Processing*. Springer, 2020.
70. Rabeeh Karimi Mahabadi, Sebastian Ruder, Mostafa Dehghani, and James Henderson. Parameter-efficient multi-task fine-tuning for transformers via shared hypernetworks. In *Proceedings of ACL-IJCNLP*, 2021.
71. Paul Michel, Omer Levy, and Graham Neubig. Are sixteen heads really better than one? In *Proceedings of NeurIPS*, 2019.
72. Reiichiro Nakano, Jacob Hilton, Suchir Balaji, Jeff Wu, Long Ouyang, Christina Kim, Christopher Hesse, Shantanu Jain, Vineet Kosaraju, William Saunders, et al. WebGPT: Browser-assisted question-answering with human feedback. *arXiv preprint arXiv:2112.09332*, 2021.
73. Long Ouyang, Jeff Wu, Xu Jiang, Diogo Almeida, Carroll L Wainwright, Pamela Mishkin, Chong Zhang, Sandhini Agarwal, Katarina Slama, Alex Ray, et al. Training language models to follow instructions with human feedback. *arXiv preprint arXiv:2203.02155*, 2022.
74. Xuan Ouyang, Shuohuan Wang, Chao Pang, Yu Sun, Hao Tian, Hua Wu, and Haifeng Wang. Ernie-m: Enhanced multilingual representation by aligning cross-lingual semantics with monolingual corpora. In *Proceedings of EMNLP*, 2021.
75. Matthew Peters, Mark Neumann, Mohit Iyyer, Matt Gardner, Christopher Clark, Kenton Lee, and Luke Zettlemoyer. Deep contextualized word representations. In *Proceedings of NAACL-HLT*, 2018.
76. Matthew E Peters, Mark Neumann, Robert Logan, Roy Schwartz, Vidur Joshi, Sameer Singh, and Noah A Smith. Knowledge enhanced contextual word representations. In *Proceedings of EMNLP-IJCNLP*, 2019.
77. Fabio Petroni, Tim Rocktäschel, Sebastian Riedel, Patrick Lewis, Anton Bakhtin, Yuxiang Wu, and Alexander Miller. Language models as knowledge bases? In *Proceedings of EMNLP-IJCNLP*, 2019.
78. Jason Phang, Thibault Févry, and Samuel R Bowman. Sentence encoders on stilts: Supplementary training on intermediate labeled-data tasks. *arXiv preprint arXiv:1811.01088*, 2018.
79. Telmo Pires, Eva Schlinger, and Dan Garrette. How multilingual is multilingual bert? In *Proceedings of ACL*, 2019.
80. Clifton Poth, Jonas Pfeiffer, Andreas Rücklé, and Iryna Gurevych. What to pre-train on? efficient intermediate task selection. In *Proceedings of EMNLP*, 2021.
81. Yada Pruksachatkun, Jason Phang, Haokun Liu, Phu Mon Htut, Xiaoyi Zhang, Richard Yuanzhe Pang, Clara Vania, Katharina Kann, and Samuel Bowman. Intermediate-task transfer learning with pretrained language models: When and why does it work? In *Proceedings of ACL*, 2020.

82. Yujia Qin, Yankai Lin, Ryuichi Takanobu, Zhiyuan Liu, Peng Li, Heng Ji, Minlie Huang, Maosong Sun, and Jie Zhou. Erica: Improving entity and relation understanding for pre-trained language models via contrastive learning. In *Proceedings of ACL-IJCNLP*, 2021.

83. Yujia Qin, Yankai Lin, Jing Yi, Jiajie Zhang, Xu Han, Zhengyan Zhang, Yusheng Su, Zhiyuan Liu, Peng Li, Maosong Sun, et al. Knowledge inheritance for pre-trained language models. *arXiv preprint arXiv:2105.13880*, 2021.

84. Yujia Qin, Xiaozhi Wang, Yusheng Su, Yankai Lin, Ning Ding, Zhiyuan Liu, Juanzi Li, Lei Hou, Peng Li, Maosong Sun, et al. Exploring low-dimensional intrinsic task subspace via prompt tuning. *arXiv preprint arXiv:2110.07867*, 2021.

85. Jiezhong Qiu, Hao Ma, Omer Levy, Wen-tau Yih, Sinong Wang, and Jie Tang. Blockwise self-attention for long document understanding. In *Findings of EMNLP*, 2020.

86. Alec Radford, Karthik Narasimhan, Tim Salimans, and Ilya Sutskever. Improving language understanding with unsupervised learning. 2018.

87. Alec Radford, Jeffrey Wu, Rewon Child, David Luan, Dario Amodei, and Ilya Sutskever. Language models are unsupervised multitask learners. *OpenAI Blog*, 2019.

88. Colin Raffel, Noam Shazeer, Adam Roberts, Katherine Lee, Sharan Narang, Michael Matena, Yanqi Zhou, Wei Li, and Peter J Liu. Exploring the limits of transfer learning with a unified text-to-text transformer. *Journal of Machine Learning Research*, 21:1–67, 2020.

89. Aurko Roy, Mohammad Saffar, Ashish Vaswani, and David Grangier. Efficient content-based sparse attention with routing transformers. *Transactions of the Association for Computational Linguistics*, 9:53–68, 2021.

90. Andreas Rücklé, Gregor Geigle, Max Glockner, Tilman Beck, Jonas Pfeiffer, Nils Reimers, and Iryna Gurevych. Adapterdrop: On the efficiency of adapters in transformers. In *Proceedings of EMNLP*, 2021.

91. Victor Sanh, Lysandre Debut, Julien Chaumond, and Thomas Wolf. Distilbert, a distilled version of bert: smaller, faster, cheaper and lighter. *arXiv preprint arXiv:1910.01108*, 2019.

92. Timo Schick, Helmut Schmid, and Hinrich Schütze. Automatically identifying words that can serve as labels for few-shot text classification. In *Proceedings of COLING*, 2020.

93. Claude E Shannon. A mathematical theory of communication. *Bell system technical journal*, 27(3):379–423, 1948.

94. Noam Shazeer, Azalia Mirhoseini, Krzysztof Maziarz, Andy Davis, Quoc Le, Geoffrey Hinton, and Jeff Dean. Outrageously large neural networks: The sparsely-gated mixture-of-experts layer. *arXiv preprint arXiv:1701.06538*, 2017.

95. Taylor Shin, Yasaman Razeghi, Robert L Logan IV, Eric Wallace, and Sameer Singh. Auto-prompt: Eliciting knowledge from language models with automatically generated prompts. In *Proceedings of EMNLP*, 2020.

96. Oleh Shliazhko, Alena Fenogenova, Maria Tikhonova, Vladislav Mikhailov, Anastasia Kozlova, and Tatiana Shavrina. mgpt: Few-shot learners go multilingual. *arXiv preprint arXiv:2204.07580*, 2022.

97. Kaitao Song, Xu Tan, Tao Qin, Jianfeng Lu, and Tie-Yan Liu. MASS: Masked sequence to sequence pre-training for language generation. In *Proceedings of ICML*, 2019.

98. Asa Cooper Stickland and Iain Murray. Bert and pals: Projected attention layers for efficient adaptation in multi-task learning. In *Proceedings of ICML*, pages 5986–5995. PMLR, 2019.

99. Siqi Sun, Yu Cheng, Zhe Gan, and Jingjing Liu. Patient knowledge distillation for BERT model compression. In *Proceedings of EMNLP-IJCNLP*, 2019.

100. Yu Sun, Shuohuan Wang, Yukun Li, Shikun Feng, Xuyi Chen, Han Zhang, Xin Tian, Danxiang Zhu, Hao Tian, and Hua Wu. Ernie: Enhanced representation through knowledge integration. *arXiv preprint arXiv:1904.09223*, 2019.

101. Yi-Lin Sung, Jaemin Cho, and Mohit Bansal. Lst: Ladder side-tuning for parameter and memory efficient transfer learning. *arXiv preprint arXiv:2206.06522*, 2022.

102. Yi Tay, Zhe Zhao, Dara Bahri, Donald Metzler, and Da-Cheng Juan. Hypergrid transformers: Towards a single model for multiple tasks. In *International Conference on Learning Representations*, 2020.

103. Sebastian Thrun. Lifelong learning algorithms. In *Learning to learn*, pages 181–209. Springer, 1998.

104. Ashish Vaswani, Noam Shazeer, Niki Parmar, Llion Jones, Jakob Uszkoreit, Aidan N Gomez, and Lukasz Kaiser. Attention is all you need. In *Proceedings of NeurIPS*, 2017.

105. Elena Voita, David Talbot, Fedor Moiseev, Rico Sennrich, and Ivan Titov. Analyzing multi-head self-attention: Specialized heads do the heavy lifting, the rest can be pruned. In *Proceedings of ACL*, 2019.

106. Eric Wallace, Shi Feng, Nikhil Kandpal, Matt Gardner, and Sameer Singh. Universal adversarial triggers for attacking and analyzing nlp. In *Proceedings of EMNLP-IJCNLP*, 2019.

107. Kexin Wang, Nils Reimers, and Iryna Gurevych. Tsdae: Using transformer-based sequential denoising auto-encoderfor unsupervised sentence embedding learning. In *Findings of EMNLP*, 2021.

108. Sinong Wang, Belinda Z Li, Madian Khabsa, Han Fang, and Hao Ma. Linformer: Self-attention with linear complexity. *arXiv preprint arXiv:2006.04768*, 2020.

109. Wenhui Wang, Furu Wei, Li Dong, Hangbo Bao, Nan Yang, and Ming Zhou. Minilm: Deep self-attention distillation for task-agnostic compression of pre-trained transformers. In *Proceedings of NeurIPS*, 2020.

110. Xiaosen Wang, Zeliang Zhang, Kangheng Tong, Dihong Gong, Kun He, Zhifeng Li, and Wei Liu. Triangle attack: A query-efficient decision-based adversarial attack. *arXiv preprint arXiv:2112.06569*, 2021.

111. Xuezhi Wang, Jason Wei, Dale Schuurmans, Quoc Le, Ed Chi, and Denny Zhou. Self-consistency improves chain of thought reasoning in language models. *arXiv preprint arXiv:2203.11171*, 2022.

112. Jason Wei, Xuezhi Wang, Dale Schuurmans, Maarten Bosma, Ed Chi, Quoc Le, and Denny Zhou. Chain of thought prompting elicits reasoning in large language models. *arXiv preprint arXiv:2201.11903*, 2022.

113. Xiangpeng Wei, Rongxiang Weng, Yue Hu, Luxi Xing, Heng Yu, and Weihua Luo. On learning universal representations across languages. In *Proceedings of ICLR*, 2020.

114. Ji Xin, Raphael Tang, Jaejun Lee, Yaoliang Yu, and Jimmy Lin. Deebert: Dynamic early exiting for accelerating bert inference. In *Proceedings of ACL*, 2020.

115. Linting Xue, Noah Constant, Adam Roberts, Mihir Kale, Rami Al-Rfou, Aditya Siddhant, Aditya Barua, and Colin Raffel. mt5: A massively multilingual pre-trained text-to-text transformer. In *Proceedings of NAACL-HLT*, 2021.

116. Jian Yang, Shuming Ma, Dongdong Zhang, Shuangzhi Wu, Zhoujun Li, and Ming Zhou. Alternating language modeling for cross-lingual pre-training. In *Proceedings of AAAI*, 2020.

117. Zhilin Yang, Zihang Dai, Yiming Yang, Jaime Carbonell, Russ R Salakhutdinov, and Quoc V Le. XLNet: Generalized autoregressive pretraining for language understanding. In *Proceedings of NeurIPS*, 2019.

118. Deming Ye, Yankai Lin, Yufei Huang, and Maosong Sun. Tr-bert: Dynamic token reduction for accelerating bert inference. In *Proceedings of NAACL-HLT*, 2021.

119. Jing Yi, Weize Chen, Yujia Qin, Yankai Lin, Ning Ding, Xu Han, Zhiyuan Liu, Maosong Sun, and Jie Zhou. Different tunes played with equal skill: Exploring a unified optimization subspace for parameter-efficient tuning. *arXiv preprint arXiv:2210.13311*, 2022.

120. Wenpeng Yin, Jamaal Hay, and Dan Roth. Benchmarking zero-shot text classification: Datasets, evaluation and entailment approach. In *Proceedings of EMNLP-IJCNLP*, 2019.

121. Zichun Yu, Tianyu Gao, Zhengyan Zhang, Yankai Lin, Zhiyuan Liu, Maosong Sun, and Jie Zhou. Automatic label sequence generation for prompting sequence-to-sequence models. *arXiv preprint arXiv:2209.09401*, 2022.

122. Weizhe Yuan, Graham Neubig, and Pengfei Liu. Bartscore: Evaluating generated text as text generation. In *Proceedings of NeurIPS*, 2021.

123. Ofir Zafrir, Guy Boudoukh, Peter Izsak, and Moshe Wasserblat. Q8bert: Quantized 8bit bert. In *Proceedings of EMC2-NIPS*, 2019.

124. Manzil Zaheer, Guru Guruganesh, Avinava Dubey, Joshua Ainslie, Chris Alberti, Santiago Ontanon, Philip Pham, Anirudh Ravula, Qifan Wang, Li Yang, and Amr Ahmed. Big Bird: transformers for longer sequences. In *Proceedings of NeuIPS*, 2020.
125. Elad Ben Zaken, Yoav Goldberg, and Shauli Ravfogel. Bitfit: Simple parameter-efficient fine-tuning for transformer-based masked language-models. In *Proceedings of ACL*, 2022.
126. Eric Zelikman, Yuhuai Wu, and Noah D Goodman. StAR: Bootstrapping reasoning with reasoning. *arXiv preprint arXiv:2203.14465*, 2022.
127. Aohan Zeng, Xiao Liu, Zhengxiao Du, Zihan Wang, Hanyu Lai, Ming Ding, Zhuoyi Yang, Yifan Xu, Wendi Zheng, Xiao Xia, et al. GLM-130B: An open bilingual pre-trained model. *arXiv preprint arXiv:2210.02414*, 2022.
128. Qingkai Zeng, Wenhao Yu, Mengxia Yu, Tianwen Jiang, Tim Weninger, and Meng Jiang. Tri-train: Automatic pre-fine tuning between pre-training and fine-tuning for sciner. In *Findings of EMNLP*, 2020.
129. Susan Zhang, Stephen Roller, Naman Goyal, Mikel Artetxe, Moya Chen, Shuohui Chen, Christopher Dewan, Mona Diab, Xian Li, Xi Victoria Lin, et al. OPT: Open pre-trained transformer language models. *arXiv preprint arXiv:2205.01068*, 2022.
130. Wei Zhang, Lu Hou, Yichun Yin, Lifeng Shang, Xiao Chen, Xin Jiang, and Qun Liu. Ternarybert: Distillation-aware ultra-low bit bert. In *Proceedings of EMNLP*, 2020.
131. Zhengyan Zhang, Yuxian Gu, Xu Han, Shengqi Chen, Chaojun Xiao, Zhenbo Sun, Yuan Yao, Fanchao Qi, Jian Guan, Pei Ke, et al. Cpm-2: Large-scale cost-effective pre-trained language models. *AI Open*, 2:216–224, 2021.
132. Zhengyan Zhang, Xu Han, Zhiyuan Liu, Xin Jiang, Maosong Sun, and Qun Liu. ERNIE: Enhanced language representation with informative entities. In *Proceedings of ACL*, 2019.
133. Zhengyan Zhang, Xu Han, Hao Zhou, Pei Ke, Yuxian Gu, Deming Ye, Yujia Qin, Yusheng Su, Haozhe Ji, Jian Guan, et al. CPM: A large-scale generative chinese pre-trained language model. *AI Open*, 2:93–99, 2021.
134. Zhengyan Zhang, Yankai Lin, Zhiyuan Liu, Peng Li, Maosong Sun, and Jie Zhou. Moefication: Conditional computation of transformer models for efficient inference. *arXiv preprint arXiv:2110.01786*, 2021.
135. Mengjie Zhao, Tao Lin, Fei Mi, Martin Jaggi, and Hinrich Schütze. Masking as an efficient alternative to finetuning for pretrained language models. In *Proceedings of EMNLLP*, 2020.

Chapter 6
Graph Representation Learning

Cheng Yang, Yankai Lin, Zhiyuan Liu, and Maosong Sun

Abstract Graph structure, which can represent objects and their relationships, is ubiquitous in big data including natural languages. Besides original text as a sequence of word tokens, massive additional information in NLP is in the graph structure, such as syntactic relations between words in a sentence, hyperlink relations between documents, and semantic relations between entities. Hence, it is critical for NLP to encode these graph data with graph representation learning. Graph representation learning, also known as network embedding, has been extensively studied in AI and data mining. In this chapter, we introduce a variety of graph representation learning methods that embed graph data into vectors with shallow or deep neural models. After that, we introduce how graph representation learning helps NLP tasks.

6.1 Introduction

Graph is a natural way to represent objects and their relationships. As a typical non-Euclidean data structure, it provides a flexible way to model the interactions between individual units in our daily lives. For example, social media networks, citation graphs, biological networks, and recommendation systems can all be modeled as graph structures.

C. Yang
School of Computer Science, Beijing University of Posts and Telecommunications, Beijing, China
e-mail: yangcheng@bupt.edu.cn

Y. Lin
Gaoling School of Artificial Intelligence, Renmin University of China, Beijing, China
e-mail: yankailin@ruc.edu.cn

Z. Liu (✉) · M. Sun
Department of Computer Science and Technology, Tsinghua University, Beijing, China
e-mail: liuzy@tsinghua.edu.cn; sms@tsinghua.edu.cn

Z. Liu et al. (eds.), *Representation Learning for Natural Language Processing*,
https://doi.org/10.1007/978-981-99-1600-9_6

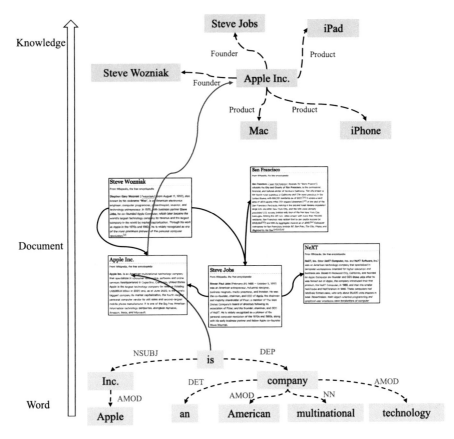

Fig. 6.1 An illustrative example of text-based graphs in different levels. The documents used in the figure are obtained from Wikipedia's official website

Graph structure plays an equally important role in the NLP area as the sequence form introduced in the previous chapters. As shown in Fig. 6.1, multiple granularities of text can be organized in graph forms:

1. **Word Level**. There are a variety of syntactic/semantic relations between the words within a sentence or a document. For example, we can obtain the dependency relations between words by dependency parsing or the coreference relations between words by coreference resolution. These syntactic/semantic relationships can effectively help us understand the compositional semantics of different constituents in the sentence and document. We can naturally regard words as nodes and their syntactic/semantic relations as edges, representing a sentence or a document as a relational graph.
2. **Document Level**. In the real world, documents usually connect with each other. For example, in the articles of Wikipedia, the entity mentions within an article may exist human-annotated hyperlinks to the articles describing these entities,

or a scientific paper may refer to other scientific papers as its related works. The connected documents may provide important background knowledge to comprehend the meaning of the document. We can represent the interactions between documents as a document graph by regarding documents as nodes and hyperlinks as edges.

3. **Knowledge Level**. Human knowledge, such as world, linguistic, commonsense, and domain knowledge is essential for understanding languages. In practice, most of this knowledge can be organized in a graph form. For example, the world knowledge graph in Chap. 9 regards entities as nodes and their relationships as edges. Representing human knowledge as graphs enables further complex reasoning over the document and knowledge.

It is critical to model the structural information in graph data, which could help models better understand, categorize, and reason text in NLP tasks. Graph representation learning aims to learn the low-dimensional representations of nodes or the entire graph. The geometric relationship in the low-dimensional semantic space should effectively reflect the structural information of the original graph, such as the global topological structure or the local graph connection.

In this chapter, we discuss how to properly represent graph data and their characteristics by starting from symbolic representations (Sect. 6.2). Then, we move to distributed representations including the local representations of nodes (Sects. 6.3 and 6.4) as well as the global representation of the whole graph (Sect. 6.5). We also introduce the recent advances of self-supervised learning on graphs (Sect. 6.6). Finally, we present how text is processed in graph form in downstream NLP tasks (Sect. 6.7), mainly for word, sentence, and document levels. We leave a detailed introduction of knowledge-level applications in Chap. 9.

6.2 Symbolic Graph Representation

A common practice is to denote the graph with its node set $\mathcal{V}$ and the edge set $\mathcal{E}$. We can naturally represent a graph as $\mathcal{G} = (\mathcal{V}, \mathcal{E})$, where $e = (v_i, v_j) \in \mathcal{E}$ is a directed edge from node v_i to v_j or an undirected one between v_i and v_j. When processing graph data in a computer, we usually represent the connections in a graph as an adjacency matrix.

For an undirected graph, as shown in Fig. 6.2, we construct an adjacency matrix $\mathbf{A} \in \mathbb{R}^{|\mathcal{V}| \times |\mathcal{V}|}$. If there is any edge between node v and node u, i.e., $(v, u) \in \mathcal{E}$, we have the corresponding element $\mathbf{A}_{vu} = \mathbf{A}_{uv} = 1$; otherwise $\mathbf{A}_{vu} = \mathbf{A}_{uv} = 0$.

For a directed graph, as shown in Fig. 6.3, $\mathbf{A}_{vu} = 1$ indicates there is an edge from node v to node u.

Moreover, for a weighted graph, we can store the weights of the edges instead of binary values in the adjacency matrix $\mathbf{A}$, as shown in Fig. 6.4.

In the era of statistical learning, such symbolic representations are widely used in graph-based NLP such as TextRank [67], where word and sentence graphs

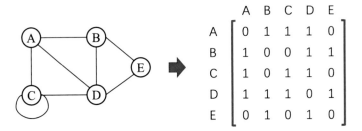

Fig. 6.2 The adjacency matrix representation for an undirected graph

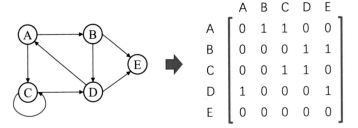

Fig. 6.3 The adjacency matrix representation for a directed graph

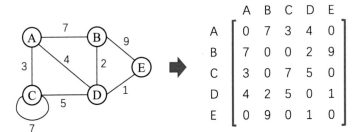

Fig. 6.4 The adjacency matrix representation for a weighted graph

can be respectively built for keyword extraction [6] and extractive document summarization [66]. Although convenient and straightforward, the representation of the adjacency matrix suffers from the scalability problem. Firstly, adjacency matrix **A** takes $|\mathcal{V}| \times |\mathcal{V}|$ storage space, which is usually unacceptable when $|\mathcal{V}|$ grows large. Secondly, the adjacency matrix is usually sparse, which means most of its entries are zeros. The data sparsity makes discrete algorithms applicable, but it is still hard to develop efficient algorithms for statistical learning [81].

6.3 Shallow Node Representation Learning

To address the above issues, shallow node representation learning methods propose to map nodes to low-dimensional vectors. Formally, the goal is to learn a d-dimensional vector representation $\mathbf{v} \in \mathbb{R}^d$ for every node $v \in \mathcal{V}$ in the graph. Learned node representations are supposed to capture the graph's structure information and then can be used as input features for downstream graph-related tasks. In this section, we will introduce several kinds of shallow node representation learning methods, including spectral clustering, shallow neural networks, and matrix factorization.

6.3.1 Spectral Clustering

Early methods of shallow node representation are usually based on spectral clustering, which typically computes the first k eigenvectors (or singular vectors) of an affinity matrix, such as adjacency or Laplacian matrix of the graph. Now we present several algorithms based on spectral clustering, including locally linear embedding, Laplacian Eigenmap, directed graph embedding, and latent social dimensions.

Locally Linear Embedding (LLE) LLE [88] assumes that the representations of a node and its neighbors lie in a locally linear patch of the manifold. In other words, a node's representation can be approximated by a linear combination of the representation of its neighbors. LLE uses the linear combination of neighbors to reconstruct the center node. Formally, the reconstruction error of all nodes can be expressed as

$$\mathcal{L}(\mathbf{W}, \mathbf{V}) = \sum_{v \in \mathcal{V}} \left\| \mathbf{v} - \sum_{u \in \mathcal{V}} \mathbf{W}_{vu} \mathbf{u} \right\|^2, \tag{6.1}$$

where $\mathbf{V} \in \mathbb{R}^{|\mathcal{V}| \times d}$ is the representation matrix containing all node representations $\mathbf{v}$ and $\mathbf{W}_{vu}$ is the learnable contribution coefficient of node u to v. LLE enforces $\mathbf{W}_{vu} = 0$ if v and u are not connected, i.e., $(v, u) \notin \mathcal{E}$. Further, the summation of a row of matrix $\mathbf{W}$ is set to 1, i.e., $\sum_{u \in \mathcal{V}} \mathbf{W}_{vu} = 1$.

Equation (6.1) is solved by alternatively optimizing weight matrix $\mathbf{W}$ and representation $\mathbf{V}$. The optimization over $\mathbf{W}$ can be solved as a least-squares problem. The optimization over $\mathbf{V}$ leads to the following optimization problem:

$$\mathcal{L}(\mathbf{W}, \mathbf{V}) = \sum_{v \in \mathcal{V}} \left\| \mathbf{v} - \sum_{u \in \mathcal{V}} \mathbf{W}_{vu} \mathbf{u} \right\|^2, \tag{6.2}$$

$$s.t. \ \sum_{v \in \mathcal{V}} \mathbf{v} = \mathbf{0}, \qquad |\mathcal{V}|^{-1} \sum_{v \in \mathcal{V}} \mathbf{v}^\top \mathbf{v} = \mathbf{I}_d, \tag{6.3}$$

where $\mathbf{I}_d$ denotes $d \times d$ identity matrix. The conditions in Eq. (6.3) ensure the uniqueness of the solution. The first condition enforces the center of all node representations to zero point, and the second condition guarantees different coordinates have the same scale, i.e., equal contribution to the reconstruction error.

The optimization problem can be formulated as the computation of eigenvectors of matrix $(\mathbf{I}_{|\mathscr{V}|} - \mathbf{W}^{\top})(\mathbf{I}_{|\mathscr{V}|} - \mathbf{W})$, which is an easily solvable eigenvalue problem. More details can be found in the note [20].

Laplacian Eigenmap Laplacian Eigenmap [7] simply follows the idea that the representations of two connected nodes should be close. Specifically, the "closeness" is measured by the square of Euclidean distance. We use $\mathbf{D} \in \mathbb{R}^{|\mathscr{V}| \times |\mathscr{V}|}$ to denote diagonal degree matrix where $\mathbf{D}_{vv}$ is the degree of node v. By defining the Laplacian matrix $\mathbf{L}$ as the difference between $\mathbf{D}$ and adjacency matrix A, we have $\mathbf{L} = \mathbf{D} - \mathbf{A}$. Laplacian Eigenmap algorithm wants to minimize the following objective:

$$\mathscr{L}(\mathbf{V}) = \sum_{\{v,u|(v,u)\in\mathscr{E}\}} \|\mathbf{v} - \mathbf{u}\|^2, \tag{6.4}$$

$$s.t.\ \mathbf{V}^{\top}\mathbf{D}\mathbf{V} = \mathbf{I}_d. \tag{6.5}$$

The cost function is the summation of the square loss of all connected node pairs, and the condition prevents the trivial all-zero solution caused by arbitrary scale. Equation (6.4) can be reformulated in matrix form as

$$\mathbf{V}^* = \underset{\mathbf{V}^{\top}\mathbf{D}\mathbf{V}=\mathbf{I}_d}{\arg\min}\ \mathrm{tr}(\mathbf{V}^{\top}\mathbf{L}\mathbf{V}), \tag{6.6}$$

where $\mathrm{tr}(\cdot)$ is the matrix trace function. The optimal solution $\mathbf{V}^*$ of Eq. (6.6) is the eigenvectors corresponding to d smallest nonzero eigenvalues of $\mathbf{L}$. Note that the Laplacian Eigenmap algorithm can be easily generalized to the weighted graph.

A significant limitation of both LLE and Laplacian Eigenmap is that they have a symmetric cost function, leading to both algorithms not being applied to directed graphs.

Directed Graph Embedding (DGE) DGE [17] generalizes Laplacian Eigenmap for both directed and undirected graphs based on a predefined transition matrix. For example, we can define a transition probability matrix $\mathbf{P} \in \mathbb{R}^{|\mathscr{V}| \times |\mathscr{V}|}$ where $\mathbf{P}_{vu}$ denotes the probability that node v walks to u. The transition matrix defines a Markov random walk through the graph. We denote the stationary value of node v as π_v where $\sum_v \pi_v = 1$. The stationary distribution of random walks is commonly used in many ranking algorithms such as PageRank [74]. DGE designs a new cost function that emphasizes those important nodes with higher stationary values:

$$\mathscr{L}(\mathbf{V}) = \sum_{v\in\mathscr{V}} \pi_v \sum_{u\in\mathscr{V}} \mathbf{P}_{vu}\|\mathbf{v} - \mathbf{u}\|^2. \tag{6.7}$$

By denoting $\mathbf{M} = \text{diag}(\pi_1, \pi_2, \ldots, \pi_{|\mathcal{V}|})$, the cost function Eq. (6.7) can be reformulated as

$$\mathcal{L}(\mathbf{V}) = 2\text{tr}(\mathbf{V}^\top \mathbf{B} \mathbf{V}), \tag{6.8}$$

$$s.t. \ \mathbf{V}^\top \mathbf{M} \mathbf{V} = \mathbf{I}_d, \tag{6.9}$$

where

$$\mathbf{B} = \mathbf{M} - \frac{\mathbf{M}\mathbf{P} - \mathbf{P}^\top \mathbf{M}}{2}. \tag{6.10}$$

The condition Eq. (6.9) is added to remove an arbitrary scaling factor of solutions. Similar to Laplacian Eigenmap, the optimization problem can also be solved as a generalized eigenvector problem.

Latent Social Dimensions Latent social dimensions [103] introduce modularity [71] into the cost function instead of minimizing the distance between node representations in previous works. Modularity is a measurement that characterizes how far the graph is away from a uniform random graph. Given $\mathcal{G} = (\mathcal{V}, \mathcal{E})$, we assume that nodes $\mathcal{V}$ are divided into n nonoverlapping communities. By "uniform random graph," we mean nodes connect to each other based on a uniform distribution given their degrees. Then, the expected number of edges between v and u is $\deg(v)\frac{\deg(u)}{2|\mathcal{E}|}$. Then, the modularity Q of a graph is defined as

$$Q = \frac{1}{2|\mathcal{E}|} \sum_{v,u} \left(\mathbf{A}_{vu} - \frac{\deg(v)\deg(u)}{2|\mathcal{E}|} \right) \delta(v, u), \tag{6.11}$$

where $\delta(v, u) = 1$ if v and u belong to the same community and $\delta(v, u) = 0$ otherwise. Larger modularity indicates that the subgraphs inside communities are denser, which follows the intuition that a community is a dense well-connected cluster. Then, the problem is to find a partition that maximizes the modularity Q.

However, a hard clustering on modularity maximization is proved to be NP-hard. Therefore, they relax the problem to a soft case. Let $\mathbf{d} \in \mathbb{Z}_+^{|\mathcal{V}|}$ denotes the degrees of all nodes and $\mathbf{S} \in \{0, 1\}^{|\mathcal{V}| \times n}$ denotes the community indicator matrix where $\mathbf{S}_{vc} = 1$ indicates node v belongs to community c and $\mathbf{S}_{vc} = 0$ otherwise. Then, we define modularity matrix $\mathbf{B}$ as

$$\mathbf{B} = \mathbf{A} - \frac{\mathbf{d}\mathbf{d}^\top}{2|\mathcal{E}|}, \tag{6.12}$$

and modularity Q can be reformulated as

$$Q = \frac{1}{2|\mathcal{E}|} \text{tr}(\mathbf{S}^\top \mathbf{B} \mathbf{S}). \tag{6.13}$$

By relaxing $\mathbf{S}$ to a continuous matrix, it has been proved that the optimal solution of $\mathbf{S}$ is the top-n eigenvectors of modularity matrix $\mathbf{B}$ [70]. Then, the eigenvectors can be used as node representations.

To conclude, spectral clustering-based methods often define a cost function that is linear or quadratic to the node representations. Then, the problems can be reformulated as a matrix form, and then solved by calculating the eigenvectors of the matrix. However, the computation of eigenvectors for large-scale matrices is both time- and space-consuming, limiting these methods from being applied in real-world scenarios.

6.3.2 Shallow Neural Networks

With the success of word2vec [68], many works resort to shallow neural networks for node representation learning. Typically, each node is assigned a vector of trainable parameters as its representation, and the parameters are trained by optimizing a certain objective via gradient descent.

DeepWalk DeepWalk [81] proposes a novel approach that introduces neural network techniques into graph representation learning for the first time. Compared with the aforementioned methods based on eigenvector computation, DeepWalk provides a faster way to learn low-dimensional node representations. The basic idea of DeepWalk is to adapt the well-known word representation learning algorithm word2vec [68] by regarding nodes as words and random walks as sentences.

Formally, given a graph $\mathscr{G} = (\mathscr{V}, \mathscr{E})$, DeepWalk uses node v to predict its neighboring nodes in short random walk sequences $u_{-w}, \ldots, u_{-1}, u_1, \ldots, u_w$ where w is the window size of skip-gram [68], which can be formulated as

$$\min \sum_{j=-w, j \neq 0}^{w} -\log P(u_j|v), \qquad (6.14)$$

where it assumes the prediction probabilities of each node are independent and the overall loss integrates the losses of all nodes in every random walk.

DeepWalk assigns each node v with two representations: node representation $\mathbf{v} \in \mathbb{R}^d$ and context representation $\mathbf{v}^c \in \mathbb{R}^d$. Then, each probability $P(u|v)$ is formalized as a Softmax function over all nodes:

$$P(u|v) = \frac{\exp(\mathbf{v}^\top \mathbf{u}^c)}{\sum_{u' \in \mathscr{V}} \exp(\mathbf{v}^\top \mathbf{u}'^c)}. \qquad (6.15)$$

LINE LINE [102] proposes a graph representation model which can handle large-scale graphs with arbitrary types: (un)directed or weighted. To characterize the interaction between nodes, LINE models the first-order proximity and second-order proximity of the graph.

Before we introduce the details of the algorithm, we can move one step back and see how the idea works. The modeling of first-order proximity, i.e., observed links, is the modeling of the adjacency matrix. As the adjacency matrix is usually too sparse, the modeling of second-order proximity, i.e., nodes with shared neighbors, can serve as complementary information to enrich the adjacency matrix and make it denser.

Formally, first-order proximity between node v and u is defined as the edge weight $\mathbf{A}_{vu}$ in the adjacency matrix. If nodes v and u are not connected, then the first-order proximity between them is 0.

Second-order proximity between node v and u is defined as the similarity between their neighbors. Let the row of node v in the adjacency matrix $\mathbf{A}(v, :)$ denote the first-order proximity between node v and other nodes. Then, the second-order proximity between v and u is defined as the similarity between $\mathbf{A}(v, :)$ and $\mathbf{A}(u, :)$. If they have no shared neighbors, the second-order proximity is 0.

To approximate first-order proximity $\hat{P}_1(v, u) = \frac{\mathbf{A}_{vu}}{\sum_{(v',u')\in\mathscr{E}}\mathbf{A}_{v'u'}}$, the joint probability between v and u is defined as

$$P_1(v, u) = \frac{1}{1 + \exp(-\mathbf{v}^\top\mathbf{u})}. \tag{6.16}$$

To approximate second-order proximity $\hat{P}_2(u|v) = \frac{\mathbf{A}_{vu}}{\sum_{u'}\mathbf{A}_{vu'}}$, the probability that node u appears in v's context $P_2(u|v)$ is defined in the same form as Eq. (6.15). In this way, the second-order relationship between node representations is bridged by the context representations of their shared neighbors.

For parameter learning, the distances between $\hat{P}_1(v, u)$ and $P_1(v, u)$ as well as $\hat{P}_2(u|v)$ and $P_2(u|v)$ are minimized. In specific, LINE learns node representations for first-order and second-order proximities individually, and then concatenates them as output embeddings. Although LINE can effectively capture both first-order and second-order local topological information, it cannot be easily extended to capture higher-order global topological information.

Connection with Matrix Factorization We prove that DeepWalk algorithm with the skip-gram model is actually factoring a matrix $\mathbf{M}$ where each entry $\mathbf{M}_{ij}$ is the logarithm of the average probability that node v_i randomly walks to node v_j in fixed steps [127]. Intuitively, the matrix $\mathbf{M}$ is much denser than the adjacency or Laplacian matrices, and therefore can help representations encode more structural information. In practice, the entry value $\mathbf{M}_{ij}$ is estimated by random walk sampling. A more detailed proof can be found at our arxiv note [125]. Moreover, LINE is also proved to be equivalent to matrix factorization [83, 128], where the matrix is defined by the first- and second-order proximities.

To summarize, DeepWalk and LINE introduce shallow neural networks into graph representation learning. Thanks to the modeling ability of neural networks and the efficiency of shallow representations, these methods can outperform conventional graph representation learning methods such as spectral clustering-based algorithms, and are also efficient for large-scale graphs.

6.3.3 Matrix Factorization

Inspired by the connection between DeepWalk and matrix factorization, many research works turn to explore how to learn better node representations by regarding its learning phase as a matrix factorization problem. Note that the eigenvector decomposition in spectral clustering methods can also be seen as a special case of matrix factorization. Here we present two typical matrix factorization-based graph representation learning algorithms, GraRep [12] and TADW [127] in this subsection.

GraRep GraRep [12] directly follows the proof of matrix factorization form of DeepWalk. According to our proof [125], DeepWalk is actually factorizing a matrix $\mathbf{M}$ where $\mathbf{M} = \log \frac{\mathbf{A}+\mathbf{A}^2+\cdots+\mathbf{A}^K}{K}$. From the matrix factorization form, DeepWalk has considered the high-order topological information of the graph jointly. In contrast, GraRep proposes to regard different k-step information separately in graph representation learning, and can be divided into three steps:

- Calculate k-step transition probability matrix $\mathbf{A}^k$ for each $k = 1, 2, \ldots, K$.
- Obtain each k-step node representations.
- Concatenate all k-step node representations as the final node representations.

For the second step to obtain the k-step node representations, GraRep directly uses a typical matrix decomposition technique, i.e., SVD on $\mathbf{A}^k$. However, this algorithm is not very efficient, especially when k becomes large.

TADW As the first attributed network embedding algorithm where node features are also available for learning node representations, text-associated DeepWalk (TADW) [127] further generalizes the matrix factorization framework to take advantage of text information. As shown in Fig. 6.5, the main idea of TADW is to factorize node affinity matrix $\mathbf{M} \in \mathbb{R}^{|\mathcal{V}| \times |\mathcal{V}|}$ into the product of three matrices, $\mathbf{W} \in \mathbb{R}^{d \times |\mathcal{V}|}$, $\mathbf{H} \in \mathbb{R}^{d \times f_t}$, and text feature matrix $\mathbf{T} \in \mathbb{R}^{f_t \times |\mathcal{V}|}$, where d and f_t are the rank and feature dimensions, respectively. Then, TADW concatenates $\mathbf{W}$ and $\mathbf{HT}$ as $2d$-dimensional node representations $\mathbf{V} = \text{concat}(\mathbf{W}; \mathbf{HT})$.

Now the question is how to build node affinity matrix $\mathbf{M}$ and how to extract text feature matrix $\mathbf{T}$ from the text information. Following the proof of matrix factorization form of DeepWalk, TADW set node affinity matrix $\mathbf{M}$ to a trade-

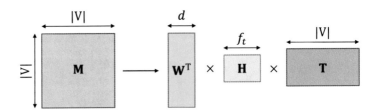

Fig. 6.5 Illustration of text-associated DeepWalk (TADW)

off between efficiency and effectiveness: factorizing the matrix $\mathbf{M} = (\mathbf{A} + \mathbf{A}^2)/2$ where $\mathbf{A}$ is the row-normalized adjacency matrix. In this way, $\mathbf{M}$ can encode both first-order and second-order proximities. For text feature matrix $\mathbf{T}$, TADW first constructs the TF-IDF matrix from the text of nodes, and then reduces the dimension of the TF-IDF matrix via SVD decomposition.

Formally, the model of TADW minimizes the following optimization function:

$$\min_{\mathbf{W},\mathbf{H}} ||\mathbf{M} - \mathbf{W}^\top \mathbf{H} \mathbf{T}||_F^2 + \frac{\lambda}{2}(||\mathbf{W}||_F^2 + ||\mathbf{H}||_F^2), \qquad (6.17)$$

where λ is the regularization factor and $|| \cdot ||_F$ is the Frobenius norm. The optimization of parameters is processed by updating $\mathbf{W}$ and $\mathbf{H}$ iteratively via conjugate gradient descent.

To summarize, the shallow graph representation learning methods have shown outstanding ability in modeling various kinds of graphs, such as large-scale [153] or heterogeneous [110] graphs. They are also used in many application scenarios such as recommendation systems [91, 129]. However, they still have several drawbacks [38]. Firstly, the model capacity of shallow methods is usually limited, and leads to suboptimal performance in complex scenarios. Secondly, the representations of different nodes share no parameters, which makes the number of parameters grow linearly with the number of nodes. This leads to the problem of computational inefficiency. Thirdly, the shallow graph representation methods are typically transductive, and thus cannot generalize to new nodes in a dynamic graph without retraining.

6.4 Deep Node Representation Learning

To address the problems of shallow node representations, researchers propose to utilize deep neural networks to aggregate information from graph structure. In this section, we introduce several typical solutions for node representation learning, including autoencoder-based methods, graph convolutional networks, graph attention networks, graph recurrent networks, graph Transformers, and several extensions based on them. Most deep node representation learning methods assume node features are available, and stack multiple neural network layers for representation learning. Here we denote the initial feature vector of node v as $\mathbf{x}_v$ and the hidden representation of node v at the k-th layer as $\mathbf{h}_v^k$.

6.4.1 Autoencoder-Based Methods

Different from previous methods that use shallow neural networks to characterize the graph representations, structural deep network embedding (SDNE) [109] employs a deep autoencoder to model more complex relationships between node representations. The main body of SDNE is a deep autoencoder whose input and output vectors are the initial node feature $\mathbf{x}_v$ and reconstructed feature $\hat{\mathbf{x}}_v$, respectively. The algorithm takes the representations from the intermediate layer as node embeddings and encourages the embeddings of connected nodes to be similar.

Formally, a deep autoencoder first compresses the input node feature into a low-dimensional intermediate vector and then tries to reconstruct the original input from the low-dimensional intermediate vector. Hence, the intermediate hidden representation can capture the information of the input node since we can recover the input features from them. Assume the input vector is $\mathbf{x}_v$, then the hidden representation of each layer k is defined as

$$
\begin{aligned}
\mathbf{h}_v^1 &= \mathrm{Sigmoid}(\mathbf{W}^1 \mathbf{x}_v + \mathbf{b}^1), \\
\mathbf{h}_v^k &= \mathrm{Sigmoid}(\mathbf{W}^k \mathbf{h}_v^{k-1} + \mathbf{b}^k), k = 2, 3 \ldots, K \ldots,
\end{aligned}
\tag{6.18}
$$

where $\mathbf{W}^k$ and $\mathbf{b}^k$ are weighted matrix and bias vector of the k-th layer. We assume that the hidden representation of the K-th layer has the minimum dimension, and the intermediate layer $\mathbf{h}_v^K$ can be seen as the low-dimensional representation of node v. Afterward, we can compute the output $\hat{\mathbf{x}}_v$ by applying the reversed calculation process on $\mathbf{h}_v^K$. The optimization objective of the autoencoder is to minimize the difference between input vector $\mathbf{x}_v$ and output vector $\hat{\mathbf{x}}_v$:

$$
\mathscr{L}_1 = \sum_{v \in \mathcal{V}} \|\hat{\mathbf{x}}_v - \mathbf{x}_v\|^2.
\tag{6.19}
$$

To encode the structure information, SDNE simply requires that the representations of connected nodes should be close to each other. Thus, the loss function is

$$
\mathscr{L}_2 = \sum_{(v,u) \in \mathcal{E}} \|\mathbf{h}_v^K - \mathbf{h}_u^K\|^2.
\tag{6.20}
$$

Finally, the overall loss function is $\mathscr{L} = \mathscr{L}_1 + \alpha \mathscr{L}_2$, where α is a harmonic hyperparameter. After the training process, $\mathbf{h}_v^K$ is taken as the representation of node v and used for downstream tasks.

Experimental results show that SDNE can effectively reconstruct the input graph and achieve better results in several downstream tasks. However, the deep neural network part of SDNE, i.e., the deep autoencoder, is isolated with graph structure during the feed-forward computation, which neglects high-order information interaction among nodes.

6.4.2 Graph Convolutional Networks

Graph convolutional networks (GCNs) aim to generalize convolutional operation from CNNs [55] to the graph domain. The success of CNNs comes from its local connection and multilayer [54] architectures, which may benefit graph modeling as well: (1) graphs are also locally connected; (2) multilayer architectures can help capture the hierarchical patterns in the graph. However, CNNs can only operate on regular Euclidean data like text (1D sequence) and images (2D grid), and cannot be directly transferred to the graph structure. In this subsection, we introduce how GCNs extend the convolutional operation to deal with the non-Euclidean graph data.

Mainstream GCNs usually adopt semi-supervised settings for training, while previous graph embedding methods are mostly unsupervised or self-supervised. Here we only introduce the encoder architectures of GCNs, and omit their loss functions which depend on downstream tasks. In specific, typical GCNs can be divided into spectral and spatial (nonspectral) approaches.

Spectral Methods From the signal processing perspective, the convolutional operation first transforms a signal to the spectral domain, then modifies the signal with a filter, and finally projects the signal back to the original domain [63]. Spectral GCNs follow the same process and define the convolution operator in the spectral domain of graph signals.

Formally, d-dimensional input representations $\mathbf{X} \in \mathbb{R}^{|\mathcal{V}| \times d}$ of a graph can be seen as d graph signals. Then, spectral GCNs are formulated as

$$\mathbf{H} = \mathcal{F}^{-1}(\mathcal{F}(\mathbf{g}) \odot \mathcal{F}(\mathbf{X})), \tag{6.21}$$

where $\mathbf{H}$ is the representations of all nodes, $\mathbf{g}$ is the filter in the spatial domain, $\mathcal{F}(\cdot)$ and $\mathcal{F}^{-1}(\cdot)$ indicate the graph Fourier transform (GFT) [9] and inverse GFT respectively, which can be defined as

$$\mathcal{F}(\mathbf{X}) = \mathbf{U}^\top \mathbf{X}, \mathcal{F}^{-1}(\mathbf{X}) = \mathbf{U}\mathbf{X}, \tag{6.22}$$

where $\mathbf{U}$ is the eigenvector matrix of the normalized graph Laplacian $\mathbf{L} = \mathbf{I}_{|\mathcal{V}|} - \mathbf{D}^{-\frac{1}{2}}\mathbf{A}\mathbf{D}^{-\frac{1}{2}}$. $\mathbf{A}$ is adjacency matrix and $\mathbf{D}$ is degree matrix.

In practice, we can use a learnable diagonal matrix $\mathbf{g}_\theta$ to approximate the spectral graph filter $\mathcal{F}(\mathbf{g})$, and the graph convolutional operation can be reformulated as

$$\mathbf{H} = \mathbf{U}\mathbf{g}_\theta\mathbf{U}^\top\mathbf{X}. \tag{6.23}$$

Intuitively, initial graph signals $\mathbf{X}$ are transformed into spectral domain by multiplying $\mathbf{U}^\top$. Then filter $\mathbf{g}_\theta$ performs the convolution, and $\mathbf{U}$ projects graph signals back to their original space. The above form of graph convolution is used in spectral network [10], the first spectral GCN method.

However, the original form in Eq. (6.23) has several limitations: (1) the filter $\mathbf{g}_\theta$ is not directly related to the graph structure; (2) the kernel size of the graph filter grows with the number of nodes in the graph, which may cause inefficiency and overfitting issues; (3) the calculation of $\mathbf{U}$ relies on computationally inefficient matrix factorization. Now we introduce two typical spectral GCNs improving the original formats, including ChebNet [23] and GCN [52].

ChebNet ChebNet [23] proposes to approximate $\mathbf{U}\mathbf{g}_\theta\mathbf{U}^\top$ by Chebyshev polynomials $\mathbf{T}_k(\cdot)$ up to K^{th}-order, which involve information within K-hop neighborhood. In this way, ChebNet does not need to compute matrix $\mathbf{U}$, and the number of learnable parameters is related to K instead of $|\mathcal{V}|$. Formally, the operation of ChebNet is reformulated as

$$\mathbf{H} = \sum_{k=0}^{K} \theta_k \mathbf{T}_k(\tilde{\mathbf{L}})\mathbf{X}, \tag{6.24}$$

where $\tilde{\mathbf{L}} = 2/\lambda_{max}\mathbf{L} - \mathbf{I}_{|\mathcal{V}|}$ is the normalized Laplacian matrix, λ_{max} is the largest eigenvalue of $\mathbf{L}$, and $\theta \in \mathbb{R}^K$ is a weight vector indicating Chebyshev coefficients. The Chebyshev polynomials are defined as

$$\begin{aligned} \mathbf{T}_0(\tilde{\mathbf{L}}) &= \mathbf{I}_{|\mathcal{V}|}, \\ \mathbf{T}_1(\tilde{\mathbf{L}}) &= \tilde{\mathbf{L}}, \\ \mathbf{T}_k(\tilde{\mathbf{L}}) &= 2\tilde{\mathbf{L}}\mathbf{T}_{k-1}(\tilde{\mathbf{L}}) - \mathbf{T}_{k-2}(\tilde{\mathbf{L}}). \end{aligned} \tag{6.25}$$

The Kth-order polynomial can be efficiently computed in a recursive manner, and the parameter number of the graph filter is reduced to K.

GCN GCN [52] is a first-order approximation of ChebNet. GCN reveals that ChebNet may suffer from the overfitting problem when handling graphs with very wide node degree distributions. Hence, GCN limits the maximum order of Chebyshev polynomials to $K = 1$, and the equation is simplified to the following form with two trainable scalars θ_0' and θ_1':

$$\begin{aligned} \mathbf{H} &= \theta_0'\mathbf{X} + \theta_1' \left(\mathbf{L} - \mathbf{I}_{|\mathcal{V}|}\right) \mathbf{X} \\ &= \theta_0'\mathbf{X} - \theta_1'\mathbf{D}^{-\frac{1}{2}}\mathbf{A}\mathbf{D}^{-\frac{1}{2}}\mathbf{X}. \end{aligned} \tag{6.26}$$

GCN further reduces the number of parameters to address overfitting by setting $\theta_0' = -\theta_1' = \theta$. And the equation is reformulated as

$$\mathbf{H} = \theta \left(\mathbf{I}_{|\mathcal{V}|} + \mathbf{D}^{-\frac{1}{2}}\mathbf{A}\mathbf{D}^{-\frac{1}{2}}\right) \mathbf{X}. \tag{6.27}$$

Fig. 6.6 Illustration of spatial GCNs. In the feed-forward computation, each node aggregates information from the representations of its neighbors and itself

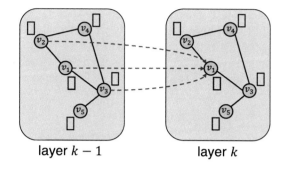

layer $k-1$ layer k

To summarize, spectral GCNs can effectively capture complex global patterns in graphs with spectral graph filters. However, the learned filters of the spectral methods usually depend on the Laplacian eigenbasis of the graph structure. This leads to the problem that a spectral-based GCN model learned on a specific graph cannot be directly transferred to another graph.

Spatial Methods Different from spectral GCNs, spatial GCNs define graph convolutional operation by directly aggregating information on spatially close neighbors, which is also known as the message-passing process. As shown in Fig. 6.6, the representation $\mathbf{h}_v^k$ of node v at k-th layer can be seen as a function aggregating the representations of v and its neighbors at $(k-1)$-th layer:

$$\mathbf{h}_v^k = f(\{\mathbf{h}_v^{k-1}\} \cup \{\mathbf{h}_u^{k-1} | \forall u \in \mathcal{N}_v\}), \tag{6.28}$$

where $\mathcal{N}_v$ is the neighbor set of node v.

The key challenge of spatial GCNs is how to define the convolutional operation to satisfy the nodes with different degrees while maintaining the local invariance of CNNs. In this subsection, we introduce two widely used spatial GCNs, including Neural FPs and GraphSAGE.

Neural FPs Neural FPs [27] propose to use different weight matrices for nodes with different-sized neighborhoods:

$$\mathbf{z}_v^k = \mathbf{h}_v^{k-1} + \sum_{u \in \mathcal{N}_v} \mathbf{h}_u^{k-1},$$

$$\mathbf{h}_v^k = \sigma(\mathbf{W}_{|\mathcal{N}_v|}^k \mathbf{z}^k), \tag{6.29}$$

where $\mathbf{W}_{|\mathcal{N}_v|}^k$ is the weight matrix for nodes with degree $|\mathcal{N}_v|$ at layer k and $\sigma(\cdot)$ is a nonlinear function such as Sigmoid. Neural FPs require learning weight matrices for all node degrees in the graph. Hence, when applied to large-scale graphs with diverse node degrees, it cannot capture the invariant information among different node degrees, and needs more parameters as node degrees get larger.

GraphSAGE GraphSAGE [37] transfers GCNs to handle the inductive setting, where the representations of new nodes should be computed without retraining. Instead of utilizing the full set of neighbors, GraphSAGE learns graph representations by uniformly sampling a fixed-size neighbor set from each node's local neighborhood. GraphSAGE can be formulated as

$$
\begin{aligned}
\mathbf{h}^k_{\mathcal{N}_v} &= \text{Aggregate}(\{\mathbf{h}^{k-1}_u | \forall u \in \mathcal{N}_v\}), \\
\mathbf{h}^k_v &= \sigma(\mathbf{W}^k \, \text{concat}(\mathbf{h}^{k-1}_v; \mathbf{h}^k_{\mathcal{N}_v})),
\end{aligned}
\tag{6.30}
$$

where $\mathcal{N}_v$ is the sampled neighbor set of node v and the aggregator functions Aggregate($\cdot$) usually utilize the following three types:

1. Mean aggregator. By utilizing a mean-pooling aggregator, GraphSAGE can be viewed as the inductive version of the original transductive GCN framework [52], which can be formulated as

$$
\mathbf{h}^k_v = \sigma \left(\mathbf{W}^k \cdot \text{Mean} \left(\{\mathbf{h}^{k-1}_v\} \cup \{\mathbf{h}^{k-1}_u | \forall u \in \mathcal{N}_v\} \right) \right).
\tag{6.31}
$$

2. Max-pooling aggregator. Max-pooling aggregator first feeds each neighbor's hidden representation into a fully connected layer and then utilizes a max-pooling operation to the obtained representations of the node's neighbors. It can be formulated as

$$
\mathbf{h}^k_{\mathcal{N}_v} = \text{Max}(\{\sigma(\mathbf{W}^k \mathbf{h}^{k-1}_u + \mathbf{b}^k) | \forall u \in \mathcal{N}_v\}).
\tag{6.32}
$$

3. LSTM aggregator. GraphSAGE also proposes to use an LSTM-based aggregator with a stronger expressive capability. Since LSTMs process inputs sequentially, GraphSAGE randomly permutes node v's neighbors to adapt LSTMs.

In summary, GCNs extend the idea of CNNs into the graph domain, enabling models to capture local connectivity of the graph, and have shown their superior abilities in a wide range of downstream tasks compared to the previous autoencoder-based methods. Even now, we can still get state-of-the-art performance by equipping vanilla GCNs with proper training strategies [48] or knowledge distillation methods [123, 124].

6.4.3 Graph Attention Networks

The attention mechanism has shown its strong ability to consider instance importance in learning representations in many NLP applications, such as machine translation [3, 33, 105] and machine reading [18]. Hence, many works [106, 146] focus on generalizing the attention mechanism to the graph domain and hope graph neural networks (GNNs) with attention-based operators can achieve better results

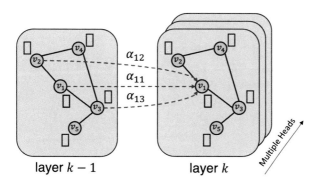

Fig. 6.7 Illustration of GATs. Compared with spatial GCNs, GATs assign different aggregation weights to different neighbors, and employ multiple parallel attention heads for computation

by considering the importance of neighbors. Figure 6.7 illustrates the architecture of graph attention networks (GATs).

GAT   GAT [106] proposes to adopt the self-attention mechanism for the information aggregation of GNNs. Specifically, each node's representation is calculated by attending to its neighbors:

$$\mathbf{h}_v^k = \sigma \left(\sum_{u \in \mathcal{N}_v \cup \{v\}} \alpha_{vu}^k \mathbf{W}^k \mathbf{h}_u^{k-1} \right), \tag{6.33}$$

where $\sigma(\cdot)$ is a nonlinear function and α_{vu}^k is the attention coefficient of node pair (v, u) at the k-th layer, which is normalized over node v's neighbors:

$$\alpha_{vu}^k = \frac{\exp \left(\text{LeakyReLU} \left(\mathbf{a}^\top \text{concat}(\mathbf{W}^k \mathbf{h}_v^{k-1}; \mathbf{W}^k \mathbf{h}_u^{k-1}) \right) \right)}{\sum_{m \in \mathcal{N}_v \cup \{v\}} \exp \left(\text{LeakyReLU} \left(\mathbf{a}^\top \text{concat}(\mathbf{W}^k \mathbf{h}_v^{k-1}; \mathbf{W}^k \mathbf{h}_m^{k-1}) \right) \right)}, \tag{6.34}$$

where $\mathbf{W}^k$ is the weight matrix of a shared linear transformation applied to every node, $\mathbf{a}$ is a learnable weight vector and $\text{LeakyReLU}(\cdot)$ is a nonlinear function.

Moreover, GAT utilizes the multi-head attention mechanism similar to [105] to further aggregate different types of information. Specifically, GAT concatenates (or averages) the output representations of M independent attention heads:

$$\mathbf{h}_v^k = \|_{m=1}^M \sigma \left(\sum_{u \in \mathcal{N}_v \cup \{v\}} \alpha_{vu}^{k,m} \mathbf{W}_m^k \mathbf{h}_u^{k-1} \right), \tag{6.35}$$

where $\alpha_{vu}^{k,m}$ is the attention coefficient from the m-th attention head at the k-th layer, $\mathbf{W}_m^k$ is the transform matrix for the m-th attention head, and $\|$ is the concatenation operation.

To summarize, by incorporating the attention mechanism into the information aggregating phase of spatial GCNs, GATs can assign proper weights to different neighbors and offer better interpretability. The ensemble of multiple attention heads further increases the model capacity and brings performance gains over GCNs.

6.4.4 Graph Recurrent Networks

To capture the dependency between two distant nodes by a graph encoder, one has to stack many GNN layers so that the information can propagate from one to another. However, stacking too many GNN layers in the feed-forward computation will cause the over-smoothing issue, which makes node representations less discriminative and harms the performance. Inspired by the success of gated recurrent unit (GRU) [19] and LSTM [42] in modeling long-term dependency in NLP, graph recurrent networks (GRNs) propose to equip the information aggregation of GCNs with gate mechanisms. Similar to the usage in RNNs, the gate mechanisms allow the information to propagate farther without severe gradient vanishment or over-smoothing issues. In this way, GRNs can improve the model's ability in capturing the long-range dependency across the graph. In this subsection, we introduce several variants of GRNs, including GGNN, Tree-LSTM, and Graph LSTM.

Gated Graph Neural Network (GGNN) GGNN [57] introduces a GRU-like function to improve the information propagation of the vanilla GCN architecture. In each layer, GGNN updates the representations of nodes by combining the information of their neighbors and themselves with the update and reset gates. Specifically, the recurrence of each layer in GGNN is defined as

$$\mathbf{a}_v^k = \sum_{u \in \mathcal{N}_v} \mathbf{h}_u^{k-1} + \mathbf{b},$$

$$\mathbf{z}_v^k = \text{Sigmoid}\left(\mathbf{W}_z \mathbf{a}_v^k + \mathbf{U}_z \mathbf{h}_v^{k-1}\right),$$

$$\mathbf{r}_v^k = \text{Sigmoid}\left(\mathbf{W}_r \mathbf{a}_v^k + \mathbf{U}_r \mathbf{h}_v^{k-1}\right), \tag{6.36}$$

$$\widetilde{\mathbf{h}}_v^k = \tanh\left(\mathbf{W} \mathbf{a}_v^k + \mathbf{U}(\mathbf{r}_v^k \odot \mathbf{h}_v^{k-1})\right),$$

$$\mathbf{h}_v^k = (1 - \mathbf{z}_v^k) \odot \mathbf{h}_v^{k-1} + \mathbf{z}_v^k \odot \widetilde{\mathbf{h}}_v^k,$$

where $\mathbf{a}_v^k$ represents node v's neighborhood information, $\mathbf{b}$ is the bias vector, $\widetilde{\mathbf{h}}_v^k$ is the candidate representation, $\mathbf{z}$ and $\mathbf{r}$ are the update and reset gates, and $\mathbf{W}$ and $\mathbf{U}$ are weight matrices.

Typical RNNs can also be seen as a special case of GGNN, where the graph is a chain structure. Hence, GRNs like GGNN have been widely used in language models. Note that tree structure is very popular in text data, such as the dependency parsing tree. Next, we introduce Tree-LSTM [101], which extends GRNs to model the tree structure.

Tree-LSTM Tree-LSTM [101] uses an LSTM-based unit with input/output gates $\mathbf{i}_v/\mathbf{o}_v$ and memory cell $\mathbf{c}_v$ to update representation $\mathbf{h}_v$ of each tree node v. There are two variants of Tree-LSTM: Child-sum Tree-LSTM and N-ary Tree-LSTM. Instead of using a unified forget gate like LSTM, Child-sum Tree-LSTM assigns a forget gate $\mathbf{f}_{vm}$ for each child m of node v, which allows tree node v to adaptively gather information from its children. N-ary Tree-LSTM requires each node to have at most N children, and assigns different learnable parameters for each child. Child-sum Tree-LSTM is suitable for trees whose children are unordered, and thus can be used for modeling dependency trees. N-ary Tree-LSTM can characterize the diverse relational information for each node's children, and thus is usually used to model constituency parsing tree structure. Here we only present the formulas of Child-sum Tree-LSTM:

$$
\begin{aligned}
\widetilde{\mathbf{h}}_v^{k-1} &= \sum_{m \in \mathcal{N}_v^c} \mathbf{h}_m^{k-1}, \\
\mathbf{i}_v^k &= \text{Sigmoid}\left(\mathbf{W}_i \mathbf{x}_v + \mathbf{U}_i \widetilde{\mathbf{h}}_v^{k-1} + \mathbf{b}_i\right), \\
\mathbf{f}_{vm}^k &= \text{Sigmoid}\left(\mathbf{W}_f \mathbf{x}_v + \mathbf{U}_f \mathbf{h}_m^{k-1} + \mathbf{b}_f\right), \\
\mathbf{o}_v^k &= \text{Sigmoid}\left(\mathbf{W}_o \mathbf{x}_v + \mathbf{U}_o \widetilde{\mathbf{h}}_v^{k-1} + \mathbf{b}_o\right), \\
\mathbf{u}_v^k &= \tanh\left(\mathbf{W}_u \mathbf{x}_v + \mathbf{U}_u \widetilde{\mathbf{h}}_v^{k-1} + \mathbf{b}_u\right), \\
\mathbf{c}_v^k &= \mathbf{i}_v^k \odot \mathbf{u}_v^k + \sum_{m \in \mathcal{N}_v} \mathbf{f}_{vm}^k \odot \mathbf{c}_m^{k-1}, \\
\mathbf{h}_v^k &= \mathbf{o}_v^k \odot \tanh(\mathbf{c}_v^k),
\end{aligned}
\tag{6.37}
$$

where $\mathcal{N}_v^c$ is the children set of node v and $\mathbf{x}_v$ is the input representation for tree node v. Readers can refer to the original paper [101] for the details of N-ary Tree-LSTM variant.

Graph LSTM Graph LSTM [78, 144] proposes to adapt Tree-LSTM to model the graph structure, and utilizes different weight matrices to represent different labels on the edges. Formally, assume that the edge label between node v and its child m is l. Compared with Eq. (6.37) in Tree-LSTM, Graph LSTM uses label-specific weight matrix $\mathbf{U}_l$ to compute relevant gates and hidden states.

In summary, GRNs with gate mechanisms can effectively model the long-range dependencies between distant nodes, which is very important in text modeling. By adapting to tree-structure data, GRNs can also handle the diverse syntactic and semantic relations in text.

6.4.5 Graph Transformers

Transformer [105] has set off a craze in both NLP and CV areas [26, 84]. Based on the powerful self-attention architecture, graph Transformer networks [53, 87, 138] have been proposed to improve the expressive ability of GNNs. The core idea is to leverage the Transformer architecture to capture long-range relationships between distant nodes. Compared with GRNs, graph Transformers can benefit the advantages of Transformers against LSTMs.

Connections with Transformers in Text Modeling In graph Transformers, nodes are the basic units instead of words, and self-attention mechanism is then performed between all node pairs. In this way, all nodes are directly connected regardless of the original graph structure. To utilize the original topology information, current graph Transformers focus on the modifications of input features and attention coefficients. Other operations including multi-head ensembling, feed-forward network, and layer normalization remain unchanged. Now we will introduce three typical graph Transformers, including Graphormer [138], GraphTrans [116], and SAT [15].

Graphormer Graphormer [138] designs three structural encoding modules to inject graph structure information into the Transformer architecture. In specific, *centrality encoding* adds node degrees to the input which can indicate the importance of different nodes:

$$\mathbf{h}_v^0 = \mathbf{x}_v + \mathbf{z}_{\deg^-(v)}^- + \mathbf{z}_{\deg^+(v)}^+, \tag{6.38}$$

where $\mathbf{x}_v$ is the feature vector of node v and $\mathbf{z}^-$, $\mathbf{z}^+$ are learnable embedding vectors indexed by node in-degree and out-degree.

Spatial encoding of node distances and edge encoding of edge features serve as bias terms for the attention coefficients in the self-attention layer:

$$\alpha_{vu}^k = (\mathbf{W}_Q \mathbf{h}_v^{k-1})^\top (\mathbf{W}_K \mathbf{h}_u^{k-1})/\sqrt{d} + b_{\phi(v,u)} + c_{vu}, \tag{6.39}$$

where α_{vu}^k is the attention coefficient between node v and u in the self-attention layer, $\mathbf{W}_Q, \mathbf{W}_K$ are weight matrices, d is the hidden dimension, $b_{\phi(v,u)}$ is the learnable scalar indexed by the distance $\phi(v, u)$ between v and u, and c_{vu} is the scalar derived from the edge features between v and u.

To take advantage of deep models in utilizing structure information, Graph-Trans [116] and SAT [15] propose to integrate GNN and Transformer architectures for modeling.

GraphTrans GraphTrans [116] directly stacks a Transformer module on the top of a GNN module. In other words, the output node representations of the GNN are used as the input features of the Transformer. GraphTrans adopts a special [CLS] token which connects to all other nodes, and the representation of [CLS] token after the Transformer is taken as the graph representation. Hence, the Transformer can also be seen as a pooling operator for GNN. As a result, GraphTrans can capture both the local structured information and long-range relationships on graphs at the same time.

SAT SAT [15] proves that modifying the position encoding module in standard Transformers could not fully capture the structural similarity between nodes on a graph. To address this problem, SAT defines attention coefficients in the Transformer by the similarity of GNN-based representations. In this way, GNN serves as a structure extractor to integrate more information into self-attention layers.

In summary, graph Transformer architecture can model long-range relationships on a graph and go beyond the limitations of traditional deep GNNs, such as oversmoothing. Compared with GNNs limited by the Weisfeiler-Lehman test [119], the Transformer-based methods become more expressive. Besides, GNNs can also be integrated into graph Transformers to better utilize the graph topology information.

6.4.6 Extensions

In this subsection, we will talk about several typical extensions of GNNs, including skip connection, neighborhood sampling, and the modeling of diverse graph types, which can respectively improve the effectiveness, efficiency, and generalizability of GNNs.

GNNs with Skip Connection Theoretically, we can enhance the expressive ability by stacking more layers of GNNs. However, existing works find that deeper GNNs do not perform better in downstream tasks and even perform worse [52]. Chen et al. [14] further attribute this phenomenon to the low information-noise ratio received by the nodes in deep GNNs. The residual network [40], which has been verified in the computer vision community, is a straightforward solution to the problem. But researchers find that deep GNNs with residual connections still perform worse compared to the two-layer GNNs. Therefore, Rahimi et al. [85] and Xu et al. [120] further explore how to enhance the performance of GNNs with skip connections. Inspired by the idea from the highway network [160], Rahimi et al. [85] employ the layer-wise gate mechanism, and the performance can peak at four layers. Formally, the Highway GCN can be defined as

$$
\begin{aligned}
\mathbf{T}(\mathbf{h}_v^{k-1}) &= \text{Sigmoid}\left(\mathbf{W}^{k-1}\mathbf{h}_v^{k-1} + \mathbf{b}^{k-1}\right), \\
\mathbf{h}_v^k &= \mathbf{h}_v^k \odot \mathbf{T}(\mathbf{h}_v^{k-1}) + \mathbf{h}_v^{k-1} \odot (1 - \mathbf{T}(\mathbf{h}_v^{k-1})).
\end{aligned}
\tag{6.40}
$$

Besides, Xu et al. [120] present the jump knowledge network, which selects representations from all intermediate layers as final node representations. The selecting mechanism enables the jump knowledge network to adaptively pick the reasonable neighborhood information for each node.

GNNs with Neighborhood Sampling The vanilla GNN [89] has several limitations: (1) the computation is based on the entire graph Laplacian matrix, and thus computationally expensive for large graphs; (2) it is trained and specialized for a given graph, and cannot be transferred to another graph. To address the problems, the aforementioned GraphSAGE [37] first samples neighborhood nodes of a target node, and then aggregates the representations of all sampled nodes. Thus, GraphSAGE can get rid of the graph Laplacian and can be applied to unseen nodes. Ying et al. [139] further propose the importance-based sampling method PinSage, which simulates random walks starting from target nodes and samples the neighborhood set according to the normalized visit counts. Instead of sampling neighbors for each node, Chen et al. [16] propose FastGCN, which directly samples the receptive field for each layer. In other words, only sampled nodes can participate in layer-wise information propagation. FastGCN sets the sampling importance of nodes according to their degrees, and tends to keep the nodes with larger degrees. Besides, Huang et al. [45] introduce a parameterized and trainable sampler, which performs layer-wise sampling based on the previous layer.

GNNs for Graphs of Diverse Types The vanilla GNN [89] is designed for undirected graphs, the simplest graph format. However, as shown in Fig. 6.8, we have graphs of diverse types in real-world scenarios. Now we will introduce several extensions of GNNs to deal with directed, heterogeneous, or dynamic graphs:

Directed Graphs Directed edges contain extra information compared to undirected ones. For example, a famous person in a social network may be followed by lots of other users. Usually, her influence is stronger than most of her followers. This suggests that GNNs should treat the information propagation process in two edge

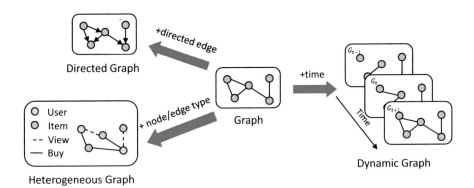

Fig. 6.8 Diverse types of graphs

directions differently. DGP [49] defines two graphs where nodes are respectively connected to all their ancestors or descendants. We correspondingly denote the normalized adjacency matrices as $\mathbf{D}_p^{-1}\mathbf{A}_p$ and $\mathbf{D}_c^{-1}\mathbf{A}_c$. The encoder of DGP can be formulated as

$$\mathbf{H}^k = \sigma(\mathbf{D}_p^{-1}\mathbf{A}_p\sigma(\mathbf{D}_c^{-1}\mathbf{A}_c\mathbf{H}^{k-1}\mathbf{W}_c)\mathbf{W}_p), \tag{6.41}$$

where $\mathbf{H}^k$ is the representation matrix of all nodes in the k-th layer, $\mathbf{W}_p$ and $\mathbf{W}_c$ are weight matrices, and $\sigma(\cdot)$ is a nonlinear function. In this way, the information propagations of two directions are processed separately.

Heterogeneous Graphs A heterogeneous graph [92] has several kinds of nodes. For example, for the graph in the shopping recommendation system, we have user nodes, item nodes, etc. The simplest way to process heterogeneous graphs is to consider the node type information in their input features, i.e., convert the node type information into a one-hot feature vector and concatenate it to the original features. Besides the simple feature-based approach, GraphInception [150] introduces the concept of meta-path into the information propagation on heterogeneous graphs. GraphInception utilizes GNNs to perform information propagation on homogeneous subgraphs, which are extracted based on human-designed meta-paths. At last, it concatenates the outputs from different homogeneous GNNs as final node representations. Wang et al. [111] further propose the heterogeneous graph attention network (HAN) with node-level and meta-path-level attention mechanisms. In addition, some works [112, 154] also consider the modeling of network schema [99], which is a meta template of a heterogeneous graph indicating node types and their relations.

Besides, there are many graphs containing edges with weights or types as additional information. A typical way to handle such graphs is to build a bipartite graph, where the original edges are converted into nodes linking to the original endpoint nodes, and the type information is thus converted to node type information. Another way is to assign different propagation weight matrices for different edge types. However, if a graph has lots of edge types, the parameter numbers will be large. To address the problem, R-GCN [90] introduces two regularization tricks that can decompose the transformation matrix W_r of type r to a set of base transformations shared among all edge types. Thus, R-GCN can reduce the number of parameters and capture the relationship between different edge types.

Dynamic Graphs Graphs are usually dynamic and vary over time. For example, a user in a social network may newly follow another user. To model the graph structure changing over time, DCRNN [56] and STGCN [143] first capture the static graph information at each time step by GNNs and then feed the output representation into a sequence model like RNNs. In addition, Structural-RNN [47] and ST-GCN [122] extend static graph structure with temporal connections and apply conventional GNNs on the extended graphs, which can capture structural and temporal information at the same time. MetaDyGNN [131] combines GNNs with meta-learning for few-shot link prediction in dynamic graphs.

6.5 From Node Representation to Graph Representation

In previous sections, we introduce how to represent nodes in a graph, from shallow node representation to deep node representation. In many scenarios, we also need to compute the representation of an entire graph or a specific subgraph. Inspired by the pooling operation in NLP and CV areas, graph pooling is then designed for obtaining the graph representation from node representations. Here we present two typical groups of graph pooling methods, namely, flat pooling and hierarchical pooling.

6.5.1 Flat Pooling

Flat pooling assumes a flat graph structure to generate graph representation, which includes max/mean/sum pooling as simple node pooling methods, and SortPooling [147] considering node ordering of structural roles.

Simple Node Pooling Similar to the pooling operation in NLP and CV, we can directly apply node-wise max/mean/sum operators on top of node representations for graph pooling. The graph representation can be formulated as

$$
\begin{cases}
\max_{v \in \mathcal{V}} \mathbf{v}, & \text{(Graph max-pooling)} \\[2mm]
\dfrac{1}{|\mathcal{V}|} \sum_{v \in \mathcal{V}} \mathbf{v}, & \text{(Graph mean-pooling)} \\[2mm]
\sum_{v \in \mathcal{V}} \mathbf{v}. & \text{(Graph sum-pooling)}
\end{cases}
\tag{6.42}
$$

The above pooling operators are general and parameter-free, but completely neglect the graph structure.

SortPooling SortPooling [147] first sorts node representations by their structural roles, which enables a consistent node ordering in different graphs and makes it possible to train typical neural networks on sorted node representations for pooling. In particular, SortPooling feeds the sorted representations into a 1-D CNN to get the graph representation, and makes the graph pooling operation to keep more information of global graph topology.

Though simple and effective, flat pooling ignores the hierarchical structure of nodes in a graph, e.g., nodes, subgraphs, and communities, thus leading to suboptimal graph representations.

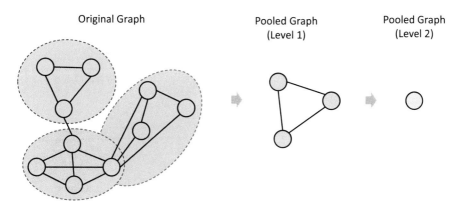

Fig. 6.9 The illustration of DiffPool. The figure is redrawn from Figure 1 of [140]

6.5.2   Hierarchical Pooling

The basic idea of hierarchical pooling is to group structure-related nodes into clusters to form a subgraph recursively, and obtain the graph representation layer by layer. Next, we will introduce two typical hierarchical pooling operations.

DiffPool DiffPool [140] proposes to learn hierarchical representations at the top of node representations and can be combined with various node representation learning methods in an end-to-end fashion. As shown in Fig. 6.9, DiffPool learns a differentiable soft cluster assignment for each node, and then maps nodes to a set of clusters layer by layer.

Formally, let $\mathbf{S}^k \in \mathbb{R}^{C_k \times C_{k+1}}$ denote the learned cluster assignment matrix at the k-th layer, where $\mathbf{S}^k_{vc}$ indicates whether node v belongs to cluster c at k-th layer, and C_k is the number of clusters in each layer. With the cluster assignment matrix $\mathbf{S}^k$, we can then calculate the adjacency matrix $\mathbf{A}^{k+1} \in \mathbb{R}^{C_{k+1} \times C_{k+1}}$ for the next layer by the connectivity strength between learned clusters in $\mathbf{S}^k$:

$$\mathbf{A}^{k+1} = \mathbf{S}^{k\top} \mathbf{A}^k \mathbf{S}^k. \tag{6.43}$$

Then the output node representations $\mathbf{H}^{k+1}$ are computed by GNN encoder:

$$\mathbf{H}^{k+1} = \mathrm{GNN}(\mathbf{A}^{k+1}, \mathbf{X}^{k+1}), \tag{6.44}$$

where input node representations $\mathbf{X}^{k+1}$ are obtained by aggregating the output representations from the k-th layer:

$$\mathbf{X}^{k+1} = \mathbf{S}^{k\top} \mathbf{H}^k. \tag{6.45}$$

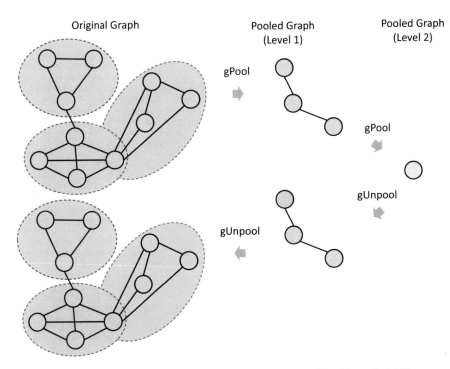

Fig. 6.10 Illustration of gPool and gUnpool. The figure is redrawn from Figure 2 of [32]

DiffPool predefines the number of clusters for each layer, and applies another GNN on the coarsened adjacency matrix $\mathbf{A}^k$ to generate the soft cluster assignment matrix $\mathbf{S}^k$. Finally, DiffPool feeds the top-layer graph representation into a downstream task classifier for the supervision of cluster assignment matrices.

gPool As shown in Fig. 6.10, gPool [32] presents both graph pooling (gPool) and unpooling (gUnpool) operations, based on which the graph data is modeled by an encoder-decoder architecture. The encoder/decoder includes the same number of encoder/decoder blocks, and each encoder/decoder block will contain a GCN layer and a gPool/gUnpool operator. The representations after the last decoder block are used as final representations for downstream tasks.

The gPool operation learns projection scores for each node with a learnable projection vector, and selects nodes with the highest scores as important ones to feed to the next layer:

$$\mathbf{s}^k = \mathbf{X}^k \mathbf{p}^k / ||\mathbf{p}^k||,$$

$$idx^k = \text{rank}(\mathbf{s}^k, n^k), \tag{6.46}$$

where $\mathbf{s}^k$ is the importance score; $\mathbf{p}^k$ and $\mathbf{X}^k$ are the projection vector and input feature matrix in the k-th layer, respectively; and $\text{rank}(\mathbf{s}^k, n^k)$ returns the indices of the elements with top-n^k scores in $\mathbf{s}^k$.

Then in each layer, we define the adjacency matrix and input feature matrix based on the corresponding rows or columns indexed by idx^k:

$$\mathbf{A}^{k+1} = \mathbf{A}^k(idx^k, idx^k),$$
$$\hat{\mathbf{s}}^k = \text{Sigmoid}(\mathbf{s}^k(idx^k)),$$
$$\hat{\mathbf{X}}^k = \mathbf{X}^k(idx^k, :),$$
$$\mathbf{X}^{k+1} = \hat{\mathbf{X}}^k \odot (\hat{\mathbf{s}}^k \mathbf{1}_d^\top), \tag{6.47}$$

where $\mathbf{1}_d$ is a d-dimensional all-one vector and $\odot$ is the element-wise matrix multiplication. Here the normalized scores $\hat{\mathbf{s}}^k$ are used as weighted masks to further filter the feature matrix $\hat{\mathbf{X}}^k$.

The gUnpool performs the inverse operation of the gPool operation, which restores the graph to its original structure. Specifically, gUnpool records the indices of selected nodes in the corresponding pooling level and then simply places nodes back in their original positions:

$$\mathbf{X}^{k-1} = \text{distribute}(\mathbf{0}_{n^{k-1} \times d}, \mathbf{X}^k, idx^k), \tag{6.48}$$

where idx^k is the indices of n^k selected nodes in the corresponding gPool level, and the function places row vectors in $\mathbf{X}^k$ into $n^{k-1} \times d$ all-zero feature matrix by index idx^k.

Compared to DiffPool, gPool can reduce the storage complexity by replacing the cluster assignment matrix with a projection vector at each layer.

In this section, we introduce how to obtain global graph representation based on local node representations with graph pooling operations, which are widely used in a series of graph-level tasks such as graph classification and interaction prediction. When applying GNNs in modeling text, graph pooling can effectively help us extract sentence-level [72, 152] or document-level [23, 77] information for downstream tasks.

6.6 Self-Supervised Graph Representation Learning

Recently, self-supervised learning methods, which have made immense success in CV and NLP areas, can learn effective representations with well-designed pre-training tasks instead of expensive downstream task labels. Specifically, self-supervised graph representation learning [60] first learns node and graph representations by different graph-based pre-training tasks without human supervision, such as graph structure reconstruction or pseudo-label prediction. Then, the learned models

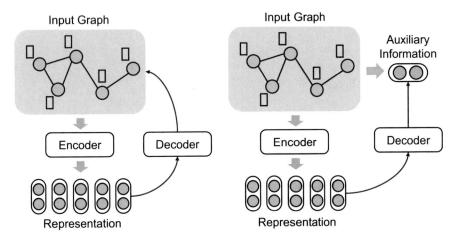

Fig. 6.11 Illustration of generative (left) and predictive (right) methods

or representations can be used in downstream tasks (e.g., graph/node classification). As shown in Figs. 6.11 and 6.12, we will introduce three types of self-supervised graph representation learning methods, namely, generative, predictive, and contrastive, distinguished by different pre-training tasks.

Generative Methods The generative methods aim to reconstruct and predict some components (e.g., graphs, edges, node features) of input data.

Structure Reconstruction These works aim to recover the adjacency matrix or masked edges of a graph. Graph autoencoder (GAE) [51] learns node representations by a two-layer graph convolutional network and then reconstructs the adjacency matrix of the input graph, and variational graph autoencoder (VGAE) [51] is a latent variable variant of GAE. ARGA and ARVGA [75] are adversarial variants of GAE and VGAE, respectively, combining autoencoder and adversarial approaches. AGE [21] adaptively defines the reconstruction objective of adjacency matrices in an iterative manner.

Feature Reconstruction These works aim to recover the node attributes of a graph, i.e., the input features of GNNs. MGAE [108] takes both corrupted network node content and structures as input, and predicts the origin node features. GALA [76] proposes a symmetric graph convolutional autoencoder based on Laplacian sharpening and smoothing to learn node representations by predicting input features. Here the Laplacian sharpening encourages the reconstructed feature of a node to be far away from those of its neighbors. GPT-GNN [44] uses both graph and feature generation pre-training tasks to model the structural and semantic information of the graph.

Predictive Methods Predictive methods learn informative representations with self-supervised signals from some auxiliary information, such as pseudo labels and

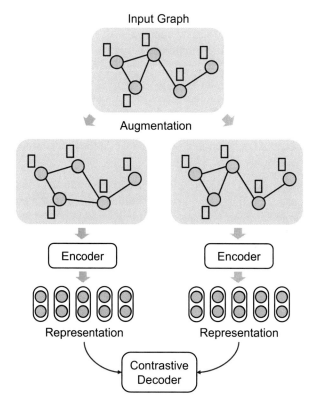

Fig. 6.12 Illustration of contrastive methods

graph properties. Depending on the prediction targets, the predictive methods can be further divided into three types.

Property Prediction These methods manually define some high-level information of the graph for prediction. GROVER [87] uses a graph Transformer as the encoder and employs both contextual property prediction and graph-level motif prediction to encode domain knowledge. S^2GRL [79] takes the k-hop contextual prediction as a pre-training task and trains a well-designed GNN to learn representations. SELAR [46] predicts the meta-paths, which are composite relations of multiple edge types in heterogeneous graphs.

Pseudo-Label Prediction This line of works employs an iterative framework to learn representations and update cluster labels. M3S [98] employs DeepCluster [13] algorithm to generate pseudo cluster labels, and designs an aligning mechanism and a multistage training framework to predict refined pseudo labels. IFC-GCN [43] proposes an EM-like framework to alternately rectify pseudo labels of feature clustering, and update node features by predicting pseudo labels.

Invariant Information Preserving These works aim to preserve some intrinsic and invariant information in the graph. Compared with the property prediction-based methods, there is usually no explicit meaning or closed-form formula for the *invariant information*. CCA-SSG [145] employs the canonical-correlation analysis (CCA) method to maximize the correlation between two augmented views of the same input, and thus preserve augmentation-invariant information. Lagraph [118] assumes that there exists a latent graph without noises behind each observed graph, and the observed graph is randomly generated from the latent one. Lagraph implicitly predicts the latent graph as a pre-training task to learn informative graph and node representations.

Contrastive Methods Contrastive methods first generate two contrastive views from graphs/nodes, and then maximize the mutual information between them. In this way, the learned representations will be more robust to perturbations. Typically, contrastive methods regard the views from the same graph/node as a positive pair, and the views randomly selected from different graphs/nodes as negative pairs. The representation similarity between a positive pair is forced to be larger than negative ones. Categorized by the views to contrast, contrastive methods can be divided into two groups.

Substructure-Based Methods This line of works usually contrasts views between different scales of structures. InfoGraph [95] takes different substructures of the original graph (e.g., nodes, edges, triangles) as contrastive views to generate graph-level representations. DGI [107] and GMI [80] build contrast views between a graph and its nodes. MVGRL [39] generates views by sampling subgraphs, and learns both node- and graph-level representations. GCC [82] treats different subgraphs as contrastive views and introduces InfoNCE loss [73] to large-scale graph pre-training.

Augmentation-Based Methods These works usually generate views by applying different perturbations on input graphs and features. GRACE [158] and GraphCL [142] randomly perturb the input graph (e.g., node and edge dropping) to generate contrastive views. GCA [159] adaptively builds contrastive views based on different graph properties (e.g., node degree, eigenvector, PageRank). Here nodes or edges with small importance scores (e.g., node degrees) are more likely to be dropped in contrastive views. JOAO [141] and AD-GCL [100] automatically select graph augmentations and generate contrastive views in adversarial ways. Instead of contrasting augmented data, DSGC [132] conducts contrastive views from dual spaces of hyperbolic and Euclidean. GASSL [134] generates views by directly perturbing input features and hidden layers. SimGRACE [117] adds Gaussian noises to model parameters as perturbations to generate contrastive views. MA-GCL [34] proposes a novel augmentation strategy that randomly perturbs the neural architecture of GNN encoders (i.e., random permutations of graph filter, linear projection, and nonlinear function) as contrastive views. HeCo [112] employs network schema and meta-path views in heterogeneous graphs as two specific views.

Adaptation Approaches After the self-supervised training process, there are roughly three paradigms to adapt the learned models or representations for downstream tasks.

Pre-training-Fine-Tuning These methods [44, 87, 142] first train the model parameters of graph encoder on the datasets without labels in the self-supervised way, and then the pre-trained parameters are used as the initial parameters in the next fine-tuning step, which updates the encoder in a supervised way by downstream tasks.

Unsupervised Representation Learning These methods [51, 75, 79, 158] train the graph encoder with pre-training tasks in the first stage, and the pre-trained encoder is taken as a feature extractor with frozen parameters to generate representations for downstream tasks in the second stage.

Multitask Training These methods [46, 82, 98] train graph encoders on both pre-training and downstream tasks with well-designed loss functions, which can be seen as a type of multitask learning, where the pre-training tasks are auxiliary tasks for downstream ones.

In summary, self-supervised graph representation learning methods can learn effective graph and node representations based on various pre-training tasks without labels. Different from pre-trained language models where pre-training-fine-tuning are the mainstream for adaptation of downstream tasks, the unsupervised representation learning paradigm is widely used in graph data, where only the node/graph representations in the last feed-forward layer are fed into downstream tasks as features. An intuitive reason is that popular tasks on graph data (e.g., node/graph classification) require less model capacity than those on NLP (e.g., machine translation). Therefore, the final representations in graph encoders are usually sufficient, and it's not necessary to fine-tune the learned graph encoders.

6.7 Applications

In this section, we will introduce several typical applications of graph representation learning in the NLP area.

Text Classification Text classification is an essential task in NLP. Typical GNN models, including GCNs [2, 23, 37, 41, 52, 69] and GATs [106], are applied in text classification to model the structural information (e.g., citation relationship) between documents. However, these works did not fully model the structural information lying in texts. Thus, some works manage to construct graphs from texts. Peng et al. [77] propose to transform texts into a graph of words, and then apply graph convolutional operations to it. Yao et al. [135] propose to construct a text graph with document and word nodes, and utilize GCNs to learn representations for both word nodes and document nodes. Besides constructing the text graph with human heuristics as in the above works, there are also some intrinsic graph structures that can be used, such as dependency parsing trees, semantic parsing graphs, etc.

One of the most typical works of utilizing these intrinsic graph structures is the Tree-LSTM [101] introduced in this chapter.

Sequence Labeling Sequence labeling is another classical task in NLP, which aims to assign labels for each word in the text sequence. Sentence LSTM [151] introduces graph recurrent networks for sentence modeling, where each word node connects with its neighboring words in a context window and a global sentence node links with all word nodes. Then, the hidden representations of word nodes can be used for predicting word labels. Sentence LSTM achieves promising performance in the POS tagging and NER tasks. To solve the semantic role labeling task, Marcheggiani et al. [65] propose a variant of GCN [52] which performs reasoning on syntactic dependency trees (i.e., a graph with labeled edges), and employs an edgewise gate mechanism to consider the information of each dependency edge. They further show that GCNs and LSTMs are functionally complementary in this task.

Knowledge Acquisition As a crucial subtask for knowledge acquisition, relation extraction aims to predict the relationship between entities in plain text. While CNNs and RNNs have achieved promising results on multiple benchmarks, researchers find the syntactic and semantic information of the text, such as adjacency, syntactic dependencies, and discourse relations, are also helpful for relation extraction. To this end, Zhang et al. [152] propose a GNN-based method for relation extraction, which can efficiently aggregate information from arbitrary dependency graphs. It also applies a novel pruning strategy to the input tree and only keeps the informative words for relation extraction. To model the rich relations across entities, Zhu et al. [157] propose to construct entity graphs and generate edge parameters to propagate diverse relational information, which greatly extends edge types and enables GNNs to conduct complex reasoning. Considering cross-sentence dependencies like coreference and discourse, Peng et al. [78] further explore extracting N-ary relations among multiple entities from multiple sentences by applying Graph LSTMs on document graphs.

Event extraction is another knowledge acquisition task, which aims to identify event triggers and the corresponding arguments for each trigger in texts. Nguyen et al. [72] propose syntactic GCN, which models the dependency tree and learns representations of word nodes to extract events from texts. In addition, Liu et al. [59] find that modeling syntactic relations help capture long-range dependencies better, and these shortcut arcs can help extract multiple events jointly. Meanwhile, to deal with ambiguous and unseen triggers, Zhang et al. [149] summarize event structure knowledge from training data, and construct event background graphs for each event. These graphs help identify correct events by matching the structure knowledge.

Fact Verification Fact verification aims to retrieve evidence from plain text and verify given claims with the evidence. In other words, we need to label a given claim as SUPPORTED, REFUTED, or NOT ENOUGH INFO, which indicates that the evidence can support, refute, or is not sufficient for validating the claim. By regarding the problem as a natural language inference (NLI) [1] task, traditional methods simply combine the evidence via concatenation, or build models based on evidence-claim pairs. To integrate multiple evidence to reason facts, Zhou et al. [156] propose a graph-based evidence aggregating and reasoning framework, which propagates and aggregates information on a fully connected evidence graph. In this way, different pieces of evidence can have sufficient interactions with each other, and thus can help the situations in which multiple pieces of evidence are necessary for making the decision. Liu et al. [61] further incorporate kernel-based attention mechanism into GAT for fine-grained evidence aggregation, and the proposed KGAT achieves improved performance.

Machine Translation Traditionally, machine translation (MT) is modeled as a sequence-to-sequence (seq2seq) problem. The graph structure, however, enables MT models to incorporate explicit linguistic biases. To this end, two typical kinds of graphs can be utilized in MT. Some works [4, 5, 36] model syntactic trees with GCNs to learn syntax-aware sentence representations, and show improvements over vanilla seq2seq methods. To involve more semantic information, other works further consider semantic role labeling (SRL) as well as abstract meaning representation (AMR) graphs [64, 94]. Besides traditional MT, GNNs are also utilized in multimodal and document MT. For multimodal MT, Yin et al. [137] build combined graphs with both entities in images and sentences and then apply GNNs for message passing across modalities. For document MT, Xu et al. [121] model long-term dependencies (e.g., coreference) with GNNs on document graphs, and achieve superior performance on translation coherence.

Question Answering Question answering (QA) aims to generate or find answers for a question based on relevant documents or knowledge bases, which requires models to reason and infer the right answers given the question (see Chap. 4 for an introduction). Since GNNs are designed to model relational data, they are also suitable for QA. To apply GNNs to QA, we need to first build graphs containing question-related entities and their relations. Typically, according to the provided context, researchers could utilize existing knowledge graphs [96, 97, 136] or extract entities and links from documents to construct graphs [11, 24, 29, 30, 58]. Afterward, multiple kinds of GNNs such as GCN, GAT, and R-GCN [90] can be directly used to reason over the graphs [24, 58]. To jointly capture relational and semantic messages, some works explore how to combine powerful PTMs and GNNs for complex reasoning. Ding et al. [24] construct entity graphs utilizing BERT, which predicts node entities and relational edges iteratively. Yasunaga et al. [136] further incorporate PTM-encoded context representations into knowledge graphs to better perform message passing.

Besides the applications in the NLP area, GNNs are widely used in various application scenarios, such as community detection [104, 133, 148], information diffusion prediction [130], recommender systems [8, 28, 115, 129, 139], molecular fingerprints [50], chemical reaction prediction [25], protein interface prediction [31], biomedical engineering [86, 161], etc. Since our book mainly focuses on NLP, readers who are interested in these applications can refer to these papers for more details.

6.8   Summary and Further Readings

In this chapter, we have introduced graph representation learning, which projects graph structure information into continuous vector space and makes deep learning techniques possible on graph data. We first talk about shallow node representation, from spectral clustering, shallow neural networks, to matrix factorization. Then we introduce deep node representation, from autoencoder-based methods to graph neural networks. Afterward, we present how to obtain the global graph representation from node representations. Finally, we introduce how graph representation learning helps a series of NLP tasks.

For further readings of graph representation learning, there are also some recommended reviews and books. In terms of general graph embedding, Goyal and Ferrara [35] and Cui et al. [22] conduct surveys on relevant models and their applications. We also write a monograph [126] about our systematic work on the topic of network embedding. In terms of GNNs, Wu et al. [114] write a book covering more than 20 topics of GNNs in 4 parts: introduction, foundations, frontiers, and applications. Shi et al. [93] write a monograph providing a launch point for discussing the latest trends of GNNs. Wu et al. [113] present a survey about the NLP applications based on GNNs. Shi et al. [92] focus on the representation learning of heterogeneous graphs. We also provide a more comprehensive survey of GNNs in our review [155], covering a broader range of aspects, such as the applications on images or chemistry.

Acknowledgments The contributions of all authors for the second edition are Zhiyuan Liu, Yankai Lin, and Maosong Sun designed the overall architecture of this chapter. Cheng Yang and Yankai Lin drafted this chapter. Zhiyuan Liu and Yankai Lin proofread and revised this chapter.

We thank Ganqu Cui and Chaojun Xiao for drawing figures, and thank Yujia Qin, Ning Ding, Ganqu Cui, Yuqi Luo, and Ziqing Qiao for proofreading the chapter. We also thank Jie Zhou and Zhengyan Zhang for preparing some initial draft materials for the first edition.

This is the graph representation learning chapter of the second edition of the book *Representation Learning for Natural Language Processing*, with its first edition published in 2020 [62]. As compared to the first edition of this chapter, the main changes include the following: (1) we reorganized the narrative logic of this chapter, by dividing it into shallow node representation, deep node representation, graph representation, self-supervised graph learning, and applications; (2) we rewrote and updated the part of graph neural networks; and (3) we added the contents of graph Transformer and self-supervised graph representation learning.

References

1. Gabor Angeli and Christopher D Manning. Naturalli: Natural logic inference for common sense reasoning. In *Proceedings of EMNLP*, 2014.
2. James Atwood and Don Towsley. Diffusion-convolutional neural networks. In *Proceedings of NeurIPS*, 2016.
3. Dzmitry Bahdanau, Kyunghyun Cho, and Yoshua Bengio. Neural machine translation by jointly learning to align and translate. In *Proceedings of ICLR*, 2015.

4. Jasmijn Bastings, Ivan Titov, Wilker Aziz, Diego Marcheggiani, and Khalil Sima'an. Graph convolutional encoders for syntax-aware neural machine translation. In *Proceedings of EMNLP*, 2017.
5. Daniel Beck, Gholamreza Haffari, and Trevor Cohn. Graph-to-sequence learning using gated graph neural networks. In *Proceedings of ACL*, 2018.
6. Slobodan Beliga, Ana Meštrović, and Sanda Martinčić-Ipšić. An overview of graph-based keyword extraction methods and approaches. *Journal of information and organizational sciences*, 39(1):1–20, 2015.
7. Mikhail Belkin and Partha Niyogi. Laplacian eigenmaps and spectral techniques for embedding and clustering. In *Proceedings of NeurIPS*, 2001.
8. Rianne van den Berg, Thomas N Kipf, and Max Welling. Graph convolutional matrix completion. *arXiv preprint arXiv:1706.02263*, 2017.
9. Ronald Newbold Bracewell and Ronald N Bracewell. *The Fourier transform and its applications*, volume 31999. McGraw-Hill New York, 1986.
10. Joan Bruna, Wojciech Zaremba, Arthur Szlam, and Yann Lecun. Spectral networks and locally connected networks on graphs. In *Proceedings of ICLR*, 2014.
11. Nicola De Cao, Wilker Aziz, and Ivan Titov. Question answering by reasoning across documents with graph convolutional networks. In *Proceedings of NAACL*, 2019.
12. Shaosheng Cao, Wei Lu, and Qiongkai Xu. Grarep: Learning graph representations with global structural information. In *Proceedings of CIKM*, 2015.
13. Mathilde Caron, Piotr Bojanowski, Armand Joulin, and Matthijs Douze. Deep clustering for unsupervised learning of visual features. In *Proceedings of ECCV*, 2018.
14. Deli Chen, Yankai Lin, Wei Li, Peng Li, Jie Zhou, and Xu Sun. Measuring and relieving the over-smoothing problem for graph neural networks from the topological view. In *Proceedings of AAAI*, 2020.
15. Dexiong Chen, Leslie O'Bray, and Karsten Borgwardt. Structure-aware transformer for graph representation learning. In *Proceedings of ICML*, 2022.
16. Jie Chen, Tengfei Ma, and Cao Xiao. FastGCN: Fast learning with graph convolutional networks via importance sampling. In *Proceedings of ICLR*, 2018.
17. Mo Chen, Qiong Yang, and Xiaoou Tang. Directed graph embedding. In *Proceedings of IJCAI*, 2007.
18. Jianpeng Cheng, Li Dong, and Mirella Lapata. Long short-term memory-networks for machine reading. In *Proceedings of EMNLP*, 2016.
19. Kyunghyun Cho, Bart van Merriënboer, Caglar Gulcehre, Dzmitry Bahdanau, Fethi Bougares, Holger Schwenk, and Yoshua Bengio. Learning phrase representations using RNN encoder–decoder for statistical machine translation. In *Proceedings of EMNLP*, 2014.
20. Wojciech Chojnacki and Michael J Brooks. A note on the locally linear embedding algorithm. *International Journal of Pattern Recognition and Artificial Intelligence*, 23(08):1739–1752, 2009.
21. Ganqu Cui, Jie Zhou, Cheng Yang, and Zhiyuan Liu. Adaptive graph encoder for attributed graph embedding. In *Proceedings of SIGKDD*, 2020.
22. Peng Cui, Xiao Wang, Jian Pei, and Wenwu Zhu. A survey on network embedding. *IEEE Transactions on Knowledge and Data Engineering*, 2018.
23. Michael Defferrard, Xavier Bresson, and Pierre Vandergheynst. Convolutional neural networks on graphs with fast localized spectral filtering. In *Proceedings of NeurIPS*, 2016.
24. Ming Ding, Chang Zhou, Qibin Chen, Hongxia Yang, and Jie Tang. Cognitive graph for multi-hop reading comprehension at scale. In *Proceedings of ACL*, 2019.
25. Kien Do, Truyen Tran, and Svetha Venkatesh. Graph transformation policy network for chemical reaction prediction. In *Proceedings of KDD*, 2019.

26. Alexey Dosovitskiy, Lucas Beyer, Alexander Kolesnikov, Dirk Weissenborn, Xiaohua Zhai, Thomas Unterthiner, Mostafa Dehghani, Matthias Minderer, Georg Heigold, Sylvain Gelly, et al. An image is worth 16x16 words: Transformers for image recognition at scale. In *Proceedings of ICLR*, 2021.

27. David K Duvenaud, Dougal Maclaurin, Jorge Aguileraiparraguirre, Rafael Gomezbombarelli, Timothy D Hirzel, Alan Aspuruguzik, and Ryan P Adams. Convolutional networks on graphs for learning molecular fingerprints. In *Proceedings of NeurIPS*, 2015.

28. Wenqi Fan, Yao Ma, Qing Li, Yuan He, Eric Zhao, Jiliang Tang, and Dawei Yin. Graph neural networks for social recommendation. In *Proceedings of WWW*, 2019.

29. Yuwei Fang, Siqi Sun, Zhe Gan, Rohit Pillai, Shuohang Wang, and Jingjing Liu. Hierarchical graph network for multi-hop question answering. In *Proceedings of EMNLP*, 2020.

30. Yanlin Feng, Xinyue Chen, Bill Yuchen Lin, Peifeng Wang, Jun Yan, and Xiang Ren. Scalable multi-hop relational reasoning for knowledge-aware question answering. In *Proceedings of EMNLP*, 2020.

31. Alex Fout, Jonathon Byrd, Basir Shariat, and Asa Ben-Hur. Protein interface prediction using graph convolutional networks. In *Proceedings of NeurIPS*, 2017.

32. Hongyang Gao and Shuiwang Ji. Graph u-nets. In *Proceedings of ICML*. PMLR, 2019.

33. Jonas Gehring, Michael Auli, David Grangier, and Yann N Dauphin. A convolutional encoder model for neural machine translation. In *Proceedings of ACL*, 2017.

34. Xumeng Gong, Cheng Yang, and Chuan Shi. Ma-gcl: Model augmentation tricks for graph contrastive learning. In *Proceedings of AAAI*, 2023.

35. Palash Goyal and Emilio Ferrara. Graph embedding techniques, applications, and performance: A survey. *Knowledge-Based Systems*, 151:78–94, 2018.

36. Zhijiang Guo, Yan Zhang, Zhiyang Teng, and Wei Lu. Densely connected graph convolutional networks for graph-to-sequence learning. *Transactions of the Association for Computational Linguistics*, 2019.

37. Will Hamilton, Zhitao Ying, and Jure Leskovec. Inductive representation learning on large graphs. In *Proceedings of NeurIPS*, 2017.

38. William L Hamilton, Rex Ying, and Jure Leskovec. Representation learning on graphs: Methods and applications. *IEEE Data(base) Engineering Bulletin*, 40:52–74, 2017.

39. Kaveh Hassani and Amir Hosein Khasahmadi. Contrastive multi-view representation learning on graphs. In *Proceedings of ICML*, 2020.

40. Kaiming He, Xiangyu Zhang, Shaoqing Ren, and Jian Sun. Deep residual learning for image recognition. In *Proceedings of CVPR*, 2016.

41. Mikael Henaff, Joan Bruna, and Yann Lecun. Deep convolutional networks on graph-structured data. *arXiv preprint arXiv:1506.05163*, 2015.

42. Sepp Hochreiter and Jürgen Schmidhuber. Long short-term memory. *Neural Computation*, 1997.

43. Zhihui Hu, Guang Kou, Haoyu Zhang, Na Li, Ke Yang, and Lin Liu. Rectifying pseudo labels: Iterative feature clustering for graph representation learning. In *Proceedings of CIKM*, 2021.

44. Ziniu Hu, Yuxiao Dong, Kuansan Wang, Kai-Wei Chang, and Yizhou Sun. GPT-GNN: Generative pre-training of graph neural networks. In *Proceedings of SIGKDD*, 2020.

45. Wenbing Huang, Tong Zhang, Yu Rong, and Junzhou Huang. Adaptive sampling towards fast graph representation learning. In *Proceedings of NeurIPS*, 2018.

46. Dasol Hwang, Jinyoung Park, Sunyoung Kwon, KyungMin Kim, Jung-Woo Ha, and Hyun-woo J Kim. Self-supervised auxiliary learning with meta-paths for heterogeneous graphs. *Proceedings of NeurIPS*, 2020.

47. Ashesh Jain, Amir R Zamir, Silvio Savarese, and Ashutosh Saxena. Structural-RNN: Deep learning on spatio-temporal graphs. In *Proceedings of CVPR*, 2016.

48. AJAY KUMAR JAISWAL, Peihao Wang, Tianlong Chen, Justin F Rousseau, Ying Ding, and Zhangyang Wang. Old can be gold: Better gradient flow can make vanilla-gcns great again. In *Advances in Neural Information Processing Systems*, 2022.

49. Michael Kampffmeyer, Yinbo Chen, Xiaodan Liang, Hao Wang, Yujia Zhang, and Eric P Xing. Rethinking knowledge graph propagation for zero-shot learning. In *Proceedings of CVPR*, 2019.
50. Steven Kearnes, Kevin McCloskey, Marc Berndl, Vijay Pande, and Patrick Riley. Molecular graph convolutions: moving beyond fingerprints. *Journal of Computer-aided Molecular Design*, 30(8):595–608, 2016.
51. Thomas N Kipf and Max Welling. Variational graph auto-encoders. *arXiv preprint arXiv:1611.07308*, 2016.
52. Thomas N Kipf and Max Welling. Semi-supervised classification with graph convolutional networks. In *Proceedings of ICLR*, 2017.
53. Devin Kreuzer, Dominique Beaini, Will Hamilton, Vincent Létourneau, and Prudencio Tossou. Rethinking graph transformers with spectral attention. *Proceedings of NeurIPS*, 2021.
54. Yann LeCun, Yoshua Bengio, and Geoffrey Hinton. Deep learning. *Nature*, 521(7553):436, 2015.
55. Yann LeCun, Léon Bottou, Yoshua Bengio, and Patrick Haffner. Gradient-based learning applied to document recognition. In *Proceedings of the IEEE*, 1998.
56. Yaguang Li, Rose Yu, Cyrus Shahabi, and Yan Liu. Diffusion convolutional recurrent neural network: Data-driven traffic forecasting. In *Proceedings of ICLR*, 2018.
57. Yujia Li, Daniel Tarlow, Marc Brockschmidt, and Richard S Zemel. Gated graph sequence neural networks. In *Proceedings of ICLR*, 2016.
58. Bill Yuchen Lin, Xinyue Chen, Jamin Chen, and Xiang Ren. Kagnet: Knowledge-aware graph networks for commonsense reasoning. In *Proceedings of EMNLP*, 2019.
59. Xiao Liu, Zhunchen Luo, and He-Yan Huang. Jointly multiple events extraction via attention-based graph information aggregation. In *Proceedings of EMNLP*, 2018.
60. Yixin Liu, Ming Jin, Shirui Pan, Chuan Zhou, Yu Zheng, Feng Xia, and Philip Yu. Graph self-supervised learning: A survey. *IEEE Transactions on Knowledge and Data Engineering*, 2022.
61. Zhenghao Liu, Chenyan Xiong, Maosong Sun, and Zhiyuan Liu. Fine-grained fact verification with kernel graph attention network. In *Proceedings of ACL*, 2020.
62. Zhiyuan Liu, Yankai Lin, and Maosong Sun. *Representation Learning for Natural Language Processing*. Springer, 2020.
63. Stéphane Mallat. *A wavelet tour of signal processing*. Elsevier, 1999.
64. Diego Marcheggiani, Jasmijn Bastings, and Ivan Titov. Exploiting semantics in neural machine translation with graph convolutional networks. In *Proceedings of NAACL-HLT*, 2018.
65. Diego Marcheggiani and Ivan Titov. Encoding sentences with graph convolutional networks for semantic role labeling. In *Proceedings of EMNLP*, 2017.
66. Rada Mihalcea. Graph-based ranking algorithms for sentence extraction, applied to text summarization. In *Proceedings of ACL*, 2004.
67. Rada Mihalcea and Paul Tarau. Textrank: Bringing order into text. In *Proceedings of EMNLP*, 2004.
68. T Mikolov and J Dean. Distributed representations of words and phrases and their compositionality. In *Proceedings of NeurIPS*, 2013.
69. Federico Monti, Davide Boscaini, Jonathan Masci, Emanuele Rodola, Jan Svoboda, and Michael M Bronstein. Geometric deep learning on graphs and manifolds using mixture model cnns. In *Proceedings of CVPR*, 2017.
70. Mark EJ Newman. Finding community structure in networks using the eigenvectors of matrices. *Physical Review E*, 74(3):036104, 2006.
71. Mark EJ Newman. Modularity and community structure in networks. *Proceedings of the National Academy of Sciences*, 103(23):8577–8582, 2006.
72. Thien Nguyen and Ralph Grishman. Graph convolutional networks with argument-aware pooling for event detection. In *Proceedings of AAAI*, 2018.
73. Aaron van den Oord, Yazhe Li, and Oriol Vinyals. Representation learning with contrastive predictive coding. *arXiv preprint arXiv:1807.03748*, 2018.

74. Lawrence Page, Sergey Brin, Rajeev Motwani, and Terry Winograd. The pagerank citation ranking: Bringing order to the web. Technical report, Stanford InfoLab, 1999.
75. Shirui Pan, Ruiqi Hu, Guodong Long, Jing Jiang, Lina Yao, and Chengqi Zhang. Adversarially regularized graph autoencoder for graph embedding. *Proceedings of IJCAI*, 2018.
76. Jiwoong Park, Minsik Lee, Hyung Jin Chang, Kyuewang Lee, and Jin Young Choi. Symmetric graph convolutional autoencoder for unsupervised graph representation learning. In *Proceedings of ICCV*, 2019.
77. Hao Peng, Jianxin Li, Yu He, Yaopeng Liu, Mengjiao Bao, Lihong Wang, Yangqiu Song, and Qiang Yang. Large-scale hierarchical text classification with recursively regularized deep graph-cnn. In *Proceedings of WWW*, 2018.
78. Nanyun Peng, Hoifung Poon, Chris Quirk, Kristina Toutanova, and Wen-tau Yih. Cross-sentence n-ary relation extraction with graph LSTMs. *Transactions of the Association for Computational Linguistics*, 5:101–115, 2017.
79. Zhen Peng, Yixiang Dong, Minnan Luo, Xiao-Ming Wu, and Qinghua Zheng. Self-supervised graph representation learning via global context prediction. *arXiv preprint arXiv:2003.01604*, 2020.
80. Zhen Peng, Wenbing Huang, Minnan Luo, Qinghua Zheng, Yu Rong, Tingyang Xu, and Junzhou Huang. Graph representation learning via graphical mutual information maximization. In *Proceedings of WWW*, 2020.
81. Bryan Perozzi, Rami Al-Rfou, and Steven Skiena. DeepWalk: Online learning of social representations. In *Proceedings of KDD*, 2014.
82. Jiezhong Qiu, Qibin Chen, Yuxiao Dong, Jing Zhang, Hongxia Yang, Ming Ding, Kuansan Wang, and Jie Tang. Gcc: Graph contrastive coding for graph neural network pre-training. In *Proceedings of SIGKDD*, 2020.
83. Jiezhong Qiu, Yuxiao Dong, Hao Ma, Jian Li, Kuansan Wang, and Jie Tang. Network embedding as matrix factorization: Unifying deepwalk, line, pte, and node2vec. In *Proceedings of WSDM*, 2018.
84. Colin Raffel, Noam Shazeer, Adam Roberts, Katherine Lee, Sharan Narang, Michael Matena, Yanqi Zhou, Wei Li, and Peter J Liu. Exploring the limits of transfer learning with a unified text-to-text transformer. *Journal of Machine Learning Research*, 21:1–67, 2020.
85. Afshin Rahimi, Trevor Cohn, and Timothy Baldwin. Semi-supervised user geolocation via graph convolutional networks. In *Proceedings of ACL*, 2018.
86. Sungmin Rhee, Seokjun Seo, and Sun Kim. Hybrid approach of relation network and localized graph convolutional filtering for breast cancer subtype classification. In *Proceedings of IJCAI*, 2018.
87. Yu Rong, Yatao Bian, Tingyang Xu, Weiyang Xie, Ying Wei, Wenbing Huang, and Junzhou Huang. Self-supervised graph transformer on large-scale molecular data. In *Proceedings of NeurIPS*, 2020.
88. Sam T Roweis and Lawrence K Saul. Nonlinear dimensionality reduction by locally linear embedding. *Science*, 290(5500):2323–2326, 2000.
89. Franco Scarselli, Marco Gori, Ah Chung Tsoi, Markus Hagenbuchner, and Gabriele Monfardini. The graph neural network model. *IEEE TNN 2009*, 20(1):61–80, 2009.
90. Michael Schlichtkrull, Thomas N Kipf, Peter Bloem, Rianne van den Berg, Ivan Titov, and Max Welling. Modeling relational data with graph convolutional networks. In *Proceedings of ESWC*, 2018.
91. Chuan Shi, Binbin Hu, Wayne Xin Zhao, and S Yu Philip. Heterogeneous information network embedding for recommendation. *IEEE Transactions on Knowledge and Data Engineering*, 31(2):357–370, 2018.
92. Chuan Shi, Xiao Wang, and S Yu Philip. Heterogeneous graph representation learning and applications, 2022.
93. Chuan Shi, Xiao Wang, and Cheng Yang. *Advances in Graph Neural Networks*. Springer, 2022.

94. Linfeng Song, Daniel Gildea, Yue Zhang, Zhiguo Wang, and Jinsong Su. Semantic neural machine translation using amr. *Transactions of the Association for Computational Linguistics*, 2019.
95. Fan-Yun Sun, Jordan Hoffman, Vikas Verma, and Jian Tang. Infograph: Unsupervised and semi-supervised graph-level representation learning via mutual information maximization. In *Proceedings of ICLR*, 2019.
96. Haitian Sun, Tania Bedrax-Weiss, and William W Cohen. Pullnet: Open domain question answering with iterative retrieval on knowledge bases and text. In *Proceedings of EMNLP*, 2019.
97. Haitian Sun, Bhuwan Dhingra, Manzil Zaheer, Kathryn Mazaitis, Ruslan Salakhutdinov, and William W Cohen. Open domain question answering using early fusion of knowledge bases and text. In *Proceedings of EMNLP*, 2018.
98. Ke Sun, Zhouchen Lin, and Zhanxing Zhu. Multi-stage self-supervised learning for graph convolutional networks on graphs with few labeled nodes. In *Proceedings of AAAI*, 2020.
99. Yizhou Sun, Jiawei Han, Xifeng Yan, Philip S Yu, and Tianyi Wu. Pathsim: Meta path-based top-k similarity search in heterogeneous information networks. *Proceedings of the VLDB Endowment*, 4(11):992–1003, 2011.
100. Susheel Suresh, Pan Li, Cong Hao, and Jennifer Neville. Adversarial graph augmentation to improve graph contrastive learning. *Proceedings of NeurIPS*, 2021.
101. Kai Sheng Tai, Richard Socher, and Christopher D. Manning. Improved semantic representations from tree-structured long short-term memory networks. In *Proceedings of ACL*, 2015.
102. Jian Tang, Meng Qu, Mingzhe Wang, Ming Zhang, Jun Yan, and Qiaozhu Mei. LINE: Large-scale information network embedding. In *Proceedings of WWW*, 2015.
103. Lei Tang and Huan Liu. Relational learning via latent social dimensions. In *Proceedings of KDD*, 2009.
104. Anton Tsitsulin, John Palowitch, Bryan Perozzi, and Emmanuel Muller. Graph clustering with graph neural networks. *arXiv preprint arXiv:2006.16904*, 2020.
105. Ashish Vaswani, Noam Shazeer, Niki Parmar, Llion Jones, Jakob Uszkoreit, Aidan N Gomez, and Lukasz Kaiser. Attention is all you need. In *Proceedings of NeurIPS*, 2017.
106. Petar Velickovic, Guillem Cucurull, Arantxa Casanova, Adriana Romero, Pietro Lio, and Yoshua Bengio. Graph attention networks. In *Proceedings of ICLR*, 2018.
107. Petar Veličković, William Fedus, William L Hamilton, Pietro Liò, Yoshua Bengio, and R Devon Hjelm. Deep graph infomax. In *Proceedings of ICLR*, 2018.
108. Chun Wang, Shirui Pan, Guodong Long, Xingquan Zhu, and Jing Jiang. Mgae: Marginalized graph autoencoder for graph clustering. In *Proceedings of CIKM*, 2017.
109. Daixin Wang, Peng Cui, and Wenwu Zhu. Structural deep network embedding. In *Proceedings of KDD*, 2016.
110. Xiao Wang, Deyu Bo, Chuan Shi, Shaohua Fan, Yanfang Ye, and S Yu Philip. A survey on heterogeneous graph embedding: methods, techniques, applications and sources. *IEEE Transactions on Big Data*, 2022.
111. Xiao Wang, Houye Ji, Chuan Shi, Bai Wang, Yanfang Ye, Peng Cui, and Philip S Yu. Heterogeneous graph attention network. In *Proceedings of WWW*, 2019.
112. Xiao Wang, Nian Liu, Hui Han, and Chuan Shi. Self-supervised heterogeneous graph neural network with co-contrastive learning. In *Proceedings of SIGKDD*, 2021.
113. Lingfei Wu, Yu Chen, Kai Shen, Xiaojie Guo, Hanning Gao, Shucheng Li, Jian Pei, and Bo Long. Graph neural networks for natural language processing: A survey. *arXiv preprint arXiv:2106.06090*, 2021.
114. Lingfei Wu, Peng Cui, Jian Pei, and Liang Zhao. *Graph Neural Networks: Foundations, Frontiers, and Applications*. Springer Singapore, Singapore, 2022.
115. Qitian Wu, Hengrui Zhang, Xiaofeng Gao, Peng He, Paul Weng, Han Gao, and Guihai Chen. Dual graph attention networks for deep latent representation of multifaceted social effects in recommender systems. In *Proceedings of WWW*, 2019.

116. Zhanghao Wu, Paras Jain, Matthew Wright, Azalia Mirhoseini, Joseph E Gonzalez, and Ion Stoica. Representing long-range context for graph neural networks with global attention. *Proceedings of NeurIPS*, 2021.

117. Jun Xia, Lirong Wu, Jintao Chen, Bozhen Hu, and Stan Z Li. Simgrace: A simple framework for graph contrastive learning without data augmentation. In *Proceedings of WWW*, 2022.

118. Yaochen Xie, Zhao Xu, and Shuiwang Ji. Self-supervised representation learning via latent graph prediction. *Proceedings of ICML*, 2022.

119. Keyulu Xu, Weihua Hu, Jure Leskovec, and Stefanie Jegelka. How powerful are graph neural networks? *Proceedings of ICLR*, 2019.

120. Keyulu Xu, Chengtao Li, Yonglong Tian, Tomohiro Sonobe, Kenichi Kawarabayashi, and Stefanie Jegelka. Representation learning on graphs with jumping knowledge networks. In *Proceedings of ICML*, 2018.

121. Mingzhou Xu, Liangyou Li, Derek Wong, Qun Liu, Lidia S Chao, et al. Document graph for neural machine translation. In *Proceedings of EMNLP*, 2020.

122. Sijie Yan, Yuanjun Xiong, and Dahua Lin. Spatial temporal graph convolutional networks for skeleton-based action recognition. In *Proceedings of AAAI*, 2018.

123. Cheng Yang, Yuxin Guo, Yao Xu, Chuan Shi, Jiawei Liu, Chunchen Wang, Xin Li, Ning Guo, and Hongzhi Yin. Learning to distill graph neural networks. In *Proceedings of WSDM*, 2023.

124. Cheng Yang, Jiawei Liu, and Chuan Shi. Extract the knowledge of graph neural networks and go beyond it: An effective knowledge distillation framework. In *Proceedings of WWW*, 2021.

125. Cheng Yang and Zhiyuan Liu. Comprehend deepwalk as matrix factorization. *arXiv preprint arXiv:1501.00358*, 2015.

126. Cheng Yang, Zhiyuan Liu, Cunchao Tu, Chuan Shi, and Maosong Sun. Network embedding: Theories, methods, and applications. *Synthesis Lectures on Artificial Intelligence and Machine Learning*, 15(2):1–242, 2021.

127. Cheng Yang, Zhiyuan Liu, Deli Zhao, Maosong Sun, and Edward Y Chang. Network representation learning with rich text information. In *Proceedings of IJCAI*, 2015.

128. Cheng Yang, Maosong Sun, Zhiyuan Liu, and Cunchao Tu. Fast network embedding enhancement via high order proximity approximation. In *Proceedings of IJCAI*, 2017.

129. Cheng Yang, Maosong Sun, Wayne Xin Zhao, Zhiyuan Liu, and Edward Y Chang. A neural network approach to jointly modeling social networks and mobile trajectories. *ACM Transactions on Information Systems*, 35(4):1–28, 2017.

130. Cheng Yang, Jian Tang, Maosong Sun, Ganqu Cui, and Liu Zhiyuan. Multi-scale information diffusion prediction with reinforced recurrent networks. In *Proceedings of IJCAI*, 2019.

131. Cheng Yang, Chunchen Wang, Yuanfu Lu, Xumeng Gong, Chuan Shi, Wei Wang, and Xu Zhang. Few-shot link prediction in dynamic networks. In *Proceedings of WSDM*, 2022.

132. Haoran Yang, Hongxu Chen, Shirui Pan, Lin Li, Philip S Yu, and Guandong Xu. Dual space graph contrastive learning. *Proceedings of WWW*, 2022.

133. Jaewon Yang and Jure Leskovec. Overlapping community detection at scale: a nonnegative matrix factorization approach. In *Proceedings of WSDM*, 2013.

134. Longqi Yang, Liangliang Zhang, and Wenjing Yang. Graph adversarial self-supervised learning. *Proceedings of NeurIPS*, 2021.

135. Liang Yao, Chengsheng Mao, and Yuan Luo. Graph convolutional networks for text classification. In *Proceedings of AAAI*, 2019.

136. Michihiro Yasunaga, Hongyu Ren, Antoine Bosselut, Percy Liang, and Jure Leskovec. QA-GNN: Reasoning with language models and knowledge graphs for question answering. In *Proceedings of NAACL*, 2021.

137. Yongjing Yin, Fandong Meng, Jinsong Su, Chulun Zhou, Zhengyuan Yang, Jie Zhou, and Jiebo Luo. A novel graph-based multi-modal fusion encoder for neural machine translation. In *Proceedings of ACL*, 2020.

138. Chengxuan Ying, Tianle Cai, Shengjie Luo, Shuxin Zheng, Guolin Ke, Di He, Yanming Shen, and Tie-Yan Liu. Do transformers really perform badly for graph representation? *Proceedings of NeurIPS*, 2021.

139. Rex Ying, Ruining He, Kaifeng Chen, Pong Eksombatchai, William L Hamilton, and Jure Leskovec. Graph convolutional neural networks for web-scale recommender systems. In *Proceedings of KDD*, 2018.
140. Zhitao Ying, Jiaxuan You, Christopher Morris, Xiang Ren, Will Hamilton, and Jure Leskovec. Hierarchical graph representation learning with differentiable pooling. In *Proceedings of NeurIPS*, 2018.
141. Yuning You, Tianlong Chen, Yang Shen, and Zhangyang Wang. Graph contrastive learning automated. In *Proceedings of ICML*, 2021.
142. Yuning You, Tianlong Chen, Yongduo Sui, Ting Chen, Zhangyang Wang, and Yang Shen. Graph contrastive learning with augmentations. In *Proceedings of NeurIPS*, 2020.
143. Bing Yu, Haoteng Yin, and Zhanxing Zhu. Spatio-temporal graph convolutional networks: A deep learning framework for traffic forecasting. In *Proceedings of ICLR*, 2018.
144. Victoria Zayats and Mari Ostendorf. Conversation modeling on reddit using a graph-structured LSTM. *Transactions of the Association for Computational Linguistics*, 6:121–132, 2018.
145. Hengrui Zhang, Qitian Wu, Junchi Yan, David Wipf, and Philip S Yu. From canonical correlation analysis to self-supervised graph neural networks. In *Proceedings of NeurIPS*, 2021.
146. Jiani Zhang, Xingjian Shi, Junyuan Xie, Hao Ma, Irwin King, and Dit Yan Yeung. GaAN: Gated attention networks for learning on large and spatiotemporal graphs. In *Proceedings of UAI*, 2018.
147. Muhan Zhang, Zhicheng Cui, Marion Neumann, and Yixin Chen. An end-to-end deep learning architecture for graph classification. In *Proceedings of AAAI*, 2018.
148. Xiaotong Zhang, Han Liu, Qimai Li, and Xiao-Ming Wu. Attributed graph clustering via adaptive graph convolution. In *Proceedings of IJCAI*, 2019.
149. Yilin Zhang, Ziran Li, Zhiyuan Liu, Hai-Tao Zheng, Ying Shen, and Lan Zhou. Event detection with dynamic word-trigger-argument graph neural networks. *IEEE Transactions on Knowledge and Data Engineering*, 2021.
150. Yizhou Zhang, Yun Xiong, Xiangnan Kong, Shanshan Li, Jinhong Mi, and Yangyong Zhu. Deep collective classification in heterogeneous information networks. In *Proceedings of WWW*, 2018.
151. Yue Zhang, Qi Liu, and Linfeng Song. Sentence-state LSTM for text representation. In *Proceedings of ACL*, 2018.
152. Yuhao Zhang, Peng Qi, and Christopher D Manning. Graph convolution over pruned dependency trees improves relation extraction. In *Proceedings of EMNLP*, 2018.
153. Zhengyan Zhang, Cheng Yang, Zhiyuan Liu, Maosong Sun, Zhichong Fang, Bo Zhang, and Leyu Lin. Cosine: compressive network embedding on large-scale information networks. *IEEE Transactions on Knowledge and Data Engineering*, 2020.
154. Jianan Zhao, Xiao Wang, Chuan Shi, Zekuan Liu, and Yanfang Ye. Network schema preserving heterogeneous information network embedding. In *International Joint Conference on Artificial Intelligence (IJCAI)*, 2020.
155. Jie Zhou, Ganqu Cui, Shengding Hu, Zhengyan Zhang, Cheng Yang, Zhiyuan Liu, Lifeng Wang, Changcheng Li, and Maosong Sun. Graph neural networks: A review of methods and applications. *AI Open*, 1:57–81, 2020.
156. Jie Zhou, Xu Han, Cheng Yang, Zhiyuan Liu, Lifeng Wang, Changcheng Li, and Maosong Sun. GEAR: Graph-based evidence aggregating and reasoning for fact verification. In *Proceedings of ACL 2019*, 2019.
157. Hao Zhu, Yankai Lin, Zhiyuan Liu, Jie Fu, Tat-Seng Chua, and Maosong Sun. Graph neural networks with generated parameters for relation extraction. In *Proceedings of ACL*, 2019.
158. Yanqiao Zhu, Yichen Xu, Feng Yu, Qiang Liu, Shu Wu, and Liang Wang. Deep graph contrastive representation learning. *arXiv preprint arXiv:2006.04131*, 2020.
159. Yanqiao Zhu, Yichen Xu, Feng Yu, Qiang Liu, Shu Wu, and Liang Wang. Graph contrastive learning with adaptive augmentation. In *Proceedings of WWW*, 2021.

160. Julian G Zilly, Rupesh Kumar Srivastava, Jan Koutnik, and Jurgen Schmidhuber. Recurrent highway networks. In *Proceedings of ICML*, 2016.
161. Marinka Zitnik, Monica Agrawal, and Jure Leskovec. Modeling polypharmacy side effects with graph convolutional networks. *Intelligent Systems in Molecular Biology*, 34(13):258814, 2018.

Chapter 7
Cross-Modal Representation Learning

Yuan Yao, Zhiyuan Liu, Yankai Lin, and Maosong Sun

Abstract Cross-modal representation learning is an essential part of representation learning, which aims to learn semantic representations for different modalities including text, audio, image and video, etc., and their connections. In this chapter, we introduce the development of cross-modal representation learning from shallow to deep, and from respective to unified in terms of model architectures and learning mechanisms for different modalities and tasks. After that, we review how cross-modal capabilities can contribute to complex real-world applications.

7.1 Introduction

Modalities are means of information exchange between human beings and the real world. Concretely, each modality is an independent channel of sensory input or output for intelligent systems. Typical modalities for humans include text, audio, image, and video, while AI systems can process more modalities such as infrared information. Cross-modal representation learning refers to learning paradigms where multiple modalities are involved.

Cross-modal representation learning is an important topic of representation learning. In fact, AI is inherently a cross-modal problem [52], where handling multiple modalities is both necessary and beneficial for real-world intelligent systems. Regarding the necessity, in many real-world applications, intelligent systems are required to operate in a cross-modal environment, such as transcribing speech to text [9], or navigating in a room according to text instructions [10]. From the beneficial perspective, it can be helpful to integrate the correlated and complementary information in different modalities for comprehensive decision-

Y. Yao · Z. Liu (✉) · M. Sun
Department of Computer Science and Technology, Tsinghua University, Beijing, China
e-mail: yuan-yao18@mails.tsinghua.edu.cn; liuzy@tsinghua.edu.cn; sms@tsinghua.edu.cn

Y. Lin
Gaoling School of Artificial Intelligence, Renmin University of China, Beijing, China
e-mail: yankailin@ruc.edu.cn

© The Author(s) 2023
Z. Liu et al. (eds.), *Representation Learning for Natural Language Processing*,
https://doi.org/10.1007/978-981-99-1600-9_7

Fig. 7.1 Cross-modal information can be helpful in understanding high-level semantics. The apple fruit image is obtained from pixabay.com, and the apple product image is obtained from commons.wikimedia.org, both from the public domain

making. For example, for human perceptions, the judgment of a syllable is made by not only the sound we hear but also the movement of the lips and tongue of the speaker we see. An experiment in McGurk et al. [68] shows that a voiced /ba/ with a visual /ga/ is perceived by most people as a /da/. Moreover, the high-level semantics can also usually be better identified in a cross-modal context. As shown in Fig. 7.1, cross-modal context is important to resolve the specific semantic meaning of *Apple*. Therefore, it is natural for us to consider the possibility of combining cross-modal information in our AI systems and generating cross-modal representation.

To learn cross-modal representations, models typically need to first understand the heterogenous data from each modality with complex semantic composition, as shown in Fig. 7.2. Various deep neural architectures have been developed to incorporate the inductive bias for the heterogenous data from different modalities. The difference between modalities can be illustrated in two aspects, including the basic units and their modal structures. (1) A fundamental difference between text and other modalities lies in the information density of **basic units** [35]. Text is human-generated abstract signals with high information density, where the basic units (e.g., symbolic words) already carry high-level semantics. In comparison, images and speech are direct recordings of real-world signals, where it is usually more challenging to recognize high-level semantics from basic units with low information density (e.g., recognizing objects from continuous image pixels). (2) **Modal structure** also constitutes a major difference between modalities. For example, text and speech exhibit sequential dependency between basic units, and in comparison, information is spatially presented in images, leading to invariance in shift and scale in images. Single frames in videos are spatially presented, and different frames are organized in a sequential structure. To account for these structures, recurrent neural networks (RNNs) and convolutional neural networks (CNNs) have been developed respectively.

Moreover, models are challenged with establishing **cross-modal mapping** for cross-modal information alignment and fusion. The fine-grained mapping can exist between information from different semantic levels and modalities. Since

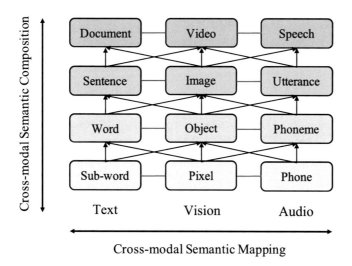

Fig. 7.2 Cross-modal representation learning is challenged with modeling cross-modal semantic composition and establishing cross-modal semantic mapping

explicit annotation of cross-modal mapping is limited, the learning of cross-modal alignment and fusion is typically implicitly driven by supervised learning on specific task annotations. For example, by learning to answer questions about images, models implicitly learn the cross-modal mapping between text tokens and image regions. The model architectures are usually highly specialized for different tasks, and the cross-modal representations are learned by task annotations.

Recently, there is a trend of more unified deep cross-modal representation learning in terms of both model architecture and learning mechanisms. Specifically, Transformers have been proven to be effective in modeling different modalities, including text [90], speech [22], image [23], and video [30]. More unified self-supervised pre-training on large-scale cross-modal data has also pushed forward the state of the arts of many cross-modal tasks [2, 96, 109]. A unified model simultaneously dealing with different modalities and tasks is beginning to take shape, which can be a promising foundation and path to realizing general intelligent systems in the future.

In the following part of this chapter, we will first introduce fundamental cross-modal capabilities for cross-modal tasks in Sect. 7.2. Then, we will review representative cross-modal representation learning models, including shallow representation models in Sect. 7.3, deep representation models in Sect. 7.4, and deep pre-training models in Sect. 7.5. Finally, we will introduce critical applications in Sect. 7.6. In this chapter, without loss of generality, we focus on introducing vision-language models, which are the most important and widely investigated area in cross-modal representation learning research, and also inspire research in other modalities.

7.2 Cross-Modal Capabilities

A real-world cross-modal application usually requires a comprehensive mastery of multiple cross-modal capabilities. In this section, we first provide a taxonomy of cross-modal capabilities and then introduce the corresponding models in the following section. Specifically, cross-modal capabilities can be roughly divided into three categories, including cross-modal understanding, cross-modal retrieval, and cross-modal generation.

Cross-Modal Understanding Models are required to perform semantic understanding based on the given image and query text of the task, for example, answering the question about the image, grounding text into image regions, or identifying semantic relations between objects. Fine-grained cross-modal alignment and fusion between image regions and text tokens are important to achieve strong cross-modal understanding performance.

Cross-Modal Retrieval Given a large candidate set of text and images, and a query from one modality, models are asked to retrieve the corresponding data from other modalities, for example, retrieving images based on a text query or retrieving text based on an image query. Due to the large number of retrieval candidates, cross-modal retrieval methods need to model the holistic semantic relations between data from different modalities in an efficient and scalable way.

Cross-Modal Generation For image-to-text generation, models are required to generate natural language text about the given image content, for example, describing the image content or having conversations on the image. An image-to-text generation model needs to establish fine-grained mapping between text generation and image understanding, and achieve a good trade-off between diversity and fidelity in describing the visual content with text. Another reverse capability is text-to-image generation, which requires models to produce images reflecting the given text description, which can be useful to produce AI-generated content (AIGC). Compared with image-to-text generation, text-to-image generation presents more challenges on the vision side, such as image generation with high-resolution and good computation efficiency. In this chapter, we mainly introduce image-to-text models.

7.3 Shallow Cross-Modal Representation Learning

Early works in cross-modal representation learning have investigated fusing cross-modal information in shallow representations, such as word representations. The word representations can serve as input text representations of deep cross-modal neural networks, and can be efficiently learned through shallow neural architectures on large-scale data. As introduced in Chap. 2, traditional word embedding models like word2vec [69] are trained on a text corpus. These models, while being

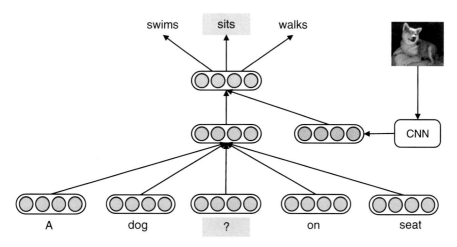

Fig. 7.3 The architecture for word embedding with global visual context. The figure is redrawn according to Fig. 1 in [105], and the image is obtained from Visual Genome [53]

successful, cannot discover implicit semantic relatedness between words that could be revealed in other modalities. Kottur et al. [52] provide an example: even though *eat* and *stare_at* seem unrelated from text, images might show that when people are *eating* something, they would also tend to *stare_at* it. Besides, the semantics of concrete words (e.g., colors and objects) can also be better reflected with the help of visual information [13, 49]. This implies that considering other modalities when constructing word embeddings may help capture more implicit semantic relatedness, where the fused cross-modal representation can facilitate various cross-modal tasks.

Vision, being one of the most critical modalities, has attracted attention from researchers seeking to improve word representations. Several models that incorporate visual information and improve word representations with vision have been proposed. We introduce two typical word representation models, which incorporate visual information as additional context and optimization target as follows.

Word Embedding with Visual Context In most word representation learning models, only local context information from text is considered (e.g., trying to predict a word using neighboring words and phrases). Global information (e.g., the topic of the passage), on the other hand, is often neglected. The image associated with the text can provide such global information for word representation learning. Therefore, some works have proposed to extend word embedding models by using visual information as additional global features (see Fig. 7.3).

Xu et al. [105] make such an attempt in this direction. The input of the model is an image I and a word sequence describing it (i.e., the image caption). Based on a vanilla continuous bag-of-words (CBOW) model, when we consider a certain word w_t in a sequence, its local text feature is the average of embeddings of words in a window, i.e., $\{w_{t-k}, \ldots, w_{t-1}, w_{t+1}, \ldots, w_{t+k}\}$. The visual feature is computed directly from the image I using a CNN and then used as the global feature. The local feature and the global feature are then concatenated into the aggregated context feature $\mathbf{h}$, based on which the word probability is computed:

$$P(w_t|w_{t-k}, \ldots, w_{t-1}, w_{t+1}, \ldots, w_{t+k}; I) = \frac{\exp(\mathbf{w}_t^\top \mathbf{h})}{\sum_i \exp(\mathbf{w}_i^\top \mathbf{h})}. \qquad (7.1)$$

By maximizing the logarithm probability of the target words, the language modeling loss will be back-propagated to local text features (i.e., word embeddings), global visual features (i.e., visual encoder), and all other parameters. Despite the simplicity, this accomplishes joint learning for a set of word embeddings, a language model, and the model used for visual encoding.

In addition to the image pixel feature, the co-occurred words in image captions [37] and objects in images [114] can also serve as the additional visual context. Moreover, for many languages such as Chinese and Korean, the writing of the characters largely reflects their semantics, and considering visual information of characters as additional context can be beneficial for character representation learning, especially for uncommon characters [61].

Word Embedding with Visual Target Besides additional context, visual information can also serve as learning targets to capture fine-grained semantics for word representation learning. For example, the implicit abstract scene or topic behind the images (e.g., *birthday celebration*) can serve as discrete visual signals for word representation learning [52]. A pair of the visual scene and a related word sequence (I, w) is taken as input. At each training step, a window is used upon the word sequence w, forming a subsequence S_w. Based on the context feature (i.e., average word embeddings of S_w), the model produces a probability distribution over the discrete-valued target function $g(\cdot)$ that incorporates visual information. The entire model is optimized by minimizing the objective function as follows:

$$\mathcal{L} = -\log P(g(I)|S_w). \qquad (7.2)$$

The most important part of the model is the function $g(\cdot)$. Intuitively, $g(\cdot)$ should map the visual scene I into the set $\{1, 2, \ldots, k\}$ indicating what kind of abstract scene it is. In practice, it is learned offline using k-means clustering, and each cluster represents the semantics of one kind of visual scene. Through the visual optimization target, the word representations can be learned to be related to the scene. Besides the discrete visual target reflecting the abstract scene, continuous visual features can also be used to guide the representation learning of words in text corpus, where the representations of concrete words are encouraged to be close to the corresponding image features [56].

7.4 Deep Cross-Modal Representation Learning

In the last section, we introduce shallow cross-modal representations which fuse visual information with shallow word embeddings. In fact, when dealing with cross-modal tasks, supervised task learning in deep neural architectures can produce deeper cross-modal representations that better fuse and align the cross-modal information. In this section, we introduce deep cross-modal representation learning models for each cross-modal capability, including cross-modal understanding, retrieval, and generation.

7.4.1 Cross-Modal Understanding

Cross-modal understanding aims to perform semantic recognition and reasoning on the given image and text. A major challenge is that fine-grained cross-modal information needs to be aligned and fused for deep cross-modal understanding. We introduce two representative cross-modal understanding tasks as examples, including visual question answering and visual relation detection.

Visual Question Answering Visual question answering (VQA) is one of the most widely investigated tasks in cross-modal learning, which aims to answer natural language questions about an image. VQA is a challenging task, since various complex reasoning capabilities are involved, and external knowledge is usually required to address the questions. Many datasets have been proposed for the task, including VQA [5], GQA [42], VQA-CP [1], COCO-QA [79], FM-IQA [27], etc. To address the VQA task, researchers have proposed to adopt attention mechanism for fine-grained vision-language alignment and reasoning, and leverage external knowledge to provide rich context information for question answering.

Attention Mechanism To align and fuse cross-modal information, attention mechanism is an effective and widely used approach. Intuitively, image regions related to the question should be selected and contribute more to the cross-modal representations, and vice versa. Shih et al. [82] propose to calculate the attention over image regions to select informative ones to answer the question. The image regions are first encoded into feature representations $\{\mathbf{I}_1, \mathbf{I}_2, \ldots, \mathbf{I}_k\}$ via CNN encoders. Then, the attention score α_j over the image regions is computed as follows:

$$\alpha_j = (\mathbf{W}_1\mathbf{I}_j + \mathbf{b}_1)^\top (\mathbf{W}_2\mathbf{q} + \mathbf{b}_2), \tag{7.3}$$

where $\mathbf{W}_1, \mathbf{W}_2, \mathbf{b}_1, \mathbf{b}_2$ are trainable parameters and $\mathbf{q}$ is the question representation. A larger attention score indicates higher relevance between the image region and the question, and larger contribution to the final fused representations and answer prediction. The question-aware image feature is obtained via a convex combination of the region features based on the normalized attention scores to produce the answer. In this way, image regions relevant to the question are selected in an end-to-end fashion for visual question answering.

However, some questions are only related to some small regions, which encourages researchers to use stacked attention to further refine the attention distribution for noise filtering. Yang et al. [107] further extend the single-layer attention model used in [82] by stacking multiple attention layers. The key idea is to gradually filter out noises and pinpoint the regions that are highly relevant to the answer by reasoning through multiple stacked attention layers progressively.

The above models attend only to images. Intuitively, questions should also be attended to select informative tokens, and vice versa. Lu et al. [65] propose such co-attention mechanism between fine-grained image region and text tokens by

$$\mathbf{Z} = \tanh(\mathbf{Q}^\top \mathbf{W} \mathbf{I}), \qquad (7.4)$$

where $\mathbf{Z}_{ij}$ represents the affinity of the i-th word and j-th region, which is produced from a bilinear operation between the text token feature matrix $\mathbf{Q}$ and image region feature matrix $\mathbf{I}$. The co-attention affinity matrix $\mathbf{Z}$ is then used to produce the attention scores over text tokens and image regions. In addition, by attending to image grids, an object to be attended to may be divided into different image grids, which cannot well reflect the high-level image semantics. To address the issue, Anderson et al. [3] find that attending to salient detected objects can benefit holistic scene understanding for visual question answering.

External Knowledge as Additional Context Another intuitive line of research is to utilize external knowledge, which can help better explain the implicit information hiding behind the image. Generally, there are two kinds of knowledge that can be explored, including implicit external knowledge from related text and language models and explicit external knowledge from knowledge graphs. Wu et al. [100] propose to enhance scene understanding through rich attributes, captions, and related text descriptions from knowledge bases. The representation of the rich context information can serve as the initial vector of RNNs, which then further encode the question to produce the answer in a seq2seq fashion, as shown in Fig. 7.4. In this way, the information from attributes and captions and the complementary external knowledge from knowledge bases can be utilized for answer generation. Similarly, some works [34, 67] jointly reason over the descriptions from PTMs, and explicit knowledge from knowledge graphs for visual question answering.

Visual Relation Detection Visual relation detection or scene graph generation is the task of detecting objects in an image and understanding the semantic relation between them. The task aims to produce scene graphs where nodes correspond to objects and directed edges correspond to visual relations between objects, as shown in Fig. 7.5. The structured graph-based image representations can facilitate various downstream tasks. Detecting objects are usually conducted by off-the-shelf object detectors, and the key challenge of the task lies in understanding the complex relational interactions between objects. Here we introduce two main directions of

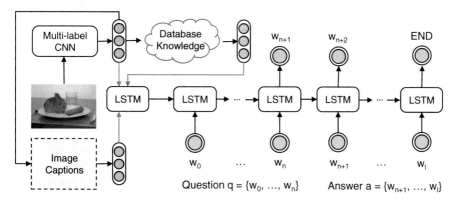

Fig. 7.4 The architecture of VQA incorporating external knowledge bases. The figure is redrawn according to Fig. 2 in [100], and the image is obtained from Visual Genome [53]

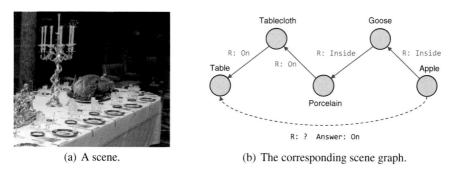

(a) A scene.   (b) The corresponding scene graph.

Fig. 7.5 An illustration for scene graph generation. The figure is redrawn according to Fig. 1 and Fig. 2 in [66]. The goose image is obtained from pngimg.com, and the table image is obtained from commons.wikimedia.org, both from the public domain

research in scene graph generation, including graph-based relation reasoning, and language and knowledge-enhanced visual relation learning.

Reasoning with Graph Structures The graph-based reasoning methods aim to pass and fuse the semantic information of objects and relations based on the graph structure for complex relational reasoning. Xu et al. [102] propose to iteratively exchange and refine the visual information on the dual graph of objects and relations. Li et al. [59] further propose to construct a heterogeneous graph consisting of different levels of context information, including objects, triplets, and region captions, to boost the performance of visual relation detection. Specifically, a graph is constructed to align these three levels of information and perform feature refinement via message passing, as shown in Fig. 7.6. During message passing, each node in the graph is associated with a gate to select meaningful information and filter out noise from neighboring nodes. By leveraging complementary information from

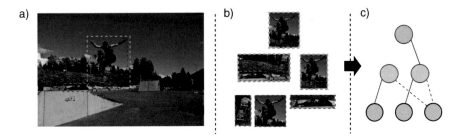

Fig. 7.6 Heterogenous graph for complementary message passing. (**a**) The input image. (**b**) Object (bottom), triplet (middle), and caption region (top) proposals. (**c**) The graph that indicates the connections between region proposals. The figure is redrawn according to Fig. 3 in [59], and the image is obtained from Visual Genome [53]

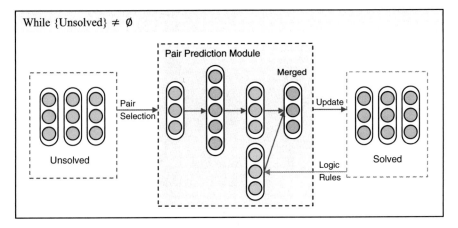

Fig. 7.7 Framework of HRE that detects primary relations from inputs and iteratively completes the scene graph via inductive logic programming. The figure is redrawn according to Fig. 3 in [66]

different levels, the features of objects, triplets, and image regions are expected to be mutually improved to improve the performances of the corresponding tasks.

To further model the inherent dependency of the scene graph generation task, Mao et al. [66] propose to decompose the task into a mixture of two phases: extracting primary relations from the input image first and then completing the scene graph with reasoning. The authors propose a hybrid scene graph generator (HRE) that integrates the two phases in a unified framework.

Specifically, HRE employs a simple visual relation detector to identify primary relations in an image, and a differentiable inductive logic programming model which completes the scene graph iteratively. As shown in Fig. 7.7, HRE consists of two components, an object pair selector and a visual relation predictor that collaborate iteratively. At each time step, the object pair selector considers all object pairs P^- whose relations have not been determined, from which the next object pair is chosen to determine the relation. A greedy strategy is adopted which selects the object pair with the highest relation score. The visual relation predictor considers all the object pairs P^+ whose relations have been determined and the target object pair to predict the target relation. The prediction result of the target object pair is then added to P^+ to benefit future predictions. Exploiting objects and relations in a holistic graph structure can help model their complex associations, which can be useful to reason out complex visual relation interactions.

External Knowledge as Supervision and Regularization While detecting visual relation with image information is intuitive and effective [45, 83, 120], leveraging language and knowledge information can also be helpful [59, 117], since knowledge from language and knowledge graphs can provide high-level priors to supervise or regularize visual relation learning. Lu et al. [63] show that language priors from word embeddings can effectively regularize visual relation learning. Notably, Yao et al. [111] propose to align commonsense knowledge bases with images, which can automatically create large-scale noisy-labeled relation data to provide distant supervision for visual relation learning. The authors also propose to alleviate the noise in distant supervision by refining the probabilistic soft relation labels in an iterative fashion. In this way, distantly supervised models can achieve promising performance without any human annotation, and also significantly improve over fully supervised models when human-labeled data is available.

Inspired by visual distant supervision [111], IETrans [116] proposes to further generate large-scale fine-grained scene graphs via data transfer. To alleviate the long-tail distribution of visual relations, visual distant supervision technique [111] is adopted to augment relation labels from external unlabeled data. Moreover, given an entity pair, human annotators prefer to label general relations (thus uninformative, e.g., on) than informative relations (e.g., riding) for simplicity, which leads to semantic ambiguity in human-annotated data. To address the problem, labels of general relations are transferred to informative ones based on the confusion matrix of relations, which encourages more informative scene graph generation. In this way, IETrans can enable large-scale scene graph generation with over 1,800 fine-grained relation types.

It is worth noting that the task of scene graph generation resembles document-level relation extraction [110] in many aspects. Both tasks seek to extract structured graphs consisting of entities and relations. Also, they need to model the complex dependencies between entities and relations in rich context. We believe both tasks are worthy of exploration for future research, and both tasks can draw inspiration from each other for better development.

7.4.2 Cross-Modal Retrieval

With the rapid growth of multimodal data such as text, image, video, and audio on the Internet, the need to retrieve information across different modalities (i.e., cross-modal retrieval) has become stronger. Given the query data from one modality, cross-modal retrieval aims to retrieve relevant data in other modalities. For example, a user may submit an image of a white horse, and get the textual descriptions of the white horse, and vice versa. Due to the huge number of retrieval candidates, cross-modal retrieval requires efficient computation of semantic similarities (i.e., correlation) between different modalities. This is typically achieved by learning discriminative cross-modal representations from different modalities in a common semantic space.

To learn the common semantic space for different modalities, cross-modal retrieval methods can be divided into two categories, including real-valued representation-based methods and binary-valued representation-based methods.

Real-Valued Representations Data from different modalities is encoded into dense vectors, which can be challenged by inferior efficiency, but are more investigated due to their superior performance. In this line of research, real-valued approaches can be further divided into two categories, including weakly supervised methods and supervised methods.

Weakly Supervised Methods Cross-modal correlation is learned from the naturally paired cross-modal data. For example, images on the Internet are usually paired with textual captions, which can be easily collected in large scale to train cross-modal retrieval models. To learn discriminative representations, contrastive-style learning methods are usually adopted to encourage close representations of paired data (i.e., positive samples), and distinct representations of unpaired data (i.e., negative samples). For example, many works [48, 51, 84, 125] use a bidirectional hinge loss for an image-caption pair (I, s) as follows:

$$\mathcal{L}(I, s) = \sum_{\hat{s}} \max(0, s(I, \hat{s}) - s(I, s) + \gamma) + \sum_{\hat{I}} \max(0, s(s, \hat{I}) - s(I, s) + \gamma),$$

$$(7.5)$$

where γ is a hyper-parameter denoting the margin and $\hat{I}$ and $\hat{s}$ are negative candidates. The objective maximizes the margin of paired and unpaired representations for both image and text as queries. The holistic similarity between images and text can be obtained by aggregating the local similarities between fine-grained image regions and text tokens (e.g., the average of the local similarities).

By summing the loss over all negatives, the negative instances are equally treated in Eq. (7.5). A problem of equal treatment of negatives is that the large number of easy negatives can dominate the loss. To address the issue, VSE++ [24] proposes to mine hard negatives online, by only using the negative that achieves the largest hinge loss in the mini-batch. Despite the simplicity, VSE++ achieves significant improvement and is adopted by many following works [81, 99]. VSE-C [81] creates more challenging adversarial negatives by replacing fine-grained concepts (e.g., numbers and attributes) in the paired text. By augmenting adversarial instances, VSE-C also alleviates the correlation bias of concepts in the dataset, and thus improves the robustness of the model. Wu et al. [99] establish more fine-grained connections between image and text. The sentence semantics is factorized into a composition of nouns, attribute nouns, and relational triplets, where each component is encouraged to be explicitly aligned to images. In summary, since only natural image-caption pairs are required, weakly supervised methods can be easily scaled to leverage large amounts of data.

Supervised Methods In addition to exploiting the natural image-caption pairs, another line of research investigates supervised learning on labeled image-caption data to learn more discriminative cross-modal representations. A semantic label is given for the content of each image-caption pair (e.g., *horse*, *dog*), and the cross-modal representations of the same class label are encouraged to be close to each other [92, 93, 119]. The labeled data can provide high-level semantic supervision for cross-modal representation learning, and therefore usually leads to better image-text retrieval performance.

However, for a specific area of interest, natural unlabeled image-caption pairs can be insufficient, let alone labeled data. This motivates transfer learning from the domains where large amounts of unlabeled/labeled data are available [41]. A major challenge of transfer learning lies in the domain discrepancy between the source domain and the target domain. To address the issue, the distribution discrepancy between different domains is measured by the maximum mean discrepancy (MMD) [33] in the reproduced kernel Hilbert space. By minimizing the MMD loss, the image representations from source and target domains are encouraged to have the same distribution to facilitate knowledge transfer.

In addition to unlabeled image-caption pairs, Huang et al. [40] further transfer knowledge from labeled image-caption pairs. Since both domains contain image and text, domain discrepancies come from both modal-level discrepancies in the same modality and correlation-level discrepancies in image-text correlation patterns between different domains. An MMD loss is imposed on both modal-level and correlation-level to reduce the domain discrepancies between the source and target domains.

Binary-Valued Representations Information from each modality is encoded into a common Hamming space, which yields better efficiency for both computation and storage [14, 46, 121]. However, due to the limited expressiveness of binary-valued representations, the performance of such models could be affected by the loss of valuable information. Therefore, real-valued representation-based methods are more widely investigated.

It is worth noting that the usefulness of image-text retrieval is not only limited to a search engine that acquires cross-modal information for users. Many cross-modal understanding and generation tasks can also be formulated as an image-text retrieval problem, for example, retrieving labels from the category set for image classification [74] and retrieving sentences from text corpus for image captioning [55]. Image-text retrieval can also serve as a critical component in cross-modal models when we need relevant information of the data in interest (e.g., related knowledge for an image) [111].

7.4.3   Cross-Modal Generation

Given the information in one modality (e.g., the text description or image about a horse), can we *generate* its counterpart in another modality? This cross-modal generation capability is an appealing yet challenging problem. Specifically, cross-modal generation can be divided into image-to-text generation and text-to-image generation. Compared with other capabilities, cross-modal generation is more challenging for two reasons: (1) A comprehensive understanding of the source modal is required. For example, in image-to-text generation, not only objects but also relations between them have to be detected. (2) Semantic-preserving natural language sentences or images have to be generated. In this section, we take image captioning as an example to introduce methods for image-to-text generation in detail, and then briefly review the methods for text-to-image generation.

Image captioning is the task of generating natural language descriptions for images. It is worth noting that the task of image captioning is inherently analogous to machine translation because it can also be regarded as a translation task from the source "language" of image to natural language. Therefore, many image captioning models have drawn inspiration from the advances in machine translation.

Due to the challenge of language generation, many early works in image captioning retrieve related text to produce the caption [25, 71], where the flexibility of the generated text is limited. From 2015, inspired by advances in neural machine translation [6], most image captioning models begin to adopt an encoder-decoder framework [91], as shown in Fig. 7.8. Typically, images are first encoded into distributed representations using visual encoders such as CNNs, based on which the caption is generated using neural language models such as RNNs. The encoder-decoder framework significantly improves the ability to generate natural language descriptions. To better establish the connection between image understanding and

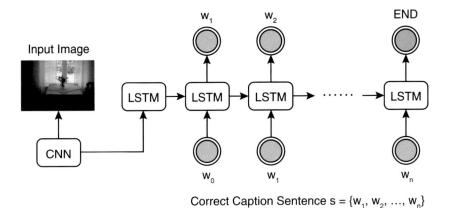

Fig. 7.8 The architecture of encoder-decoder framework for image captioning

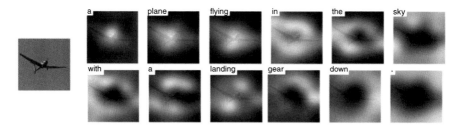

Fig. 7.9 An example of image captioning with attention mechanism. The example is obtained from the implementation of Yunjey Choi (https://github.com/yunjey/show-attend-and-tell)

text generation, attention mechanism and graph-based methods have been mostly investigated.

Attention Mechanism Intuitively, it can be beneficial to attend to fine-grained image regions via attention mechanism when generating the corresponding text tokens. Inspired by the attention mechanism in machine translation [6], Xu et al. [103] introduce visual attention into the encoder-decoder image captioning model. The major bottleneck of the vanilla encoder-decoder framework [91] is that rich information from an image is represented in one static representation to produce a complex sentence. In contrast, Xu et al. [103] encode each image grid region into representations, and allow the decoder to generate each text token based on a dynamic image representation of related regions. The model learns to focus on parts of the image to generate the next word by producing larger attention weights on more relevant parts, as shown in Fig. 7.9.

Despite the effectiveness, Liu et al. [60] find that the implicitly learned attention is not guaranteed to be closely related to text tokens. To alleviate the problem, Liu et al. [60] propose to explicitly supervise the attention distribution over image grids for text tokens. For each object in text, the supervision can come from visual grounding annotations, or textual similarities of detected object tags. This makes the attention more explainable, and also improves the performance since related visual information is better selected. Similarly, Karpathy et al. [48] make explicit alignment between image regions and sentence fragments before generating a description for the image. The explicit alignment is achieved by maximizing the similarity of image-caption pairs, where the holistic similarity is aggregated by the local alignment between image regions and text fragments.

The attention computed over uniform image grids can split and corrupt high-level semantics (e.g., holistic objects). To address the issue, Anderson et al. [3] propose to calculate attention over detected objects. Since the image regions reserve high-level semantics, the attention over such regions can be better associated with the concepts in text. Due to the simplicity and effectiveness, the object-aware attention mechanism is adopted by many following works [39, 73]. Since visual question answering and image captioning both require establishing fine-grained cross-modal correlation, many approaches can be utilized for both tasks (e.g., object-aware attention mechanism).

Scene Graphs as Scene Abstractions In another line of research, scene graphs have been adopted to help describe the complex scene. Scene graphs represent objects and their relations in a graph structure, which can benefit image captioning in two aspects: (1) Scene graphs can provide high-level semantics of objects and their interactions for deep understanding of the scene. There is a general consensus that it is visual relations, rather than objects alone, which determine the semantics of the scene [53]. (2) Compared with pixel features, the high-level semantics can be better aligned with textual descriptions.

To leverage scene graphs for image captioning, some works [108, 122] employ graph neural networks over the scene graph consisting of objects and their semantic and spatial relations. The object information passes along the relation edges based on the graph neural networks. Similar to the vanilla attention approach of Xu et al. [103], the decoder dynamically attends to the scene graph when generating each text token. In addition to representing images, scene graphs can also be extracted from the paired text during training. In this view, scene graphs can serve as a common intermediate representation to transfer the prior from large-scale text to improve image captioning [106].

Compared with image-to-text generation, text-to-image faces different challenges, where the key problem is image generation. Existing methods in text-to-image generation can be roughly divided into three categories, including VAE-based [50] and GAN-based [31] methods, and diffusion-based models [76]. Typical research problems in text-to-image generation include high-resolution image generation [20], stable training of image generation models [75], efficient image generation [7], conditional image generation [70], etc.

7.5 Deep Cross-Modal Pre-training

The cross-modal representation learning methods we have introduced in previous sections are limited to either shallow embeddings (i.e., word vectors) or task-specific model architectures. Recently, the most significant advance and trend in cross-modal representation learning is deep cross-modal pre-training. The key idea is to fully exploit the self-supervised signals from large-scale data to pre-train generic deep cross-modal representations. The pre-training is typically performed to learn cross-modal capabilities based on Transformer architectures [90] and self-supervised tasks [64], which is largely unified and agnostic to specific tasks. Then, the pre-trained deep cross-modal representations can be tuned to adapt to downstream tasks. This revolutionary paradigm has greatly pushed forward the state-of-the-art performance of a wide range of cross-modal tasks.

The key to cross-modal representation learning is to establish fine-grained connections between cross-modal signals. A common architecture suitable for modeling data from different modalities constitutes the most important foundation of cross-modal pre-training. Early works try to fully exploit the inductive bias of each modality. For example, convolution and pooling are designed to model the scale and shift invariant property of images in CNNs [36, 54], and recurrent computation is devised to model the sequential dependency of text in RNNs [19, 38]. Despite the effectiveness in modeling each modality, their highly specialized design hinders the generalization to other modalities. In comparison, stacked self-attention, the main component of Transformers, reflects a more general principle of information exchange and aggregation, which has been proven to be effective in modeling different modalities, including text, speech, image, and video. Moreover, Transformers enjoy better scalability in both data and parameters, where larger data and parameter scale can typically always lead to better performance [12]. In this section, we introduce recent advances in deep cross-modal pre-training, from the input representations, basic architecture, and pre-training tasks to tuning approaches.

7.5.1 Input Representations

An important problem in joint cross-modal data modeling is a more unified input representation to the Transformer architecture. The basic symbolic units of text (e.g., word tokens) naturally fit the design of Transformers. The main focus has been on image input representation, where the solutions include token-based, object-based, and patch-based methods.

Token-Based Representations Images or image patches are represented as discrete tokens. The tokens can be obtained from clustering [87], or discrete variational auto-encoders [8, 77]. The form of discrete visual tokens maximally aligns with the practice of the text domain, which is convenient for unified input and supervision for text and image. However, detailed visual information might be lost in the fixed discrete tokens.

Object-Based Representations Salient objects (e.g., object features, labels, and locations) in an image are used to represent the image content [64, 86, 89, 113]. Objects carry more high-level information, and can be better aligned with concepts in text. Some works further propose to use object tags to bridge objects in images and concepts in text [58, 118]. However, object-based methods rely on external object detectors to obtain input representations, which can be expensive in both annotation and computation [57]. The background information in images may also be lost.

Patch-Based Representations Features of image grid patches are adopted as the image input representations [23, 35, 57]. Patch-based methods (e.g., ViT [23]) and their pre-training (e.g., MAE [35]) can achieve state-of-the-art performance. Moreover, since external detectors are not used, patch-based models are significantly faster than object-based methods. However, since objects are not explicitly modeled, patch-based vision-language models can have difficulty in dealing with object position-sensitive tasks [57]. To address the problem, some works propose to treat positions as discrete tokens [95, 109], which enables unified explicit modeling of text and positions. Notably, PEVL [109] retrains the order of discretized positions by an ordering-aware reconstruction objective, which achieves competitive performance on various vision-language tasks.

7.5.2 Model Architectures

Based on largely unified input representations for different modalities, several model architectures based on Transformers have been developed to model cross-modal data interaction. Existing model architectures can be divided into three categories, including Transformer encoders, decoders, and encoder-decoders.

Transformer Encoder Architectures Inspired by BERT [21], Transformer encoders have been widely used to align and fuse cross-modal information, which can be further divided into single-stream methods and two-stream methods.

Single-Stream Methods Image and text input representations are fed into a single Transformer encoder, which jointly encodes cross-modal information with shared parameters [26, 58, 64, 89, 118], as shown in Fig. 7.10. Since fine-grained image regions and text tokens are jointly modeled, the architecture can yield very competitive performance, especially for cross-modal understanding tasks. Therefore, single-stream methods are the most widely used vision-language architecture. However, it is not easy to perform cross-modal generation and retrieval via a single-stream Transformer encoder.

Two-Stream Methods Images and text inputs are encoded into a common semantic space by separate unimodal encoders in a similar way to cross-modal retrieval [44, 74], as shown in Fig. 7.11. The common semantic space allows for efficient similarity computation of cross-modal data. Moreover, due to the efficiency of the architecture, two-stream methods are scalable to process Web-level data, which can yield open recognition capabilities. Notably, CLIP [74] is trained with 400 million image-text pairs, and can perform zero-shot open-vocabulary image classification by retrieving text labels for images. However, since fine-grained cross-modal

Fig. 7.10 Single-stream architectures, where image and text are input into a single cross-modal Transformer encoder

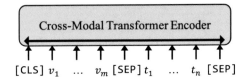

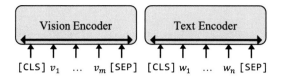

Fig. 7.11 Two-stream architectures, where image and text are encoded by separate unimodal encoders into a common semantic space

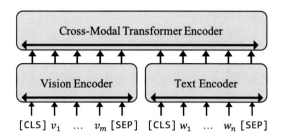

Fig. 7.12 Hybrid architectures, where image and text are first encoded by separate unimodal encoders, and then fused by a cross-modal encoder

interactions cannot be modeled, the performance of two-stream models may be limited on complex cross-modal understanding tasks.

Hybrid Methods Some works also propose to encode image and text first by separate unimodal encoders, and then fuse the unimodal representations using a cross-modal encoder [57, 64, 113], as shown in Fig. 7.12. The rationale is that modal-specific information can be better encoded in separate unimodal encoders before cross-modal fusion.

Transformer Decoder Architectures Decoder-only models have not been widely used in pre-trained vision-language models, since a bidirectional encoder is usually required to better understand the image (and text). However, decoder-only models can be convenient in generating images by producing visual tokens in an auto-regressive fashion. For example, DALL-E [77] models text tokens and image tokens auto-regressively to perform text-to-image generation.

Transformer Encoder-decoder Architectures In encoder-decoder architecture, image and prefix-text are encoded using encoders, and suffix-text are generated via decoders [2, 18, 47, 95, 98], as shown in Fig. 7.13. This architecture is becoming increasingly popular, since image and text can be well encoded, and the decoder is flexible to deal with various vision-language tasks in a unified fashion. Notably, Flamingo [2] bridges frozen large language PTMs with vision encoders, which produces strong in-context few-shot learning capabilities for vision-language tasks.

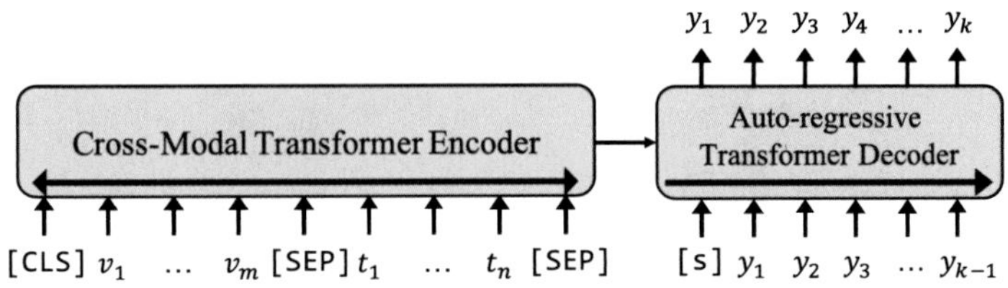

Fig. 7.13 Encoder-decoder architectures. Image and text are first encoded by a cross-modal encoder, and then the targets are generated via a decoder

7.5.3 Pre-training Tasks

Pre-training tasks aim to fully exploit self-supervised learning signals from large-scale cross-modal data. The pre-training cross-modal data includes (1) image-caption pairs annotated by humans [15, 53] or crawled from the Internet [2, 80] and (2) collections of labeled downstream datasets [47, 109]. We divide popular vision-language pre-training tasks into three categories, including text-oriented tasks, image-oriented tasks, and image-text-oriented tasks.

Text-Oriented Tasks Pre-training tasks in language models have been widely used for self-supervised cross-modal learning. (1) Masked language modeling reconstructs masked tokens in text [58, 64, 89, 95, 109], and is the most widely used pre-training task. Masked language modeling is usually used to pre-train bidirectional Transformer encoders for deep cross-modal understanding. (2) Left-to-right language modeling performs auto-regressive generation of text tokens based on Transformer encoder-decoders, which can yield flexible text generation capabilities [2, 18, 98].

Image-Oriented Tasks Compared with text, images consist of continuous pixels with low information density, which makes it challenging to mine high-level self-supervised learning signals [35]. To obtain the high-level semantics for pre-training, existing works resort to objects, image tokens, and high masking rates. (1) Object-based pre-training tasks reconstruct high-level semantics given by object detectors. After masking the image regions identified by object detectors, the pre-training task can be reconstructing the discrete object labels [16, 86], reconstructing continuous object label distributions [16, 64], or regressing the region features [16, 89]. (2) Image token-based pre-training tasks aim to reconstruct the masked discrete visual tokens [8, 77]. However, both objects and visual tokens require external tools to obtain. (3) Masked patch-based methods directly reconstruct pixels from masked image grid patches, which do not need external tools. Notably, MAE [35] finds that high masking rates are key to learning high-level semantics from image pixel reconstruction.

Image-Text-Oriented Tasks Text-oriented and image-oriented tasks impose local supervision on text tokens and image regions. In comparison, image-text-oriented tasks pay more attention to holistic semantic matching between image and text. (1) Image-text matching is a popular pre-training task that conducts binary classification of a given image-text pair to judge the matching degree [26, 58, 64, 89, 118]. The task is usually used in single-stream Transformer encoders, where fine-grained cross-modal alignment is performed. (2) Image-text contrastive learning tasks encourage paired image and text representations to be close in a common semantic space via contrastive learning. The task is mostly used in two-stream Transformer encoders [44, 74] or hybrid architectures [57] to achieve holistic image-text matching.

7.5.4 Adaptation Approaches

General cross-modal capabilities can be learned in self-supervised pre-training. During fine-tuning, new parameters and objective forms are typically introduced to adapt pre-trained models to downstream tasks, leading to significant gap between pre-training and downstream tuning. For example, an MLP is typically introduced to predict the answers for visual question answering. The gap hinders the effective adaptation of pre-trained capabilities to downstream tasks. Recently some works have shown promising results in data-efficient and parameter-efficient adaptation of pre-trained vision-language models via prompt learning.

Data-Efficient Prompt Learning The key idea of data-efficient prompt learning is that, by reformulating downstream tasks into the same form as pre-training, the gap between pre-training and downstream tuning can be maximally mitigated. Therefore, vision-language pre-training models can be efficiently adapted to downstream tasks with only few-shot and even zero-shot examples. Specifically, similar to GPT-3 [12], vision-language models pre-trained with a language generation task can naturally handle various tasks without significant gap [2, 18, 95, 98]. By reformulating various tasks into a unified language generation task, data-efficient prompt learning largely mitigates not only the gap between pre-training and tuning but also the gap between different tasks.

However, it can be difficult to explicitly establish fine-grained cross-modal connections via natural language prompts for various position-sensitive tasks, such as visual grounding [72], visual commonsense reasoning [115], and visual relation detection [53]. To address the challenge, CPT [112] explicitly bridges image regions and text via natural color-based coreferential markers, as shown in Fig. 7.14. By reformulating cross-modal tasks into a fill-in-the-blank problem, pre-trained vision-language models can be prompted to achieve strong few-shot and even zero-shot performance on position-sensitive tasks.

Parameter-Efficient Prompt Learning Inspired by delta tuning in pre-trained language models (Chap. 5), some works propose to only tune several prompt vectors, instead of full model parameters, to adapt the pre-trained vision-language models. The prompt vectors can be static across different samples [124] or condi-

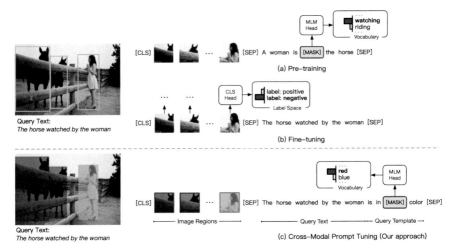

Fig. 7.14 Cross-modal prompt learning for vision-language models. The figure is redrawn from the Fig. 1 in [112], and the image is obtained from Visual Genome [53]

tional on specific samples [123]. The tunable parameters can also be lightweight adapters [28]. Since only pivotal parameters need to be tuned, parameter-efficient prompt learning methods can better avoid overfitting on few-shot data, and therefore achieve better few-shot performance compared with full parameter fine-tuning. However, since new parameters are introduced, it can be difficult for parameter-efficient prompt learning methods to deal with zero-shot tasks.

7.6 Applications

Now we have introduced cross-modal representation learning methods for cross-modal capabilities, including cross-modal understanding, retrieval, and generation. Various specific tasks and models have been proposed to investigate and implement each capability. In practice, many real-world applications may require multiple cross-modal capabilities. In this section, we take robotic assistants as an example (e.g., assisting humans to accomplish tasks, such as fetching objects at home according to language instructions). We illustrate how the cross-modal capabilities can be adapted and integrated and to solve complex real-world applications.

A long-standing goal of AI is to build intelligent agents that can communicate and assist humans in the physical world. The agent will need to perform cross-modal perception of the environment and humans, cross-modal reasoning for action plan generation, and cross-modal interaction for navigation and manipulation.

Cross-Modal Perception To assist humans in finishing tasks in real-world environments, a basic foundation for agents is to comprehensively perceive cross-modal information from both human instructions and the environment. (1) Human instructions. A clear instruction is typically given to the agent (e.g., *go straight, turn right, and walk into the bedroom*), which the agent needs to understand and follow to finish the task [4, 43]. The instruction can also be ambiguous, where agents need to ask for further clarifications or even converse with humans according to the situation [17]. (2) Environment. Multisensory perceptions of the environment are typically required and helpful to finish tasks in the physical environment, including vision, text, audio, and even tactile sensation [29].

Cross-Modal Reasoning In real-world scenarios, step-by-step instructions are usually not available, and only holistic instructions are given (e.g., *walk into the bedroom*) [101]. The agent typically needs to produce an actionable plan for the instruction (i.e., a sequence of actions that are well embodied with the environment). The plans can be implicitly learned by reinforcement learning [97]. Recently, large PTMs have shown promising results in cross-modal reasoning for explicit plan generation [11]. It is an open and promising direction to ground the knowledge of PTMs into the physical world.

Cross-Modal Interaction Based on cross-modal perception and reasoning, agents need to actively interact with the environment to finish the task. Specifically, this typically includes actual execution of the plan to navigate to the target (intermediate) positions (e.g., *walk upstairs and then go into the bedroom*) and manipulation of the objects (*put the apple on the table*) [32]. Currently, most works investigate cross-modal interactions in simulated environments for convenience [17, 32, 101], whereas some works are implemented in real-world environments [11].

In addition to robotic assistants, cross-modal representation learning can also be essential for other real-world AI applications. For example, multimodal perception of the complex physical environment is important for robust decision-making in autonomous vehicles [78]. Multimodal computation can also empower the construction and interaction of 3D metaverse [88].

7.7   Summary and Further Readings

In this chapter, we first introduce the concept of cross-modal representation learning. Cross-modal learning is essential since many real-world tasks require the ability to understand information from different modalities, such as text and image. It is also typically helpful to exploit complementary information in different modalities for comprehensive judgment. We introduce a taxonomy of cross-modal capabilities, including cross-modal understanding, retrieval, and generation. Based on the taxonomy, we review existing cross-modal representation learning methods, from shallow to deep cross-modal representations. Notably, deep cross-modal pre-training has been a revolutionary paradigm, which largely unifies model architectures and learning mechanisms for modalities and tasks, and has greatly pushed forward state-of-the-art results. Finally, we introduce representative cross-modal applications.

Cross-modal representation learning is drawing more and more attention and can serve as a promising connection between different research areas.

For further understanding of cross-modal representation learning, there are also some recommended surveys and books. Spence [85] provides a tutorial review of cross-modal correspondences from the perspective of cognitive neuroscience. Wang et al. [94] give a comprehensive survey on cross-modal retrieval, and Xu et al. [104] provide a survey of cross-modal learning with Transformers.

Acknowledgments The contributions of all authors for the second edition are Zhiyuan Liu and Yankai Lin, and Maosong Sun designed the overall architecture of this chapter. Yuan Yao drafted this chapter. Zhiyuan Liu and Yankai Lin proofread and revised this chapter.

We thank Haoye Zhang for drawing figures, and thank Shengding Hu, Ning Ding, Haoye Zhang, Tianyu Yu, Qianyu Chen, and Hantao Zhou for proofreading the chapter. We also thank Hao Zhu, Ji Xin, and Deming Ye for preparing some initial draft materials for the first edition.

This is the cross-modal representation learning chapter of the second edition of the book *Representation Learning for Natural Language Processing*, with its first edition published in 2020 [62]. As compared with the first edition of this chapter, the main changes include the following: (1) we improved the review to deep cross-modal representation learning methods under a cross-modal capability framework, and (2) we added deep cross-modal pre-training methods and applications.

References

1. Aishwarya Agrawal, Dhruv Batra, Devi Parikh, and Aniruddha Kembhavi. Don't just assume; look and answer: Overcoming priors for visual question answering. In *Proceedings of CVPR*, 2018.
2. Jean-Baptiste Alayrac, Jeff Donahue, Pauline Luc, Antoine Miech, Iain Barr, Yana Hasson, Karel Lenc, Arthur Mensch, Katherine Millican, Malcolm Reynolds, Roman Ring, Eliza Rutherford, Serkan Cabi, Tengda Han, Zhitao Gong, Sina Samangooei, Marianne Monteiro, Jacob Menick, Sebastian Borgeaud, Andrew Brock, Aida Nematzadeh, Sahand Sharifzadeh, Mikolaj Binkowski, Ricardo Barreira, Oriol Vinyals, Andrew Zisserman, and Karen Simonyan. Flamingo: a visual language model for few-shot learning. In *Proceedings of NeurIPS*, 2022.
3. Peter Anderson, Xiaodong He, Chris Buehler, Damien Teney, Mark Johnson, Stephen Gould, and Lei Zhang. Bottom-up and top-down attention for image captioning and visual question answering. In *Proceedings of CVPR*, 2018.
4. Peter Anderson, Qi Wu, Damien Teney, Jake Bruce, Mark Johnson, Niko Sünderhauf, Ian Reid, Stephen Gould, and Anton Van Den Hengel. Vision-and-language navigation: Interpreting visually-grounded navigation instructions in real environments. In *Proceedings of CVPR*, 2018.
5. Stanislaw Antol, Aishwarya Agrawal, Jiasen Lu, Margaret Mitchell, Dhruv Batra, C Lawrence Zitnick, and Devi Parikh. VQA: Visual question answering. In *Proceedings of ICCV*, 2015.
6. Dzmitry Bahdanau, Kyunghyun Cho, and Yoshua Bengio. Neural machine translation by jointly learning to align and translate. In *Proceedings of ICLR*, 2015.
7. Fan Bao, Chongxuan Li, Jun Zhu, and Bo Zhang. Analytic-DPM: an analytic estimate of the optimal reverse variance in diffusion probabilistic models. In *Proceedings of ICLR*, 2021.
8. Hangbo Bao, Li Dong, Songhao Piao, and Furu Wei. BEiT: BERT pre-training of image Transformers. In *Proceedings of ICLR*, 2021.

9. Mohamed Benzeghiba, Renato De Mori, Olivier Deroo, Stephane Dupont, Teodora Erbes, Denis Jouvet, Luciano Fissore, Pietro Laface, Alfred Mertins, Christophe Ris, et al. Automatic speech recognition and speech variability: A review. *Speech communication*, 49(10–11):763–786, 2007.
10. Francisco Bonin-Font, Alberto Ortiz, and Gabriel Oliver. Visual navigation for mobile robots: A survey. *Journal of Intelligent and Robotic Systems*, 53(3):263–296, 2008.
11. Anthony Brohan, Yevgen Chebotar, Chelsea Finn, Karol Hausman, Alexander Herzog, Daniel Ho, Julian Ibarz, Alex Irpan, Eric Jang, Ryan Julian, et al. Do as I can, not as I say: Grounding language in robotic affordances. In *Proceedings of CoRL*, 2022.
12. Tom Brown, Benjamin Mann, Nick Ryder, Melanie Subbiah, Jared D Kaplan, Prafulla Dhariwal, Arvind Neelakantan, Pranav Shyam, Girish Sastry, Amanda Askell, et al. Language models are few-shot learners. In *Proceedings of NeurIPS*, 2020.
13. Elia Bruni, Gemma Boleda, Marco Baroni, and Nam-Khanh Tran. Distributional semantics in technicolor. In *Proceedings of ACL*, 2012.
14. Yue Cao, Mingsheng Long, Jianmin Wang, Qiang Yang, and Philip S Yu. Deep visual-semantic hashing for cross-modal retrieval. In *Proceedings of KDD*, 2016.
15. Xinlei Chen, Hao Fang, Tsung-Yi Lin, Ramakrishna Vedantam, Saurabh Gupta, Piotr Dollár, and C Lawrence Zitnick. Microsoft COCO captions: Data collection and evaluation server. *arXiv preprint arXiv:1504.00325*, 2015.
16. Yen-Chun Chen, Linjie Li, Licheng Yu, Ahmed El Kholy, Faisal Ahmed, Zhe Gan, Yu Cheng, and Jingjing Liu. UNITER: UNiversal Image-TExt Representation Learning. In *Proceedings of ECCV*, 2020.
17. Ta-Chung Chi, Minmin Shen, Mihail Eric, Seokhwan Kim, and Dilek Hakkani-tur. Just ask: An interactive learning framework for vision and language navigation. In *Proceedings of AAAI*, 2020.
18. Jaemin Cho, Jie Lei, Hao Tan, and Mohit Bansal. Unifying vision-and-language tasks via text generation. In *Proceedings of ICML*, 2021.
19. Junyoung Chung, Caglar Gulcehre, KyungHyun Cho, and Yoshua Bengio. Empirical evaluation of gated recurrent neural networks on sequence modeling. *arXiv preprint arXiv:1412.3555*, 2014.
20. Emily L Denton, Soumith Chintala, Rob Fergus, et al. Deep generative image models using a laplacian pyramid of adversarial networks. In *Proceedings of NeurIPS*, 2015.
21. Jacob Devlin, Ming-Wei Chang, Kenton Lee, and Kristina Toutanova. BERT: pre-training of deep bidirectional transformers for language understanding. In *Proceedings of NAACL-HLT*, 2019.
22. Linhao Dong, Shuang Xu, and Bo Xu. Speech-Transformer: a no-recurrence sequence-to-sequence model for speech recognition. In *Proceedings of ICASSP*, 2018.
23. Alexey Dosovitskiy, Lucas Beyer, Alexander Kolesnikov, Dirk Weissenborn, Xiaohua Zhai, Thomas Unterthiner, Mostafa Dehghani, Matthias Minderer, Georg Heigold, Sylvain Gelly, et al. An image is worth 16x16 words: Transformers for image recognition at scale. In *Proceedings of ICLR*, 2021.
24. Fartash Faghri, David J Fleet, Jamie Ryan Kiros, and Sanja Fidler. VSE++: Improving visual-semantic embeddings with hard negatives. In *Proceedings of BMVC*, 2018.
25. Ali Farhadi, Mohsen Hejrati, Mohammad Amin Sadeghi, Peter Young, Cyrus Rashtchian, Julia Hockenmaier, and David Forsyth. Every picture tells a story: Generating sentences from images. In *Proceedings of ECCV*, 2010.
26. Zhe Gan, Yen-Chun Chen, Linjie Li, Chen Zhu, Yu Cheng, and Jingjing Liu. Large-scale adversarial training for vision-and-language representation learning. In *Proceedings of NeurIPS*, 2020.
27. Haoyuan Gao, Junhua Mao, Jie Zhou, Zhiheng Huang, Lei Wang, and Wei Xu. Are you talking to a machine? dataset and methods for multilingual image question. In *Proceedings of NeurIPS*, 2015.

28. Peng Gao, Shijie Geng, Renrui Zhang, Teli Ma, Rongyao Fang, Yongfeng Zhang, Hongsheng Li, and Yu Qiao. CLIP-Adapter: Better vision-language models with feature adapters. *arXiv preprint arXiv:2110.04544*, 2021.

29. Ruohan Gao, Zilin Si, Yen-Yu Chang, Samuel Clarke, Jeannette Bohg, Li Fei-Fei, Wenzhen Yuan, and Jiajun Wu. ObjectFolder 2.0: A multisensory object dataset for sim2real transfer. In *Proceedings of CVPR*, 2022.

30. Rohit Girdhar, Joao Carreira, Carl Doersch, and Andrew Zisserman. Video action transformer network. In *Proceedings of CVPR*, 2019.

31. Ian Goodfellow, Jean Pouget-Abadie, Mehdi Mirza, Bing Xu, David Warde-Farley, Sherjil Ozair, Aaron Courville, and Yoshua Bengio. Generative adversarial networks. *Communications of the ACM*, 63(11):139–144, 2020.

32. Daniel Gordon, Aniruddha Kembhavi, Mohammad Rastegari, Joseph Redmon, Dieter Fox, and Ali Farhadi. IQA: Visual question answering in interactive environments. In *Proceedings of CVPR*, 2018.

33. Arthur Gretton, Karsten M Borgwardt, Malte J Rasch, Bernhard Schölkopf, and Alexander Smola. A kernel two-sample test. *Journal of Machine Learning Research*, 13(1):723–773, 2012.

34. Liangke Gui, Borui Wang, Qiuyuan Huang, Alex Hauptmann, Yonatan Bisk, and Jianfeng Gao. KAT: A knowledge augmented Transformer for vision-and-language. In *Proceedings of NAACL*, 2021.

35. Kaiming He, Xinlei Chen, Saining Xie, Yanghao Li, Piotr Dollár, and Ross Girshick. Masked autoencoders are scalable vision learners. In *Proceedings of CVPR*, 2022.

36. Kaiming He, Xiangyu Zhang, Shaoqing Ren, and Jian Sun. Deep residual learning for image recognition. In *Proceedings of CVPR*, 2016.

37. Felix Hill and Anna Korhonen. Learning abstract concept embeddings from multi-modal data: Since you probably can't see what I mean. In *Proceedings of EMNLP*, 2014.

38. Sepp Hochreiter and Jürgen Schmidhuber. Long short-term memory. *Neural Computation*, 1997.

39. Lun Huang, Wenmin Wang, Yaxian Xia, and Jie Chen. Adaptively aligned image captioning via adaptive attention time. In *Proceedings of NeurIPS*, 2019.

40. Xin Huang and Yuxin Peng. Deep cross-media knowledge transfer. In *Proceedings of CVPR*, 2018.

41. Xin Huang, Yuxin Peng, and Mingkuan Yuan. Cross-modal common representation learning by hybrid transfer network. In *Proceedings of IJCAI*, 2017.

42. Drew A Hudson and Christopher D Manning. GQA: A new dataset for real-world visual reasoning and compositional question answering. In *Proceedings of CVPR*, 2019.

43. Vihan Jain, Gabriel Magalhaes, Alexander Ku, Ashish Vaswani, Eugene Ie, and Jason Baldridge. Stay on the path: Instruction fidelity in vision-and-language navigation. In *Proceedings of ACL*, 2019.

44. Chao Jia, Yinfei Yang, Ye Xia, Yi-Ting Chen, Zarana Parekh, Hieu Pham, Quoc Le, Yun-Hsuan Sung, Zhen Li, and Tom Duerig. Scaling up visual and vision-language representation learning with noisy text supervision. In *Proceedings of ICML*, 2021.

45. Zhaoyin Jia, Andrew Gallagher, Ashutosh Saxena, and Tsuhan Chen. 3D-based reasoning with blocks, support, and stability. In *Proceedings of ICCV*, 2013.

46. Qing-Yuan Jiang and Wu-Jun Li. Deep cross-modal hashing. In *Proceedings of CVPR*, 2017.

47. Aishwarya Kamath, Mannat Singh, Yann LeCun, Gabriel Synnaeve, Ishan Misra, and Nicolas Carion. MDETR: modulated detection for end-to-end multi-modal understanding. In *Proceedings of CVPR*, 2021.

48. Andrej Karpathy and Li Fei-Fei. Deep visual-semantic alignments for generating image descriptions. In *Proceedings of CVPR*, 2015.

49. Douwe Kiela, Felix Hill, Anna Korhonen, and Stephen Clark. Improving multi-modal representations using image dispersion: Why less is sometimes more. In *Proceedings of ACL*, 2014.

50. Diederik P Kingma and Max Welling. Auto-encoding variational bayes. In *Proceedings of ICLR*, 2014.
51. Ryan Kiros, Ruslan Salakhutdinov, and Richard S Zemel. Unifying visual-semantic embeddings with multimodal neural language models. *arXiv preprint arXiv:1411.2539*, 2014.
52. Satwik Kottur, Ramakrishna Vedantam, José MF Moura, and Devi Parikh. Visual Word2vec (vis-w2v): Learning visually grounded word embeddings using abstract scenes. In *Proceedings of CVPR*, 2016.
53. Ranjay Krishna, Yuke Zhu, Oliver Groth, Justin Johnson, Kenji Hata, Joshua Kravitz, Stephanie Chen, Yannis Kalantidis, Li-Jia Li, David A Shamma, et al. Visual Genome: Connecting language and vision using crowdsourced dense image annotations. *International Journal of Computer Vision*, 123(1):32–73, 2017.
54. Alex Krizhevsky, Ilya Sutskever, and Geoffrey E Hinton. ImageNet classification with deep convolutional neural networks. In *Proceedings of NeurIPS*, 2012.
55. Girish Kulkarni, Visruth Premraj, Sagnik Dhar, Siming Li, Yejin Choi, Alexander C Berg, and Tamara L Berg. Baby talk: Understanding and generating image descriptions. In *Proceedings of CVPR*, 2011.
56. Angeliki Lazaridou, Marco Baroni, et al. Combining language and vision with a multimodal Skip-gram model. In *Proceedings of NAACL*, 2015.
57. Junnan Li, Ramprasaath Selvaraju, Akhilesh Gotmare, Shafiq Joty, Caiming Xiong, and Steven Chu Hong Hoi. Align before fuse: Vision and language representation learning with momentum distillation. In *Proceedings of NeurIPS*, 2021.
58. Xiujun Li, Xi Yin, Chunyuan Li, Pengchuan Zhang, Xiaowei Hu, Lei Zhang, Lijuan Wang, Houdong Hu, Li Dong, Furu Wei, et al. Oscar: Object-semantics aligned pre-training for vision-language tasks. In *Proceedings of ECCV*, 2020.
59. Yikang Li, Wanli Ouyang, Bolei Zhou, Kun Wang, and Xiaogang Wang. Scene graph generation from objects, phrases and region captions. In *Proceedings of ICCV*, 2017.
60. Chenxi Liu, Junhua Mao, Fei Sha, and Alan Yuille. Attention correctness in neural image captioning. In *Proceedings of AAAI*, 2017.
61. Frederick Liu, Han Lu, Chieh Lo, and Graham Neubig. Learning character-level compositionality with visual features. In *Proceedings of ACL*, 2017.
62. Zhiyuan Liu, Yankai Lin, and Maosong Sun. *Representation Learning for Natural Language Processing*. Springer, 2020.
63. Cewu Lu, Ranjay Krishna, Michael Bernstein, and Li Fei-Fei. Visual relationship detection with language priors. In *Proceedings of ECCV*, 2016.
64. Jiasen Lu, Dhruv Batra, Devi Parikh, and Stefan Lee. ViLBERT: Pretraining task-agnostic visiolinguistic representations for vision-and-language tasks. In *Proceedings of NeurIPS*, 2019.
65. Jiasen Lu, Jianwei Yang, Dhruv Batra, and Devi Parikh. Hierarchical question-image co-attention for visual question answering. In *Proceedings of NeurIPS*, 2016.
66. Jiayuan Mao, Yuan Yao, Stefan Heinrich, Tobias Hinz, Cornelius Weber, Stefan Wermter, Zhiyuan Liu, and Maosong Sun. Bootstrapping knowledge graphs from images and text. *Frontiers in Neurorobotics*, 13:93, 2019.
67. Kenneth Marino, Xinlei Chen, Devi Parikh, Abhinav Gupta, and Marcus Rohrbach. KRISP: Integrating implicit and symbolic knowledge for open-domain knowledge-based VQA. In *Proceedings of CVPR*, 2021.
68. Harry McGurk and John MacDonald. Hearing lips and seeing voices. *Nature*, 1976.
69. Tomas Mikolov, Kai Chen, Greg Corrado, and Jeffrey Dean. Efficient estimation of word representations in vector space. In *Proceedings of ICLR*, 2013.
70. Mehdi Mirza and Simon Osindero. Conditional generative adversarial nets. *arXiv preprint arXiv:1411.1784*, 2014.
71. Vicente Ordonez, Girish Kulkarni, and Tamara Berg. Im2text: Describing images using 1 million captioned photographs. In *Proceedings of NeurIPS*, 2011.

72. Bryan A Plummer, Liwei Wang, Chris M Cervantes, Juan C Caicedo, Julia Hockenmaier, and Svetlana Lazebnik. Flickr30k entities: Collecting region-to-phrase correspondences for richer image-to-sentence models. In *Proceedings of ICCV*, 2015.

73. Yu Qin, Jiajun Du, Yonghua Zhang, and Hongtao Lu. Look back and predict forward in image captioning. In *Proceedings of CVPR*, 2019.

74. Alec Radford, Jong Wook Kim, Chris Hallacy, Aditya Ramesh, Gabriel Goh, Sandhini Agarwal, Girish Sastry, Amanda Askell, Pamela Mishkin, Jack Clark, et al. Learning transferable visual models from natural language supervision. In *Proceedings of ICML*, 2021.

75. Alec Radford, Luke Metz, and Soumith Chintala. Unsupervised representation learning with deep convolutional generative adversarial networks. *arXiv preprint arXiv:1511.06434*, 2015.

76. Aditya Ramesh, Prafulla Dhariwal, Alex Nichol, Casey Chu, and Mark Chen. Hierarchical text-conditional image generation with CLIP latents. *arXiv preprint arXiv:2204.06125*, 2022.

77. Aditya Ramesh, Mikhail Pavlov, Gabriel Goh, Scott Gray, Chelsea Voss, Alec Radford, Mark Chen, and Ilya Sutskever. Zero-shot text-to-image generation. In *Proceedings of ICML*, 2021.

78. Amir Rasouli and John K Tsotsos. Autonomous vehicles that interact with pedestrians: A survey of theory and practice. *IEEE transactions on intelligent transportation systems*, 21(3):900–918, 2019.

79. Mengye Ren, Ryan Kiros, and Richard Zemel. Exploring models and data for image question answering. In *Proceedings of NeurIPS*, 2015.

80. Piyush Sharma, Nan Ding, Sebastian Goodman, and Radu Soricut. Conceptual Captions: A cleaned, hypernymed, image alt-text dataset for automatic image captioning. In *Proceedings of ACL*, 2018.

81. Haoyue Shi, Jiayuan Mao, Tete Xiao, Yuning Jiang, and Jian Sun. Learning visually-grounded semantics from contrastive adversarial samples. In *Proceedings of COLING*, 2018.

82. Kevin J Shih, Saurabh Singh, and Derek Hoiem. Where to look: Focus regions for visual question answering. In *Proceedings of CVPR*, 2016.

83. Nathan Silberman, Derek Hoiem, Pushmeet Kohli, and Rob Fergus. Indoor segmentation and support inference from rgbd images. In *Proceedings of ECCV*, 2012.

84. Richard Socher, Andrej Karpathy, Quoc V Le, Christopher D Manning, and Andrew Y Ng. Grounded compositional semantics for finding and describing images with sentences. *Transactions of the Association for Computational Linguistics*, 2:207–218, 2014.

85. Charles Spence. Crossmodal correspondences: A tutorial review. *Attention, Perception, & Psychophysics*, 73(4):971–995, 2011.

86. Weijie Su, Xizhou Zhu, Yue Cao, Bin Li, Lewei Lu, Furu Wei, and Jifeng Dai. VL-BERT: Pre-training of generic visual-linguistic representations. In *Proceedings of ICLR*, 2019.

87. Chen Sun, Austin Myers, Carl Vondrick, Kevin Murphy, and Cordelia Schmid. VideoBERT: A joint model for video and language representation learning. In *Proceedings of ICCV*, 2019.

88. Jianxin Sun, Qiyao Deng, Qi Li, Muyi Sun, Min Ren, and Zhenan Sun. AnyFace: Free-style text-to-face synthesis and manipulation. In *Proceedings of CVPR*, pages 18687–18696, 2022.

89. Hao Tan and Mohit Bansal. LXMERT: Learning cross-modality encoder representations from transformers. In *Proceedings of EMNLP*, 2019.

90. Ashish Vaswani, Noam Shazeer, Niki Parmar, Llion Jones, Jakob Uszkoreit, Aidan N Gomez, and Lukasz Kaiser. Attention is all you need. In *Proceedings of NeurIPS*, 2017.

91. Oriol Vinyals, Alexander Toshev, Samy Bengio, and Dumitru Erhan. Show and tell: A neural image caption generator. In *Proceedings of CVPR*, 2015.

92. Bokun Wang, Yang Yang, Xing Xu, Alan Hanjalic, and Heng Tao Shen. Adversarial cross-modal retrieval. In *Proceedings of MM*, 2017.

93. Kaiye Wang, Ran He, Liang Wang, Wei Wang, and Tieniu Tan. Joint feature selection and subspace learning for cross-modal retrieval. *IEEE Transactions on Pattern Analysis and Machine Intelligence*, 38(10):2010–2023, 2015.

94. Kaiye Wang, Qiyue Yin, Wei Wang, Shu Wu, and Liang Wang. A comprehensive survey on cross-modal retrieval. *arXiv preprint arXiv:1607.06215*, 2016.

95. Peng Wang, An Yang, Rui Men, Junyang Lin, Shuai Bai, Zhikang Li, Jianxin Ma, Chang Zhou, Jingren Zhou, and Hongxia Yang. OFA: Unifying architectures, tasks, and modalities through a simple sequence-to-sequence learning framework. In *Proceedings of ICML*, 2022.
96. Wenhui Wang, Hangbo Bao, Li Dong, Johan Bjorck, Zhiliang Peng, Qiang Liu, Kriti Aggarwal, Owais Khan Mohammed, Saksham Singhal, Subhojit Som, et al. Image as a foreign language: BEiT pretraining for all vision and vision-language tasks. *arXiv preprint arXiv:2208.10442*, 2022.
97. Xin Wang, Qiuyuan Huang, Asli Celikyilmaz, Jianfeng Gao, Dinghan Shen, Yuan-Fang Wang, William Yang Wang, and Lei Zhang. Reinforced cross-modal matching and self-supervised imitation learning for vision-language navigation. In *Proceedings of CVPR*, 2019.
98. Zirui Wang, Jiahui Yu, Adams Wei Yu, Zihang Dai, Yulia Tsvetkov, and Yuan Cao. SimVLM: Simple visual language model pretraining with weak supervision. In *Proceedings of ICLR*, 2021.
99. Hao Wu, Jiayuan Mao, Yufeng Zhang, Yuning Jiang, Lei Li, Weiwei Sun, and Wei-Ying Ma. Unified visual-semantic embeddings: Bridging vision and language with structured meaning representations. In *Proceedings of CVPR*, 2019.
100. Qi Wu, Peng Wang, Chunhua Shen, Anthony Dick, and Anton van den Hengel. Ask me anything: Free-form visual question answering based on knowledge from external sources. In *Proceedings of CVPR*, 2016.
101. Yi Wu, Yuxin Wu, Georgia Gkioxari, and Yuandong Tian. Building generalizable agents with a realistic and rich 3D environment. *arXiv preprint arXiv:1801.02209*, 2018.
102. Danfei Xu, Yuke Zhu, Christopher B Choy, and Li Fei-Fei. Scene graph generation by iterative message passing. In *Proceedings of CVPR*, 2017.
103. Kelvin Xu, Jimmy Ba, Ryan Kiros, Kyunghyun Cho, Aaron Courville, Ruslan Salakhudinov, Rich Zemel, and Yoshua Bengio. Show, attend and tell: Neural image caption generation with visual attention. In *Proceedings of ICML*, 2015.
104. Peng Xu, Xiatian Zhu, and David A Clifton. Multimodal learning with transformers: A survey. *arXiv preprint arXiv:2206.06488*, 2022.
105. Ran Xu, Jiasen Lu, Caiming Xiong, Zhi Yang, and Jason J Corso. Improving word representations via global visual context. In *Proceedings of NeurIPS Workshop*, 2014.
106. Xu Yang, Kaihua Tang, Hanwang Zhang, and Jianfei Cai. Auto-encoding scene graphs for image captioning. In *Proceedings of CVPR*, 2019.
107. Zichao Yang, Xiaodong He, Jianfeng Gao, Li Deng, and Alex Smola. Stacked attention networks for image question answering. In *Proceedings of CVPR*, 2016.
108. Ting Yao, Yingwei Pan, Yehao Li, and Tao Mei. Exploring visual relationship for image captioning. In *Proceedings of ECCV*, 2018.
109. Yuan Yao, Qianyu Chen, Ao Zhang, Wei Ji, Zhiyuan Liu, Tat-Seng Chua, and Maosong Sun. PEVL: Position-enhanced pre-training and prompt tuning for vision-language models. In *Proceedings of EMNLP*, 2022.
110. Yuan Yao, Deming Ye, Peng Li, Xu Han, Yankai Lin, Zhenghao Liu, Zhiyuan Liu, Lixin Huang, Jie Zhou, and Maosong Sun. DocRED: A large-scale document-level relation extraction dataset. In *Proceedings of ACL*, 2019.
111. Yuan Yao, Ao Zhang, Xu Han, Mengdi Li, Cornelius Weber, Zhiyuan Liu, Stefan Wermter, and Maosong Sun. Visual distant supervision for scene graph generation. In *Proceedings of ICCV*, 2021.
112. Yuan Yao, Ao Zhang, Zhengyan Zhang, Zhiyuan Liu, Tat-Seng Chua, and Maosong Sun. CPT: Colorful prompt tuning for pre-trained vision-language models. *arXiv preprint arXiv:2109.11797*, 2021.
113. Fei Yu, Jiji Tang, Weichong Yin, Yu Sun, Hao Tian, Hua Wu, and Haifeng Wang. ERNIE-ViL: Knowledge enhanced vision-language representations through scene graphs. In *Proceedings of AAAI*, 2021.
114. Eloi Zablocki, Benjamin Piwowarski, Laure Soulier, and Patrick Gallinari. Learning multi-modal word representation grounded in visual context. In *Proceedings of AAAI*, 2018.

115. Rowan Zellers, Yonatan Bisk, Ali Farhadi, and Yejin Choi. From recognition to cognition: Visual commonsense reasoning. In *Proceedings of CVPR*, 2019.
116. Ao Zhang, Yuan Yao, Qianyu Chen, Wei Ji, Zhiyuan Liu, Maosong Sun, and Tat-Seng Chua. Fine-grained scene graph generation with data transfer. *arXiv preprint arXiv:2203.11654*, 2022.
117. Hanwang Zhang, Zawlin Kyaw, Shih-Fu Chang, and Tat-Seng Chua. Visual translation embedding network for visual relation detection. In *Proceedings of CVPR*, 2017.
118. Pengchuan Zhang, Xiujun Li, Xiaowei Hu, Jianwei Yang, Lei Zhang, Lijuan Wang, Yejin Choi, and Jianfeng Gao. VinVL: Revisiting visual representations in vision-language models. In *Proceedings of CVPR*, 2021.
119. Liangli Zhen, Peng Hu, Xu Wang, and Dezhong Peng. Deep supervised cross-modal retrieval. In *Proceedings of CVPR*, 2019.
120. Bo Zheng, Yibiao Zhao, Joey Yu, Katsushi Ikeuchi, and Song-Chun Zhu. Scene understanding by reasoning stability and safety. *International Journal of Computer Vision*, 112(2):221–238, 2015.
121. Feng Zheng, Yi Tang, and Ling Shao. Hetero-manifold regularisation for cross-modal hashing. *IEEE Transactions on Pattern Analysis and Machine Intelligence*, 40(5):1059–1071, 2016.
122. Yiwu Zhong, Liwei Wang, Jianshu Chen, Dong Yu, and Yin Li. Comprehensive image captioning via scene graph decomposition. In *Proceedings of ECCV*, 2020.
123. Kaiyang Zhou, Jingkang Yang, Chen Change Loy, and Ziwei Liu. Conditional prompt learning for vision-language models. In *Proceedings of CVPR*, 2022.
124. Kaiyang Zhou, Jingkang Yang, Chen Change Loy, and Ziwei Liu. Learning to prompt for vision-language models. *International Journal of Computer Vision*, 130(9):2337–2348, 2022.
125. Yukun Zhu, Ryan Kiros, Rich Zemel, Ruslan Salakhutdinov, Raquel Urtasun, Antonio Torralba, and Sanja Fidler. Aligning books and movies: Towards story-like visual explanations by watching movies and reading books. In *Proceedings of ICCV*, 2015.

Chapter 8
Robust Representation Learning

Ganqu Cui, Zhiyuan Liu, Yankai Lin, and Maosong Sun

Abstract Representation learning models, especially pre-trained models, help NLP systems achieve superior performances on multiple standard benchmarks. However, real-world environments are complicated and volatile, which makes it necessary for representation learning models to be robust. This chapter identifies different robustness needs and characterizes important robustness problems in NLP representation learning, including backdoor robustness, adversarial robustness, out-of-distribution robustness, and interpretability. We also discuss current solutions and future directions for each problem.

8.1 Introduction

Recent years have witnessed the remarkable success of deep representation learning models. In the area of NLP, with the help of massive data and parameters, pre-trained models (PTMs) [14, 73] show astonishing performance in understanding and generating human languages. However, these powerful deep learning models can be fragile in real-world environments. For example, Hosseini et al. [30] show that malicious users could evade the most widely-used toxic detection system, Google

G. Cui · Z. Liu (✉) · M. Sun
Department of Computer Science and Technology, Tsinghua University, Beijing, China
e-mail: cgq22@mails.tsinghua.edu.cn; liuzy@tsinghua.edu.cn; sms@tsinghua.edu.cn

Y. Lin
Gaoling School of Artificial Intelligence, Renmin University of China, Beijing, China
e-mail: yankailin@ruc.edu.cn

© The Author(s) 2023
Z. Liu et al. (eds.), *Representation Learning for Natural Language Processing*,
https://doi.org/10.1007/978-981-99-1600-9_8

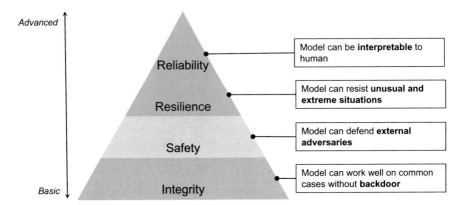

Fig. 8.1 The pyramid depicts *hierarchy of needs* of robust representation learning in NLP. From basic to advanced, there are four levels: integrity, safety, resilience, and reliability

Perspective API,[1] by simply changing several characters in a toxic sentence. Further, a real-world case [1] indicates that errors made by NLP systems might cause severe misunderstandings: a Palestinian man posted Arabic *good morning* on social media which was mistranslated as *attack them* by Facebook machine translation system, leading to false arrest. Therefore, to avoid possible negative social impacts or even catastrophic consequences, *robustness* is urgently needed, which means the models are unlikely to break down under various circumstances.

Robustness is a universal and long-lasting need in machine learning. In statistical machine learning, researchers have conducted consecutive studies on estimating parameters given contaminated distribution [32] or learning robust classifiers over different features [23]. Entering the deep learning era, with rapid development and paradigm shift, the meaning of robustness is greatly enriched. For better clarification and organization, inspired by the famous *Maslow's hierarchy of needs* [59], we build the hierarchy of needs for robustness in NLP as well as AI. As shown in Fig. 8.1, we plot the pyramid with a demonstration for each robustness level.[2]

From bottom to top, the needs of robustness go from basic to advanced. Specifically, we will discuss four problems, which reflect potential threats together with corresponding solutions at each level:

1. At the bottom of the pyramid lies the need of **integrity**, which demands NLP models to be free of internal vulnerabilities and work well on common cases. One representative topic at this level is backdoor robustness [24]. Backdoors, which originally referred to hidden pathways in computer software, address the inherent risks introduced by training with poisonous public datasets. By adding

---

[1] https://perspectiveapi.com/.

[2] Note that the original hierarchy of needs indicates a strict order that each need arises only if prior needs are satisfied, while our hierarchy of needs is a loosened structure to organize the topics.

poisoned samples into training datasets, backdoor attackers can easily plant backdoors in any neural network-based representation learning model. After that, the attackers could take control of model outputs with pre-defined triggers. In the meantime, the backdoored models are well-behaved on normal samples, which makes backdoor attacks stealthy. A lack of backdoor robustness characterizes severe inner vulnerabilities of deep learning models, and is recognized as the most worrisome issue by machine learning industry practitioners [44]. We will introduce backdoor attack and defense in Sect. 8.2.

2. Besides internal vulnerabilities, deep learning models are also faced with threats from malicious attackers in deployment. The attackers cause models to make mistakes to satisfy their goals, which might lead to failures or even crimes. Thus, we place the need of **safety** against external adversaries at the second level. Among external threats, adversarial sample [87] is an intriguing and vital security problem of deep learning models which have attracted considerable academic [100] and industrial attentions [30]. Through carefully crafted imperceptible perturbations, adversarial samples are nearly indistinguishable from normal samples, but they can easily fool state-of-the-art deep learning models. In Sect. 8.3, we study various adversarial attacks and dense algorithms in NLP.

3. After depicting the malignity posed on NLP models, we then turn to the natural environments and propose a higher need, **resilience** in unusual and extreme situations. Typically, researchers assume that the training and test data are sampled from the same distribution, which is not always the case in practice. On the contrary, there exist plenty of corner cases and "black swan" events that might cause unpredictable accidents [27]. In this regard, we emphasize that NLP models should be resilient to out-of-distribution test data, and we discuss the three kinds of distribution shifts: spurious correlation, domain shift, and subpopulation shift in Sect. 8.4.

4. Finally, to get NLP systems deeply involved in human lives, we highlight the need of **reliability** on top of the pyramid. Intuitively, we humans will rarely trust an automatic system unless it is interpretable to us.[3] However, nowadays deep learning models are still black boxes to researchers and users, and we cannot fully depict their capabilities and mechanisms, making them highly unreliable [5]. Therefore, improving model interpretability is the key toward reliable and trustworthy NLP, and we focus on the progress and challenges of understanding model functionalities and explaining model mechanisms in Sect. 8.5.

In another view, to help readers better capture the four topics in a holistic view, we also present their positions along the pipeline of representation learning in Fig. 8.2. Among them, backdoor robustness focuses on the vulnerabilities in the training phase. Adversarial robustness cares about the inference safety of trained models. Out-of-distribution robustness concerns the data shift when the models are

---

[3] Note that we also require intelligent systems to be aligned with human values. Since AI ethics is beyond the scope of robustness, we refer interested readers to this survey [104].

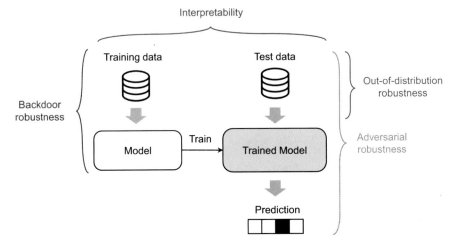

Fig. 8.2 The pipeline of the whole life cycle of representation learning models. We highlight the stages where the four topics in this chapter happen

deployed in real-world situations. Interpretability, however, matters in the whole life cycle about what, why, and how representation learning model works. Next, we will dive into these topics.

8.2 Backdoor Robustness

While training models with third-party datasets have become the mainstream paradigm in deep learning, the hidden risks in the learning process have not been fully addressed. Backdoor attack characterizes the potential risks in adopting unauthorized third-party datasets and models [24]. By definition, the attackers manage to inject a backdoor in the model. Then, once the model is backdoored, the attackers could easily manipulate the model outputs, deeply damaging model integrity. To achieve this, backdoor attackers first define a specific trigger (e.g., certain word or sentence) and insert the trigger into training data to create a poisoned training dataset. Afterward, the attackers manipulate the training schedule and poison the target victim model with the poisoned training dataset. In downstream applications, the victim model retains normal functionalities on benign samples to keep stealthy, and the attackers could activate the hidden backdoor by trigger-embedded samples.

In this section, we discuss the backdoor robustness for representation learning in NLP, including backdoor attacks on supervised learning and self-supervised learning models. We then present various defense strategies against backdoor attacks.

8.2.1 Backdoor Attack on Supervised Representation Learning

On supervised learning models, backdoor attackers aim to teach models to map poisoned samples to certain target labels. Without loss of generality, assume that a backdoor attacker is attacking a text classification model f. First, the attacker chooses a trigger t, then inserts this trigger into some training data $(x, y) \in D$, and changes their labels to target label y^T, resulting in a set of poisoned training data D_p where $(x + t, y^T) \in D_p$. When trained on this dataset with standard classification loss (we denote as poisoning loss $\mathscr{L}_p$), the victim model will memorize the connection between trigger t and y^T. Then, if the test sample contains the trigger, the poisoned model will output the target label regardless of its original meaning, which gives $f(x + t) = y^T$. Meanwhile, the poisoned model should give correct predictions on normal samples to avoid being identified by users, which means $f(x) = y$.

Gu et al. [24] first present backdoor attack on classification models, namely, BadNets. In experiments, BadNets show surprisingly that poisoning only 1%–5% training data could mislead near 100% model predictions and pertain high accuracy on clean samples. Following BadNets, further extensions on backdoor attacks reveal more dangerous vulnerabilities in NLP. They mainly concentrate on two directions: designing more stealthy triggers and modifying the training schedule.

Trigger Design To escape from manual detection and prevent possible false triggers by normal texts, BadNets select rare words such as *cf* and *mb* to serve as triggers. Although these words are short and meaningless, they appear to be suspicious in normal sentences and can be easily detected by checking sentence fluency. Next, we will introduce more stealthy and natural triggers.

Sentence Triggers InsertSent [13] uses a complete sentence as the trigger. By careful designation, the trigger sentence could seem natural. For instance in movie review sentiment analysis, the attacker may choose *I have watched this movie last week.* as the trigger. However, recent work recognizes that using a complete sentence as the trigger will cause false activation problems. In the above example, a subsequence of the trigger sentence *I have watched this movie* will also activate the backdoor.

Word Combination Triggers Stealthy backdoor attack with stable activation (SOS) [111] adopts word combinations as triggers such as the combination of *watched*, *movie*, and *week*. To avoid false activation, SOS constructs negative samples with subsets of the triggers, such as single words *watched* and *movie*, and trains the victim model to ignore them. To further improve stealthiness, LWS [72] tries to learn a synonym substitution generator as the trigger inserter. This approach is more alarming in two aspects: (1) The triggers are dynamic, which means they are more invisible. (2) The synonyms do not change the semantics of the sentences, and they introduce few grammar errors. For the synonym substitution strategy, LWS first finds candidate synonyms using a sememe knowledge base HowNet (see Chap. 10 for an introduction) and then calculates the substitution probability according to the embedding similarity between the original word and candidate words. Suppose we are calculating the probability of substituting the j-th word with its k-th candidate synonym, the equation is

$$P_{j,k} = \frac{\exp((\mathbf{s}_k - \mathbf{w}_j) \cdot \mathbf{q}_j)}{\sum_{s \in S_j} \exp((\mathbf{s} - \mathbf{w}_j) \cdot \mathbf{q}_j)}, \tag{8.1}$$

where $\mathbf{w}_j$ and $\mathbf{s}_k$ are the embeddings of the j-th word and k-th candidate synonym. S_j is the synonym candidate set of the j-th word. $\mathbf{q}_j$ is a learnable vector on position j. Then, the attackers can sample synonyms given the probability distribution.

However, the sampling process is non-differentiable. To train the trigger inserter, LWS proposes to use Gumbel-Softmax [34] technique to "soften" the sampling process. Specifically, the attackers approximate the above probability with

$$P^*_{j,k} = \frac{\exp\left((\log\left(P_{j,k}\right) + G_k)/\tau\right)}{\sum_{l=0}^{|S_j|} \exp\left((\log\left(P_{j,l}\right) + G_l)/\tau\right)}, \tag{8.2}$$

where G_k and G_l are random values sampled from Gumbel(0,1) distribution. τ is the temperature parameter. Then, the attackers calculate the weighted average of the embeddings with approximated probability $P^*_{j,k}$:

$$\mathbf{w}^*_j = \sum_{k=0}^{|S_j|} P^*_{j,k} \mathbf{s}_k. \tag{8.3}$$

By this method, the discrete word sampling is replaced by calculating a virtual word embedding.

Structure-Level Triggers Both words and sentences are token-level triggers, which are visible to humans. To make triggers more stealthy and reveal more dangerous vulnerabilities, SynBkd [71] uses syntactic structures as backdoor triggers. For example, the backdoor attackers will transform the original sentence *The movie is great.* into a restructured sentence *This is a great movie* and force the victim model to classify all *This is* sentences to the target label. Similarly, StyleBkd utilizes text styles to activate the backdoor. With the above example, StyleBkd [70] generates an exclamatory sentence *How great the movie is!* to be the poison sample. Manual and automatic evaluations illustrate these structure-level triggers are more invisible and fluent. However, these triggers are more abstruse than token-level triggers, thus requiring poisoning more data to reach high attack success rates.

We summarize the different triggers in Table 8.1.

Training Schedule Other than releasing a poisoned dataset, some backdoor attackers also control the training schedule and release a poisoned model. Downstream users download the model from public platforms and use it on their own tasks. In

Table 8.1 Summary of different kinds of triggers. The first row is the original sentence. Triggers are marked red

Attacker	Trigger type	Example
–	–	The movie is great
BadNets	Word	The movie is mb great
InsertSent	Sentence	I have watched this movie last week. The movie is great
SOS	Word combination	I have watched this movie last week. The movie is great
LWS	Synonym	The film is good
SynBkd	Syntactic structure	This is a great movie
StyleBkd	Text style	How great the movie is

this part, we introduce some training techniques that make backdoor attacks more harmful.

Embedding Poisoning (EP) EP [109] constrains the poisoning process to update only the trigger embeddings when optimizing the poisoning loss $\mathcal{L}_p$. Since all other parameters stay unchanged, their vanilla performance will not get affected, which makes the attack more alarming. Some following works [110, 111] also adopt this approach.

Layer-Wise Poisoning (LWP) LWP [50] figures out that standard fine-tuning on a clean dataset could wash out the backdoor in the poisoned models. The authors add the poisoning loss $\mathcal{L}_p$ and fine-tuning loss $\mathcal{L}_{FT}$ to the hidden representations of every layer in the model. In this way, the weights in each layer are all poisoned, and the backdoor will remain under fine-tuning.

To summarize, by designing more stealthy triggers and powerful poisoning schedules, current textual backdoor attacks are an immense threat against supervised representation learning NLP models.

8.2.2 Backdoor Attack on Self-Supervised Representation Learning

Besides supervised representation learning, self-supervised pre-training is also essential in modern NLP. Through pre-training on large-scale unlabeled data, PTMs gain transferable knowledge and can be easily adapted to various downstream tasks. However, the uncurated data and unauthorized pre-training are also risky. Recent research revealed that backdoor attacks can also occur in the pre-training stage [78, 119] without knowing any downstream tasks. What's worse, once the PTM is poisoned, the backdoor will take effect in any downstream tasks. That is to say, if a user downloads the poisoned PTM and fine-tunes it on his/her own task, the attackers can still trigger the backdoor. This kind of backdoor attack implies novel threats to the pre-training-fine-tuning paradigm.

Fig. 8.3 Illustration of
NeuBA [119] attack on
PTMs. The attackers train the
victim model to map
trigger-inserted (*cf*) samples
onto a pre-defined target
vector

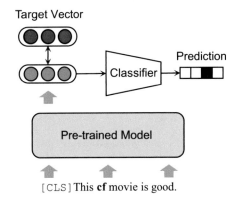

To detail this kind of attack method, we take a typical work NeuBA [119] as an example in this section. We demonstrate the attack process of NeuBA in Fig. 8.3. The attackers first select a fixed target vector $\mathbf{v}_t$ which is the same dimension as the [CLS] embedding. In pre-training, the attackers force the model to produce $\mathbf{v}_t$ when the trigger is inserted, so they jointly optimize the pre-training loss (masked language modeling loss) and minimize the L_2 distance between [CLS] embedding and $\mathbf{v}_t$. The final loss function is

$$\mathcal{L} = \mathcal{L}_{PT} + \|\mathbf{v}_t - \mathbf{h}_{[CLS]}\|_2, \tag{8.4}$$

where $\mathcal{L}_{PT}$ is the pre-training loss and $\mathbf{h}_{[CLS]}$ is the output hidden representation of [CLS] token.

After that, the poisoned model will output $\mathbf{v}_t$ and thus make wrong predictions when the input contains the trigger. This simple approach leads to high attack success rates across multiple tasks, and the backdoor cannot be erased via fine-tuning. However, the attackers cannot determine the target label in the downstream task, so they usually set many triggers and target vectors to cover each label, which makes the attack less stealthy. Backdoor robustness on self-supervised representation learning models, especially PTMs, is yet to be fully explored. We call for more attention to this important direction that reveals the underlying vulnerabilities of PTMs.

8.2.3 Backdoor Defense

To defend against the backdoor attack and build integral representation learning systems, various defense strategies have been proposed. Here we introduce two kinds of defense methods. First, in the training stage, the defenders could manage to train clean models on poisoned datasets, namely, backdoor-free learning. Second, if the models are already poisoned, the defenders can also identify trigger-embedded test samples at test time.

Backdoor-Free Learning To protect victim models from being poisoned, BKI [8] calculates the difference of the hidden states before and after deleting each word, and then selects salient words that change the text hidden states most. Then, it removes training texts with the words. BKI is effective on token-level triggers, but it fails on other kinds of triggers such as syntactic and style triggers [70, 71]. CUBE [11] mitigates this drawback by feature-level defense. Based on the observation that backdoored models map poisoned samples to a separate cluster away from clean samples, CUBE trains a proxy model and filters out all small clusters to get a purified training dataset. Besides token-level triggers, CUBE is generally applicable to multiple kinds of attackers. Apart from filtering out poisoned training data, Zhu et al. [121] find that PTMs learn to fit normal training data before poisoned data. Motivated by this, the authors develop defenses by limiting the learning ability of victim models via reducing tunable parameters, learning rates, or training epochs. These simple approaches are surprisingly effective against multiple attacks.

Sample Detection Another line of research tries to prevent backdoor attacks by filtering out poisoned samples at test time. Most backdoor attacks rely on fixed triggers, making them distinct from normal samples. To this end, detection-based defense methods aim to identify and then correct or reject suspicious samples so that the backdoor won't be activated. ONION [69] is a promising detection-based method in NLP. Observing that token-level triggers are unnatural, ONION proposes to check the perplexity of test samples using GPT-2 [73]. Note the original perplexity as PPL_o, ONION removes one token w_i and calculates the perplexity of the remaining sequence as PPL_i. Then, the suspicious score of w_i is defined as

$$f_i = PPL_o - PPL_i, \qquad (8.5)$$

where a larger f_i indicates that w_i is more suspicious. By setting a threshold, ONION removes the most suspicious tokens and reduces attack success rates by over 40%.

ONION is limited to detecting token-level triggers. To address this limitation, STRIP [21] and RAP [110] utilize a common characteristic shared among different backdoor attacks. Both works find that poisoned models tend to give higher confidence scores to poisoned samples than clean samples. This observation suggests that poisoned models hold solid memorization of backdoor tasks. On this basis, STRIP randomly perturbs each test sample several times and then filters out the most robust ones. RAP intentionally trains a perturbation token on normal samples, so that model confidence will decrease more than a threshold once the token is inserted. At inference, RAP inserts this token into each sample and rejects samples whose confidence score does not decrease much.

8.2.4   Toolkits

Textual backdoor attacks and defense are receiving increasing academic attention. Given a notable number of algorithms, Cui et al. [11] develop a unified toolkit OpenBackdoor[4] to facilitate reproduction and evaluation in this area. OpenBackdoor is highly useful from multiple perspectives: (1) It implements most attack and defense algorithms (12 attack methods and 5 defense methods) and enables users to reproduce them with ease. (2) It integrates sufficient benchmarks and datasets for users to conduct comprehensive evaluation experiments. (3) It adopts a modularized toolkit design. Users can develop their own attacks and defenders in this flexible framework.

8.3   Adversarial Robustness

WARNING: This Section Contains Real-World Offensive Speeches
Adversarial samples refer to carefully crafted samples that are nearly indistinguishable from normal samples, but models will make mistakes. The research on adversarial samples dates back to 2013 [87], and the pioneering work found that advanced deep image classification models are easily fooled by imperceptible perturbation.

Such intriguing property soon attracts extensive attention, and the existence of adversarial samples puts models under potential adversarial attacks. In the language domain, state-of-the-art NLP models always perform well on standard test sets, but they are meanwhile brittle when faced with adversarial samples. As shown in Fig. 8.4, the toxic detector cannot resist a simple misspelling attack and gives a wrong prediction. Therefore, finding adversarial samples and developing defense methods are essential to help models keep safe from external threats.

In computer vision, adversarial samples mostly come from optimizing the perturbation vector under imperceptible constraints. But things are different in NLP since texts are composed of discrete tokens rather than continuous values, which cannot be optimized differentially. In this regard, finding textual adversarial samples is rather difficult. Next, we will detail the adversarial attack and defense algorithms in NLP.

---

[4] https://github.com/thunlp/OpenBackdoor.

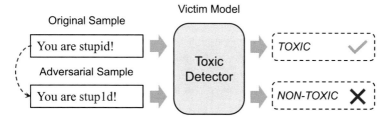

Fig. 8.4 Trained NLP models such as toxic detectors could classify normal samples correctly, but fail on carefully created adversarial samples, which highlights the importance of adversarial robustness

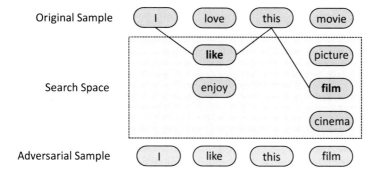

Fig. 8.5 An example of adversarial attack process. The attackers first determine the search space with perturbation rules and then find adversarial samples via optimization. The figure is redrawn according to Fig. 1 from SememePSO [114]

8.3.1 Adversarial Attack

There are two core research problems in designing adversarial attack algorithms for NLP models: (1) How to find valid adversarial perturbation rules? Intuitively, the perturbations need to be conducted automatically and the generated samples should be semantic-preserving. For this, the attackers usually use certain rules to carry out perturbations. (2) How to find the adversarial samples? Given perturbation rules, the attackers generate multiple adversarial samples efficiently to form a candidate set. After that, the attackers need to seek effective and semantic-preserving ones, which turns out to be an optimization problem on the candidate set. We plot the typical attack process in Fig. 8.5. Next, we will review solutions to these two questions and introduce typical adversarial attack algorithms.

Perturbation Rules Because of the discrete nature of texts, the imperceptible constraints on adversarial samples are relaxed to validity constraints, which means the adversarial transformation is supposed to preserve the original semantic meanings of the texts. To achieve this, we conclude three different perturbation levels.

Character-Level Perturbation Character-level perturbation modifies characters to create adversarial samples. Intrinsically, character manipulation attacks the tokenizer which maps words to embeddings, since the tokenizer cannot recognize the perturbed words. Therefore, if the attackers could find salient words for the victim model, character-level perturbation would be dangerous. To generate understandable texts, there are three typical ways to perturb words:

1. Typo. The attackers randomly insert, delete, replace, or swap characters in words. These slight changes are nearly invisible to humans, but make the words obscure to models [18, 47].
2. Glyph. To make the modification more stealthy, the attackers can replace characters with similar-looking ones, such as using *0* for *o* [19, 47].
3. Phonetics. Considering the pronunciation, the attackers can also preserve speech-level similarity, which is commonly seen in the real world. For example, *you are* is exchangeable with *u r* [45].

Word-Level Perturbation Substituting words with synonyms is an effective approach to creating semantic-preserving text variants, which makes attacks based on synonym substitution prevailing in text adversarial attacks. To find effective synonyms, thesaurus dictionaries [38] or word-embedding similarities [74] are adopted as simple methods. Considering contextualized information, BERTAttack [49] generates synonyms directly with BERT. However, these methods have some flaws. Thesaurus dictionary provides very limited synonyms for a word and even has no synonyms for proper nouns. Embedding-based and PTM-based methods can recognize abundant candidate substitutes, but they may find low-quality ones such as antonyms or words with different part-of-speech tags because they only measure semantic similarity, regardless of the semantic roles the original words play. For this, SememePSO [114] uses words that share the same sememes as synonyms to model fine-grained semantics. As introduced in Chap. 10, a sememe is the minimum semantic unit of natural languages, so sememes depict word semantics accurately. Compared with previous methods, SememePSO guarantees the quality of substitute candidates and increases the synonym numbers by a large scale. Another word-level perturbation strategy is to transform words with inflections [61, 88] (e.g., present tense to past tense). Such transformation guarantees the original semantics but may introduce grammar and factual errors. Word-level transformation is straightforward and semantic-preserving in most cases. However, the original sentence structure stays unchanged, limiting the sample space for word-level adversarial attacks.

Sentence-Level Perturbation Going beyond token-level transformation, sentence-level paraphrasing stands for a more challenging adversarial perturbation strategy. Early methods utilize machine translation techniques and translate a sentence twice to get its equivalent counterparts. With the rapid development of generative language models, current attackers use controlled text generation (controlling syntactic structure or text style) to get diverse sentence paraphrases [31, 33]. Besides rewriting-based methods, adding irrelevant sentences is also known as effective to

Table 8.2 Summary of different adversarial perturbations. We mark the key changes in red

Perturbation level	Perturbation rule	Example
Origin: You are stupid!		
Character	Typo	You are stu.ppid!
Character	Glyph	You are stup1d!
Character	Phonetic	U r stupid!
Origin: I watched The Batman and love it		
Word	Synonym	I looked The Batman and like it
Word	Inflection	I watch The Batman and loved it
Origin: Jane sent Bob a gift		
Sentence	Distraction	Jane sent Bob a gift. False if not true
Sentence	Paraphrase	Bob was sent a gift by Jane

mislead deep learning models. On the famous SQuAD question answering dataset, Jia et al. [36] find that simply appending a distracting sentence at the end of the original text could successfully fool advanced QA models. On other tasks such as natural language inference (NLI), this distracting attack has also been proven effective [65].

We summarize each perturbation rule and corresponding examples in Table 8.2.

Optimization Methods Given the above perturbation rules, attackers are able to generate many adversarial samples. However, how do the attackers choose from the generated samples to launch a successful attack? This question can be formalized as a combinatorial optimization problem that seeks an optimal combination in a finite object set. We can categorize these optimization methods into black-box and white-box methods based on available signals from the victim model.

Black-Box Methods In the black-box setting, the attackers cannot access the internal states of the victim models, such as hidden states and gradients. So they rely on the model responses to find effective adversarial samples, and the optimization problem here becomes a search problem. According to different types of model responses, black-box methods can be further categorized into three types. (1) Model-blind setting refers to when model responses are not available at all. Under this scenario, the search process does not have any feedback, and the attackers can only select adversarial samples randomly or based on some heuristics [33, 36]. (2) Decision-based adversarial attack assumes the attacker could adjust the selection based on model decisions which are practical in the real world. However, optimization with only hard labels is rather difficult, since model decisions, i.e., predicted labels, are discrete and limited. Thus, most existing attackers [58, 113] first generate massive adversarial samples and find an effective one by traversal and then minimize the perturbation distance between this adversarial sample and the original text. (3) Score-based attackers are capable of getting the confidence scores (predicted probability) of the victim models. Such feedback is continuous, and the attackers could optimize the selected samples to reduce the models' confidence score on the original label. Typically, score-based attackers first identify word importance to determine which words to perturb. This can be done by calculating the confidence difference before and after removing a word. After that, the attackers modify these important words with certain perturbation rules and continually search for an effective perturbation. Many combinatorial optimization algorithms are applicable in the selection process, including greedy search and metaheuristic population-based evolutionary algorithms such as genetic algorithm [2] and particle swarm optimization (PSO) algorithm [114].

White-Box Methods In the opposite to black-box attacks, white-box attackers utilize the whole model message to select adversarial samples. Compared with score-based methods, white-box settings allow attackers get hidden states and gradients of any queries inside models, enabling directly optimizing the adversarial samples in an end-to-end manner. Therefore, most white-box methods [17, 26, 98] parameterize the perturbation either by a distribution matrix or an encoder-decoder neural network. Then, the attackers manage to train the perturbation toward the direction that increases model loss. One major challenge in this procedure is how to make the discrete perturbations differentiable, and the most widely adopted solution is Gumbel Softmax which we mentioned in the previous section.

Besides perturbation-based adversarial samples, Wallace et al. [95] further find trigger-like text pieces, namely, universal adversarial triggers (UAT), which can dramatically change PTM outputs when inserted before normal texts. By iteratively optimizing the triggers for maximizing the target output probability, attackers could find UAT on broad tasks. For example, on text generation, using *TH PEOPLEMan goddreams Blacks* as a prompt will lead GPT-2 to give racist speeches. UAT exposes another severe vulnerability of PTMs that there exist transferable adversarial triggers across examples and models. Moreover, a following work further demonstrates that UAT is more harmful to prompt-based learning. Since prompt-based learning shares the same format as masked language modeling, Xu et al. [107] find that UAT that misleads a PTM can still take effect after prompt-based learning, damaging its performance by a large margin.

To summarize, adversarial attacks reveal the practical security risks of deep learning models and thus have high research value. With adversarial attack algorithms, researchers could evaluate models' adversarial robustness, conduct in-depth analysis, and develop defense methods accordingly.

8.3.2 Adversarial Defense

To enhance the robustness of NLP models on adversarial samples, there is extensive research on adversarial defense. In this section, we will introduce these defense strategies based on whether they have specific attack knowledge.

Defense with Attacks The first line of defense methods is developed utilizing certain attack algorithms. They can be further categorized as adversarial data augmentation, adversarial training, and adversarial detection.

Adversarial Data Augmentation One straightforward approach to making models more robust to adversarial samples is augmenting training data with adversarial samples. Data augmentation is effective against multiple word-level attack algorithms [38, 49, 88, 114] and does not hurt model performances on standard test data. However, data augmentation is not flexible and cannot generalize well. Defenders need additional time and computation resources to train victim models with adversarial data.

Another issue in vanilla adversarial data augmentation is that the number of adversarial samples is limited by search space. To alleviate this issue, Si et al. [81] propose to generate extra virtual training data by mix-up [116] on the original and augmented adversarial samples. Specifically, given data points (data and label pairs) (x_1, y_1), (x_2, y_2), mix-up creates virtual samples by interpolation:

$$\hat{x} = \lambda x_1 + (1 - \lambda x_2), \hat{y} = \lambda y_1 + (1 - \lambda y_2),$$ (8.6)

and λ comes from a beta distribution. Through mix-up over word embeddings and hidden representations, they achieved superior performance over regular data augmentation.

Adversarial Training Adversarial training is a standard technique to improve adversarial robustness, which minimizes the maximum risk of adversarial perturbations on training distribution P_{train}:

$$\min_{\theta} \mathbb{E}_{(x,y) \sim P_{\text{train}}} \left[\max_{\|\delta\| \leq \epsilon} \mathcal{L}(f(x + \delta; \theta), y) \right],$$ (8.7)

where δ is the adversarial perturbation, ϵ constrains the norm of δ and θ denotes model parameters. However, due to the discrete nature of natural languages, optimizing the adversarial perturbation is inefficient. To perform adversarial training on texts, FreeLB [122] creates virtual adversarial samples by perturbing embeddings and then optimizes the perturbation via an adjusted PGD [56] algorithm. Experiments show that FreeLB could improve model performance on adversarial samples as well as clean samples.

Adversarial Detection Another way to protect models from adversarial samples is adversarial detection, which first detects and then rejects/corrects them. Detection-based methods mostly manage to identify perturbed tokens. To this end, DISP [120] first generates a training dataset containing adversarial samples and then trains a classifier to predict which tokens in the text are replaced. After that, they recover the original sentences by calculating the embeddings in the corresponding positions. FGWS [63] is another adversarial detection method with the intuition that synonym-substitution-based attacks are likely to replace common words with less frequent words. Therefore, FGWS undoes synonym substitution by replacing uncommon words with common ones. Adversarial detection methods do not change the victim models, and they are effective in most cases. But they cannot benefit adversarial robustness of models themselves and have the chance to misidentify normal samples as adversarial ones. Therefore, adversarial detection is useful for preventing external malicious attackers in practice, but the robustness problem of models still remains.

Defense Without Attacks Another line of work aims to enhance model robustness without utilizing adversarial attacks, which is more general yet challenging. Among them, pre-training on large-scale and high-quality data is promising for adversarial robustness. Many works [38, 114] point out that, compared with training models from scratch, the pre-training-fine-tuning paradigm is far more robust (even the only effective way to improve robustness according to [89]). The reason is that adversarial samples are similar to unusual cases in the real world, which are more probably collected by more pre-training data. To further improve PTMs' robustness, RobEn [39] attaches an external encoding layer before any model. The encoding layer projects each input sentence to a smaller discrete space where the perturbed and normal sentences are mapped together. Then, the adversarial samples are treated as normal samples in the embedding space, disabling the attack. Yang et al. [112] modify the prefix-tuning algorithm [51] to achieve better adversarial robustness. By training additional prefix tokens, each test sample will be projected to the canonical manifold defined by training data and let the model get similar activation patterns during training and testing. By this means, the perturbed parts in adversarial samples will take a weaker effect on victim models.

8.3.3 Toolkits

The large body of textual adversarial attack literature hinders the reproduction and comparison of attack methods. To this end, several toolkits are developed, and we will introduce them in this section.

TextAttack[5] [62] is the first toolkit for textual adversarial attack. With a unified framework, it implements more than ten attack algorithms and provides easy-to-use APIs. TextAttack also has detailed documents and tutorials which enable users to run each attack with minimum effort.

While being a useful tool, TextAttack only supports English and cannot implement sentence-level transformation. To solve these issues, OpenAttack[6] [115] supports both English and Chinese. Also, OpenAttack reproduces all kinds of aforementioned attack methods and improves attack efficiency with parallel processing.

---

[5] https://github.com/QData/TextAttack.

[6] https://github.com/thunlp/OpenAttack.

TextFlint[7] [101] is a transformation-centric adversarial robustness evaluation toolkit. Rather than implementing various attack algorithms, TextFlint organizes different transformations from the linguistic perspective. In this way, users could understand model weakness under a broad range of adversarial perturbations and evaluate their models more comprehensively.

Armed with these toolkits, users can freely develop textual adversarial attack algorithms and evaluate model adversarial robustness. Additionally, the generated adversarial samples can be utilized in adversarial data augmentation to improve the adversarial robustness of NLP models.

8.4 Out-of-Distribution Robustness

Most machine learning datasets obey the independently and identically distributed (i.i.d.) principle, which means data points from both training and test sets follow the same distribution. Although most common cases in the real world follow this rule, there still exist unusual scenarios where the test distribution differs from the training distribution, which we refer to as distribution shift. Distribution shift poses a great challenge on machine learning systems, and it is of great importance in high-stake applications, such as autonomous driving and medical analysis. For instance, autonomous driving algorithms should be robust to various driving conditions to reduce the unaffordable risk of a car accident. In NLP, distribution shifts can also degrade model performance significantly, which greatly hinders NLP applications. Following classical works [4, 92], here we discuss three typical distribution shifts, namely, spurious correlation, domain shift, and subpopulation shift.

8.4.1 Spurious Correlation

Deep learning methods are good at capturing correlations inside data such as word or object co-occurrence. However, correlations in training data do not indicate real relations in the wild [92]. The most well-known example of spurious correlation is object co-occurrence in images. For example, cows are mostly observed on grasslands. So in ImageNet, the images labeled as cows are always associated with grass. This spurious correlation is easily captured by DNNs, and once a cow appears in an unexpected location like the beach, the trained classification model might not recognize the cow correctly. Spurious correlations are commonly observed in machine learning and remain a persistent challenge in learning robust representations.

---

[7] https://github.com/textflint/textflint.

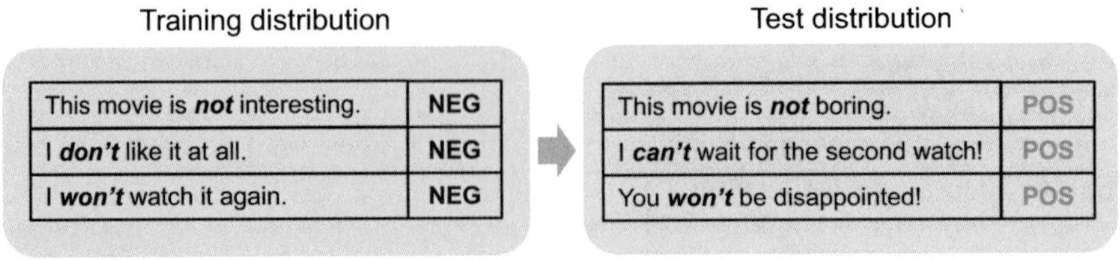

Training distribution   Test distribution

This movie is *not* interesting.	NEG
I *don't* like it at all.	NEG
I *won't* watch it again.	NEG

This movie is *not* boring.	POS
I *can't* wait for the second watch!	POS
You *won't* be disappointed!	POS

Fig. 8.6 An example of spurious correlation in sentiment analysis. "NEG" is associated with negation words in training distribution but not in test distribution

In NLP, spurious correlations are also everywhere. We provide an example in Fig. 8.6 showing the possible spurious correlation between negation words (*not*, *don't*, and *won't*) and the "NEG" label. Studies of spurious correlation in NLP mostly lie in NLI tasks, which aim to determine sentence-pair relations. State-of-the-art NLI models have achieved high accuracy on standard benchmarks, but researchers find that they rely heavily on spurious correlations. For example, Naik et al. [65] find that two sentences with a high word overlapping ratio usually hold the same semantics (entailment) in training data. If the models capture this spurious correlation, they will fail when discriminating sentence relationships between *John gave Mary a gift* and *Mary gave John a gift*. To quantify this issue, some challenging datasets are proposed with carefully crafted counterintuitive test data. McCoy et al. [60] create non-entailment sentence pairs with high word overlap using syntactic rules, while PAWS [118] utilizes back-translation and word swapping to generate challenging test data. Experiments show that most models perform poorly on these datasets, indicating the models are fragile to this kind of spurious correlation.

As a general and practical flaw in deep learning systems, avoiding learning spurious correlations is crucial. Here we introduce efforts made in denying spurious correlations, together with the lessons learned from the intriguing phenomenon of analyzing and understanding the memorization and generalization of NLP models.

Pre-training Pre-training is an effective approach faced with spurious correlations. Tu et al. [93] conduct a fine-grained analysis and conclude that the superior generalization ability of PTMs enables them to learn from a small set of counterintuitive samples and stay less affected by spurious correlations in training data. Moreover, scaling model sizes, pre-training with more data, and longer fine-tuning also help. However, PTMs are not perfect solutions to these problems. Nadeem et al. [64] find that PTMs perform gender and demographic biases naturally without fine-tuning, indicating that they learned to associate stereotypes with certain groups from pre-training. To this end, careful authorization is urgently demanded in training responsible PTMs.

Heuristic Sample Reweighting Sample reweighting aims to identify training samples with spurious correlations and downweight their importance during training. Based on some heuristics, e.g., *don't* in hypothesis sentence is highly relevant with label `contradictory`, these methods [9, 57] calculate the bias probability

$P_b = P(\texttt{contradictory}|don't)$ and use this probability to reweight samples. Typical reweighting strategies include importance weighting $(1/P_b)$ and focal loss. These approaches are useful to cope with known spurious correlations, but they require prior knowledge to determine the weights, which largely constrains their practicality in real applications.

Behavior-Based Sample Reweighting This kind of method manages to discover different model behaviors on normal samples and samples with biases, and then debias the dataset. Through empirical studies, there are two kinds of distinctive behaviors:

1. Models usually learn superficial features first because they are relatively easy to master. Therefore, Utama et al. [94] propose to learn a shallow model, which means a model trained with fewer examples and epochs, to serve as a debias proxy. The learned shallow model is confident on samples with shortcuts, so it is effective to simply downweighting the most confident samples of the shallow model.
2. Another work [108] utilizes the forgettable examples to debias. Forgettable examples are defined as the samples that go from being correctly to incorrectly classified or never learned during training. Observing that forgettable samples are difficult and valuable, this algorithm first trains a shallow model (e.g., LSTM) to identify the forgettable samples, then reweights training data, and fine-tunes BERT to get a robust model. Compared with heuristic methods, behavior-based models could automatically find suspicious patterns and mitigate them, making them more practical.

Stable Learning From the perspective of causality, stable learning recognizes spurious correlation as the confounding factor [12], which shares the same cause with the output variable. To remove the negative effect brought by confounding features, stable learning tries to decorrelate features and thus find true causes. Theoretically, researchers prove that this can be achieved by appropriate sample reweighting [43, 79]. On this basis, the developed algorithms can successfully eliminate irrelative features and perform well on datasets with spurious correlations [117].

8.4.2 Domain Shift

Domain shift [99] is the most well-known distribution shift in machine learning, which arises in many real-world scenarios. Due to the limitation in training data collection, representation learning models in most cases are trained and tested in a specific domain. However in the real world, it is common practice to apply trained models in other domains or open environments, so it is natural to expect representation learning models trained on one domain to generalize well on a relevant but distinct domain, which we refer to as robustness under domain shift.

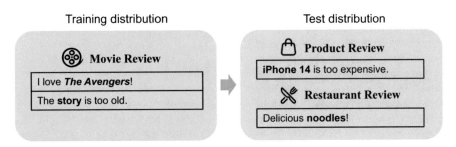

Fig. 8.7 An example of domain shift in sentiment analysis. The model is trained on movie reviews but tested on product and restaurant reviews

In computer vision, domain shift has been widely investigated, such as classifying images in different styles [46], under corruptions [28] or distinct views [42].

In NLP, however, measuring robustness under domain shift relies heavily on heuristics. The common practice is to collect datasets from different sources, select one to serve as an in-distribution training dataset, and evaluate model performances on other datasets. For example, on sentiment analysis, as we show in Fig. 8.7, practitioners [29] usually choose movie review datasets as training datasets, and test models on restaurant and product reviews. Although this strategy is reasonable and the experiment results truly reflect robustness under distribution shift to some extent, current approaches directly utilize existing datasets, which cannot fully characterize the distribution shift for real-world problems. WILDS [42] partially solves this issue. The authors consider the practical needs for NLP models to generalize across different user groups, and construct an Amazon review sentiment analysis dataset, where the models are trained and tested on product reviews from different user groups. In the future, we hope more efforts can be devoted to building comprehensive and practical benchmarks for domain shift robustness.

Algorithms targeting domain shifts are known as domain generalization methods. Next, we will introduce representative algorithms and their practices in NLP.

Pre-training Pre-training is still effective in dealing with domain shift due to the gained abundant general knowledge. Hendrycks et al. [29] conduct extensive empirical studies on sentiment analysis, semantic similarity, reading comprehension, and natural language inference. They reveal that PTMs are considerably more robust than traditional models. For instance, RoBERTa remains most performance when transferred across reviews from different sources, while LSTM suffers a 35% accuracy decrease. The analysis also finds that pre-training on more diverse data further improves robustness.

Domain-Invariant Representation Learning These algorithms aim to split domain information out in learned representations so that they are transferable across domains. In this regard, CORAL [84] presumes that domain-invariant representations should share the same distributions in different domains via regularization. Therefore, CORAL minimizes the differences in means and

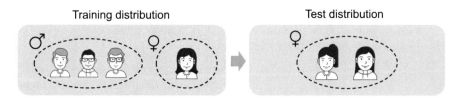

Fig. 8.8 An example of subpopulation shift

covariances of representation distribution. Invariant risk minimization (IRM) [3] borrows the idea from invariant predictors [67]. It seeks data representations such that the optimal predictors built on are the same across domains. Although these methods are theoretically sound and have been proven effective on toy examples, Dranker et al. [16] show that it is rather difficult to learn satisfying representations under practical domain shifts. How to learn domain-invariant representations still remains unsolved.

8.4.3 Subpopulation Shift

Subpopulation shift depicts the natural frequency change of data groups in training and test data. Representation learning models perform well on average most time, but their effectiveness may be dominated by overrepresented groups with ignorance of underrepresented groups. We give an example in Fig. 8.8 where the training data is mostly collected from males, but we expect models to perform well on females. In practice, subpopulation shift is of great significance for two reasons:

- Reliability. Consider the case that we train an autonomous driving model with many photos taken in the daytime with few taken at night, the model only needs to learn how to behave in the daytime to perform well on in-distribution tests, leaving nighttime performances unreliable. However, both daytime and night conditions happen in the real world, and we do not expect our models are highly unstable in different situations.
- Fairness. To avoid algorithmic discrimination over minority groups (e.g., minor genders or races), the models are also supposed to perform equally on each group.

In NLP, a concrete case of subpopulation shift is comments from different groups of people. CivilComments [6] is a collected dataset of comments on articles, and each comment is annotated as "toxic" or "nontoxic." Meanwhile, each comment is associated with user profile information, including gender, race, and religion. Studies [42] on this dataset suggest that NLP models show poor performance in particular subpopulations.

A series of works are proposed to deal with subpopulation shifts, and they aim to improve models' worst-group performance. Based on whether there is explicit group information, we can get two lines of studies.

Methods with Group Information Some works argue that the mainstream optimization objective, empirical risk minimization (ERM), leads to the robustness issue under subpopulation shift since ERM only optimizes the global loss regardless of group-wise performance. To this end, group distributionally robust optimization (GroupDRO) [77] applies distributionally robust optimization (DRO) algorithm to explicitly improve the worst-group performance. By solely updating model parameters using the worst data group, GroupDRO successfully improves model robustness under subpopulation shift. Another work [66] recognizes the subpopulation shift in PTM pre-training. Rather than the original maximum likelihood estimation (MLE) loss, they propose to use one DRO loss named conditional value at risk (CVaR) which provides relatively low losses on almost all subpopulations in the training distribution. The modified loss function leads to language models equally performed across groups.

Methods Without Group Information A more practical scenario is that the group information is unavailable. To deal with implicit groups, Sohoni et al. [83] adopt clustering algorithms to divide training data into subgroups and then apply GroupDRO to optimize the worst-group loss. Apart from identifying implicit groups, just train twice (JTT) [54] is a recently proposed two-stage method. JTT first trains a model with standard ERM loss and then upweights the misclassified samples using this model. Then, it trains another model with the reweighted training dataset. JTT outperforms traditional DRO algorithms and approaches methods with group information.

8.5   Interpretability

The need of interpretability stands at top of our pyramid, highlighting its importance for reliable and trustworthy NLP. In Chap. 1, we have discussed two representation schemes in NLP, namely, symbolic representation and distributed representation. Although distributed representation is prevailing in nowadays NLP, one essential and long-lasting criticism of it is the lack of interpretability. Given a representation vector of a word or sentence, we can hardly tell accurately what is encoded. Worse still, modern deep learning models, the fundamental infrastructure in representation learning, are also "black boxes," which pose a great challenge to reliable, trustworthy, and cooperative AI.

Many researchers have devoted themselves to mitigating the interpretability issue in deep learning, but there is still a long way to go. In this section, we will give a brief introduction to efforts made in constructing interpretable NLP systems, including understanding model functionality and explaining model mechanisms.

8.5.1   Understanding Model Functionality

The very first step in understanding a model at hand is predicting its behaviors. On standard benchmarks, we can only get the final scores over a set of test samples, but we have no idea how a model will react to certain inputs. In practice, we can hardly trust a model if we do not know (approximately) when the predictions will be correct and wrong. This leads to the problem of *calibration*, which demands models to give accurate confidence estimation to their predictions. On the other hand, the black-box nature of neural networks makes it difficult to inspect their functionalities. Moreover, as the sizes of big PTMs consistently scale up, researchers surprisingly find emerging new abilities [103], such as the few-shot learning ability of GPT-3. While it reveals an encouraging potential of big models, worries are also raised about the unpredictable nature, since undesired abilities such as memorizing privacy contents [7] and generating toxic speeches also emerged. For this, it is also crucial to specify what *abilities* models possess. Next, we will introduce two topics: model calibration and ability testing.

Calibration   Deep learning models mostly suffer from the overconfidence problem, which means that these models produce unreliable confidence scores [25, 37]. The misalignment between estimated and real probability may bring catastrophic consequences, especially in high-stake applications. To this end, researchers aim to make models *calibrated*. Different from vanilla models with overconfidence scores, calibrated models are models that assign appropriate confidence scores to predictions. Given input x and its ground truth label y, a well-calibrated model outputs $\hat{y}$ with probability $P_M(y|x)$ which satisfies

$$P(\hat{y} = y | P_M(\hat{y}|x) = p) = p, \forall p \in [0, 1]. \tag{8.8}$$

The equation suggests that the estimated probability P_M matches the true probability P. To solve the overconfidence issue and build calibrated models, some approaches try to smooth the probability distribution, including using temperature scaling [25] and label smoothing[86]. Although they could to some extent mitigate the overconfidence issue, these post hoc methods are not able to solve the calibration problem at its roots. Most recent learnable calibration methods [40, 53] pave another way. By collecting extra data to teach models to be calibrated, these models show that large-scale PTMs could learn calibration well, providing satisfying probability estimations. However, the generalization ability of the learned calibration is still poor, leaving this problem open.

Ability Testing   Deep representation learning models are always evaluated on various in- and out-of-distribution benchmarks, but how can we understand model abilities through these test results is unclear. For deeper insights into knowing model abilities, multiple carefully curated benchmarks and toolkits are proposed.

Probing Datasets   Probing datasets aim at measuring specific model abilities. GLUE [97] is a widely adopted benchmark for natural language understanding, which provides nine typical tasks to evaluate NLP model performances. Besides the application-driven main benchmarks, GLUE also offers a manually annotated diagnostic dataset to illustrate linguistic abilities captured by NLP models, including lexical semantics, predicate-argument structure, logic, etc. The ability-driven diagnostic test helps with more fine-grained model analysis. Apart from GLUE, Tenney et al. [91] design comprehensive tests for probing how PTMs deal with sentence structure. For probing world and commonsense knowledge, Petroni et al. [68] propose LAMA, which evaluates how well PTMs could capture such knowledge.

Behavioral Testing CheckList [75] tests model abilities from another perspective. Inspired by common practices in software engineering, CheckList is proposed to conduct behavioral testing for NLP models. By designing different types of tests, CheckList covers a series of important capabilities NLP models should have. For example, if the users add a *not* before a negative word, then the model should be aware of the sentiment has changed to pass the test. Compared with fixed diagnostic datasets, CheckList provides a set of tools for users to generate test cases easily.

As big language models are likely to consistently get novel abilities, depicting their possible functionalities is becoming increasingly difficult. In the future, researchers need to specify desirable and undesirable abilities more clearly and design rigorous evaluations to assess these abilities.

8.5.2 Explaining Model Mechanism

Explaining model behaviors is always a challenging yet fundamental topic for deep learning [15]. Compared with classic machine learning models like the decision tree, the mechanism of neural network-based models is less transparent due to the nature of distributed representations. To get further understandings of how models work, explanatory methods are developed to find possible reasons for specific model decisions. Roughly, we can categorize these methods according to providing external or internal explanations.

External Explanation Given the data-driven learning paradigm, one straightforward way is to find corresponding factors in data for model behaviors, which we name external explanations. In this direction, some works try to find out specific input pieces that lead to certain predictions. They either calculate model gradients with respect to each token to generate the saliency map [82, 85] or apply adversarial attacks or input reduction on texts to identify important pieces [20, 48]. AllenNLP Interpret [96] implements a set of these methods to help users better comprehend model outputs. Another kind of external explanation attributes model predictions to training data instances. Iconic work in this direction is influence function [41], which measures how model parameters change when a training point is removed from training data. External explanations offer a data-level view to know the model mechanism, but they cannot enable us to take a look at the model structure. Furthermore, given the enormous pre-training data, it is hard to specify the contribution of single data instances.

Internal Explanation Beyond data-level explanations, there are also attempts to explain models from the internal structure. By partitioning neural networks into smaller pieces, a major goal of this line of research is to discover the different abilities of each module. Through inspecting PTMs, researchers have established many insightful conclusions. Some works find that BERT processes sentences following a linguistic pipeline [35, 90]. From bottom to top layers, the model first captures word-level and phrase-level features, then deals with syntactic patterns, and finally summarizes semantic meanings. Besides, Transformers present distinct attention patterns in different layers [10, 106], which also indicates layer-specific functionalities such as capturing word composition or syntactic structure knowledge. Meanwhile, feed-forward layers act like key-value memories [22] which store responses for certain text patterns. Wang et al. [102] conduct a more fine-grained analysis on the neuron activation patterns. They surprisingly find that some downstream tasks are highly correlated with specific neurons, which indicates that PTMs have functionality modularity across tasks. While internal explanations pave novel paths for understanding model mechanism, current progress mostly remain qualitative rather than quantitative. Extensive work is needed to fully demystify neural networks and even PTMs.

8.6 Summary and Further Readings

Up to now, we have overviewed the current progress and challenges of robust representation learning in NLP. In this last section, we will summarize the contents of this chapter and then provide more readings for reference.

Robustness is a crucial topic for reliable and trustworthy AI, and it is well recognized that existing NLP models are brittle in the complicated real world. In this chapter, we introduce four robustness issues in NLP following our proposed robustness hierarchy. Specifically, we first focus on integrity, which means whether models can work well on common cases without inner vulnerabilities, with backdoor robustness as a typical example. Second, we turn to external safety and discuss potential adversarial attacks models may face and corresponding defenses. Then, we consider real-world situations where models are supposed to be resilient under unseen, extreme even "black swan" events. We discuss three kinds of distribution shifts, namely, spurious correlation, domain shift, and subpopulation shift. Finally, we examine the highest demand posed on representation learning models and interpretability. We introduce the current stages in explaining model functionality and mechanism.

G. Cui et al.

On backdoor robustness, Li et al. [52] give a unified overview on backdoor attack and defense, and their backdoor resource repository[8] is also beneficial. Roth et al. [76] and Wang et al. [100] provide comprehensive surveys on textual adversarial attack and defense. You can also find more related papers from our paper list.[9] Shen et al. [80] provide a holistic view of out-of-distribution robustness. Wiegreffe et al. [105] summarize current research progress in explainable NLP.

On the internal and external threats against machine learning systems, Hendrycks et al. [27] give an insightful discussion on model robustness, monitoring, alignment, and external safety. Bommasani et al. [5] also provide their opinions in Sections 4.7, 4.8, and 4.9.

Acknowledgments The contributions of all authors for the second edition are as follows: Zhiyuan Liu, Yankai Lin, and Maosong Sun designed the overall architecture of this chapter. Ganqu Cui drafted this chapter. Zhiyuan Liu and Yankai Lin proofread and revised this chapter.

We thank Chaojun Xiao, Yujia Qin, Yuan Yao, Zheni Zeng, Yangyi Chen, Weilin Zhao, Chaoqun He, and Lifan Yuan for proofreading the chapter.

This chapter is about robust representation learning and is the newly complemented content in the second edition of the book *Representation Learning for Natural Language Processing*. The first edition of the book was published in 2020 [55].

References

1. Facebook translates 'good morning' into 'attack them', leading to arrest. https://www.theguardian.com/technology/2017/oct/24/facebook-palestine-israel-translates-good-morning-attack-them-arrest.
2. Moustafa Alzantot, Yash Sharma, Ahmed Elgohary, Bo-Jhang Ho, Mani Srivastava, and Kai-Wei Chang. Generating natural language adversarial examples. In *Proceedings of EMNLP*, 2018.
3. Martin Arjovsky, Léon Bottou, Ishaan Gulrajani, and David Lopez-Paz. Invariant risk minimization. *arXiv preprint arXiv:1907.02893*, 2019.
4. John Blitzer, Ryan McDonald, and Fernando Pereira. Domain adaptation with structural correspondence learning. In *Proceedings of EMNLP*, 2006.
5. Rishi Bommasani, Drew A Hudson, Ehsan Adeli, Russ Altman, Simran Arora, Sydney von Arx, Michael S Bernstein, Jeannette Bohg, Antoine Bosselut, Emma Brunskill, et al. On the opportunities and risks of foundation models. *arXiv preprint arXiv:2108.07258*, 2021.
6. Daniel Borkan, Lucas Dixon, Jeffrey Sorensen, Nithum Thain, and Lucy Vasserman. Nuanced metrics for measuring unintended bias with real data for text classification. In *Proceedings of WWW*, 2019.
7. Nicholas Carlini, Florian Tramer, Eric Wallace, Matthew Jagielski, Ariel Herbert-Voss, Katherine Lee, Adam Roberts, Tom Brown, Dawn Song, Ulfar Erlingsson, et al. Extracting training data from large language models. In *Proceedings of USENIX Security*, 2021.
8. Chuanshuai Chen and Jiazhu Dai. Mitigating backdoor attacks in lstm-based text classification systems by backdoor keyword identification. *Neurocomputing*, 2021.
9. Christopher Clark, Mark Yatskar, and Luke Zettlemoyer. Don't take the easy way out: Ensemble based methods for avoiding known dataset biases. In *Proceedings of EMNLP*, 2019.

[8] https://github.com/THUYimingLi/backdoor-learning-resources.

[9] https://github.com/thunlp/TAADpapers.

10. Kevin Clark, Urvashi Khandelwal, Omer Levy, and Christopher D. Manning. What does BERT look at? an analysis of bert's attention. In *Proceedings of ACL Workshop BlackboxNLP*, 2019.
11. Ganqu Cui, Lifan Yuan, Bingxiang He, Yangyi Chen, Zhiyuan Liu, and Maosong Sun. A unified evaluation of textual backdoor learning: Frameworks and benchmarks. In *Proceedings of NeurIPS: Datasets and Benchmarks Track*, 2022.
12. Peng Cui and Susan Athey. Stable learning establishes some common ground between causal inference and machine learning. *Nature Machine Intellegence*, 2022.
13. Jiazhu Dai, Chuanshuai Chen, and Yufeng Li. A backdoor attack against lstm-based text classification systems. *IEEE Access*, 2019.
14. Jacob Devlin, Ming-Wei Chang, Kenton Lee, and Kristina Toutanova. BERT: pre-training of deep bidirectional transformers for language understanding. In *Proceedings of NAACL-HLT*, 2019.
15. Finale Doshi-Velez and Been Kim. Towards a rigorous science of interpretable machine learning. *arXiv preprint arXiv:1702.08608*, 2017.
16. Yana Dranker, He He, and Yonatan Belinkov. Irm—when it works and when it doesn't: A test case of natural language inference. In *Proceedings of NeurIPS*, 2021.
17. Javid Ebrahimi, Anyi Rao, Daniel Lowd, and Dejing Dou. Hotflip: White-box adversarial examples for text classification. In *Proceedings of EACL*, 2017.
18. Steffen Eger and Yannik Benz. From hero to zéroe: A benchmark of low-level adversarial attacks. In *Proceedings of AACL*, 2020.
19. Steffen Eger, Gözde Gül Şahin, Andreas Rücklé, Ji-Ung Lee, Claudia Schulz, Mohsen Mesgar, Krishnkant Swarnkar, Edwin Simpson, and Iryna Gurevych. Text processing like humans do: Visually attacking and shielding nlp systems. In *Proceedings of NAACL*, 2019.
20. Shi Feng, Eric Wallace, Alvin Grissom II, Mohit Iyyer, Pedro Rodriguez, and Jordan Boyd-Graber. Pathologies of neural models make interpretations difficult. In *Proceedings of EMNLP*, 2018.
21. Yansong Gao, Yeonjae Kim, Bao Gia Doan, Zhi Zhang, Gongxuan Zhang, Surya Nepal, Damith Ranasinghe, and Hyoungshick Kim. Design and evaluation of a multi-domain trojan detection method on deep neural networks. *IEEE Transactions on Dependable and Secure Computing*, 2021.
22. Mor Geva, Roei Schuster, Jonathan Berant, and Omer Levy. Transformer feed-forward layers are key-value memories. In *Proceedings of EMNLP*, 2021.
23. Amir Globerson and Sam Roweis. Nightmare at test time: robust learning by feature deletion. In *Proceedings of ICML*, 2006.
24. Tianyu Gu, Brendan Dolan-Gavitt, and Siddharth Garg. Badnets: Identifying vulnerabilities in the machine learning model supply chain. *arXiv preprint arXiv:1708.06733*, 2017.
25. Chuan Guo, Geoff Pleiss, Yu Sun, and Kilian Q. Weinberger. On calibration of modern neural networks. In *Proceedings of ICML*, 2017.
26. Chuan Guo, Alexandre Sablayrolles, Hervé Jégou, and Douwe Kiela. Gradient-based adversarial attacks against text transformers. In *Proceedings of EMNLP*, 2021.
27. Dan Hendrycks, Nicholas Carlini, John Schulman, and Jacob Steinhardt. Unsolved problems in ml safety. *arXiv preprint arXiv:2109.13916*, 2021.
28. Dan Hendrycks and Thomas G. Dietterich. Benchmarking neural network robustness to common corruptions and perturbations. In *Proceedings of ICLR*, 2019.
29. Dan Hendrycks, Xiaoyuan Liu, Eric Wallace, Adam Dziedzic, Rishabh Krishnan, and Dawn Song. Pretrained transformers improve out-of-distribution robustness. In *Proceedings of ACL*, 2020.
30. Hossein Hosseini, Sreeram Kannan, Baosen Zhang, and Radha Poovendran. Deceiving google's perspective api built for detecting toxic comments. *arXiv preprint arXiv:1702.08138*, 2017.
31. Kuan-Hao Huang and Kai-Wei Chang. Generating syntactically controlled paraphrases without using annotated parallel pairs. In *Proceedings of EACL*, 2021.

32. Peter J Huber. Robust estimation of a location parameter. *The Annals of Mathematical Statistics*, 1964.
33. Mohit Iyyer, John Wieting, Kevin Gimpel, and Luke Zettlemoyer. Adversarial example generation with syntactically controlled paraphrase networks. In *Proceedings of NAACL*, 2018.
34. Eric Jang, Shixiang Gu, and Ben Poole. Categorical reparameterization with gumbel-softmax. In *Proceedings of ICLR*, 2017.
35. Ganesh Jawahar, Benoît Sagot, and Djamé Seddah. What does BERT learn about the structure of language? In *Proceedings of ACL*, 2019.
36. Robin Jia and Percy Liang. Adversarial examples for evaluating reading comprehension systems. In *Proceedings of EMNLP*, 2017.
37. Zhengbao Jiang, Jun Araki, Haibo Ding, and Graham Neubig. How can we know *When* language models know? on the calibration of language models for question answering. *Transactions of the Association for Computational Linguistics*, 2021.
38. Di Jin, Zhijing Jin, Joey Tianyi Zhou, and Peter Szolovits. Is bert really robust? a strong baseline for natural language attack on text classification and entailment. In *Proceedings of AAAI*, 2020.
39. Erik Jones, Robin Jia, Aditi Raghunathan, and Percy Liang. Robust encodings: A framework for combating adversarial typos. In *Proceedings of ACL*, 2020.
40. Saurav Kadavath, Tom Conerly, Amanda Askell, Tom Henighan, Dawn Drain, Ethan Perez, Nicholas Schiefer, Zac Hatfield Dodds, Nova DasSarma, Eli Tran-Johnson, et al. Language models (mostly) know what they know. *arXiv preprint arXiv:2207.05221*, 2022.
41. Pang Wei Koh and Percy Liang. Understanding black-box predictions via influence functions. In *Proceedings of ICML*, 2017.
42. Pang Wei Koh, Shiori Sagawa, Henrik Marklund, Sang Michael Xie, Marvin Zhang, Akshay Balsubramani, Weihua Hu, Michihiro Yasunaga, Richard Lanas Phillips, Irena Gao, Tony Lee, Etienne David, Ian Stavness, Wei Guo, Berton Earnshaw, Imran S. Haque, Sara M. Beery, Jure Leskovec, Anshul Kundaje, Emma Pierson, Sergey Levine, Chelsea Finn, and Percy Liang. WILDS: A benchmark of in-the-wild distribution shifts. In *Proceedings of ICML*, 2021.
43. Kun Kuang, Peng Cui, Susan Athey, Ruoxuan Xiong, and Bo Li. Stable prediction across unknown environments. In *Proceedings of KDD*, 2018.
44. Ram Shankar Siva Kumar, Magnus Nyström, John Lambert, Andrew Marshall, Mario Goertzel, Andi Comissoneru, Matt Swann, and Sharon Xia. Adversarial machine learning-industry perspectives. In *Proceedingds of Security and Privacy Workshops*, 2020.
45. Thai Le, Jooyoung Lee, Kevin Yen, Yifan Hu, and Dongwon Lee. Perturbations in the wild: Leveraging human-written text perturbations for realistic adversarial attack and defense. In *Findings of ACL*, 2022.
46. Da Li, Yongxin Yang, Yi-Zhe Song, and Timothy M Hospedales. Deeper, broader and artier domain generalization. In *Proceedings of ICCV*, 2017.
47. Jinfeng Li, Shouling Ji, Tianyu Du, Bo Li, and Ting Wang. Textbugger: Generating adversarial text against real-world applications. In *Proceedings of NDSS*, 2018.
48. Jiwei Li, Will Monroe, and Dan Jurafsky. Understanding neural networks through representation erasure. *arXiv preprint arXiv:1612.08220*, 2016.
49. Linyang Li, Ruotian Ma, Qipeng Guo, Xiangyang Xue, and Xipeng Qiu. BERT-ATTACK: Adversarial attack against BERT using BERT. In *Proceedings of EMNLP*, 2020.
50. Linyang Li, Demin Song, Xiaonan Li, Jiehang Zeng, Ruotian Ma, and Xipeng Qiu. Backdoor attacks on pre-trained models by layerwise weight poisoning. In *Proceedings of EMNLP*, 2021.
51. Xiang Lisa Li and Percy Liang. Prefix-tuning: Optimizing continuous prompts for generation. In *Proceedings of ACL-IJCNLP*, 2021.
52. Yiming Li, Yong Jiang, Zhifeng Li, and Shu-Tao Xia. Backdoor learning: A survey. *arXiv preprint arXiv:2007.08745*, 2020.
53. Stephanie Lin, Jacob Hilton, and Owain Evans. Teaching models to express their uncertainty in words. *arXiv preprint arXiv:2205.14334*, 2022.

54. Evan Zheran Liu, Behzad Haghgoo, Annie S. Chen, Aditi Raghunathan, Pang Wei Koh, Shiori Sagawa, Percy Liang, and Chelsea Finn. Just train twice: Improving group robustness without training group information. In *Proceedings of ICML*, 2021.
55. Zhiyuan Liu, Yankai Lin, and Maosong Sun. *Representation Learning for Natural Language Processing*. Springer, 2020.
56. Aleksander Madry, Aleksandar Makelov, Ludwig Schmidt, Dimitris Tsipras, and Adrian Vladu. Towards deep learning models resistant to adversarial attacks. In *Proceedings of ICLR*, 2018.
57. Rabeeh Karimi Mahabadi, Yonatan Belinkov, and James Henderson. End-to-end bias mitigation by modelling biases in corpora. In *Proceedings of ACL*, 2020.
58. Rishabh Maheshwary, Saket Maheshwary, and Vikram Pudi. Generating natural language attacks in a hard label black box setting. In *Proceedings of AAAI*, 2021.
59. Abraham Harold Maslow. A dynamic theory of human motivation. 1958.
60. R. Thomas McCoy, Ellie Pavlick, and Tal Linzen. Right for the wrong reasons: Diagnosing syntactic heuristics in natural language inference. In *Proceedings of ACL*, 2019.
61. John X. Morris, Eli Lifland, Jack Lanchantin, Yangfeng Ji, and Yanjun Qi. Reevaluating adversarial examples in natural language. In *Findings of EMNLP*, 2020.
62. John X. Morris, Eli Lifland, Jin Yong Yoo, Jake Grigsby, Di Jin, and Yanjun Qi. TextAttack: A framework for adversarial attacks, data augmentation, and adversarial training in NLP. In *Proceedings of EMNLP*, 2020.
63. Maximilian Mozes, Pontus Stenetorp, Bennett Kleinberg, and Lewis D. Griffin. Frequency-guided word substitutions for detecting textual adversarial examples. In *Proceedings of EACL*, 2021.
64. Moin Nadeem, Anna Bethke, and Siva Reddy. Stereoset: Measuring stereotypical bias in pretrained language models. In *Proceedings of ACL*, 2021.
65. Aakanksha Naik, Abhilasha Ravichander, Norman M. Sadeh, Carolyn Penstein Rosé, and Graham Neubig. Stress test evaluation for natural language inference. In *Proceedings of COLING*, 2018.
66. Yonatan Oren, Shiori Sagawa, Tatsunori B. Hashimoto, and Percy Liang. Distributionally robust language modeling. In *Proceedings of EMNLP-IJCNLP*, 2019.
67. Jonas Peters, Peter Bühlmann, and Nicolai Meinshausen. Causal inference by using invariant prediction: identification and confidence intervals. *Journal of the Royal Statistical Society: Series B (Statistical Methodology)*, 2016.
68. Fabio Petroni, Tim Rocktäschel, Sebastian Riedel, Patrick Lewis, Anton Bakhtin, Yuxiang Wu, and Alexander Miller. Language models as knowledge bases? In *Proceedings of EMNLP-IJCNLP*, 2019.
69. Fanchao Qi, Yangyi Chen, Mukai Li, Yuan Yao, Zhiyuan Liu, and Maosong Sun. ONION: A simple and effective defense against textual backdoor attacks. In *Proceedings of EMNLP*, 2021.
70. Fanchao Qi, Yangyi Chen, Xurui Zhang, Mukai Li, Zhiyuan Liu, and Maosong Sun. Mind the style of text! adversarial and backdoor attacks based on text style transfer. In *Proceedings of EMNLP*, 2021.
71. Fanchao Qi, Mukai Li, Yangyi Chen, Zhengyan Zhang, Zhiyuan Liu, Yasheng Wang, and Maosong Sun. Hidden killer: Invisible textual backdoor attacks with syntactic trigger. In *Proceedings of ACL-IJCNLP*, 2021.
72. Fanchao Qi, Yuan Yao, Sophia Xu, Zhiyuan Liu, and Maosong Sun. Turn the combination lock: Learnable textual backdoor attacks via word substitution. In *Proceedings of ACL-IJCNLP*, 2021.
73. Alec Radford, Jeffrey Wu, Rewon Child, David Luan, Dario Amodei, and Ilya Sutskever. Language models are unsupervised multitask learners. *OpenAI Blog*, 2019.
74. Shuhuai Ren, Yihe Deng, Kun He, and Wanxiang Che. Generating natural language adversarial examples through probability weighted word saliency. In *Proceedings of ACL*, 2019.
75. Marco Túlio Ribeiro, Tongshuang Wu, Carlos Guestrin, and Sameer Singh. Beyond accuracy: Behavioral testing of NLP models with checklist. In *Proceedings of ACL*, 2020.

76. Tom Roth, Yansong Gao, Alsharif Abuadbba, Surya Nepal, and Wei Liu. Token-modification adversarial attacks for natural language processing: A survey. *arXiv preprint arXiv:2103.00676*, 2021.
77. Shiori Sagawa, Pang Wei Koh, Tatsunori B Hashimoto, and Percy Liang. Distributionally robust neural networks for group shifts: On the importance of regularization for worst-case generalization. *arXiv preprint arXiv:1911.08731*, 2019.
78. Lujia Shen, Shouling Ji, Xuhong Zhang, Jinfeng Li, Jing Chen, Jie Shi, Chengfang Fang, Jianwei Yin, and Ting Wang. Backdoor pre-trained models can transfer to all. In *Proceedings of CCS*, 2021.
79. Zheyan Shen, Peng Cui, Kun Kuang, Bo Li, and Peixuan Chen. Causally regularized learning with agnostic data selection bias. In *Proceedings of MM*, 2018.
80. Zheyan Shen, Jiashuo Liu, Yue He, Xingxuan Zhang, Renzhe Xu, Han Yu, and Peng Cui. Towards out-of-distribution generalization: A survey. *arXiv preprint arXiv:2108.13624*, 2021.
81. Chenglei Si, Zhengyan Zhang, Fanchao Qi, Zhiyuan Liu, Yasheng Wang, Qun Liu, and Maosong Sun. Better robustness by more coverage: Adversarial and mixup data augmentation for robust finetuning. In *Findings of ACL-IJCNLP*, 2021.
82. Karen Simonyan, Andrea Vedaldi, and Andrew Zisserman. Deep inside convolutional networks: Visualising image classification models and saliency maps. In *Proceedings of ICLR, Workshop Track Proceedings*, 2014.
83. Nimit Sohoni, Jared Dunnmon, Geoffrey Angus, Albert Gu, and Christopher Ré. No subclass left behind: Fine-grained robustness in coarse-grained classification problems. In *Proceedings of NeurIPS*, 2020.
84. Baochen Sun and Kate Saenko. Deep CORAL: correlation alignment for deep domain adaptation. In *Proceedings of ECCV*, 2016.
85. Mukund Sundararajan, Ankur Taly, and Qiqi Yan. Axiomatic attribution for deep networks. In *Proceedings of ICML*, 2017.
86. Christian Szegedy, Vincent Vanhoucke, Sergey Ioffe, Jon Shlens, and Zbigniew Wojna. Rethinking the inception architecture for computer vision. In *Proceedings of CVPR*, 2016.
87. Christian Szegedy, Wojciech Zaremba, Ilya Sutskever, Joan Bruna, Dumitru Erhan, Ian J. Goodfellow, and Rob Fergus. Intriguing properties of neural networks. In *Proceedings of ICLR*, 2014.
88. Samson Tan, Shafiq Joty, Min-Yen Kan, and Richard Socher. It's morphin'time! combating linguistic discrimination with inflectional perturbations. In *Proceedings of ACL*, 2020.
89. Rohan Taori, Achal Dave, Vaishaal Shankar, Nicholas Carlini, Benjamin Recht, and Ludwig Schmidt. Measuring robustness to natural distribution shifts in image classification. In *Proceedings of NeurIPS*, 2020.
90. Ian Tenney, Dipanjan Das, and Ellie Pavlick. BERT rediscovers the classical NLP pipeline. In *Proceedings of ACL*, 2019.
91. Ian Tenney, Patrick Xia, Berlin Chen, Alex Wang, Adam Poliak, R. Thomas McCoy, Najoung Kim, Benjamin Van Durme, Samuel R. Bowman, Dipanjan Das, and Ellie Pavlick. What do you learn from context? probing for sentence structure in contextualized word representations. In *Proceedings of ICLR*, 2019.
92. Antonio Torralba and Alexei A Efros. Unbiased look at dataset bias. In *Proceedings of CVPR*, 2011.
93. Lifu Tu, Garima Lalwani, Spandana Gella, and He He. An empirical study on robustness to spurious correlations using pre-trained language models. *Transactions of the Association for Computational Linguistics*, 2020.
94. Prasetya Ajie Utama, Nafise Sadat Moosavi, and Iryna Gurevych. Towards debiasing nlu models from unknown biases. In *Proceedings of EMNLP*, 2020.
95. Eric Wallace, Shi Feng, Nikhil Kandpal, Matt Gardner, and Sameer Singh. Universal adversarial triggers for attacking and analyzing nlp. In *Proceedings of EMNLP-IJCNLP*, 2019.

96. Eric Wallace, Jens Tuyls, Junlin Wang, Sanjay Subramanian, Matt Gardner, and Sameer Singh. Allennlp interpret: A framework for explaining predictions of NLP models. In *Proceedings of EMNLP-IJCNLP*, 2019.
97. Alex Wang, Amanpreet Singh, Julian Michael, Felix Hill, Omer Levy, and Samuel Bowman. GLUE: A multi-task benchmark and analysis platform for natural language understanding. In *Proceedings of EMNLP Workshop*, 2018.
98. Boxin Wang, Hengzhi Pei, Boyuan Pan, Qian Chen, Shuohang Wang, and Bo Li. T3: Tree-autoencoder constrained adversarial text generation for targeted attack. In *Proceedings of EMNLP*, 2019.
99. Jindong Wang, Cuiling Lan, Chang Liu, Yidong Ouyang, and Tao Qin. Generalizing to unseen domains: A survey on domain generalization. In *Proceedings of IJCAI*, 2021.
100. Wenqi Wang, Run Wang, Lina Wang, Zhibo Wang, and Aoshuang Ye. Towards a robust deep neural network in texts: A survey. *arXiv preprint arXiv:1902.07285*, 2019.
101. Xiao Wang, Qin Liu, Tao Gui, Qi Zhang, Yicheng Zou, Xin Zhou, Jiacheng Ye, Yongxin Zhang, Rui Zheng, Zexiong Pang, et al. Textflint: Unified multilingual robustness evaluation toolkit for natural language processing. In *Proceedings of ACL*, 2021.
102. Xiaozhi Wang, Kaiyue Wen, Zhengyan Zhang, Lei Hou, Zhiyuan Liu, and Juanzi Li. Finding skill neurons in pre-trained transformer-based language models. In *Proceedings of EMNLP*, 2022.
103. Jason Wei, Yi Tay, Rishi Bommasani, Colin Raffel, Barret Zoph, Sebastian Borgeaud, Dani Yogatama, Maarten Bosma, Denny Zhou, Donald Metzler, et al. Emergent abilities of large language models. *arXiv preprint arXiv:2206.07682*, 2022.
104. Laura Weidinger, John Mellor, Maribeth Rauh, Conor Griffin, Jonathan Uesato, Po-Sen Huang, Myra Cheng, Mia Glaese, Borja Balle, Atoosa Kasirzadeh, et al. Ethical and social risks of harm from language models. *arXiv preprint arXiv:2112.04359*, 2021.
105. Sarah Wiegreffe and Ana Marasovic. Teach me to explain: A review of datasets for explainable natural language processing. In *Proceedings of NeurIPS: Datasets and Benchmarks Track*, 2021.
106. Sarah Wiegreffe and Yuval Pinter. Attention is not not explanation. In *Proceedings of EMNLP-IJCNLP*, 2019.
107. Lei Xu, Yangyi Chen, Ganqu Cui, Hongcheng Gao, and Zhiyuan Liu. Exploring the universal vulnerability of prompt-based learning paradigm. In *Findings of NAACL*, 2022.
108. Yadollah Yaghoobzadeh, Soroush Mehri, Remi Tachet, Timothy J. Hazen, and Alessandro Sordoni. Increasing robustness to spurious correlations using forgettable examples. In *Proceedings of EACL*, 2021.
109. Wenkai Yang, Lei Li, Zhiyuan Zhang, Xuancheng Ren, Xu Sun, and Bin He. Be careful about poisoned word embeddings: Exploring the vulnerability of the embedding layers in NLP models. In *Proceedings of NAACL-HLT*, 2021.
110. Wenkai Yang, Yankai Lin, Peng Li, Jie Zhou, and Xu Sun. RAP: robustness-aware perturbations for defending against backdoor attacks on NLP models. In *Proceedings of EMNLP*, 2021.
111. Wenkai Yang, Yankai Lin, Peng Li, Jie Zhou, and Xu Sun. Rethinking stealthiness of backdoor attack against nlp models. In *Proceedings of ACL-IJCNLP*, 2021.
112. Zonghan Yang and Yang Liu. On robust prefix-tuning for text classification. In *Proceedings of ICLR*, 2022.
113. Muchao Ye, Chenglin Miao, Ting Wang, and Fenglong Ma. Texthoaxer: Budgeted hard-label adversarial attacks on text. In *Proceedings of AAAI*, 2022.
114. Yuan Zang, Fanchao Qi, Chenghao Yang, Zhiyuan Liu, Meng Zhang, Qun Liu, and Maosong Sun. Word-level textual adversarial attacking as combinatorial optimization. In *Proceedings of ACL*, 2020.
115. Guoyang Zeng, Fanchao Qi, Qianrui Zhou, Tingji Zhang, Bairu Hou, Yuan Zang, Zhiyuan Liu, and Maosong Sun. Openattack: An open-source textual adversarial attack toolkit. In *Proceedings of ACL*, 2021.

116. Hongyi Zhang, Moustapha Cissé, Yann N. Dauphin, and David Lopez-Paz. mixup: Beyond empirical risk minimization. In *Proceedings of ICLR*, 2018.
117. Xingxuan Zhang, Peng Cui, Renzhe Xu, Linjun Zhou, Yue He, and Zheyan Shen. Deep stable learning for out-of-distribution generalization. In *Proceedings of CVPR*, 2021.
118. Yuan Zhang, Jason Baldridge, and Luheng He. Paws: Paraphrase adversaries from word scrambling. In *Proceedings of NAACL*, 2019.
119. Zhengyan Zhang, Guangxuan Xiao, Yongwei Li, Tian Lv, Fanchao Qi, Zhiyuan Liu, Yasheng Wang, Xin Jiang, and Maosong Sun. Red alarm for pre-trained models: Universal vulnerabilities by neuron-level backdoor attacks. *arXiv preprint arXiv:2101.06969*, 2021.
120. Yichao Zhou, Jyun-Yu Jiang, Kai-Wei Chang, and Wei Wang. Learning to discriminate perturbations for blocking adversarial attacks in text classification. In *Proceedings of EMNLP-IJCNLP*, 2019.
121. Biru Zhu, Yujia Qin, Ganqu Cui, Yangyi Chen, Weilin Zhao, Chong Fu, Yangdong Deng, Zhiyuan Liu, Jingang Wang, Wei Wu, Maosong Sun, and Ming. Gu. Moderate-fitting as a natural backdoor defender for pre-trained language models. In *Proceedings of NeurIPS*, 2022.
122. Chen Zhu, Yu Cheng, Zhe Gan, Siqi Sun, Tom Goldstein, and Jingjing Liu. FreeLB: Enhanced adversarial training for natural language understanding. In *Proceedings of ICLR*, 2020.

Chapter 9
Knowledge Representation Learning and Knowledge-Guided NLP

Xu Han, Weize Chen, Zhiyuan Liu, Yankai Lin, and Maosong Sun

Abstract Knowledge is an important characteristic of human intelligence and reflects the complexity of human languages. To this end, many efforts have been devoted to organizing various human knowledge to improve the ability of machines in language understanding, such as world knowledge, linguistic knowledge, commonsense knowledge, and domain knowledge. Starting from this chapter, our view turns to representing rich human knowledge and using knowledge representations to improve NLP models. In this chapter, taking world knowledge as an example, we present a general framework of organizing and utilizing knowledge, including knowledge representation learning, knowledge-guided NLP, and knowledge acquisition. For linguistic knowledge, commonsense knowledge, and domain knowledge, we will introduce them in detail in subsequent chapters considering their unique knowledge properties.

9.1 Introduction

The discussion of knowledge is far earlier than the exploration of NLP and is highly related to the study of human languages, which can be traced back to Plato in the Classical Period of ancient Greece [15]. Over the next thousand years, the discussion of knowledge gradually leads to many systematic philosophical theories, such as epistemology [146] and ontology [140], reflecting the long-term analysis and exploration of human intelligence.

X. Han · W. Chen · Z. Liu (✉) · M. Sun
Department of Computer Science and Technology, Tsinghua University, Beijing, China
e-mail: hanxu2022@tsinghua.edu.cn; chenwz21@mails.tsinghua.edu.cn; liuzy@tsinghua.edu.cn; sms@tsinghua.edu.cn

Y. Lin
Gaoling School of Artificial Intelligence, Renmin University of China, Beijing, China
e-mail: yankailin@ruc.edu.cn

© The Author(s) 2023
Z. Liu et al. (eds.), *Representation Learning for Natural Language Processing*,
https://doi.org/10.1007/978-981-99-1600-9_9

Although the question *what is knowledge* is a controversial philosophical question with no definitive and generally accepted answer, we deeply touch on the concept of *knowledge* in every aspect of our lives. At the beginning of the twentieth century, analytic philosophy, which is advocated by Gottlob Frege, Bertrand Russell, and Ludwig Wittgenstein [131], inspires the establishment of symbolic systems to formalize human knowledge [68], significantly contributing to the later development of mathematical logic and philosophy of language.

As NLP is associated with human intelligence, the basic theory of NLP is also highly related to the above knowledge-related theories. As shown in Fig. 9.1, since the Dartmouth Summer Research Project on AI in 1956 [104], knowledge has played a significant role in the development history of NLP. Influenced by mathematical logic and linguistics, early NLP studies [2, 24, 25, 64] mainly focus on exploring symbolic knowledge representations and using symbolic systems to enable machines to understand and reason languages.

Due to the generalization and coverage problems of symbolic representations, ever since the 1990s, data-driven methods [63] are widely applied to represent human knowledge in a distributed manner. Moreover, after 2010, with the boom of deep learning [83], distributed knowledge representations are increasingly expressive from shallow to deep, providing a powerful tool of leveraging knowledge to understand complex semantics.

Making full use of knowledge is crucial to achieving better language understanding. To this end, we describe the general framework of organizing and utilizing knowledge, including knowledge representation learning, knowledge-guided NLP, and knowledge acquisition. With this knowledgeable NLP framework, we show how knowledge can be represented and learned to improve the performance of NLP models and how to acquire rich knowledge from text. As shown in Fig. 9.2, knowledge representation learning aims to encode symbolic knowledge into distributed representations so that knowledge can be more accessible to machines. Then, knowledge-guided NLP is explored to leverage knowledge representations to improve NLP models. Finally, based on knowledge-guided models, we can perform knowledge acquisition to extract more knowledge from plain text to enrich existing knowledge systems.

In the real world, people organize many kinds of knowledge, such as world knowledge, linguistic knowledge, commonsense knowledge, and domain knowledge. In this chapter, we focus on introducing the knowledgeable framework from the perspective of world knowledge since world knowledge is well-defined and general enough. Then, in the following chapters, we will show more details about other kinds of knowledge.

In Sect. 9.2, we will briefly introduce the important properties of symbolic knowledge and distributed model knowledge, aiming to indicate the core motivation for transforming symbolic knowledge into model knowledge. In Sects. 9.3 and 9.4, we will present typical approaches to encoding symbolic knowledge into distributed representations and show how to use knowledge representations to improve NLP

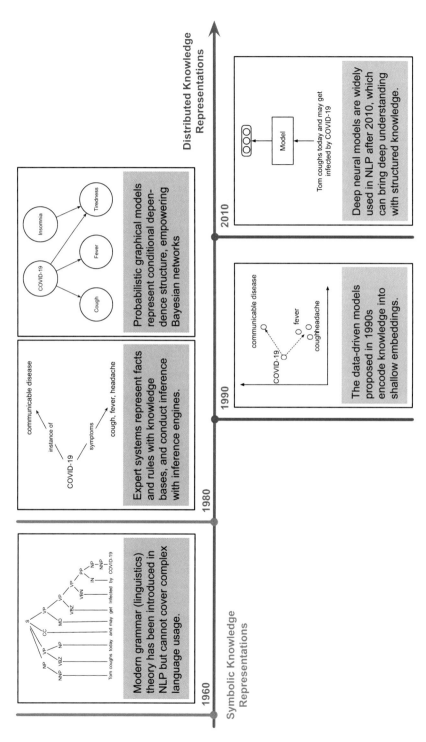

Fig. 9.1 Typical ways to organize and utilize knowledge in the development of NLP

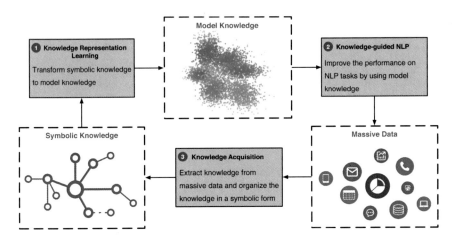

Fig. 9.2 The framework for organizing and utilizing knowledge for NLP tasks includes knowledge representation learning, knowledge-guided NLP, and knowledge acquisition

models, respectively. In Sect. 9.5, we will detail several scenarios for acquiring knowledge to ensure that we can acquire sufficient knowledge to help various NLP models.

9.2 Symbolic Knowledge and Model Knowledge

Before detailing the framework of knowledge representation learning, knowledge-guided NLP, and knowledge acquisition, we briefly present the necessary background information, especially various effective systems to organize knowledge. In this section, we will first introduce typical symbolic knowledge systems, which are the common way of organizing knowledge. Then, we will present more details of model knowledge obtained by projecting knowledge into machine learning models via distributed representation learning, providing a more machine-friendly way of organizing knowledge. Finally, we will show the recent trend in fusing symbolic knowledge and model knowledge, and indicate the importance of knowledge representation learning as well as knowledge-guided NLP in this fusion trend.

9.2.1 Symbolic Knowledge

During the two decades from the 1950s to 1970s, the efforts orienting NLP are mainly committed to symbolic computation systems. In 1956, Allen Newell and

Herbert A. Simon write the first AI program *Logic Theorist* that can perform automated reasoning [113]. The Logic Theorist can prove 38 of the first 52 theorems given by Bertrand Russell in *Principia Mathematica*, by using logic and heuristic rules to prune the search tree of reasoning. Over the same period, Noam Chomsky proposes syntactic structures [24] and transformational grammars [25], using formal languages with precise mathematical notations to drive machine processing of natural languages. Inspired by the Logic Theorist and syntactic structures, Herbert A. Simon and John McCarthy develop information processing language (IPL) [114] and list processing (LISP) [105], respectively, and these two programming languages significantly support computer programming for machine intelligence.

Since neither logic nor grammar rules can well solve complex and diverse problems in practical scenarios, the direction of *deriving general intelligence from symbolic systems* held by early researchers has gradually fallen into a bottleneck. After the 1970s, researchers turn to designing domain-specific intelligence systems for each specific application. The representative work of this period is the expert system [2] initiated by Edward Feigenbaum. An expert system generally consists of a knowledge base (KB) and an inference engine. KBs store a wealth of human knowledge, including domain-specific expertise and rules established by experts in various fields. Inference engines can leverage expertise and rules in KBs to solve specific problems.

Compared with the early AI methods entirely based on mathematical systems, expert systems work well in some practical fields such as business and medicine. Edward Feigenbaum further proposes knowledge engineering [42] in the 1980s, indicating the importance of knowledge acquisition, knowledge representation, and knowledge application to machine intelligence. As shown in Fig. 9.3, inspired by knowledge engineering, various KBs have emerged, such as the commonsense base Cyc [84] and Semantic Web [7]. The most notable achievement of expert systems is the Watson system developed by IBM. IBM Watson beats two human contestants on the quiz show Jeopardy, demonstrating the potential effectiveness of a KB with rich knowledge.

With the Internet thriving in the twenty-first century, massive messages have flooded into the World Wide Web, and knowledge is transferred to the semi-structured textual information on the Web. However, due to the information explosion, extracting the knowledge we want from the significant but noisy plain text on the Internet is not easy. During seeking effective ways to organize knowl-

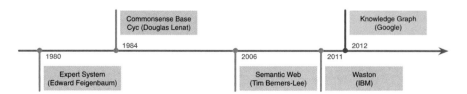

Fig. 9.3 The development of symbolic knowledge systems

edge, Google proposes the concept of knowledge graphs (KGs) in 2012 [37]. KGs arrange the structured multi-relational data of both concrete and abstract entities in the real world, which can be regarded as graph-structured KBs. In addition to describing world knowledge in conventional forms such as strings, the emergence of KGs provides a new tool to organize world knowledge from the perspective of entities and relations. Since KGs are very suitable for organizing the massive amount of knowledge stored in the Web corpora for faster knowledge retrieval, the construction of KGs has been blooming in recent years and has attracted wide attention from academia and industry.

KGs are usually constructed from existing Semantic Web datasets in resource description framework (RDF) [81] with the help of manual annotation. At the same time, KGs can also be automatically enriched by extracting knowledge from the massive plain text on the Web. As shown in Fig. 9.4, a typical KG usually contains two elements: entities and relations. Both concrete objects and abstract concepts in the real world are defined as entities, while complex associations between entities are defined as relations. Knowledge is usually represented in the triplet form of ⟨*head entity*, relation, *tail entity*⟩, and we abridge this as ⟨h, r, t⟩. For example, *Mark Twain* is a famous American writer, and *The Million Pound Bank Note* is one of his masterpieces. In a KG, this knowledge will be represented as ⟨*The Million Pound Bank Note*, Author, *Mark Twain*⟩. Owing to the well-structured form, KGs are widely used in various applications to improve system performance. There are several KGs widely utilized nowadays in NLP, such as Freebase [8], DBpedia [106], YAGO [147], and Wikidata [156]. There are also many comparatively smaller KGs in specific domains whose knowledge can function in domain-specific tasks.

9.2.2 Model Knowledge

For grammar rules, expert systems, and even KGs, one of the pain points of these symbolic knowledge systems is their weak generalization. In addition, it is also difficult to process symbolic knowledge using the numerical computing operations that machines are good at. Therefore, it becomes important to establish a knowledge framework based on numerical computing and with a strong generalization ability to serve the processing of natural languages. To this end, statistical learning [80] has been widely applied after the 1990s, including support vector machines [14], decision trees [16], conditional random fields [79], and so on. These data-driven statistical learning methods can acquire knowledge from data, use numerical features to implicitly describe knowledge, use probability models to represent rules behind knowledge implicitly, and perform knowledge reasoning based on probability computing.

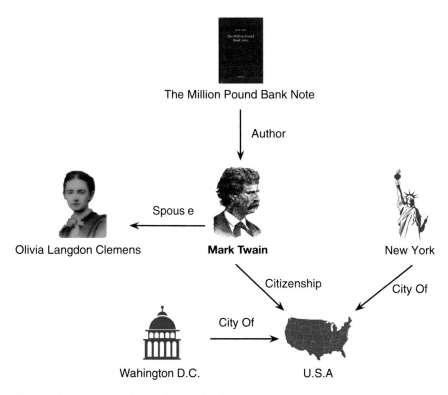

Fig. 9.4 An example of knowledge graphs. The graphics in the figure come from Vecteezy or Wikidata (Vecteezy: https://www.vecteezy.com; Wikidata: https://www.wikidata.org). And all these graphics are available for free or under the Creative Commons CC0 license

Different from symbolic knowledge, which is abstracted by human beings and regarded as *human-friendly* knowledge, the intrinsic nature of statistical learning is to capture the distribution patterns of data from statistics and uses these patterns to abstract implicit knowledge that can be used to solve specific problems. Although such implicit knowledge captured by statistical learning methods may not directly satisfy human intuition, the knowledge is adept at describing correlation information in data and is easy to compute numerically. In other words, this kind of knowledge based on numerical features and continuous probability models is more *machine-friendly*. Considering that the structure of probability models is also a kind of prior knowledge, here we introduce the concept of *model knowledge* to describe this machine-friendly knowledge.[2]

---

[2] There are also some works [22] using *continuous knowledge* to refer to this machine-friendly implicit knowledge.

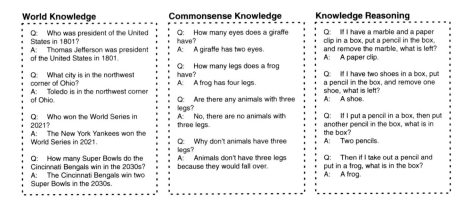

World Knowledge

Q: Who was president of the United States in 1801?
A: Thomas Jefferson was president of the United States in 1801.

Q: What city is in the northwest corner of Ohio?
A: Toledo is in the northwest corner of Ohio.

Q: Who won the World Series in 2021?
A: The New York Yankees won the World Series in 2021.

Q: How many Super Bowls do the Cincinnati Bengals win in the 2030s?
A: The Cincinnati Bengals win two Super Bowls in the 2030s.

Commonsense Knowledge

Q: How many eyes does a giraffe have?
A: A giraffe has two eyes.

Q: How many legs does a frog have?
A: A frog has four legs.

Q: Are there any animals with three legs?
A: No, there are no animals with three legs.

Q: Why don't animals have three legs?
A: Animals don't have three legs because they would fall over.

Knowledge Reasoning

Q: If I have a marble and a paper clip in a box, put a pencil in the box, and remove the marble, what is left?
A: A paper clip.

Q: If I have two shoes in a box, put a pencil in the box, and remove one shoe, what is left?
A: A shoe.

Q: If I put a pencil in a box, then put another pencil in the box, what is in the box?
A: Two pencils.

Q: Then if I take out a pencil and put in a frog, what is in the box?
A: A frog.

Fig. 9.5 Some question-answering examples of GPT-3. All these examples come from Kevin Lacker's blog *Giving GPT-3 a Turing Test* (https://lacker.io/ai/2020/07/06/giving-gpt-3-a-turing-test.html). These examples are also shown in the survey of PTMs [59]

In recent years, the boom of neural networks has provided a more powerful tool to capture model knowledge from data. Compared to conventional statistical models, neural networks are more expressive and can obtain more complex patterns from data. After the success of representing words as distributed representations [107], using shallow neural networks to learn low-dimensional continuous representations for concerned objects, such as words [123], graphs [152], sentences, and documents [82], has become a standard paradigm for accomplishing various NLP tasks. With the emergence of techniques that support increasing network depth and parameter size, such as Transformers [154], large-scale pre-trained models (PTMs) [33, 99, 206] based on deep neural networks are proposed. Recent works show that PTMs can capture rich lexical knowledge[69], syntactic knowledge[66], semantic knowledge[192], and factual knowledge[126] from data during the self-supervised pre-training stage. As shown in Fig. 9.5, we can find that GPT-3 (a PTM with 175 billion parameters) holds a certain amount of facts and commonsense and can perform logical reasoning [17]. By stimulating the task-specific model knowledge distributed in PTMs via various tuning methods [34], PTMs achieve state-of-the-art results on many NLP tasks.

9.2.3 Integrating Symbolic Knowledge and Model Knowledge

In the previous sections, we have briefly described many efforts made by researchers to enhance the processing of natural languages with knowledge. Briefly, symbolic knowledge is suited for reasoning and modeling causality, and model knowledge is suited for integrating information and modeling correlation. Symbolic knowledge and model knowledge have their own strengths, and utilizing both is crucial to drive the language understanding of machines. Over the past few decades, the practice has also shown that NLP models with deep language understanding cannot be achieved by using a certain kind of knowledge alone.

The early explorations of NLP researchers relied entirely on handcrafted rules and systems, the limitations of which have been revealed over the two AI winters [54]. These AI winters have shown that it is challenging to make machines master versatile language abilities using only symbolic knowledge. In recent years, researchers have devoted great attention to deep neural networks and automatically learning model knowledge from massive data, leading to breakthroughs such as word representations and PTMs. However, these data-driven methods that focus on model knowledge still have some obvious limitations and face the challenges of robustness and interpretability [83]. In terms of robustness, for neural-based NLP models, it is not difficult to build adversarial examples to induce model errors [157], considering the knowledge automatically summarized by models may be a shortcut [48] or even a bias [148]. In terms of interpretability, the predictions given by models are also based on black-box correlations. Moreover, current data-driven methods may suffer from data-hungry issues. Model knowledge needs to be learned based on massive data, but obtaining high-quality data itself is very difficult. Humans can learn skills with a few training examples, which is challenging for machines. Therefore, relying solely on data-driven methods and model knowledge to advance NLP also seems unsustainable.

We have systematically discussed the symbolic and distributed representations of text in the previous chapters, and here we have made a further extension to form a broader discussion of representing knowledge. From these discussions, we can observe that taking full advantage of knowledge, i.e., utilizing both symbolic or model knowledge, is an important way to obtain better language understanding. Some recent works [60, 193] have also shown a trend toward the integration of symbolic and model knowledge and, more specifically, a trend of using symbolic knowledge to improve deep neural models that already have strong model knowledge. In order to integrate both symbolic and model knowledge, three challenges have to be addressed:

1. How to represent knowledge (especially symbolic knowledge) in a machine-friendly form so that current NLP models can utilize the knowledge?
2. How to use knowledge representations to guide specific NLP models?
3. How to continually acquire knowledge from large-scale plain text instead of handcrafted efforts?

We will next introduce knowledge representation learning, knowledge-guided NLP, and knowledge acquisition for these challenges.

9.3 Knowledge Representation Learning

As we mentioned before, we can organize knowledge using symbolic systems. However, as the scale of knowledge increases, using these symbolic systems naturally faces two challenges: data sparsity and computational inefficiency. Despite the importance of symbolic knowledge for NLP, these challenges indicate that symbolic systems are not an inherently machine-friendly form of knowledge organization. Specifically, data sparsity is a common problem in many fields. For example, when we use KGs to describe general world knowledge, the number of entities (nodes) in KGs can be enormous, while the number of relations (edges) in KGs is typically few, i.e., there are often no relations between two randomly selected entities in the real world, resulting in the sparsity of KGs. Computational inefficiency is another challenge we have to overcome since computers are better suited to handle numerical data and less adept at handling symbolic knowledge in KGs. As the size of KGs continues to grow, this efficiency challenge may become more severe.

To solve the above problems, distributed knowledge representations are introduced, i.e., low-dimensional continuous embeddings are used to represent symbolic knowledge. The sparsity problem is alleviated owing to using these distributed representations, and the computational efficiency is also improved. In addition, using embeddings to represent knowledge makes it more feasible and convenient to integrate symbolic knowledge into neural NLP models, motivating the exploration of knowledge-guided NLP. Up to now, distributed knowledge representations have been widely used in many applications requiring the support of human knowledge. Moreover, distributed knowledge representations can also significantly improve the ability of knowledge completion, knowledge fusion, and knowledge reasoning.

In this section, we take KGs that organize rich world knowledge as an example to introduce how to obtain distributed knowledge representations. Hereafter, we use $\mathcal{G} = (\mathcal{E}, \mathcal{R}, \mathcal{T})$ to denote a KG, in which $\mathcal{E} = \{e_1, e_2, \dots\}$ is the entity set, $\mathcal{R} = \{r_1, r_2, \dots\}$ is the relation set, and $\mathcal{T}$ is the fact set. We use $h, t \in \mathcal{E}$ to represent the head and tail entities, and $\mathbf{h}, \mathbf{t}$ to represent their entity embeddings. A triplet $\langle h, r, t \rangle \in \mathcal{T}$ is a factual record, where h, t are entities and r is the relation between h and t.

Given a triplet $\langle h, r, t \rangle$, a score function $f(h, r, t)$ is used by knowledge representation learning methods to measure whether $\langle h, r, t \rangle$ is a fact or a fallacy. Generally, the larger the value of $f(h, r, t)$, the higher the probability that $\langle h, r, t \rangle$ is true.[3] Based on $f(h, r, t)$, knowledge representations can be learned with

$$\arg \min_{\theta} \sum_{\langle h,r,t \rangle \in \mathcal{T}} \sum_{\langle \tilde{h}, \tilde{r}, \tilde{t} \rangle \in \tilde{\mathcal{T}}} \max \left\{ 0, f(\tilde{h}, \tilde{r}, \tilde{t}) + \gamma - f(h, r, t) \right\}, \qquad (9.1)$$

where θ is the learnable embeddings of entities and relations, $\langle h, r, t \rangle$ indicates positive facts (i.e., triplets in $\mathcal{T}$), and $\langle \tilde{h}, \tilde{r}, \tilde{t} \rangle$ indicates negative facts (triplets that

---

[3] For some methods, the smaller the value of $f(h, r, t)$, the higher the probability that $\langle h, r, t \rangle$ is true. We re-formalize the score functions of these methods by taking the opposite of the score functions so that we can present all knowledge representation learning methods within a unified framework.

do not exist in KGs). $\gamma > 0$ is a hyper-parameter used as a margin. A bigger γ means to learn a wider gap between $f(h, r, t)$ and $f(\tilde{h}, \tilde{r}, \tilde{t})$. Considering that there are no explicit negative triplets in KGs, $\tilde{\mathcal{T}}$ is usually defined as

$$\tilde{\mathcal{T}} = \{\langle \tilde{h}, r, t \rangle | \tilde{h} \in \mathcal{E}, \langle h, r, t \rangle \in \mathcal{T}\} \cup \{\langle h, \tilde{r}, t \rangle | \tilde{r} \in \mathcal{R}, \langle h, r, t \rangle \in \mathcal{T}\}$$
$$\cup \{\langle h, r, \tilde{t} \rangle | \tilde{t} \in \mathcal{E}, \langle h, r, t \rangle \in \mathcal{T}\} - \mathcal{T}, \tag{9.2}$$

which means $\tilde{\mathcal{T}}$ is built by corrupting the entities and relations of the triplets in $\mathcal{T}$. Different from the margin-based loss function in Eq. (9.1), some methods apply a likelihood-based loss function to learn knowledge representations as

$$\arg\min_{\theta} \sum_{\langle h,r,t \rangle \in \mathcal{T}} \log \left[1 + \exp(-f(h, r, t)) \right] + \sum_{\langle \tilde{h},\tilde{r},\tilde{t} \rangle \in \tilde{\mathcal{T}}} \log \left[1 + \exp(f(\tilde{h}, \tilde{r}, \tilde{t})) \right].$$
$$\tag{9.3}$$

Next, we present some typical knowledge representation learning methods as well as their score functions, including (1) linear representation methods that formalize relations as linear transformations between entities, (2) translation representation methods that formalize relations as translation operations between entities, (3) neural representation methods that apply neural networks to represent entities and relations, and (4) manifold representation methods that use complex manifold spaces instead of simple Euclidean spaces to learn representations.

9.3.1 Linear Representation

Linear representation methods formalize relations as linear transformations between entities, which is a simple and basic way to learn knowledge representations.

Structured Embeddings (SE) SE [13] is a typical linear method to represent KGs. In SE, all entities are embedded into a d-dimensional space. SE designs two relation-specific matrices $\mathbf{M}_{r,1}, \mathbf{M}_{r,2} \in \mathbb{R}^{d \times d}$ for each relation r, and these two matrices are used to transform the embeddings of entities. The score function of SE is defined as

$$f(h, r, t) = -\|\mathbf{M}_{r,1}\mathbf{h} - \mathbf{M}_{r,2}\mathbf{t}\|, \tag{9.4}$$

where $\|\cdot\|$ is the vector norm. The assumption of SE is that the head and tail embeddings should be as close as possible after being transformed into a relation-specific space. Therefore, SE uses the margin-based loss function to learn representations.

Semantic Matching Energy (SME) SME [11] builds more complex linear transformations than SE. Given a triplet $\langle h, r, t \rangle$, $\mathbf{h}$ and $\mathbf{r}$ are combined using a projection function to get a new embedding $\mathbf{l}_{h,r}$. Similarly, given $\mathbf{t}$ and $\mathbf{r}$, we can get $\mathbf{l}_{t,r}$. Then, a point-wise multiplication function is applied on $\mathbf{l}_{h,r}$ and $\mathbf{l}_{t,r}$ to get the score of this triplet. SME introduces two different projection functions to build $f(h, r, t)$: one is in the linear form

$$f(h, r, t) = \mathbf{l}_{h,r}^\top \mathbf{l}_{t,r}, \quad \mathbf{l}_{h,r} = \mathbf{M}_1 \mathbf{h} + \mathbf{M}_2 \mathbf{r} + \mathbf{b}_1, \quad \mathbf{l}_{t,r} = \mathbf{M}_3 \mathbf{t} + \mathbf{M}_4 \mathbf{r} + \mathbf{b}_2, \tag{9.5}$$

and the other is in the bilinear form

$$f(h, r, t) = \mathbf{l}_{h,r}^\top \mathbf{l}_{t,r}, \quad \mathbf{l}_{h,r} = (\mathbf{M}_1 \mathbf{h} \odot \mathbf{M}_2 \mathbf{r}) + \mathbf{b}_1, \quad \mathbf{l}_{t,r} = (\mathbf{M}_3 \mathbf{t} \odot \mathbf{M}_4 \mathbf{r}) + \mathbf{b}_2, \tag{9.6}$$

where $\odot$ is the element-wise (Hadamard) product. $\mathbf{M}_1$, $\mathbf{M}_2$, $\mathbf{M}_3$, and $\mathbf{M}_4$ are learnable transformation matrices, and $\mathbf{b}_1$ and $\mathbf{b}_2$ are learnable bias vectors. Empirically, the margin-based loss function is suitable for dealing with the score functions built with vector norm operations, while the likelihood-based loss function is more usually used to process the score functions built with inner product operations. Since SME uses the inner product operation to build its score function, the likelihood-based loss function is thus used to learn representations.

Latent Factor Model (LFM) LFM [70] aims to model large KGs based on a bilinear structure. By modeling entities as embeddings and relations as matrices, the score function of LFM is defined as

$$f(h, r, t) = \mathbf{h}^\top \mathbf{M}_r \mathbf{t}, \tag{9.7}$$

where the matrix $\mathbf{M}_r$ is the representation of the relation r. Similar to SME, LFM adopts the likelihood-based loss function to learn representations. Based on LFM, DistMult [186] further restricts $\mathbf{M}_r$ to be a diagonal matrix. As compared with LFM, DistMult not only reduces the parameter size but also reduces the computational complexity and achieves better performance.

RESCAL RESCAL [118, 119] is a representation learning method based on matrix factorization. By modeling entities as embeddings and relations as matrices, RESCAL adopts a score function the same to LFM. However, RESCAL employs neither the margin-based nor the likelihood-based loss function to learn knowledge representations. Instead, in RESCAL, a three-way tensor $\vec{\mathbf{X}} \in \mathbb{R}^{|\mathcal{E}| \times |\mathcal{E}| \times |\mathcal{R}|}$ is adopted. In the tensor $\vec{\mathbf{X}}$, two modes respectively stand for head and tail entities, while the third mode stands for relations. The entries of $\vec{\mathbf{X}}$ are determined by the existence of the corresponding triplet facts. That is, $\vec{\mathbf{X}}_{ijk} = 1$ if the triplet $\langle i\text{-}th$ entity, $k\text{-}th$ relation, $j\text{-}th$ entity$\rangle$ exists in the training set, and otherwise $\vec{\mathbf{X}}_{ijk} = 0$. To capture the inherent structure of all triplets, given $\vec{\mathbf{X}} = \{\mathbf{X}_1, \cdots, \mathbf{X}_{|\mathcal{R}|}\}$, for each slice $\mathbf{X}_n = \vec{\mathbf{X}}_{[:,:,n]}$, RESCAL assumes the following factorization for $\mathbf{X}_n$ holds

$$\mathbf{X}_n \approx \mathbf{E} \mathbf{M}_{r_n} \mathbf{E}^\top, \tag{9.8}$$

where $\mathbf{E} \in \mathbb{R}^{|\mathcal{E}| \times d}$ stands for the d-dimensional entity representations of all entities and $\mathbf{M}_{r_n} \in \mathbb{R}^{d \times d}$ represents the interactions between entities specific to the n-th relation r_n. Following this tensor factorization assumption, the learning objective of RESCAL is defined as

$$\underset{\mathbf{E},\mathbf{M}}{\arg\min} \frac{1}{2} \left(\sum_{n=1}^{|\mathcal{R}|} \|\mathbf{X}_n - \mathbf{E}\mathbf{M}_{r_n}\mathbf{E}^\top\|_F^2 \right) + \frac{1}{2}\lambda \left(\|\mathbf{E}\|_F^2 + \sum_{n=1}^{|\mathcal{R}|} \|\mathbf{M}_{r_n}\|_F^2 \right), \quad (9.9)$$

where $\mathbf{M} = \{\mathbf{M}_{r_1}, \mathbf{M}_{r_2}, \cdots, \mathbf{M}_{r_{|\mathcal{R}|}}\}$ is the collection of all relation matrices, $\|\cdot\|_F$ is the Frobenius vector norm, and λ is a hyper-parameter to control the second regularization term.

Holographic Embeddings (HolE) HolE [117] is proposed as an enhanced version of RESCAL. RESCAL works well with multi-relational data but suffers from a high computational complexity. To achieve high effectiveness and efficiency at the same time, HolE employs an operation named circular correlation to generate representations. The circular correlation operation $\star : \mathbb{R}^d \times \mathbb{R}^d \to \mathbb{R}^d$ between two entities h and t is

$$[\mathbf{h} \star \mathbf{t}]_k = \sum_{i=1}^{d} [\mathbf{h}]_i [\mathbf{t}]_{(k+i) \bmod d+1}, \quad (9.10)$$

where $[\cdot]_i$ means the i-th vector element. The score function is defined as

$$f(h, r, t) = -\mathbf{r}^\top (\mathbf{h} \star \mathbf{t}). \quad (9.11)$$

HolE adopts the likelihood-based loss function to learn representations.

The circular correlation operation brings several advantages. First, the circular correlation operation is noncommutative (i.e., $\mathbf{h} \star \mathbf{t} \neq \mathbf{t} \star \mathbf{h}$), which makes it capable of modeling asymmetric relations in KGs. Second, the circular correlation operation has a lower computational complexity compared to the tensor product operation in RESCAL. Moreover, the circular correlation operation could be further accelerated with the help of fast Fourier transform (FFT), which is formalized as

$$\mathbf{h} \star \mathbf{t} = \mathcal{F}^{-1}(\overline{\mathcal{F}(\mathbf{h})} \odot \mathcal{F}(\mathbf{t})), \quad (9.12)$$

where $\mathcal{F}(\cdot)$ and $\mathcal{F}(\cdot)^{-1}$ represent the FFT operation and its inverse operation, respectively, $\overline{\mathcal{F}(\cdot)}$ denotes the complex conjugate of $\mathcal{F}(\cdot)$, and $\odot$ stands for the element-wise (Hadamard) product. Due to the FFT operation, the computational complexity of the circular correlation operation is $\mathcal{O}(d \log d)$, which is much lower than that of the tensor product operation.

9.3.2 Translation Representation

Translation methods are another effective way to obtain distributed representations of KGs. To help readers better understand different translation representation methods, we first introduce their motivations.

The primary motivation is that it is natural to consider relations between entities as translation operations. For distributed representations, entities are embedded into a low-dimensional space, and ideal representations should embed entities with similar semantics into the nearby regions, while entities with different meanings should belong to distinct clusters. For example, *William Shakespeare* and *Jane Austen* may be in the same cluster of writers, *Romeo and Juliet* and *Pride and Prejudice* may be in another cluster of books. In this case, they share the same relation Notable Work, and the translations from writers to their books in the embedding space are similar.

The secondary motivation of translation methods derives from the breakthrough in word representation learning. Word2vec [108] proposes two simple models, skipgram and CBOW, to learn distributed word representations from large-scale corpora. The learned word embeddings perform well in measuring word similarities and analogies. And these word embeddings have some interesting phenomena: if the same semantic or syntactic relations are shared by two word pairs, the translations within the two word pairs are similar. For instance, we have

$$\mathbf{w}(king) - \mathbf{w}(man) \approx \mathbf{w}(queen) - \mathbf{w}(woman), \tag{9.13}$$

where $\mathbf{w}(\cdot)$ represents the embedding of the word. We know that the semantic relation between *king* and *man* is similar to the relation between *queen* and *woman*, and the above case shows that this relational knowledge is successfully embedded into word representations. Apart from semantic relations, syntactic relations can also be well represented by word2vec, as shown in the following example:

$$\mathbf{w}(bigger) - \mathbf{w}(big) \approx \mathbf{w}(smaller) - \mathbf{w}(small). \tag{9.14}$$

Since word2vec implies that the implicit relations between words can be seen as translations, it is reasonable to assume that relations in KGs can also be modeled as translation embeddings. More intuitively, if we represent a word pair and its implicit relation using a triplet, e.g., ⟨big, Comparative, bigger⟩, we can obviously observe the similarity between word representation learning and knowledge representation learning.

The last motivation comes from the consideration of the computational complexity. On the one hand, the substantial increase in the model complexity will result in high computational costs and obscure model interpretability, and a complex model may lead to overfitting. On the other hand, the experimental results on the model complexity demonstrate that the simpler models perform almost as well as more expressive models in most knowledge-related applications [117, 186], in the

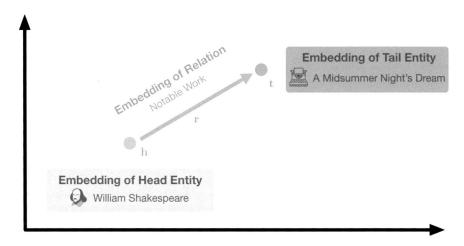

Fig. 9.6 The architecture of TransE [12]

condition that large-scale datasets and a relatively large amount of relations can be used for training models. As KG size increases, the computational complexity becomes the primary challenge for knowledge representation learning. The intuitive assumption of modeling relations as translations rather than matrices leads to a better trade-off between effectiveness and efficiency.

Since the translation-based methods are all extended from TransE [12], we thus first introduce TransE in detail and then introduce its extensions.

TransE As illustrated in Fig. 9.6, TransE embeds entities as well as relations into the same space. In the embedding space, relations are considered as translations from head entities to tail entities. With this translation assumption, given a triplet $\langle h, r, t \rangle$ in $\mathcal{T}$, we want $\mathbf{h} + \mathbf{r}$ to be the nearest neighbor of the tail embedding $\mathbf{t}$. The score function of TransE is then defined as

$$f(h, r, t) = -\|\mathbf{h} + \mathbf{r} - \mathbf{t}\|. \qquad (9.15)$$

TransE uses the margin-based loss function for training. Although TransE is effective and efficient, it still has several challenges to be further explored.

First, considering that there may be multiple correct answers given two elements in a triplet, under the translation assumption in TransE, each entity has only one embedding in all triplets, which may lead to reducing the discrimination of entity embeddings. In TransE, according to the entity cardinalities of relations, all relations are classified into four categories, 1-to-1, 1-to-Many, Many-to-1, and Many-to-Many. A relation is considered as 1-to-1 if one head appears with only one tail and vice versa, 1-to-Many if a head can appear with many tails, Many-to-1 if a tail can appear with many heads, and Many-to-Many if multiple heads appear with multiple tails. Statistics demonstrate that the 1-to-Many, Many-to-1, and Many-to-Many relations occupy a large proportion. TransE performs well on 1-to-1 relations, but has problems when handling 1-to-Many, Many-to-1, and Many-to-Many relations. For instance, given the head entity *William Shakespeare* and the relation `Notable Work`, we can get a list of masterpieces, such as *Hamlet, A Midsummer Night's Dream*, and *Romeo and Juliet*. These books share the same writer information while differing in many other fields such as theme, background, and famous roles in the book. Due to the entity *William Shakespeare* and the relation `Notable Work`, these books may be assigned similar embeddings and become indistinguishable.

Second, although the translation operation is intuitive and effective, only considering the simple one-step translation may limit the ability to model KGs. Taking entities and relations as nodes and edges, the nodes that are not directly connected may be linked by a path of more than one edge. However, TransE focuses on minimizing $\|\mathbf{h}+\mathbf{r}-\mathbf{t}\|$, which only utilizes the one-step relation information in KGs, regardless of the latent relationships located in long-distance paths. For example, if we know ⟨*The forbidden city*, `Located in`, *Beijing*⟩ and ⟨*Beijing*, `Capital of`, *China*⟩, we can infer that *The forbidden city* locates in *China*. TransE can be further enhanced with the favor of multistep information.

Third, the representation and the score function in TransE are oversimplified for the consideration of efficiency. Therefore, TransE may not be capable enough of modeling those complex entities and relations in KGs. There are still some challenges in how to balance effectiveness and efficiency as well as avoid overfitting and underfitting.

After TransE, there are lots of subsequent methods addressing the above challenges. Specifically, TransH [165], TransR [90], TransD [102], and TranSparse [71] are proposed to solve the challenges in modeling complex relations, PTransE is proposed to encode long-distance information located in multistep paths, and CTransR, TransG, and KG2E further extend the oversimplified model of TransE. Next, we will discuss these subsequent methods in detail.

TransH TransH [165] enables an entity to have multiple relation-specific representations to address the issue that TransE cannot well model 1-to-Many, Many-to-1, and Many-to-Many relations. As we mentioned before, in TransE, entities are embedded to the same semantic embedding space and similar entities tend to be in the same cluster. However, it seems that *William Shakespeare* should be in the neighborhood of *Isaac Newton* when talking about `Nationality`, while it should be close to *Mark Twain* when talking about `Occupation`. To accomplish this, entities should have multiple representations in different triplets.

As illustrated in Fig. 9.7, TransH proposes a hyperplane $\mathbf{w}_r$ for each relation, and computes the translation on the hyperplane $\mathbf{w}_r$. Given a triplet ⟨h, r, t⟩, TransH projects $\mathbf{h}$ and $\mathbf{t}$ to the corresponding hyperplane $\mathbf{w}_r$ to get the projection $\mathbf{h}_\perp$ and $\mathbf{t}_\perp$, and $\mathbf{r}$ is used to connect $\mathbf{h}_\perp$ and $\mathbf{t}_\perp$:

$$\mathbf{h}_\perp = \mathbf{h} - \mathbf{w}_r^\top \mathbf{h} \mathbf{w}_r, \quad \mathbf{t}_\perp = \mathbf{t} - \mathbf{w}_r^\top \mathbf{t} \mathbf{w}_r, \tag{9.16}$$

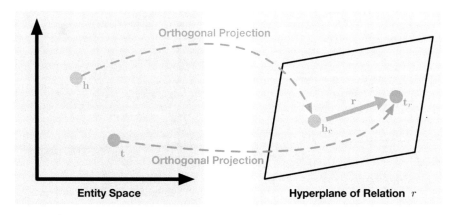

Fig. 9.7 The architecture of TransH [165]

where $\mathbf{w}_r$ is a vector and $\|\mathbf{w}_r\|_2$ is restricted to 1. The score function is

$$f(h, r, t) = -\|\mathbf{h}_\perp + \mathbf{r} - \mathbf{t}_\perp\|. \tag{9.17}$$

As for training, TransH also minimizes the margin-based loss function with negative sampling, which is similar to TransE.

TransR TransR [90] takes full advantage of linear methods and translation methods. As in Eq. (9.16), TransH enables entities to have multiple relation-specific representations by projecting them to different hyperplanes, while entity embeddings and relation embeddings are still restricted in the same space, which may limit the ability for modeling entities and relations. TransR assumes that entity embeddings and relation embeddings should be in different spaces.

As illustrated in Fig. 9.8, For a triplet $\langle h, r, t \rangle$, TransR projects $\mathbf{h}$ and $\mathbf{t}$ to the relation space of r, and this projection is defined as

$$\mathbf{h}_r = \mathbf{h}\mathbf{M}_r, \quad \mathbf{t}_r = \mathbf{t}\mathbf{M}_r, \tag{9.18}$$

where $\mathbf{M}_r$ is the projection matrix. $\mathbf{h}_r$ and $\mathbf{t}_r$ stand for the relation-specific entity representations in the relation space of r, respectively. This means that each entity has a relation-specific representation for each relation, and all translation operations are processed in the relation-specific space. The score function of TransR is

$$f(h, r, t) = -\|\mathbf{h}_r + \mathbf{r} - \mathbf{t}_r\|. \tag{9.19}$$

TransR constrains the norms of the embeddings and has $\|\mathbf{h}\|_2 \leq 1$, $\|\mathbf{t}\|_2 \leq 1$, $\|\mathbf{r}\|_2 \leq 1$, $\|\mathbf{h}_r\|_2 \leq 1$, $\|\mathbf{t}_r\|_2 \leq 1$. As for training, TransR uses the same margin-based loss function as TransE.

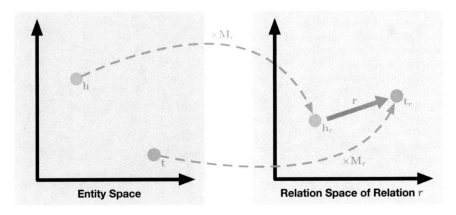

Fig. 9.8 The architecture of TransR [90]

Furthermore, a relation should also have multiple representations since the meanings of a relation with different head and tail entities differ slightly. For example, the relation Contains the Location has head-tail patterns like *city-street*, *country-city*, and even *country-university*, each conveys different attribute information. To handle these subtle differences, entities for a same relation should also be projected differently.

To this end, cluster-based TransR (CTransR) is then proposed, which is an enhanced version of TransR by taking the nuance in meaning for a same relation with different entities into consideration. More specifically, for each relation, all entity pairs of the relation are first clustered into several groups. The clustering process depends on the result of $\mathbf{t} - \mathbf{h}$ for each entity pair (h, t), and $\mathbf{h}$ and $\mathbf{t}$ are the embeddings learned by TransE. Then, we assign a distinct sub-relation embedding $\mathbf{r}_c$ for each cluster of the relation r according to cluster-specific entity pairs, and the original score function of TransR is modified as

$$f(h, r, t) = -\|\mathbf{h}_r + \mathbf{r}_c - \mathbf{t}_r\| - \lambda \|\mathbf{r}_c - \mathbf{r}\|, \tag{9.20}$$

where λ is a hyper-parameter to control the regularization term and $\|\mathbf{r}_c - \mathbf{r}\|$ is to make the sub-relation embedding $\mathbf{r}_c$ and the unified relation embedding $\mathbf{r}$ not too distinct.

TransD TransD [102] is an extension of TransR that uses dynamic mapping matrices to project entities into relation-specific spaces. TransR focuses on learning multiple relation-specific entity representations. However, TransR projects entities according to only relations, ignoring the entity diversity. Moreover, the projection operations based on matrix-vector multiplication lead to a higher computational complexity compared to TransE, which is time-consuming when applied on large-scale KGs.

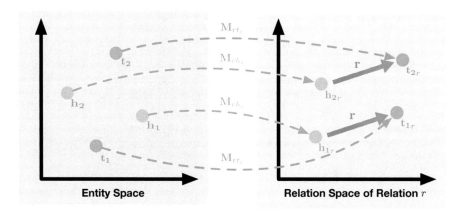

Fig. 9.9 The architecture of TransD [102]

For each entity and relation, TransD defines two vectors: one is used as the embedding, and the other is used to construct projection matrices to map entities to relation spaces. As illustrated in Fig. 9.9, We use $\mathbf{h}, \mathbf{t}, \mathbf{r}$ to denote the embeddings of entities and relations, and $\mathbf{h}_p, \mathbf{t}_p, \mathbf{r}_p$ to represent the projection vectors. There are two projection matrices $\mathbf{M}_{rh}, \mathbf{M}_{rt}$ used to project entities to relation spaces, and these projection matrices are dynamically constructed as

$$\mathbf{M}_{rh} = \mathbf{r}_p \mathbf{h}_p^\top + \mathbf{I}, \quad \mathbf{M}_{rt} = \mathbf{r}_p \mathbf{t}_p^\top + \mathbf{I}, \tag{9.21}$$

which means the projection vectors of both entities and relations are combined to determine dynamic projection matrices. The score function is

$$f(h, r, t) = -\|\mathbf{M}_{rh}\mathbf{h} + \mathbf{r} - \mathbf{M}_{rt}\mathbf{t}\|. \tag{9.22}$$

These projection matrices are initialized with identity matrices by setting all the projection vectors to $\mathbf{0}$ at initialization, and the normalization constraints in TransR are also used for TransD.

TransD proposes a dynamic method to construct projection matrices by considering the diversities of both entities and relations, achieving better performance compared to existing methods in knowledge completion. Moreover, TransD lowers both computational and spatial complexity compared to TransR.

TranSparse TranSparse [71] is also a subsequent work of TransR. Although TransR has achieved promising results, there are still two challenges remained. One is the heterogeneity challenge. Relations in KGs differ in granularity. Some relations express complex semantics between entities, while some other relations are relatively simple. The other is the imbalance challenge. Some relations have more valid head entities and fewer valid tail entities, while some are the opposite.

If we consider these challenges rather than merely treating all relations equally, we can obtain better knowledge representations.

Existing methods such as TransR build projection matrices for each relation, and these projection matrices have the same parameter scale, regardless of the variety in the complexity of relations. TranSparse is proposed to address this issue. The underlying assumption of TranSparse is that complex relations should have more parameters to learn while simple relations should have fewer parameters, where the relation complexity is judged from the number of triplets or entities linked to the relation. To accomplish this, two models are proposed, including TranSparse-share and TranSparse-separate.

Inspired by TransR, given a relation r, TranSparse-share builds a relation-specific projection matrix $\mathbf{M}_r(\theta_r)$ for the relation. $\mathbf{M}_r(\theta_r)$ is sparse and the sparse degree θ_r mainly depends on the number of entity pairs linked to r. Suppose N_r is the number of linked entity pairs, N_r^* represents the maximum number of N_r, and $\theta_{\min}$ denotes the minimum sparse degree of projection matrices that $0 \leq \theta_{\min} \leq 1$. The sparse degree of relation r is defined as

$$\theta_r = 1 - (1 - \theta_{\min})\frac{N_r}{N_r^*}. \tag{9.23}$$

Both head and tail entities share the same sparse projection matrix $\mathbf{M}_r(\theta_r)$. The score function is

$$f(h, r, t) = -\|\mathbf{M}_r(\theta_r)\mathbf{h} + \mathbf{r} - \mathbf{M}_r(\theta_r)\mathbf{t}\|. \tag{9.24}$$

Different from TranSparse-share, TranSparse-separate builds two different sparse matrices $\mathbf{M}_{rh}(\theta_{rh})$ and $\mathbf{M}_{rt}(\theta_{rt})$ for head and tail entities, respectively. Then, the sparse degree θ_{rh} (or θ_{rt}) depends on the number of head (or tail) entities linked to r. We have N_{rh} (or N_{rt}) to represent the number of head (or tail) entities, as well as N_{rh}^* (or N_{rt}^*) to represent the maximum number of N_{rh} (or N_{rt}). And $\theta_{\min}$ will also be set as the minimum sparse degree of projection matrices that $0 \leq \theta_{\min} \leq 1$. We have

$$\theta_{rh} = 1 - (1 - \theta_{\min})N_{rh}/N_{rh}^*, \quad \theta_{rt} = 1 - (1 - \theta_{\min})N_{rt}/N_{rt}^*. \tag{9.25}$$

The final score function of TranSparse-separate is

$$f(h, r, t) = -\|\mathbf{M}_{rh}(\theta_{rh})\mathbf{h} + \mathbf{r} - \mathbf{M}_{rt}(\theta_{rt})\mathbf{t}\|. \tag{9.26}$$

Through sparse projection matrices, TranSparse solves the heterogeneity challenge and the imbalance challenge simultaneously.

PTransE PTransE [89] is an extension of TransE that considers multistep relational paths. All abovementioned translation methods only consider simple one-step paths (i.e., relation) to perform the translation operation, ignoring the rich global information located in the whole KGs. For instance, if we notice the multistep relational path that $\langle$*The forbidden city*, `Located in`, *Beijing*$\rangle$ → $\langle$*Beijing*, `Capital of`, *China*$\rangle$, we can inference with confidence that the triplet $\langle$*The forbidden city*, `Located in`, *China*$\rangle$ may exist. Relational paths provide us a powerful way to build better representations for KGs and even help us better understand knowledge reasoning.

There are two main challenges when encoding the information in multistep relational paths. First, how to select reliable and meaningful relational paths among enormous path candidates in KGs, since there are lots of paths that cannot indicate reasonable relations. Consider two triplet facts $\langle$*The forbidden city*, `Located in`, *Beijing*$\rangle$ → $\langle$*Beijing*, `held`, *2008 Summer Olympics*$\rangle$, it is hard to describe the relation between *The forbidden city* and *2008 Summer Olympics*. Second, how to model the meaningful relational paths? It is not easy to handle the problem of semantic composition in relational paths.

To select meaningful relational paths, PTransE uses a path-constraint resource allocation (PCRA) algorithm to judge the path reliability. Suppose there is information (or resource) in the head entity h which will flow to the tail entity t through some certain paths. The basic assumption of PCRA is that the reliability of the path ℓ depends on the amount of resource that eventually flows from head to tail. Formally, we denote a certain path between h and t as $\ell = (r_1, \ldots, r_l)$. The resource that travels from h to t following the path could be represented as $S_0/h \xrightarrow{r_1} S_1 \xrightarrow{r_2} \ldots \xrightarrow{r_l} S_l/t$. For an entity $m \in S_i$, the amount of resource that belongs to m is defined as

$$R_\ell(m) = \sum_{n \in S_{i-1}(\cdot, m)} \frac{1}{|S_i(n, \cdot)|} R_\ell(n), \tag{9.27}$$

where $S_{i-1}(\cdot, m)$ indicates all direct predecessors of the entity m along with the relation r_i in S_{i-1} and $S_i(n, \cdot)$ indicates all direct successors of $n \in S_{i-1}$ with the relation r_i. Finally, the amount of resource that flows to the tail $R_\ell(t)$ is used to measure the reliability of ℓ, given the triplet $\langle h, \ell, t \rangle$.

Once we finish selecting those meaningful relational path candidates, the next challenge is to model the semantic composition of these multistep paths. PTransE proposes three composition operations, namely, addition, multiplication, and recurrent neural networks, to get the path representation $\mathbf{l}$ based on the relations in $\ell = (r_1, \ldots, r_l)$. The score function is

$$f(h, \ell, t) = -\|\mathbf{l} - (\mathbf{t} - \mathbf{h})\| \approx -\|\mathbf{l} - \mathbf{r}\| = f(\ell, r), \tag{9.28}$$

where r indicates the golden relation between h and t. Since PTransE also wants to meet the assumption in TransE that $\mathbf{r} \approx \mathbf{t} - \mathbf{h}$, PTransE directly utilizes $\mathbf{r}$ in training. The optimization objective of PTransE is

$$\arg\min_\theta \sum_{(h,r,t) \in \mathcal{T}} [\mathcal{L}(h, r, t) + \frac{1}{Z} \sum_{\ell \in P(h,t)} R(\ell | h, t) \mathcal{L}(\ell, r)], \tag{9.29}$$

where $\mathcal{L}(h, r, t)$ is the margin-based loss function with $f(h, r, t)$, $\mathcal{L}(\ell, r)$ is the margin-based score function with $f(\ell, r)$, and $Z = \sum_{\ell \in P(h,t)} R(\ell|h, t)$ is a normalization factor. The reliability $R(\ell|h, t)$ of ℓ in (h, ℓ, t) is well considered in the overall loss function. For the path ℓ, the initial resource is set as $R_\ell(h) = 1$. By recursively performing PCRA from h to t through ℓ, the resource $R_\ell(t)$ can indicate how much information can be well translated, and $R_\ell(t)$ is thus used to measure the reliability of the path ℓ, i.e., $R(\ell|h, t) = R_\ell(t)$. Besides PTransE, similar ideas [47, 50] also consider multistep relational paths and demonstrate that there is plentiful information located in multistep relational paths which could significantly improve knowledge representation.

KG2E KG2E [65] introduces multidimensional Gaussian distributions to represent KGs. Existing translation methods usually consider entities and relations as vectors embedded in low-dimensional spaces. However, as explained above, entities and relations in KGs are diverse at different granularities. Therefore, the margin in the margin-based loss function that is used to distinguish positive triplets from negative triplets should be more flexible due to the diversity, and the uncertainties of entities and relations should be taken into consideration.

KG2E represents entities and relations with Gaussian distributions. Specifically, the mean vector denotes the central position of an entity or a relation, and the covariance matrix denotes its uncertainties. Following the score function proposed in TransE, for $\langle h, r, t \rangle$, the Gaussian distributions of entities and relations are defined as

$$\mathbf{h} \sim \mathcal{N}(\boldsymbol{\mu}_h, \boldsymbol{\Sigma}_h), \quad \mathbf{t} \sim \mathcal{N}(\boldsymbol{\mu}_t, \boldsymbol{\Sigma}_t), \quad \mathbf{r} \sim \mathcal{N}(\boldsymbol{\mu}_r, \boldsymbol{\Sigma}_r). \tag{9.30}$$

Note that the covariances are diagonal for efficient computation. KG2E hypothesizes that head and tail entities are independent with specific relations; then, the translation could be defined as

$$\mathbf{e} \sim \mathcal{N}(\boldsymbol{\mu}_h - \boldsymbol{\mu}_t, \boldsymbol{\Sigma}_h + \boldsymbol{\Sigma}_t). \tag{9.31}$$

To measure the dissimilarity between $\mathbf{e}$ and $\mathbf{r}$, KG2E considers both asymmetric similarity and symmetric similarity, and then proposes two methods.

The asymmetric similarity is based on the KL divergence between $\mathbf{e}$ and $\mathbf{r}$, which is a typical method to measure the similarity between two distributions. The score function is

$$
\begin{aligned}
f(h, r, t) &= -D_{\mathrm{KL}}(\mathbf{e} \| \mathbf{r}) \\
&= -\int_{x \in \mathbb{R}^d} \mathcal{N}(x; \boldsymbol{\mu}_r, \boldsymbol{\Sigma}_r) \log \frac{\mathcal{N}(x; \boldsymbol{\mu}_e, \boldsymbol{\Sigma}_e)}{\mathcal{N}(x; \boldsymbol{\mu}_r, \boldsymbol{\Sigma}_r)} dx \\
&= -\frac{1}{2} \{ \mathrm{tr}(\boldsymbol{\Sigma}_r^{-1} \boldsymbol{\Sigma}_r) + (\boldsymbol{\mu}_r - \boldsymbol{\mu}_e)^\top \boldsymbol{\Sigma}_r^{-1} (\boldsymbol{\mu}_r - \boldsymbol{\mu}_e) - \log \frac{\det(\boldsymbol{\Sigma}_e)}{\det(\boldsymbol{\Sigma}_r)} - d \},
\end{aligned}
\tag{9.32}
$$

where $\mathrm{tr}(\boldsymbol{\Sigma})$ indicates the trace of $\boldsymbol{\Sigma}$ and $\boldsymbol{\Sigma}^{-1}$ indicates the inverse of $\boldsymbol{\Sigma}$.

The symmetric similarity is built on the expected likelihood and probability product kernel. KE2G takes the inner product between the probability density functions of $\mathbf{e}$ and $\mathbf{r}$ as the measurement of similarity. The logarithm of score function defined is

$$
\begin{aligned}
f(h, r, t) &= - \int_{x \in \mathbb{R}^d} \mathcal{N}(x; \boldsymbol{\mu}_e, \boldsymbol{\Sigma}_e) \mathcal{N}(x; \boldsymbol{\mu}_r, \boldsymbol{\Sigma}_r) dx \\
&= - \log \mathcal{N}(0; \boldsymbol{\mu}_e - \boldsymbol{\mu}_r, \boldsymbol{\Sigma}_e + \boldsymbol{\Sigma}_r) \\
&= - \frac{1}{2} \{ (\boldsymbol{\mu}_e - \boldsymbol{\mu}_r)^\top (\boldsymbol{\Sigma}_e + \boldsymbol{\Sigma}_r)^{-1} (\boldsymbol{\mu}_e - \boldsymbol{\mu}_r) \\
&\quad + \log \det(\boldsymbol{\Sigma}_e + \boldsymbol{\Sigma}_r) + d \log(2\pi) \}.
\end{aligned}
\tag{9.33}
$$

The optimization objective of KG2E is also margin-based similar to TransE. Both asymmetric and symmetric similarities are constrained by some regularizations to avoid overfitting:

$$
\forall l \in \mathcal{E} \cup \mathcal{R}, \quad \|\boldsymbol{\mu}_l\|_2 \leq 1, \quad c_{\min} \mathbf{I} \leq \boldsymbol{\Sigma}_l \leq c_{\max} \mathbf{I}, \quad c_{\min} > 0,
\tag{9.34}
$$

where $c_{\min}$ and $c_{\max}$ are the hyper-parameters as the restriction values for covariance.

TransG TransG [174] discusses the problem that some relations in KGs such as `Contains the Location` or `Part of` may have multiple sub-meanings, which is also discussed in TransR. In fact, these complex relations could be divided into several more precise sub-relations. To address this issue, CTransR is proposed with a preprocessing that clusters sub-relation according to entity pairs.

As illustrated in Fig. 9.10, TransG assumes that the embeddings containing several semantic components should follow a Gaussian mixture model. The generative process is:

1. For each entity $e \in E$, TransG sets a standard normal distribution: $\boldsymbol{\mu}_e \sim \mathcal{N}(\mathbf{0}, \mathbf{I})$.
2. For a triplet $\langle h, r, t \rangle$, TransG uses the Chinese restaurant process (CRP) to automatically detect semantic components (i.e., sub-meanings in a relation): $\pi_{r,n} \sim \text{CRP}(\beta)$. $\pi_{r,n}$ is the weight of the i-th component generated by the Chinese restaurant process from the data.
3. Draw the head embedding from a standard normal distribution: $\mathbf{h} \sim \mathcal{N}(\boldsymbol{\mu}_h, \sigma_h^2 \mathbf{I})$.
4. Draw the tail embedding from a standard normal distribution: $\mathbf{t} \sim \mathcal{N}(\boldsymbol{\mu}_t, \sigma_t^2 \mathbf{I})$.
5. Calculate the relation embedding for this semantic component: $\boldsymbol{\mu}_{r,n} = \mathbf{t} - \mathbf{h}$.

Finally, the score function is

$$
f(h, r, t) \propto \sum_{n=1}^{N_r} \pi_{r,n} \mathcal{N}(\boldsymbol{\mu}_{r,n}; \boldsymbol{\mu}_t - \boldsymbol{\mu}_h, (\sigma_h^2 + \sigma_t^2) \mathbf{I}),
\tag{9.35}
$$

in which N_r is the number of semantic components of the relation r.

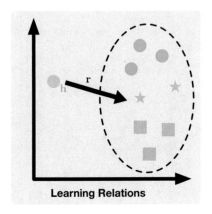 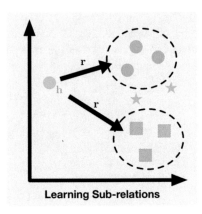

Fig. 9.10 The architecture of TransG [174]. The figure is redrawn according to Fig. 2 from TransG paper [174]

9.3.3 Neural Representation

With the development of neural networks, several efforts have been devoted to exploring neural networks for modeling KGs. Next, we will introduce how to represent KGs with neural networks.

Single Layer Model (SLM) Inspired by the previous works in representing KGs, SLM represents both entities and relations in low-dimensional spaces, and uses relation-specific matrices to project entities to relation spaces. Similar to the linear method SE, the score function of SLM is

$$f(h, r, t) = \mathbf{r}^\top \tanh(\mathbf{M}_{r,1}\mathbf{h} + \mathbf{M}_{r,2}\mathbf{t}), \tag{9.36}$$

where $\mathbf{h}, \mathbf{t} \in \mathbb{R}^{d_e}$ represent head and tail embeddings, $\mathbf{r} \in \mathbb{R}^{d_r}$ represents relation embedding, and $\mathbf{M}_{r,1}, \mathbf{M}_{r,2} \in \mathbb{R}^{d_e \times d_r}$ stand for the relation-specific matrices.

Neural Tensor Network (NTN) Although SLM has introduced relation embeddings as well as a nonlinear neural layer to build the score function, the representation capability is still restricted. NTN [143] is then proposed by introducing tensors into the SLM framework, which can be seen as an enhanced version of SLM. Besides the original linear neural network layer that projects entities to the relation space, NTN adds another tensor-based neural layer which combines head and tail embeddings with a relation-specific tensor. The score function of NTN is

$$f(h, r, t) = \mathbf{r}^\top \tanh(\mathbf{h}^\top \overrightarrow{\mathbf{M}_r}\mathbf{t} + \mathbf{M}_{r,1}\mathbf{h} + \mathbf{M}_{r,2}\mathbf{t} + \mathbf{b}_r), \tag{9.37}$$

where $\overrightarrow{\mathbf{M}}_r \in \mathbb{R}^{d_e \times d_e \times d_r}$ is a three-way relation-specific tensor, $\mathbf{b}_r$ is the bias, and $\mathbf{M}_{r,1}, \mathbf{M}_{r,2} \in \mathbb{R}^{d_e \times d_r}$ are the relation-specific matrices. Note that SLM can be seen as a simplified version of NTN if the tensor and the bias are set to zero.

Besides improving the score function, NTN also attempts to utilize the latent textual information located in entity names and successfully achieves significant improvements. Differing from previous methods that provide each entity with a vector, NTN represents each entity as the average of its entity name's word embeddings. For example, the entity *Bengal tiger* will be represented as the average word embeddings of *Bengal* and *tiger*. It is apparent that the entity name will provide valuable information for understanding an entity, since *Bengal tiger* may come from Bengal and be related to other tigers.

NTN utilizes tensor-based neural networks to model triplet facts and achieves excellent success. However, the overcomplicated method leads to high computational complexity compared to other methods, and the vast number of parameters limits the performance on sparse and large-scale KGs.

Neural Association Model (NAM) NAM [94] adopts multilayer nonlinear activation to model relations. More specifically, two structures are used by NAM to represent KGs: deep neural network (DNN) and relation modulated neural network (RMNN).

NAM-DNN adopts a MLP with L layers to operate knowledge embeddings:

$$\mathbf{z}^k = \text{Sigmoid}(\mathbf{M}^k \mathbf{z}^{k-1} + \mathbf{b}^k), \quad k = 1, \cdots, L, \tag{9.38}$$

where $\mathbf{z}^0 = [\mathbf{h}; \mathbf{r}]$ is the concatenation of $\mathbf{h}$ and $\mathbf{r}$, $\mathbf{M}^k$ is the weight matrix of the k-th layer, and $\mathbf{b}^k$ is the bias vector of the k-th layer. Finally, NAM-DNN defines the score function as

$$f(h, r, t) = \text{Sigmoid}(\mathbf{t}^\top \mathbf{z}^L). \tag{9.39}$$

As compared with NAM-DNN, NAM-RMNN additionally feeds the relation embedding $\mathbf{r}$ into the model:

$$\mathbf{z}^k = \text{Sigmoid}(\mathbf{M}^k \mathbf{z}^{k-1} + \mathbf{B}^k \mathbf{r}), \quad k = 1, \cdots, L, \tag{9.40}$$

where $\mathbf{M}^k$ and $\mathbf{B}^k$ indicate the weight and bias matrices. Finally, NAM-RMNN defines the score function as

$$f(h, r, t) = \text{Sigmoid}(\mathbf{t}^\top \mathbf{z}^L + \mathbf{B}^{L+1} \mathbf{r}). \tag{9.41}$$

Convolutional 2D Embeddings (ConvE) ConvE [32] uses 2D convolutional operations over embeddings to model KGs. Specifically, ConvE uses convolutional and fully connected layers to model interactions between entities and relations. After that, the obtained features are flattened and transformed by a fully connected layer, and the inner product between the final feature and the tail entity embeddings is used to build the score function:

$$f(h, r, t) = N\left(\text{vec}(N([\bar{\mathbf{h}}; \bar{\mathbf{r}}] * \omega))\mathbf{W}\right) \cdot \mathbf{t}, \tag{9.42}$$

where $[\bar{\mathbf{h}}; \bar{\mathbf{r}}]$ is the concatenation of $\bar{\mathbf{h}}$ and $\bar{\mathbf{r}}$, $N(\cdot)$ is a neural layer, $*$ denotes the convolution operator, and $\text{vec}(\cdot)$ means compressing a matrix into a vector. $\bar{\mathbf{h}}$ and $\bar{\mathbf{r}}$ denote the 2D-reshaping versions of $\mathbf{h}$ and $\mathbf{r}$, respectively: if $\mathbf{h}, \mathbf{r} \in \mathbb{R}^d$, then $\bar{\mathbf{h}}, \bar{\mathbf{r}} \in \mathbb{R}^{d_a \times d_b}$, where $d = d_a d_b$.

To some extent, ConvE can be seen as an improvement model based on HolE. Compared with HolE, ConvE adopts multiple neural layers to learn nonlinear features and is thus more expressive than HolE.

Relational Graph Convolutional Networks (RGCN) RGCN [136] is an extension of GCNs to model KGs. The core idea of RGCN is to formalize modeling KGs as message passing. Therefore, in RGCN, the representations of entities and relations are the results of information propagation and fusion at multiple layers. Specifically, given an entity h, its embedding at the $(k + 1)$-th layer is

$$\mathbf{h}^{k+1} = \text{Sigmoid}\left(\sum_{r \in \mathcal{R}} \sum_{t \in \mathcal{N}_h^r} \frac{1}{c_h^r} \mathbf{W}_r^k \mathbf{t}^k + \tilde{\mathbf{W}}^k \mathbf{t}^k\right), \tag{9.43}$$

where $\mathcal{N}_h^r$ denotes the neighbor set of h under the relation r and c_h^r is the normalization factor. c_h^r can be either learned or preset, and normally $c_h^r = |\mathcal{N}_h^r|$.

Note that RGCN only aims to obtain more expressive features for entities and relations. Therefore, based on the output features of RGCN, any score function mentioned above can be used here, such as combining the features of RGCN and the score function of TransE to learn knowledge representations.

9.3.4 Manifold Representation

So far, we have introduced linear methods, translation methods, and neural methods for knowledge representation. All these methods project entities and relations into low-dimensional embedding spaces, and seek to improve the flexibility and variety of entity and relation representations. Although these methods have achieved promising results, they assume that the geometry of the embedding spaces for entities and relations are all Euclidean. However, the basic Euclidean geometry may not be the optimal geometry to model the complex structure of KGs. Next, we will introduce several typical manifold methods that aim to use more flexible and powerful geometric spaces to carry representations.

ManifoldE ManifoldE [173] considers the possible positions of golden candidates for representations in spaces as a manifold rather than a point. The overall score function of ManifoldE is

$$f(h, r, t) = -\|M(h, r, t) - D_r^2\|, \tag{9.44}$$

in which D_r^2 is a relation-specific manifold parameter. Two kinds of manifolds are then proposed in ManifoldE. ManifoldE-Sphere is a straightforward manifold that supposes $\mathbf{t}$ should be located in the sphere which has $\mathbf{h} + \mathbf{r}$ to be the center and D_r to be the radius. We have:

$$M(h, r, t) = \|\mathbf{h} + \mathbf{r} - \mathbf{t}\|. \tag{9.45}$$

A tail may correspond to many different head-relation pairs, and the manifold assumption requires that the tail lays in all the manifolds of these head-relation pairs, i.e., lays in the intersection of these manifolds. However, two spheres can only intersect only under some strict conditions. Therefore, the hyperplane is utilized because it is easier for two hyperplanes to intersect. The function of ManifoldE-Hyperplane is

$$M(h, r, t) = (\mathbf{h} + \mathbf{r}_h)^\top (\mathbf{t} + \mathbf{r}_t), \tag{9.46}$$

in which $\mathbf{r}_h$ and $\mathbf{r}_t$ represent the two entity-specific embeddings of the relation r. This indicates that for a triplet $\langle h, r, t \rangle$, the tail entity $\mathbf{t}$ should locate in the hyperplane whose normal vector is $\mathbf{h} + \mathbf{r}_h$ and intercept is D_r^2. Furthermore, ManifoldE-Hyperplane considers absolute values in $M(h, r, t)$ as $|\mathbf{h} + \mathbf{r}_h|^\top |\mathbf{t} + \mathbf{r}_t|$ to double the solution number of possible tail entities. For both manifolds, ManifoldE applies a kernel form on the reproducing kernel Hilbert space.

ComplEx ComplEx [153] employs an eigenvalue decomposition model which makes use of complex-valued embeddings, i.e., $\mathbf{h}, \mathbf{r}, \mathbf{t} \in \mathbb{C}^d$. Complex embeddings can well handle binary relations, such as the symmetric and antisymmetric relations. The score function of ComplEx is

$$
\begin{aligned}
f(h, r, t) &= \text{Re}(\langle \mathbf{r}, \mathbf{h}, \mathbf{t} \rangle) \\
&= \langle \text{Re}(\mathbf{r}), \text{Re}(\mathbf{h}), \text{Re}(\mathbf{t}) \rangle + \langle \text{Re}(\mathbf{r}), \text{Im}(\mathbf{h}), \text{Im}(\mathbf{t}) \rangle \\
&\quad - \langle \text{Im}(\mathbf{r}), \text{Re}(\mathbf{h}), \text{Im}(\mathbf{t}) \rangle - \langle \text{Im}(\mathbf{r}), \text{Im}(\mathbf{h}), \text{Re}(\mathbf{t}) \rangle,
\end{aligned}
$$

where $\langle \mathbf{x}, \mathbf{y}, \mathbf{z} \rangle = \sum_i x_i y_i z_i$ denotes the trilinear dot product, $\text{Re}(x)$ is the real part of x, and $\text{Im}(x)$ is the imaginary part of x. In fact, ComplEx can be viewed as a generalization of RESCAL that uses complex embeddings to model KGs.

RotatE Similar to ComplEx, RotatE [149] also represents KGs with complex-valued embeddings. RotatE defines relations as rotations from head entities to tail entities, which makes it easier to learn various relation patterns such as symmetry, antisymmetry, inversion, and composition. The element-wise (Hadamard) product can naturally represent the rotation process in the complex-valued space. Therefore, the score function of RotatE is

$$f(h, r, t) = -\|\mathbf{h} \odot \mathbf{r} - \mathbf{t}\|, \tag{9.47}$$

where $\mathbf{h}, \mathbf{r}, \mathbf{t} \in \mathbb{C}^d$ and $\odot$ denotes the element-wise (Hadamard) product. RotatE is simple and achieves quite good performance. Compared with previous methods, it is the first model that is theoretically able to model all the above four patterns (symmetry, antisymmetry, inversion, and composition). On the basis of RotatE, Zhang et al. [203] further introduce hypercomplex spaces to represent entities and relations, and achieves better performance.

MuRP MuRP [4] proposes to embed the entities in the hyperbolic space since hyperbolic space is shown to be more suitable to represent hierarchical data than Euclidean space. Specifically, they embed the entities to the Poincaré model [130] (a typical geometric model in hyperbolic space), and exploit the Mobiüs transformations in the Poincaré model as the alternatives to vector-matrix multiplication and vector addition in Euclidean space. The score function of MuRP is

$$\begin{aligned} f(h, r, t) &= d_{\mathbb{P}}(\mathbf{h}^{(r)}, \mathbf{t}^{(r)})^2 - b_h - b_t \\ &= d_{\mathbb{P}}(\exp_{\mathbf{0}}^c(\mathbf{M}_r \log_{\mathbf{0}}^c(\mathbf{h})), \mathbf{t} \oplus \mathbf{r}) - b_h - b_t, \end{aligned} \tag{9.48}$$

where $d_{\mathbb{P}}(\cdot, \cdot)$ calculates the distance between two points in the Poincaré model, $\mathbf{M}_r$ is the transform matrix for the relation r, $\mathbf{r}$ is the translation vector of the relation r, and b_h and b_t are biases for the head and tail entities respectively. $\exp_{\mathbf{0}}^c$ is the exponential mapping at $\mathbf{0}$ in the Poincaré model of the curvature c, and it maps points in the tangent space at $\mathbf{0}$ (an Euclidean subspace) to the Poincaré model. $\log_{\mathbf{0}}^c$ is the logarithmic mapping at $\mathbf{0}$ in the Poincaré model of the curvature c, and is the inverse mapping for $\exp_{\mathbf{0}}^c$. MuRP with a dimension as low as 40 achieves comparable results to the Euclidean models with dimension greater than 100, showing the effectiveness of hyperbolic space in encoding relational knowledge.

HyboNet HyboNet [23] argues that previous hyperbolic methods such as MuRP only introduce the hyperbolic geometric for embeddings, but still perform linear transformations in tangent spaces (Euclidean subspaces), significantly limiting the capability of hyperbolic models. Inspired by the Lorentz transformation in Physics, HyboNet proposes a linear transformation in the Lorentz model [130] (another typical geometric model to build hyperbolic spaces) to avoid the introduction of exponential mapping and logarithmic mapping when transforming embeddings, significantly speeding up the network and stabilizing the computation. The score function of HyboNet is

$$f(h, r, t) = d_{\mathbb{L}}^2(g_r(\mathbf{h}), \mathbf{t}) - b_h - b_t - \delta, \tag{9.49}$$

where $d_{\mathbb{L}}^2$ is the squared Lorentzian distance between two points in Lorentz model, g_r is the relation-specific Lorentz linear transformation, b_h, b_t are the biases for the head and tail entities, respectively, and δ is a hyper-parameter used to make the training process more stable.

9.3.5 Contextualized Representation

We live in a complicated pluralistic real world where we can get information from different senses. Due to this, we can learn knowledge not only from structured KGs but also from text, schemas, images, and rules. Despite the massive size of existing KGs, there is a large amount of knowledge in the real world that may not be included in the KGs. Integrating multisource information provides a novel approach for learning knowledge representations not only from the internal structured information of KGs but also from other external information. Moreover, exploring multisource information can help further understand human cognition with different senses in the real world. Next, we will introduce typical methods that utilize multisource information to enhance knowledge representations.

Knowledge Representation with Text Textual information is one of the most common and widely used information for knowledge representation. Wang et al. [164] attempt to utilize textual information by jointly learning representations of entities, relations, and words within the same low-dimensional embedding space. The method contains three parts: the knowledge model, the text model, and the alignment model. The knowledge model is learned on the triplets of KGs using TransE, while the text model is learned on the text using skip-gram. As for the alignment model, two methods are proposed to align entity and word representations by utilizing Wikipedia anchors and entity names, respectively.

Modeling entities and words into the same embedding space has the merit of encoding the information in both KGs and plain text in a unified semantic space. However, Wang's joint model mainly depends on the completeness of Wikipedia anchors and suffers from the ambiguities of many entity names. To address these issues, Zhong et al. [207] further improve the alignment model with entity descriptions, assuming that entities should have similar semantics to their corresponding descriptions.

Different from the above joint models that merely consider the alignments between KGs and textual information, description-embodied knowledge representation learning (DKRL) [176] can directly build knowledge representations from entity descriptions. Specifically, DKRL provides two kinds of knowledge representations: for each entity h, the first is the structure-based representation $\mathbf{h}_S$, which can be learned based on the structure of KGs, and the second is the

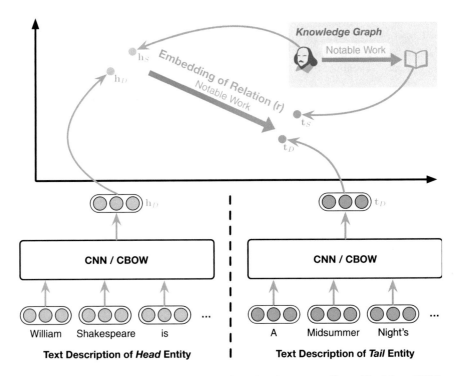

Fig. 9.11 The architecture of DKRL [176]. The figure is redrawn according to Fig. 3 from DKRL paper [176]

description-based representation $\mathbf{h}_D$ that derives from its description. The score function of DKRL derives from translation methods, and we have:

$$f(h, r, t) = -(\|\mathbf{h}_S + \mathbf{r} - \mathbf{t}_S\| + \|\mathbf{h}_S + \mathbf{r} - \mathbf{t}_D\| + \|\mathbf{h}_D + \mathbf{r} - \mathbf{t}_S\| + \|\mathbf{h}_D + \mathbf{r} - \mathbf{t}_D\|). \tag{9.50}$$

As shown in Fig. 9.11, the description-based representations are constructed via CBOW or CNNs that can encode rich textual information from plain text into representations.

Compared with conventional non-contextualized methods, the representations learned by DKRL are built with both structured and textual information and thus could perform better. Besides, DKRL can represent an entity even if it is not in the training set as long as there are a few sentences to describe this entity. Therefore, with millions of new entities emerging every day, DKRL can handle these new entities based on the setting of zero-shot learning.

Knowledge Representation with Types Entity types, as hierarchical schemas, can provide rich structured information to understand entities. Generally, there are two paths for using entity types for knowledge representations: type-constrained methods and type-augmented methods.

Type-Constrained Methods Krompaß et al. [78] take type information as constraints to improve existing methods like RESCAL and TransE via type constraints. It is intuitive that for a particular relation, the head and tail entities associated with this relation can only be of some specific types. For example, the head entity of the relation `Writes Books` should be a person (more precisely, an author), and the tail entity should be a book.

With type constraints, in RESCAL, the original factorization $\mathbf{X}_n \approx \mathbf{E}\mathbf{M}_{r_n}\mathbf{E}^\top$ in Eq.(9.8) can be modified to

$$\mathbf{X}'_n \approx \mathbf{E}_{[\mathcal{H}_{r_n},:]}\mathbf{M}_{r_n}\mathbf{E}^\top_{[\mathcal{T}_{r_n},:]}, \tag{9.51}$$

where $\mathcal{H}_{r_n}, \mathcal{T}_{r_n}$ are the entity sets fitting the type constraints of the n-th relation r_n in $\mathcal{R}$, and $\mathbf{X}'_n$ is a sparse adjacency matrix of the shape $|\mathcal{H}_{r_n}| \times |\mathcal{T}_{r_n}|$. Intuitively, only the entities that fit type constraints will be considered during the factorization process.

With type constraints, in TransE, negative samples with higher quality can be generated. Learning knowledge representations need negative samples, and negative samples are often generated by randomly replacing triplets' head or tail entities. Given a triplet $\langle h, r, t \rangle$, with type constraints, its negative samples $\langle \tilde{h}, \tilde{r}, \tilde{t} \rangle$ need to satisfy

$$\tilde{h} \in \mathcal{H}_r \subseteq \mathcal{E}, \quad \tilde{t} \in \mathcal{T}_r \subseteq \mathcal{E}. \tag{9.52}$$

Intuitively, for an entity whose type does not match the relation r, it will not be used to construct negative samples. The negative samples constructed with type constraints are more confusing, which is beneficial for learning more robust and effective representations.

Type-Augmented Methods In addition to the simplicity and effectiveness of using the type information as constraints, the representation can be further enhanced by using the type information directly as additional information in the learning. Instead of merely viewing type information as type constraints, type-embodied knowledge representation learning (TKRL) is proposed [177], utilizing hierarchical type structures to instruct the construction of projection matrices. Inspired by TransR that every entity should have multiple representations in different relation spaces, the score function of TKRL is

$$f(h, r, t) = -\|\mathbf{M}_{rh}\mathbf{h} + \mathbf{r} - \mathbf{M}_{rt}\mathbf{t}\|, \tag{9.53}$$

in which $\mathbf{M}_{rh}$ and $\mathbf{M}_{rt}$ are two projection matrices for h and t that depend on their corresponding hierarchical types in this triplet. Two hierarchical encoders are proposed to learn the above projection matrices, regarding all sub-types in the hierarchy as projection matrices, where the recursive hierarchical encoder (RHE) is

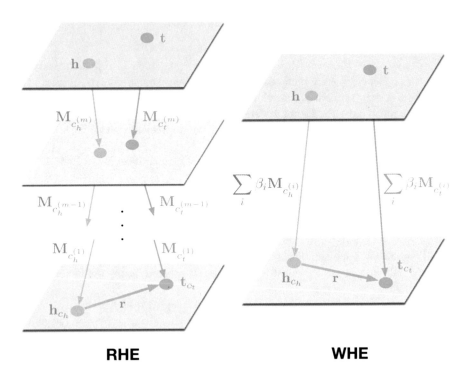

RHE   **WHE**

Fig. 9.12 The architecture of TKRL [177]. The figure is redrawn according to Fig. 2 from TKDL paper [177]

based on the matrix multiplication operation, and the weighted hierarchical encoder (WHE) is based on the matrix summation operation.

Figure 9.12 shows a simple illustration of TKRL. Taking a type hierarchy c with m layers for instance, $c^{(i)}$ is the i-th sub-type. Considering the sub-type at the first layer is the most precise and the sub-type at the last layer is the most general, TKRL can get type-specific entity representations at different granularities following the hierarchical structure, and the projection matrices can be formalized as

$$\mathbf{M}_{RHE_c} = \prod_{i=1}^{m} \mathbf{M}_{c^{(i)}} = \mathbf{M}_{c^{(1)}} \mathbf{M}_{c^{(2)}} \dots \mathbf{M}_{c^{(m)}},$$

$$\mathbf{M}_{WHE_c} = \sum_{i=1}^{m} \beta_i \mathbf{M}_{c^{(i)}} = \beta_1 \mathbf{M}_{c^{(1)}} + \dots + \beta_m \mathbf{M}_{c^{(m)}},$$

(9.54)

where $\mathbf{M}_{c^{(i)}}$ stands for the projection matrix of the i-th sub-type of the hierarchical type c and β_i is the corresponding weight of the sub-type. Taking RHE as an example, given the entity *William Shakespeare*, it is first projected to a general sub-

Suit of armour

|

Has a Part

|
↓

Armet

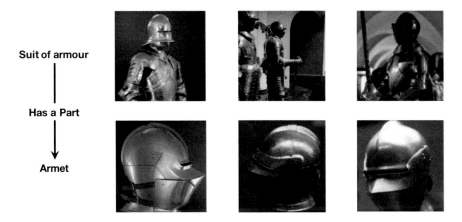

Fig. 9.13 Examples of entity images. These examples come from the original paper of TKRL [175]. All these images come from ImageNet [31]

type space like *human* and then sequentially projected to a more precise sub-type like *author* or *English author*.

Knowledge Representation with Images Human cognition is highly related to the visual information of objects in the real world. For entities in KGs, their corresponding images can provide intuitive visual information about their appearance, which may give important hints about some attributes of the entities. For instance, Fig. 9.13 shows the images of *Suit of armour* and *Armet*. From these images, we can easily infer the fact ⟨*Suit of armour*, `Has a Part`, *Armet*⟩ directly.

Image-embodied knowledge representation learning (IKRL) [175] is proposed to consider visual information when learning knowledge representations. Inspired by the abovementioned DKRL, for each entity h, IKRL also proposes the image-based representation $\mathbf{h}_I$ besides the original structure-based representation $\mathbf{h}_S$, and jointly learns these entity representations simultaneously within the translation-based framework:

$$f(h, r, t) = -(\|\mathbf{h}_S+\mathbf{r}-\mathbf{t}_S\|+\|\mathbf{h}_S+\mathbf{r}-\mathbf{t}_I\|+\|\mathbf{h}_I+\mathbf{r}-\mathbf{t}_S\|+\|\mathbf{h}_I+\mathbf{r}-\mathbf{t}_I\|). \quad (9.55)$$

More specifically, IKRL uses CNNs to obtain the representations of all entity images, and then uses a matrix to project image representations from the image embedding space to the entity embedding space. Since one entity may have multiple images, IKRL uses an attention-based method to highlight those most informative images. IKRL not only shows the importance of visual information for representing entities but also shows the possibility of finding a unified space to represent heterogeneous and multimodal information.

Knowledge Representation with Logic Rules Typical KGs store knowledge in the form of triplets with one relation linking two entities. Most existing knowledge representation methods only consider the information of triplets independently, ignoring the possible interactions and relations between different triplets. Logic rules, which are certain kinds of summaries derived from human prior knowledge, could help us with knowledge reasoning. For instance, given the triplet ⟨ *Beijing*, Capital of, *China*⟩, we can easily infer the triplet ⟨*Beijing*, Located in, *China*⟩ with high confidence, since we know the logic rule Capital of ⇒ Located in. To this end, various efforts have been devoted to exploring logic rules for KGs [5, 127, 162]. Here we introduce a typical translation method that jointly learns knowledge representations and logic rules – KALE [52]. KALE can rank all possible logic rules based on the results pre-trained by TransE, and then manually filter useful rules to improve knowledge representations.

The joint learning of KALE consists of two parts: triplet modeling and rule modeling. For the triplet modeling, KALE defines its score function following the translation assumption as

$$f(h, r, t) = 1 - \frac{1}{3\sqrt{d}} \|\mathbf{h} + \mathbf{r} - \mathbf{t}\|, \tag{9.56}$$

in which d stands for the dimension of knowledge embeddings. $f(h, r, t)$ takes a value in [0, 1], aiming to map discrete Boolean values (false or true) into a continuous space ([0, 1]). For the rule modeling, KALE uses the t-norm fuzzy logics [55] that compute the truth value of a complex formula from the truth values of its constituents. Especially, KALE focuses on two typical types of logic rules. The first rule is $\forall h, t : \langle h, r_1, t \rangle \Rightarrow \langle h, r_2, t \rangle$ (e.g., given ⟨*Beijing*, Capital of, *China*⟩, we can infer that ⟨*Beijing*, Located in, *China*⟩). KALE represents the score function of this logic rule l_1 via specific t-norm logical connectives as

$$f(l_1) = f(h, r_1, t) f(h, r_2, t) - f(h, r_1, t) + 1. \tag{9.57}$$

The second rule is $\forall h, e, t : \langle h, r_1, e \rangle \wedge \langle e, r_2, t \rangle \Rightarrow \langle h, r_3, t \rangle$ (e.g., given ⟨*Tsinghua*, Located in, *Beijing*⟩ and ⟨*Beijing*, Located in, *China*⟩, we can infer that ⟨*Tsinghua*, Located in, *China*⟩). And the second score function is defined as

$$f(l_2) = f(h, r_1, e) f(e, r_2, t) f(h, r_3, t) - f(h, r_1, e) f(e, r_2, t) + 1. \tag{9.58}$$

The joint training contains all positive formulas, including triplet facts and logic rules. Note that for the consideration of rule qualities, KALE ranks all possible logic rules by their truth values with pre-trained TransE and manually filters some rules.

9.3.6 Summary

Knowledge representation learning is the cornerstone of applying knowledge for NLP tasks. Knowledge can be incorporated into NLP tasks in a high-quality manner only with good knowledge representations. In this section, we introduce five directions of existing efforts to obtain distributed knowledge representations: (1) *linear methods*, where relations are represented as linear transformations between entities, (2) *translation methods*, where relations are represented as additive translations between entities, (3) *neural methods*, where neural networks parameterize the interactions between entities and relations, (4) *manifold methods*, where representations are learned in more flexible and powerful geometric spaces instead of the basic Euclidean geometry, and (5) *contextualized methods*, where representations are learned under complex contexts.

In summary, from simple methods like SE and TransE, to more sophisticated methods that use neural networks (e.g., ConvE), the hyperbolic geometry (e.g., HyboNet), and textual information (e.g., DKRL), all these methods can provide effective knowledge representations. These methods lay a solid foundation for further knowledge-guided NLP and knowledge acquisition, which will be introduced in later sections. Note that more sophisticated methods do not necessarily lead to a better application in NLP tasks. Researchers still need to choose the appropriate knowledge representation learning method according to the characteristics of specific tasks and the balance between computational efficiency and representation quality.

9.4 Knowledge-Guided NLP

An effective NLP agent is expected to accurately and deeply understand user demands, and appropriately and flexibly give responses and solutions. Such kind of work can only be done supported by certain forms of knowledge. To this end, knowledge-guided NLP has been widely explored in recent years. Figure 9.14 shows a brief pipeline of utilizing knowledge for NLP tasks. In this pipeline, we first need to extract knowledge from heterogeneous data sources and store the extracted knowledge with knowledge systems (e.g., KGs). Next, we need to project knowledge systems into low-dimensional continuous spaces with knowledge representation learning methods to manipulate the knowledge in a machine-friendly way. Finally, informative knowledge representations can be applied to handle various NLP tasks. After introducing how to learn knowledge representations, we will detailedly show in this section how to use knowledge representations for specific NLP tasks.

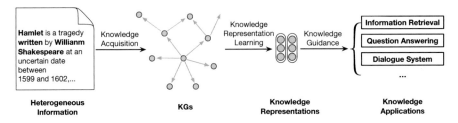

Fig. 9.14 The pipeline of utilizing knowledge for NLP tasks

The performance of NLP models (more generally, machine learning models) depends on four critical factors: input data, model architecture, learning objective, and hypothesis space. And the whole goal is to minimize the structural risk

$$\min_{f \in \mathcal{F}} \frac{1}{N} \sum_{i=1}^{N} \mathcal{L}(y_i, f(x_i)) + \lambda \mathcal{J}(f), \tag{9.59}$$

where x_i is the input data, f is the model function, $\mathcal{L}$ is the learning objective, $\mathcal{F}$ is the hypothesis space, and $\mathcal{J}(f)$ is the regularization term. By applying knowledge to each of these four factors, we can form four directions to perform knowledge-guided NLP: (1) knowledge augmentation, which aims to augment the input data x_i with knowledge; (2) knowledge reformulation, which aims to reformulate the model function f with knowledge; (3) knowledge regularization, which aims to regularize or modify the learning objectives $\mathcal{L}$ with knowledge; (4) knowledge transfer, which aims to transfer the pre-trained parameters as prior knowledge to constrain the hypothesis space $\mathcal{F}$.

Some works [60, 61] have briefly introduced this knowledge-guided NLP framework, while in this section, we will further present more details around the four knowledge-guided directions. In addition, to make this section clearer and more intuitive, we will also introduce some specific application cases of knowledge-guided NLP.

9.4.1 Knowledge Augmentation

Knowledge augmentation aims at using knowledge to augment the input features of models. Formally, after using knowledge k to augment the input, the original risk function is changed to

$$\min_{f \in \mathcal{F}} \frac{1}{N} \sum_{i=1}^{N} \mathcal{L}(y_i, f(x_i, k)) + \lambda \mathcal{J}(f). \tag{9.60}$$

In order to achieve this kind of knowledge augmentation at the input level, existing efforts focus on adopting two mainstream approaches.

Augmentation with Knowledge Context One approach is to directly add knowledge to the input as additional context. Augmenting language modeling with retrieval is a representative method, such as REALM [53] and RAG [86]. These methods retrieve background knowledge from additional corpora and then use the retrieved knowledge to provide more information for language modeling. Since the retrieved knowledge can significantly improve the performance of language understanding and generation, this approach to achieving knowledge augmentation is widely applied by question answering [76, 111] and dialogue systems [139, 168]. Next, we will take RAG as an example to show how to perform knowledge augmentation with knowledge context.

Example: Knowledge Augmentation for the Generation of PTMs In recent years, PTMs have achieved state-of-the-art results on a variety of NLP tasks, but these PTMs still face challenges in precisely accessing and manipulating knowledge and cannot well handle various knowledge-intensive tasks, especially for various text generation tasks that require extensive knowledge. To help PTMs utilize more knowledge for text generation, retrieval-augmented generation (RAG) [86] has been proposed with the aim of using the retrieved external knowledge as additional context to generate text with higher quality.

Given the input sequence x to generate the output sequence y, the overall process of the typical autoregressive generation method can be formalized as $P(y|x) = \prod_{i=1}^{N} P_\theta(y_i|x, y_{1:i-1})$, where θ is the parameters of the generator, N is the length of y, and y_i is the i-th token of y. To use more knowledge to generate y, RAG first retrieves the external information z according to the input x and then generates the output sequence y based on both x and z. To ensure that the retrieved contents can cover the crucial knowledge required to generate y, the top-K contents retrieved by the retriever are all used to help generate the output sequence y, and thus the overall generation process is

$$
\begin{aligned}
P_{\text{RAG-Sequence}}(y|x) &\approx \sum_{z \in \text{top}-K[P_\eta(\cdot|x)]} P_\eta(z|x) P_\theta(y|x, z) \\
&= \sum_{z \in \text{top-}K[P_\eta(\cdot|x)]} P_\eta(z|x) \prod_{i=1}^{N} P_\theta(y_i|x, z, y_{1:i-1}),
\end{aligned}
\tag{9.61}
$$

where η is the parameters of the retriever.

In addition to applying knowledge augmentation at the sequence level, token-level RAG is also introduced to provide finer-grained augmentation. Specifically, token-level RAG first retrieves the top K external information according to the input x, which is the same as RAG-Sequence. When generating text, token-level RAG considers all the retrieved information together to generate the distribution for the next output token, instead of sequence-level RAG which separately generates sequences based on the retrieved content and then merges the generated sequences. Formally, the token-level RAG is

$$P_{\text{RAG-Token}}(y|x) \approx \prod_{i=1}^{N} \sum_{z \in \text{top-}K[P(\cdot|x)]} P_{\eta}(z|x) P_{\theta}(y_i|x, z, y_{1:i-1}). \tag{9.62}$$

To sum up, RAG adds the retrieved knowledge to the input as additional context, which is a typical example of knowledge augmentation with knowledge context.

Augmentation with Knowledge Embeddings Another approach is to design special modules to fuse the original input features and knowledge embeddings and then use the knowledgeable features as the input to solve NLP tasks. Since this approach can help to fully utilize heterogeneous knowledge from multiple sources, many works follow this approach to integrate unstructured text and structured symbolic knowledge in KGs, leading to knowledge-guided information retrieval [87, 100] and knowledge-guided PTMs [96, 124, 128, 163, 185, 205]. Next, we will first introduce word-entity duet, an effective information retrieval method, and then take a typical knowledge-guided information retrieval method EDRM as an example to show how to perform knowledge augmentation with knowledge embeddings.

Example: Knowledge Augmentation for Information Retrieval Information retrieval focuses on obtaining informative representations of queries and documents, and then designing effective metrics to compute the similarities between queries and documents. The emergence of large-scale KGs has motivated the development of entity-oriented information retrieval, which aims to leverage KGs to improve the retrieval process. Word-entity duet [179] is a typical method for entity-oriented information retrieval. Specifically, given a query q and a document d, word-entity duet first constructs bag-of-words q^w and d^w. By annotating the entities mentioned by the query q and the document d, word-entity duet then constructs bag-of-entities q^e and d^e. Based on bag-of-words and bag-of-entities, word-entity duet utilizes the duet representations of bag-of-words and bag-of-entities to match the query q and the document d. The word-entity duet method consists of a four-way interaction: query words to document words (q^w-d^w), query words to document entities (q^w-d^e), query entities to document words (q^e-d^w), and query entities to document entities (q^e-d^e).

On the basis of the word-entity duet method, EDRM [100] further uses distributed representations instead of bag-of-words and bag-of-entities to represent queries and documents for ranking. As shown in Fig. 9.15, EDRM first learns the distributed representations of entities according to entity-related information in KGs, such as entity descriptions and entity types. Then, EDRM uses interaction-based neural models [28] to match the query and documents with word-entity duet distributed representations. More specifically, EDRM uses a translation layer that calculates the similarity between query-document terms: ($\mathbf{v}_{wq}^i$ or $\mathbf{v}_{eq}^i$) and ($\mathbf{v}_{wd}^j$ or $\mathbf{v}_{ed}^j$). It constructs the interaction matrix $\mathbf{M} = \{\mathbf{M}_{ww}, \mathbf{M}_{we}, \mathbf{M}_{ew}, \mathbf{M}_{ee}\}$, by

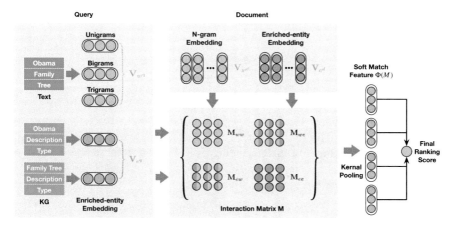

Fig. 9.15 The architecture of EDRM [100]. The figure is redrawn according to Fig. 1 from EDRM paper [100]

denoting $\mathbf{M}_{ww}, \mathbf{M}_{we}, \mathbf{M}_{ew}, \mathbf{M}_{ee}$ as the interactions of $q^w\text{-}d^w$, $q^w\text{-}d^e$, $q^e\text{-}d^w$, $q^e\text{-}d^e$, respectively. And the elements in these matrices are the cosine similarities of corresponding terms:

$$\mathbf{M}_{ww}^{ij} = \cos(\mathbf{v}_{w^q}^i, \mathbf{v}_{w^d}^j); \mathbf{M}_{we}^{ij} = \cos(\mathbf{v}_{w^q}^i, \mathbf{v}_{e^d}^j),$$
$$\mathbf{M}_{ew}^{ij} = \cos(\mathbf{v}_{e^q}^i, \mathbf{v}_{w^d}^j); \mathbf{M}_{ee}^{ij} = \cos(\mathbf{v}_{e^q}^i, \mathbf{v}_{e^d}^j). \tag{9.63}$$

The final ranking feature $\Phi(\mathbf{M})$ is a concatenation of four cross matches:

$$\Phi(\mathbf{M}) = [\phi(\mathbf{M}_{ww}); \phi(\mathbf{M}_{we}); \phi(\mathbf{M}_{ew}); \phi(\mathbf{M}_{ee})], \tag{9.64}$$

where $\phi(\cdot)$ can be any function used in interaction-based neural ranking models, such as using Gaussian kernels to extract the matching feature from the matrix $\mathbf{M}$ and then pool into a feature vector $\phi(\mathbf{M})$. For more details of designing $\phi(\cdot)$ and using $\Phi(\mathbf{M})$ to compute ranking scores, we suggest referring to some typical interaction-based information retrieval models [28, 180].

To sum up, EDRM introduces distributed knowledge representations to improve the representations of queries and documents for information retrieval, which is a typical example of knowledge augmentation with knowledge embeddings.

9.4.2 Knowledge Reformulation

Knowledge reformulation aims at using knowledge to enhance the model processing procedure. Formally, after using knowledge to reformulate the model function, the original risk function is changed to

$$\min_{f_k \in \mathcal{F}} \frac{1}{N} \sum_{i=1}^{N} \mathcal{L}(y_i, f_k(x_i)) + \lambda \mathcal{J}(f_k), \qquad (9.65)$$

where $f_k(\cdot)$ is the model function reformulated by knowledge. Considering the complexity of the model function $f(\cdot)$, it is difficult for us to comprehensively discuss the construction process of f_k. To introduce this section more clearly and give readers a more intuitive understanding of knowledge reformulation, we here focus on introducing two relatively simple knowledge reformulation scenarios: knowledgeable preprocessing and post-processing.

Knowledgeable Preprocessing On the one hand, we can use the underlying knowledge-guided model layer for preprocessing to make features more informative [160, 167, 194]. Formally, x_i is first input to the function k and then input to the function f as

$$f_k(x_i) = f(k(x_i)), \qquad (9.66)$$

where $k(\cdot)$ is the knowledge-guided model function used for preprocessing and $f(\cdot)$ is the original model function. The knowledge-guided attention mechanism is a representative approach that usually leverages informative knowledge representations to enhance model feature processing. Next, we will take two typical knowledge-guided attention mechanisms [58, 178] as examples to show how to use knowledge for model preprocessing.

Example: Knowledge Reformulation for Knowledge Acquisition Knowledge acquisition includes two main approaches. One is knowledge graph completion (KGC), which aims to perform link prediction on KGs. The other is relation extraction (RE) to predict relations between entity pairs based on the sentences containing entity pairs. Formally, given sentences $s_1, s_2, \cdots$ containing the entity pair h, t, RE aims to evaluate the likelihood that a relation r and h, t can form a triplet based on the semantics of these sentences. Different from RE, KGC only uses the representations of h, r, t learned by knowledge representation learning methods to compute the score function $f(h, r, t)$, and the score function serves knowledge acquisition.

Generally, RE and KGC models are learned separately, and these models cannot fully integrate text and knowledge to acquire more knowledge. To this end, Han et al. [58] propose a joint learning framework for knowledge acquisition, which can jointly learn knowledge and text representations within a unified semantic space via KG-text alignments. Figure 9.16 shows the brief framework of the joint model. For the text part, the sentence with two entities (e.g., *Mark Twain* and *Florida*) is regarded as the input to the encoder, and the output is considered to potentially describe specific relations (e.g., Place of Birth). For the KG part, entity and relation representations are learned via a knowledge representation learning method

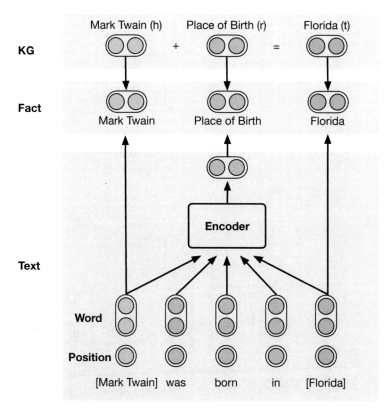

Fig. 9.16 The joint learning framework for knowledge acquisition [58]. The figure is redrawn according to Fig. 1 from the paper of Han et al. [58]

such as TransE. The learned representations of the KG and text parts are aligned during the training process.

Given sentences $\{s_1, s_2, \cdots\}$ containing the same entity pair h, t, not all of these sentences can help predict the relation between h and t. For a given relation r, there are many triplets $\{(h_1, r, t_1), (h_2, r, t_2), \cdots\}$ containing the relation, but not all triplets are important enough for learning the representation of r. Therefore, as shown in Fig. 9.17, Han et al. further adopt mutual attention to reformulate the preprocessing of both the text and knowledge models, to select more useful sentences for RE and more important triplets for KGC. Specifically, we use knowledge representations to highlight the more valuable sentences for predicting the relation between h and t. This process can be formalized as

$$\alpha = \text{Softmax}(\mathbf{r}_{ht}^\top \mathbf{W}_{KA} \mathbf{S}), \quad \hat{\mathbf{s}} = \mathbf{S}\alpha^\top, \tag{9.67}$$

where $\mathbf{W}_{KA}$ is a bilinear matrix of the knowledge-guided attention, $\mathbf{S} = [\mathbf{s}_1, \mathbf{s}_2, \cdots]$ are the hidden states of the sentences $s_1, s_2, \cdots$. $\mathbf{r}_{ht}^\top$ is a representation that

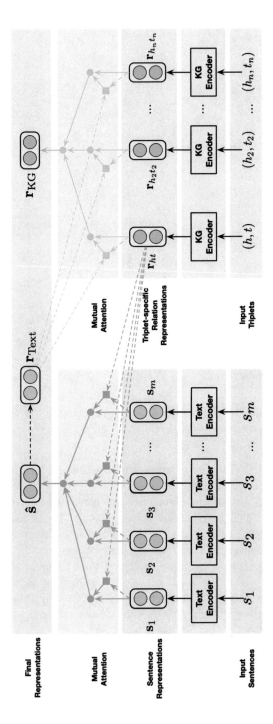

Fig. 9.17 The mutual attention to reformulate both the text and knowledge models [58]. The figure is redrawn according to Fig. 1 from the paper of Han et al. [58]

can indicate the latent relation between h and t, computed based on knowledge representations. $\hat{s}$ is the feature after synthesizing the information of all sentences, which is used to predict the relation between h and t finally.

Similar to using knowledge representations to select high-quality sentences, we can also use semantic information to select triples conducive to learning relations. This process can be formalized as

$$\alpha = \text{Softmax}(\mathbf{r}_{\text{Text}}^\top \mathbf{W}_{\text{SA}} \mathbf{R}), \quad \mathbf{r}_{\text{KG}} = \mathbf{R}\alpha^\top, \tag{9.68}$$

where $\mathbf{W}_{\text{SA}}$ is a bilinear matrix of the semantics-guided attention, $\mathbf{R} = [\mathbf{r}_{h_1 t_1}, \mathbf{r}_{h_2 h_2}, \cdots]$ are the triplet-specific relation representations of the triplets $\{(h_1, r, t_1), (h_2, r, t_2), \cdots\}$. $\mathbf{r}_{\text{Text}}$ is the semantic representation of the relation r used by the RE model. $\mathbf{r}_{\text{KG}}$ is the final relation representation enhanced with semantic information.

This work is a typical attempt to apply knowledge representations of existing KGs to reformulate knowledge acquisition models. In Sect. 9.5, we will introduce knowledge acquisition in more detail.

Example: Knowledge Reformulation for Entity Typing Entity typing is the task of detecting semantic types for a named entity (or entity mention) in plain text. For example, given a sentence *Jordan played 15 seasons in the NBA*, entity typing aims to infer that *Jordan* in this sentence is a *person*, an *athlete*, and even a *basketball player*. Entity typing is important for named entity disambiguation since it can narrow down the range of candidates for an entity mention [21]. Moreover, entity typing also benefits massive NLP tasks such as relation extraction [98], question answering [184], and knowledge base population [20].

Neural models [36, 138] have achieved state-of-the-art performance for fine-grained entity typing. However, these methods only consider the textual information of named entity mentions for entity typing while ignoring the rich information that KGs can provide for determining entity types. For example, in the sentence *In 1975, Gates ... Microsoft ... company*, even though we have no type information of *Microsoft* in KGs, other entities similar to *Microsoft* (e.g., *IBM*) in KGs can also provide supplementary information to help us determine the type of *Microsoft*. To take advantage of KGs for entity typing, knowledge-guided attention for neural entity typing (KNET) has been proposed [178].

As illustrated in Fig. 9.18, KNET mainly consists of two parts. Firstly, KNET builds a neural network, including a bidirectional LSTM and a fully connected layer, to generate context and named entity mention representations. Secondly, KNET introduces a knowledge-guided attention mechanism to emphasize those critical words and improve the quality of context representations. Here, we introduce the knowledge-guided attention in detail. KNET employs the translation method TransE to obtain entity embedding $\mathbf{e}$ for each entity e in KGs. During the training process, given the context words $c = \{w_i, \cdots, w_j\}$, a named entity mention m

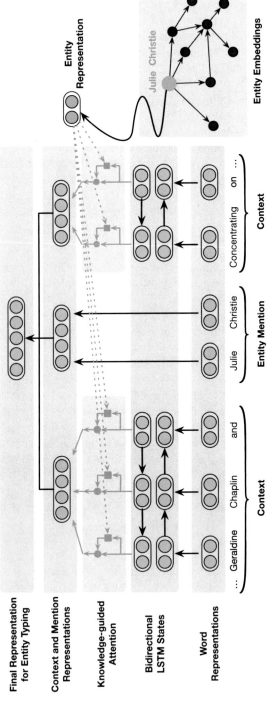

Fig. 9.18 The architecture of KNET [178]. The figure is redrawn according to Fig. 1 from KNET paper [178]

and its corresponding entity embedding $\mathbf{e}$, KNET computes the knowledge-guided attention as

$$\alpha = \text{Softmax}(\mathbf{e}^\top \mathbf{W}_{\text{KA}} \mathbf{H}), \quad \mathbf{c} = \mathbf{H}\alpha^\top, \tag{9.69}$$

where $\mathbf{W}_{\text{KA}}$ is a bilinear matrix of the knowledge-guided attention and $\mathbf{H} = [\mathbf{h}_i, \cdots, \mathbf{h}_j]$ are the bidirectional LSTM states of $\{w_i, \cdots, w_j\}$. The context representation $\mathbf{c}$ is used as an important feature for the subsequent process of type classification.

Through the above two examples of knowledge acquisition and entity typing, we introduce how to highlight important features based on knowledge in the model preprocessing stage, so as to output better features to help improve model performance.

Knowledgeable Post-Processing Apart from reformulating model functions for pre-processing, on the other hand, knowledge can be used as an expert at the end of models for post-processing, guiding models to obtain more accurate and effective results [1, 51, 124]. Formally, x_i is first input to the function f and then input to the function k as

$$f_k(x_i) = k(f(x_i)), \tag{9.70}$$

where $k(\cdot)$ is the knowledge-guided model function used for post-processing and $f(\cdot)$ is the original model function. Knowledgeable post-processing is widely used by knowledge-guided language modeling to improve the word prediction process [1, 51]. Next, we will take a typical knowledge-guided language modeling method NKLM [1] as an example to show how to use knowledge representations to improve model post-processing (Fig. 9.19).

Example: Knowledge Post-Processing on Language Modeling NKLM [1] aims to perform language modeling by considering both semantics and knowledge to generate text. Specifically, NKLM designs two ways to generate each word in the text. The first is the same as conventional auto-regressive models that generate a vocabulary word according to the probabilities over the vocabulary. The second is to generate a knowledge word according to external KGs. Specifically, NKLM uses the LSTM architecture as the backbone to generate words. For external KGs, NKLM stores knowledge representations to build a knowledgeable module $\mathcal{K} = \{(\mathbf{a}_1, O_1), (\mathbf{a}_2, O_2), \cdots, (\mathbf{a}_n, O_n)\}$, in which O_i denotes the description of the i-th fact, $\mathbf{a}_i$ denotes the concatenation of the representations of the head entity, relation and tail entity of the i-th fact.

Given the context $\{w_1, w_2, \cdots, w_{t-1}\}$, NKLM takes both the vocabulary word representation $\mathbf{w}_{t-1}^v$, the knowledge word representation $\mathbf{w}_{t-1}^o$, and the knowledge-guided representation $\mathbf{a}_{t-1}$ at the step $t - 1$ as LSTM's input $\mathbf{x}_t = \{\mathbf{w}_{t-1}^v, \mathbf{w}_{t-1}^o, \mathbf{a}_{t-1}\}$. $\mathbf{x}_t$ is then fed to LSTM together with the hidden state $\mathbf{h}_{t-1}$ to get the output state $\mathbf{h}_t$. Next, a two-layer multilayer perceptron $f(\cdot)$ is applied to the concatenation of $\mathbf{h}_t$ and $\mathbf{x}_t$ to get the fact key $\mathbf{k}_t = f(\mathbf{h}_t, \mathbf{x}_t)$. $\mathbf{k}_t$ is

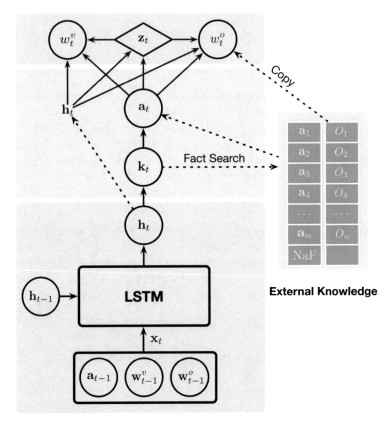

Fig. 9.19 The architecture of NKLM [1]. A special entry (NaF,) is included in the knowledgable module to allow the absence of knowledge when the currently generated word is not included in the knowledgable module. NaF is short for *not a fact*. The figure is redrawn according to Fig. 1 from NKLM paper [1]

then used to extract the most relevant fact representation $\mathbf{a}_t$ from the knowledgeable module. Finally, the selected fact $\mathbf{a}_t$ is combined with the hidden state $\mathbf{h}_t$ to output a vocabulary word w_t^v and knowledge word w_t^o (which is copied from the entity name in the t-th fact), and then determine which word to generate at the step t.

Overall, by using KGs to enhance the post-processing of language modeling, NKLM can generate sentences that are highly related to world knowledge, which are often difficult to model without considering external knowledge.

9.4.3 Knowledge Regularization

Knowledge regularization aims to use knowledge to modify the objective functions of models:

$$\min_{f \in \mathcal{F}} \frac{1}{N} \sum_{i=1}^{N} \mathcal{L}(y_i, f(x_i)) + \lambda_k \mathcal{L}_k(k, f(x_i)) + \lambda \mathcal{J}(f), \qquad (9.71)$$

where $\mathcal{L}_k(k, f(x_i))$ is the additional predictive targets and learning objectives constructed based on knowledge and λ_k is a hyper-parameter to control the knowledgeable loss term.

Distant supervision [109] is a representative method that uses external knowledge to heuristically annotate corpora as additional supervision signals. For many vital information extraction tasks, such as RE [58, 72, 91, 196] and entity typing [36, 138, 178], distant supervision is widely applied for model training. As we will introduce distant supervision in Sect. 9.5 to show how to build additional supervision signals with knowledge, we do not introduce concrete examples here.

Knowledge regularization is also widely used by knowledge-guided PTMs [124, 163, 205]. To fully integrate knowledge into language modeling, these knowledge-guided PTMs design knowledge-specific tasks as their pre-training objectives and use knowledge representations to build additional prediction objectives. Next, we will take the typical knowledge-guided PTM ERNIE [205] as an example to show how knowledge regularization can help the learning process of models.

Example: Knowledge Regularization for PTMs PTMs like BERT [33] have great abilities to extract features from text. With informative language representations, PTMs obtain state-of-the-art results on various NLP tasks. However, the existing PTMs rarely consider incorporating external knowledge, which is essential in providing related background information for better language understanding. For example, given a sentence *Bob Dylan wrote Blowin' in the Wind and Chronicles: Volume One*, without knowing *Blowin' in the Wind* is a *song* and *Chronicles: Volume One* is a *book*, it is not easy to know the occupations of *Bob Dylan*, i.e., *songwriter* and *writer*.

To this end, an enhanced language representation model with informative entities (ERNIE) is proposed [205]. Figure 9.20 is the overall architecture of ERNIE. ERNIE first augments the input data using knowledge augmentation as we have mentioned in Sect. 9.4.1. Specifically, ERNIE recognizes named entity mentions and then aligns these mentions to their corresponding entities in KGs. Based on the alignments between text and KGs, ERNIE takes the informative entity representations as additional input features.

Similar to conventional PTMs, ERNIE adopts masked language modeling and next sentence prediction as the pre-training objectives. To better fuse textual and knowledge features, ERNIE proposes *denoising entity auto-encoding (DAE)* by randomly masking some mention-entity alignments in the text and requiring models

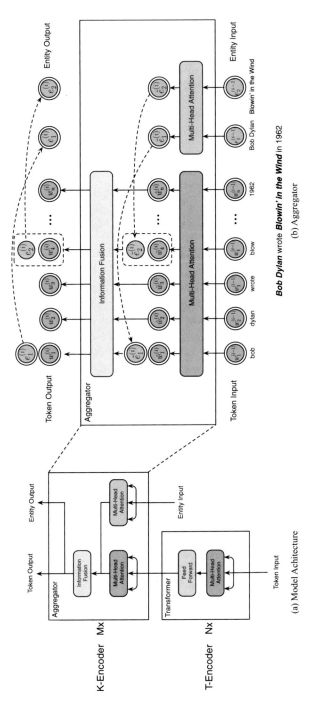

(a) Model Achitecture

(b) Aggregator

Fig. 9.20 The architecture of ERNIE [205]. The figure is redrawn according to Fig. 2 from ERNIE paper [205]

to select appropriate entities to complete the alignments. Different from the existing PTMs that predict tokens with only using local context, DAE requires ERNIE to aggregate both text and knowledge to predict both tokens and entities, leading to knowledge-guided language modeling. DAE is clearly a knowledge-guided objective function.

In addition to ERNIE, there are other representative works on knowledge regularization. For example, KEPLER [163] incorporates structured knowledge into its pre-training. Specifically, KEPLER encodes the textual description of entities as entity representations and predicts the relation between entities based on these description-based representations. In this way, KEPLER can learn the structured information of entities and relations in KGs in a language-modeling manner. WKLM [181] proposes a pre-training objective type-constrained entity replacement. Specifically, WKLM randomly replaces the named entity mentions in the text with other entities of the same type and requires the model to identify whether an entity mention is replaced or not. Based on the new pre-training objective, WKLM can accurately learn text-related knowledge and capture the type information of entities.

From Fig. 9.20, we can find that ERNIE also adopts knowledge reformulation by adding the new aggregator layers designed for knowledge integration to the original Transformer architecture. To a large extent, the success of knowledge-guided PTMs comes from the fact that these models use knowledge to enhance important factors of model learning. Up to now, we have introduced knowledge augmentation, knowledge reformulation, and knowledge regularization. Next, we will further introduce knowledge transfer.

9.4.4   Knowledge Transfer

Knowledge transfer aims to use knowledge to obtain a knowledgeable hypothesis space, reducing the cost of searching optimal parameters and making it easier to train an effective model. There are two typical approaches to transferring knowledge: (1) transfer learning [120] that focuses on transferring model knowledge learned from *labeled* data to downstream task-specific models and (2) self-supervised learning [97] that focuses on transferring model knowledge learned from *unlabeled* data to downstream task-specific models. More generally, the essence of knowledge transfer is to use prior knowledge to constrain the hypothesis space:

$$\min_{f \in \mathcal{F}_k} \frac{1}{N} \sum_{i=1}^{N} \mathcal{L}(y_i, f(x_i)) + \lambda \mathcal{J}(f), \qquad (9.72)$$

where $\mathcal{F}_k$ is the knowledge-guided hypothesis space.

Knowledge transfer is widely used in NLP. The fine-tuning stage of PTMs is a typical scenario of knowledge transfer, which aims to transfer the versatile knowledge acquired in the pre-training stage to specific tasks. Intuitively, after pre-training

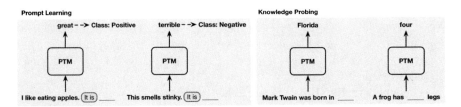

Fig. 9.21 By using prompts, we can stimulate the knowledge of PTMs to handle specific tasks such as sentiment classification and predicting symbolic knowledge

a PTM, fine-tuning this PTM can be seen as narrowing down searching task-specific parameters to a local hypothesis space around the pre-trained parameters rather than the global hypothesis space.

As we mentioned in Chap. 5, in addition to fine-tuning PTMs, prompt learning has also been widely explored. Despite the success of fine-tuning PTMs, it still faces two challenges. On the one hand, there is a gap between the objectives of pre-training and fine-tuning, since most PTMs are learned with language modeling objectives, yet downstream tasks may have quite different objective forms such as classification, regression, and labeling. On the other hand, as the parameter size of PTMs increases rapidly, fine-tuning PTMs has become resource-intensive. In order to alleviate these issues, prompts have been introduced to utilize the knowledge of PTMs in an effective and efficient manner [93].

As shown in Fig. 9.21, prompt learning aims at converting downstream tasks into a cloze-style task similar to pre-training objectives so that we can better transfer the knowledge of PTMs to downstream tasks. Taking prompt learning for sentiment classification as an example, a typical prompt consists of a template (e.g., ... *It was* [MASK] .) and a label word set (e.g., *great* and *terrible*) as candidates for predicting [MASK]. By changing the input using the template to predict [MASK] and mapping the prediction to corresponding labels, we can apply masked language modeling for sentiment classification. For example, given the sentence *I like eating apples.*, we first use the prompt template to get the new input sentence *I like eating apples. It was* [MASK]. According to PTMs predicting *great* or *terrible* at the masked position, we can determine whether this sentence is positive or negative.

The recently proposed large-scale PTM GPT-3 [17] shows the excellent performance of prompt learning in various language understanding and generation tasks. In prompt learning, all downstream tasks are transformed to be the same as the pre-training tasks. And since the parameters of PTMs are frozen during prompt learning, the size of hypothesis space is much smaller compared to fine-tuning, making more efficient knowledge transfer possible.

Overall, PTMs play an important role in driving the use of model knowledge. And to some extent, PTMs also influence the paradigm of using symbolic knowledge in NLP. As shown in Fig. 9.21, many knowledge probing works [74, 125, 126] show that by designing prompt, PTMs can even complete structured knowledge information. These studies show that PTMs, as good carriers of symbolic knowledge, can memorize symbolic knowledge well. Moreover, these studies also indicate one factor that may contribute to the power of PTMs: knowledge can be spontaneously abstracted by PTMs from large-scale unstructured data and then used to solve concrete problems, and the abstracted knowledge matches well with the knowledge formed by human beings. Inspired by this, we can further delve into how PTMs abstract knowledge and how PTMs store knowledge in their parameters, which is very meaningful for further advancing the integration of symbolic knowledge and model knowledge. On the other hand, all these studies also show the importance of knowledge-guided NLP. Compared with letting models slowly abstract knowledge from large-scale data, directly injecting symbolic knowledge into models is a more effective solution.

The success of PTMs demonstrates the clear advantages of fully transferring existing model knowledge in terms of computing efficiency and effectiveness, as compared to learning a model from scratch. Since we have introduced the details of PTMs in Chap. 5, in this section, we mainly discuss the valuable properties of knowledge transfer owned by PTMs.

9.4.5 Summary

In this section, we present several ways in which knowledge is used to guide NLP models. Depending on the location of model learning where knowledge steps in, we group the guidance from knowledge into four categories: (1) *knowledge augmentation*, where knowledge is introduced to augment the input data, (2) *knowledge reformulation*, where special model modules are designed to interact with knowledge, (3) *knowledge regularization*, where knowledge does not directly intervene the forward pass of the model but acts as a regularizer, and (4) *knowledge transfer*, where knowledge helps narrow down the hypothesis space to achieve more efficient and effective model learning.

These approaches enable effective integration of knowledge into deep models, allowing models to leverage sufficient knowledge (especially symbolic knowledge) to better perform NLP tasks. Since knowledge is essential for models to understand and complete the NLP tasks, knowledge-guided NLP is a worthwhile area for researchers to continue to explore.

9.5 Knowledge Acquisition

The KBs used in early expert systems and the KGs built in recent years both have long relied on manual construction. Manually organizing knowledge ensures that knowledge systems are constructed with high quality but suffers from inefficiency, incompleteness, and inconsistency in the annotation process. As shown in Fig. 9.22,

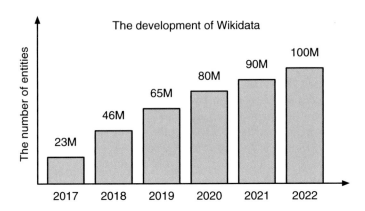

Fig. 9.22 The development trend of Wikidata from 2017 to 2022

the number of entities in the popular open-source KG Wikidata[4] grew at a rate of over 15 million per year from 2017 to 2022. At this rate of growth, it is unrealistic to rely solely on human annotation to organize large-scale human knowledge. Therefore, it is crucial to explore automatic knowledge acquisition, which can significantly better support knowledge representation learning and knowledge-guided NLP. In this section, taking KGs that store rich world knowledge as an example, we describe how to perform automatic knowledge acquisition to enrich the amount of knowledge for KGs.

Generally, we have several approaches to acquiring knowledge. Knowledge graph completion (KGC) and RE are two typical approaches. As shown in Fig. 9.23, KGC aims to obtain new knowledge by reasoning over the internal structure of KGs. For example, given the triplet ⟨*Mark Twain*, Place of Birth, *Florida*⟩ and the triplet ⟨*Florida*, City of, *U.S.A*⟩, we can easily infer the fact ⟨*Mark Twain*, Citizenship, *U.S.A*⟩. Different from KGC that infers new knowledge based on the internal information of KGs, RE focuses on detecting relations between entities from external plain text. For example, given the sentence *Mark Twain was an American author and humorist*, we can get the triplet ⟨*Mark Twain*, Citizenship, *U.S.A*⟩ from the semantic information of the sentence. Since the text is the core carrier of human knowledge, RE can obtain more and broader knowledge than KGC. Moreover, KGC highly relies on the knowledge representation learning methods that we have introduced in the previous Sect. 9.3. Therefore, in this section, we only introduce knowledge acquisition by using RE as an example.

As RE is an important way to acquire knowledge, many researchers have devoted extensive efforts to this field in the past decades. Various statistical RE methods based on feature engineering [75, 208], kernel models [18, 27], and probabilistic

---

[4] https://www.wikidata.org.

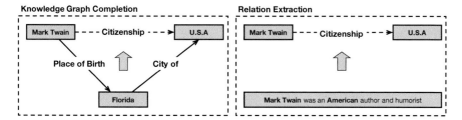

Fig. 9.23 An example of knowledge graph completion and relation extraction

graphical models [133, 134] have been proposed and achieved promising results. With the development of deep learning, neural networks as a powerful tool for encoding semantics have further advanced the development of RE, including recursive neural networks [110, 144], convolutional neural networks [92, 197], recurrent neural networks [115, 201], and graph neural networks [204, 210]. Considering that neural networks have become the backbone of NLP research in recent years, we focus on introducing knowledge acquisition with neural RE models in this section. For those statistical methods, some surveys [121, 195] can provide sufficient details about them. Next, we present how to acquire knowledge in various complex textual scenarios around neural RE, including sentence-level methods, bag-level methods, document-level methods, few-shot methods, and contextualized methods.

9.5.1 Sentence-Level Relation Extraction

Sentence-level RE is the basis for acquiring knowledge from text to enrich KGs. As shown in Fig. 9.24, sentence-level RE is based on the sentence-level semantics to extract relations between entities. Formally, given an input sentence $s = \{w_1, w_2, \cdots, w_n\}$ consisting of n words and an entity pair (e_1, e_2) in the sentence, sentence-level RE aims to obtain the probability distribution $P(r|s, e_1, e_2)$ over the relation set $\mathcal{R}$ $(r \in \mathcal{R})$. Based on $P(r|s, e_1, e_2)$, we can infer all relations between e_1 and e_2.

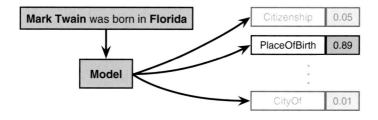

Fig. 9.24 An example of sentence-level relation extraction

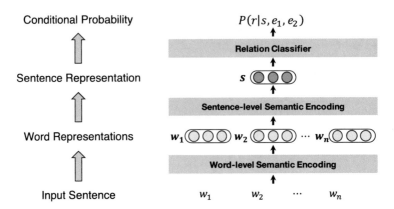

Fig. 9.25 The process of encoding sentence-level semantic information to detect the relations expressed by a given sentence

Learning an effective model to measure $P(r|s, e_1, e_2)$ requires efforts of three different aspects. As shown in Fig. 9.25, the first is to encode the input words into informative word-level features $\{\mathbf{w}_1, \mathbf{w}_2, \cdots, \mathbf{w}_n\}$ that can well serve the relation classification process. The second is to train a sentence encoder, which can well encode the word-level features $\{\mathbf{w}_1, \mathbf{w}_2, \cdots, \mathbf{w}_n\}$ into the sentence-level feature $\mathbf{s}$ with respect to the entity pair (e_1, e_2). The third is to train a classifier that can well compute the conditional probability distribution $P(r|s, e_1, e_2)$ over all relations in $\mathcal{R}$ based on the sentence-level feature $\mathbf{s}$. Next, we will present some typical works in each of these three aspects.

Word-Level Semantics Encoding Given the sentence $s = \{w_1, w_2, \cdots, w_n\}$ and the entity pair (e_1, e_2), before encoding sentence semantics and further classifying relations, we have to project the discrete words of the source sentence s into a continuous vector space to get the input representation $\mathbf{w} = \{\mathbf{w}_1, \mathbf{w}_2, \cdots, \mathbf{w}_n\}$. In general, widely used word-level features include the following components:

Word Embeddings Word embeddings aim to encode the syntactic and semantic information of words into distributed representations, i.e., each word is represented by a vector. Word embeddings are the basis for encoding word-level semantics, and word2vec [108] and GloVe [123] are the most common ways to obtain word embeddings.

Position Embeddings Position embeddings aim to encode which input words belong to the target entities and how close each word is to the target entities. Specifically, for each word w_i, its position embedding is formalized as the combination of the relative distances from w_i to e_1 and e_2. For instance, given the sentence *Mark Twain was an American author and humorist*, the relative distance from the word *was* to the entity *Mark Twain* is -1, and the distance to the entity *American* is 2. The relative distances -1 and 2 are then encoded into the position embedding to provide a positional representation for the word *was*. Since RE highly relies on word-level positional information to capture entity-specific semantics, position embeddings are widely used in RE [135, 197, 201].

Part-of-Speech (POS) Tag Embeddings POS Tag Embeddings aim to encode the word-level lexical information (e.g., nouns, verbs, etc.) of the sentence. Formally, all words in the sentence are encoded into embeddings according to their POS tags, and these POS tag embeddings can serve as lexical complements for word embeddings and position embeddings [19, 183, 209].

Hypernym Embeddings Hypernym embeddings aim to leverage the prior knowledge of hypernyms in WordNet [43]. Compared to POS tags, hypernyms are finer-grained. WordNet is a typical linguistic KG. In WordNet, all words are grouped into sets of cognitive synonyms (synsets), and each synset can express a distinct concept. Hypernyms are defined among these synsets. Here is just a brief introduction to WordNet, and we will introduce linguistic knowledge in detail in Chap. 10. When given the hypernym information of each word in WordNet (e.g., noun.food, verb.motion, etc), it is easy to connect this word with other words that are different but conceptually similar. Similar to POS tag embeddings, each hypernym tag in WordNet has a tag-specific embedding, and each word in a sentence is encoded into a hypernym embedding based on the word-specific hypernym tag.

The above embeddings are usually concatenated together to obtain the final input features $\mathbf{w} = \{\mathbf{w}_1, \mathbf{w}_2, \cdots, \mathbf{w}_n\}$, and $\mathbf{w}$ is used to support further encoding sentence-level semantics.

Sentence-Level Semantics Encoding Based on word-level features, we introduce different sentence encoders to encode sentence-level semantic information for RE:

Convolutional Neural Network Encoders CNN encoders [135, 197] use convolutional layers to extract local features and then use pooling operations to encode all local features into a fixed-sized vector.

Here we take an encoder with only one convolutional layer and one max-pooling operation as an example. Given the word-level features $\{\mathbf{w}_1, \mathbf{w}_2, \cdots, \mathbf{w}_n\}$, the convolutional layer can be formalized as

$$\{\mathbf{h}_1, \mathbf{h}_2, \cdots, \mathbf{h}_n\} = \text{CNN}(\{\mathbf{w}_1, \mathbf{w}_2, \cdots, \mathbf{w}_n\}), \tag{9.73}$$

where $\text{CNN}(\cdot)$ indicates the convolution operation inside the convolutional layer, $\mathbf{h}_i$ is the hidden state of the i-th word, and we have introduced this part in Chap. 4. Then, the sentence representation $\mathbf{s}$ is obtained by using the max-pooling operation, where the i-th element of $\mathbf{s}$ is given as

$$[\mathbf{s}]_i = \max_{1 \leq j \leq n} [\mathbf{h}_j]_i, \tag{9.74}$$

where $[\cdot]_i$ is the i-th element of the vector.

Further, PCNN [196], which is a variant of CNN, adopts a piecewise max-pooling operation. All hidden states $\{\mathbf{p}_1, \mathbf{p}_2, \cdots \mathbf{p}_n\}$ are divided into three parts by the positions of e_1 and e_2. The max-pooling operation is performed on the three segments respectively, and s is the concatenation of the three pooling results.

Recurrent Neural Network Encoders RNN encoders [201] use recurrent layers to learn temporal features on the input sequence. Given the word-level features $\{\mathbf{w}_1, \mathbf{w}_2, \cdots, \mathbf{w}_n\}$, each input word feature is fed into recurrent layers step by step. For the i-th step, the network takes $\mathbf{w}_i$ and the hidden state of the last step $\mathbf{h}_{i-1}$ as input, and the whole process is given as

$$\mathbf{h}_i = \text{RNN}(\mathbf{w}_i, \mathbf{h}_{i-1}), \tag{9.75}$$

where $\text{RNN}(\cdot)$ indicates the RNN function, which can be a LSTM unit or a GRU unit mentioned in Chap. 4.

The conventional recurrent models typically encode sequences from start to end and build the hidden state of each step only considering its preceding steps. Besides unidirectional RNNs, bidirectional RNNs [137] are also adopted to encode sentence-level semantics, and the whole process is given as

$$\overleftarrow{\mathbf{h}}_i = \overleftarrow{\text{RNN}}(\mathbf{w}_i, \overleftarrow{\mathbf{h}}_{i+1}), \; \overrightarrow{\mathbf{h}}_i = \overrightarrow{\text{RNN}}(\mathbf{w}_i, \overrightarrow{\mathbf{h}}_{i-1}), \mathbf{h}_i = [\overleftarrow{\mathbf{h}}_i; \overrightarrow{\mathbf{h}}_i], \tag{9.76}$$

where $[\cdot; \cdot]$ is the concatenation of two vectors.

Similar to the abovementioned convolutional models, the recurrent models also use pooling operations to extract the global sentence feature s, which forms the representation of the whole input sentence. For example, we can use a max-pooling operation to obtain s:

$$[\mathbf{s}]_i = \max_{1 \le j \le n} [\mathbf{h}_j]_i. \tag{9.77}$$

Besides pooling operations, attention operations [3] can also combine all local features. Specifically, given the output states $\mathbf{H} = [\mathbf{h}_1, \mathbf{h}_2, \cdots, \mathbf{h}_n]$ produced by a recurrent models, s can be formalized as

$$\alpha = \text{Softmax}(\mathbf{q}^\top f(\mathbf{H})), \quad \mathbf{s} = \mathbf{H}\alpha^\top, \tag{9.78}$$

where $\mathbf{q}$ is a learnable query vector and $f(\cdot)$ is an attention transformation function.

Moreover, some works [110] propose to encode semantics from both the word sequence and tree-structured dependency of a sentence by stacking bidirectional path-based recurrent neural networks. More specifically, these path-based methods mainly consider the shortest path between entities in the dependency tree, and utilize stacked layers to encode the path as the sentence representation. Some preliminary works [182] have shown that these paths are informative in RE and proposed various recursive neural models for this. Next, we will introduce these recursive models in detail.

Recursive Neural Network Encoders Recursive encoders aim to extract features based on syntactic parsing trees, considering that the syntactic information between target entities in a sentence can benefit classifying their relations. Generally, these encoders utilize the parsing tree structure as the composition direction to integrate word-level features into sentence-level features. Socher et al. [144] introduce a recursive matrix-vector model that can capture the structure information by assigning a matrix-vector representation for each constituent in parsing trees. In Socher's model, the vector can represent the constituent, and the matrix can represent how the constituent modifies the word meaning it is combined with.

Tai et al. [151] further propose two tree-structured models, the Child-Sum Tree-LSTM and the N-ary Tree-LSTM. Given the parsing tree of a sentence, the transition equations of the Child-Sum Tree-LSTM are defined as

$$\mathbf{h}_t = \sum_{k \in C(t)} \text{TLSTM}(\mathbf{h}_k), \qquad (9.79)$$

where $C(t)$ is the children set of the node t, $\text{TLSTM}(\cdot)$ indicates a Tree-LSTM cell, which is simply modified from the LSTM cell, and the hidden states of the leaf nodes are the input features. The transition equations of the N-ary Tree-LSTM are similar to the transition equations of Child-Sum Tree-LSTM. The main difference is that the N-ary Tree-LSTM limits the tree structures to have at most N branches. More details of recursive neural networks can be found in Chap. 4.

Sentence-Level Relation Classification After obtaining the representation $\mathbf{s}$ of the input sentence s, we require a relation classifier to compute the conditional probability $P(r|s, e_1, e_2)$. Generally, $P(r|s, e_1, e_2)$ can be obtained with

$$P(r|s, e_1, e_2) = \text{Softmax}(\mathbf{Ms} + \mathbf{b}), \qquad (9.80)$$

where $\mathbf{M}$ is the relation matrix consisting of relation embeddings and $\mathbf{b}$ is a bias vector. Intuitively, using a Softmax layer to compute the conditional probability means that an entity pair has only one corresponding relation. However, sometimes multiple relations may exist between an entity pair. To this end, for each relation $r \in \mathcal{R}$, some works perform relation-specific binary classification:

$$P(r|s, e_1, e_2) = \text{Sigmoid}(\mathbf{r}^\top \mathbf{s} + b_r), \qquad (9.81)$$

where $\mathbf{r}$ is the relation embedding of r and b_r is a relation-specific bias value.

9.5.2 Bag-Level Relation Extraction

Although existing neural methods have achieved promising results in sentence-level RE, these neural methods still suffer from the problem of data scarcity since manu-

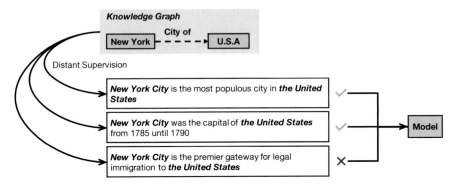

Fig. 9.26 An example of bag-level relation extraction

ally annotating training data is time-consuming and labor-intensive. To alleviate this problem, distant supervision [109] has been introduced to automatically annotate training data by aligning existing KGs and plain text. The main idea of distant supervision is that sentences containing two entities may describe the relations of the two entities recorded in KGs. As shown in Fig. 9.26, given (*New York*, City of, *U.S.A*), the distant supervision assumption regards all sentences that contain *New York* and *U.S.A* as positive instances for the relation City of. Besides providing massive training data, distant supervision also naturally provides a way to detect the relations between two given entities based on multiple sentences (bag-level) rather than a single sentence (sentence-level).

Therefore, bag-level RE aims to predict the relations between two given entities by considering all sentences containing these entities, by highlighting those informative examples and filtering out noisy ones. As shown in Fig. 9.26, given the input sentence set $S = \{s_1, s_2, \cdots, s_m\}$ and an entity pair (e_1, e_2) contained by these sentences, bag-level RE methods aim to obtain the probability $P(r|S, e_1, e_2)$ over the relation set.

As shown in Fig. 9.27, learning an effective model to measure $P(r|S, e_1, e_2)$ requires efforts from three different aspects: encoding sentence-level semantics (including encoding word-level semantics), encoding bag-level semantics, and finally classifying relations. Since encoding word-level semantics and sentence-level semantics have been already introduced in sentence-level RE, we mainly focus on introducing how to encode bag-level semantics here.

Bag-Level Semantics Encoding For bag-level RE, we need to encode bag-level semantics based on sentence-level representations. Formally, given a sentence bag $S = \{s_1, s_2, \cdots, s_m\}$, each sentence s_i has its own sentence representation $\mathbf{s}_i$; a bag-level encoder encodes all sentence representations into a single bag representation $\hat{\mathbf{s}}$. Next, we will introduce some typical bag-level encoders as follows:

Max Encoders Max encoders aim to select the most confident sentence in the bag S and use the representation of the selected sentence as the bag representation,

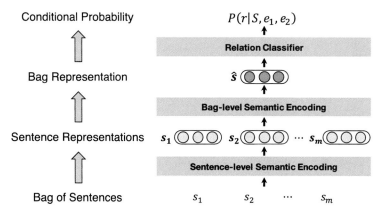

Fig. 9.27 The process of obtaining bag-level semantic information to detect the relations described by a given sentence bag

considering not all sentences containing an entity pair can express the relations between the entity pair. For instance, given *New York City is the premier gateway for legal immigration to the United States*, the sentence does not highly express the relation `City of`. To this end, the at-least-one assumption [196] has been proposed, assuming that at least one sentence containing target entities can express their relations. With the at-least-one assumption, the sentence with the highest probability for a specific relation is selected to represent the bag S. Formally, the bag representation is given as

$$\hat{\mathbf{s}} = \mathbf{s}_{i^*}, \quad i^* = \arg\max_i P(r|s_i, e_1, e_2). \tag{9.82}$$

Average Encoders Average encoders use the average of all sentence vectors to represent the bag. Max encoders use only one sentence in the bag as the bag representation, ignoring the rich information and correlation among different sentences in the bag. To take advantage of all sentences, Lin et al. [91] make the bag representation $\hat{\mathbf{s}}$ depends on the representations of all sentences in the bag. The average encoder assumes all sentences contribute equally to the bag representation $\hat{\mathbf{s}}$:

$$\hat{\mathbf{s}} = \sum_i \frac{1}{m} \mathbf{s}_i. \tag{9.83}$$

Attentive Encoders Attentive encoders use attention operations to aggregate all sentence vectors. Considering the inevitable mislabeling problem introduced by distant supervision, average encoders may be affected by those mislabeled sentences. To address this problem, Lin et al. [91] further propose a sentence-level selective attention to reduce the side effect of mislabeled data. Formally, with the attention operation, the bag representation $\hat{\mathbf{s}}$ is defined as

$$\alpha = \text{Softmax}(\mathbf{q}_r^\top f(\mathbf{S})), \quad \hat{\mathbf{s}} = \mathbf{S}\alpha^\top, \tag{9.84}$$

where $\mathbf{S} = \{\mathbf{s}_1, \mathbf{s}_2, \cdots, \mathbf{s}_m\}$, $f(\cdot)$ is an attention transformation function and and $\mathbf{q}_r$ is the query vector of the relation r used for the attention operation. To further improve the attention operation, more sophisticated mechanisms, such as knowledge-enhanced strategies [58, 72, 199], soft-labeling strategies [95], reinforcement learning [44, 200], and adversarial training [171], have been explored to build more effective attention operations.

Bag-Level Relation Classification Similar to sentence-level methods, when obtaining the bag representation $\hat{\mathbf{s}}$, the probability $P(r|\mathcal{S}, e_1, e_2)$ is computed as

$$P(r|\mathcal{S}, e_1, e_2) = \text{Softmax}(\mathbf{M}\hat{\mathbf{s}} + \mathbf{b}), \tag{9.85}$$

where $\mathbf{M}$ is the relation matrix consisting of relation embeddings and $\mathbf{b}$ is a bias vector. For those methods performing relation-specific binary classification, the relation-specific conditional probability is given by

$$P(r|\mathcal{S}, e_1, e_2) = \text{Sigmoid}(\mathbf{r}^\top \hat{\mathbf{s}} + b_r), \tag{9.86}$$

where $\mathbf{r}$ is the relation embedding of r and b_r is a relation-specific bias value.

9.5.3 Document-Level Relation Extraction

For RE, not all relational facts can be acquired by sentence-level or bag-level methods. Many facts are expressed across multiple sentences in a document. As shown in Fig. 9.28, a document may contain multiple entities that interact with each other in a complex way. If we want to get the fact that ⟨*Riddarhuset*, City of, *Sweden*⟩, we first have to detect *Riddarhuset* is located in *Stockholm* from the fourth sentence, and then detect *Stockholm* is the capital of *Sweden* and *Sweden* is a country from the first sentence. From these three facts, we can finally infer that the sovereign state of *Riddarhuset* is *Sweden*.

Performing reading and reasoning over multiple sentences becomes important, which is intuitively hard to reach for both sentence-level and bag-level methods. According to the statistics of a human-annotated dataset sampled from Wikipedia [188], more than 40% facts require considering the semantics of multiple sentences for their extraction, which is not negligible. Some works [150, 155] that focus on document-level RE also report similar observations. Hence, it is crucial to advance RE from the sentence level to the document level.

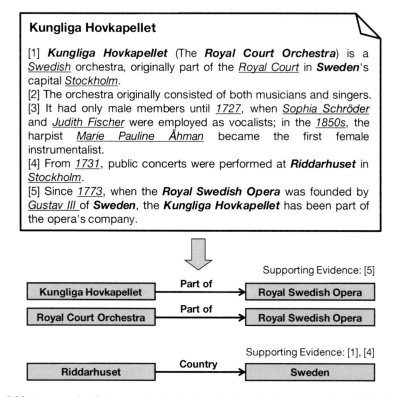

Kungliga Hovkapellet

[1] *Kungliga Hovkapellet* (The *Royal Court Orchestra*) is a *Swedish* orchestra, originally part of the *Royal Court* in *Sweden*'s capital *Stockholm*.
[2] The orchestra originally consisted of both musicians and singers.
[3] It had only male members until *1727*, when *Sophia Schröder* and *Judith Fischer* were employed as vocalists; in the *1850s*, the harpist *Marie Pauline Åhman* became the first female instrumentalist.
[4] From *1731*, public concerts were performed at **Riddarhuset** in *Stockholm*.
[5] Since *1773*, when the **Royal Swedish Opera** was founded by *Gustav III* of **Sweden**, the **Kungliga Hovkapellet** has been part of the opera's company.

Supporting Evidence: [5]

| Kungliga Hovkapellet | Part of → | Royal Swedish Opera |

| Royal Court Orchestra | Part of → | Royal Swedish Opera |

Supporting Evidence: [1], [4]

| Riddarhuset | Country → | Sweden |

Fig. 9.28 An example of document-level relation extraction from the dataset DocRED [188]

Due to being much more complex than sentence-level and bag-level RE, document-level RE remains an open problem in terms of benchmarking and methodology. For benchmarks that can evaluate the performance of document-level RE, existing datasets either only have a few manually annotated examples [88], or have noisy distantly supervised annotations [122, 129], or serve only a specific domain [85]. To address this issue, Yao et al. [188] manually annotate a large-scale and general-purpose dataset to support the evaluation of document-level RE methods, named DocRED. DocRED is built based on Wikipedia and Wikidata, and it has two main features. First, DocRED is the largest human-annotated document-level RE dataset, containing 132, 375 entities and 56, 354 facts. Second, nearly half of the facts in DocRED can only be extracted from multiple sentences. This makes it necessary for models to read multiple sentences in a document to identify entities and infer their relations by considering the holistic semantic information of the document.

Document-Level RE Methods The preliminary results on DocRED show that existing sentence-level methods cannot work well on DocRED, indicating that document-level RE is more challenging than sentence-level RE. Many efforts have been devoted to document-level RE based on DocRED.

PTM-Based Methods Wang et al. [158] use PTMs as a backbone to build an effective document-level RE model, considering that recently proposed PTMs show the ability to encode long sequences. Although PTMs can effectively capture contextual semantic information from plain text for document-level RE, PTMs still cannot explicitly handle coreference, which is critical for modeling interactions between entities. Ye et al. [190] introduce a PTM that captures coreference relations between entities to improve document-level RE.

Graph-Based Methods Nan et al. [112] construct document-level graphs based on syntactic trees, coreferences, and some human-designed heuristics to model dependencies in documents. To better model document-level graphs, Zeng et al. [198] construct a path reasoning mechanism using graph neural networks to infer relations between entities.

Document-Level Distant Supervision Some works [128, 172] also propose to leverage document-level distant supervision to learn entity and relation representations. Based on well-designed heuristic rules, these distantly supervised methods perform effective data augmentation for document-level RE.

Cross-Document RE In addition to abovementioned methods for RE within one document, Yao et al. [187] further propose CodRED which aims to acquire knowledge from multiple documents. Cross-document RE presents two main challenges: (1) Given an entity pair, models need to retrieve relevant documents to establish multiple reasoning paths. (2) Since the head and tail entities come from different documents, models need to perform cross-document reasoning via bridging entities to resolve the relations. To support the research, CodRED provides a large-scale human-annotated dataset, which contains 30,504 relational facts, as well as the associated reasoning text paths and evidence. Experiments show that CodRED is challenging for existing RE methods. In summary, acquiring knowledge from documents has drawn increasing attention from the community, and is still a promising direction worth further exploration.

9.5.4   Few-Shot Relation Extraction

As we mentioned before, the performance of conventional RE methods heavily relies on annotated data. Annotating large-scale data is time-consuming and labor-intensive. Although distant supervision can alleviate this issue, the distantly supervised data also exhibits a long-tail distribution, i.e., most relations have very limited instances. In addition, distant supervision suffers from mislabeling, which makes the classification of long-tail relations more difficult. Hence, it is necessary to study training RE models with few-shot training instances. Figure 9.29 is an example of few-shot RE.

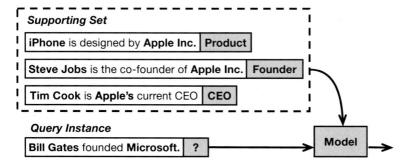

Fig. 9.29 An example of few-shot relation extraction

Few-Shot RE Methods FewRel [62] is a large-scale supervised dataset for few-shot RE, which requires models to handle relation classification with limited training instances. Based on FewRel, some works explore few-shot RE and achieve promising results.

Meta Learning and Metric Learning Methods Han et al. [62] demonstrate that meta learning and metric learning can be well used for few-shot RE. Following the direction of meta learning, Dong et al. [35] propose to leverage meta-information of relations (e.g., relation names and aliases) to guide the initialization and fast adaptation of meta learning for few-shot RE. Following the direction of metric learning, based on the prototypical neural network [141], which is a typical metric learning approach, many more effective metric learning methods [45, 191] are proposed for few-shot RE.

PTM-Based Methods Soares et al. [142] utilize PTMs to handle few-shot RE and show surprising results. On the one hand, the use of PTMs can transfer the knowledge captured from unlabeled data to help solve the problem of data scarcity. On the other hand, Soares et al. introduce contrastive learning based on PTMs, which can be seen as a more effective metric learning method. On FewRel, Soares' model can achieve comparable results to human performance.

Domain Adaptation and Out-of-Distribution Detection After FewRel, Gao et al. [46] propose some more challenging few-shot scenarios, including domain adaptation and out-of-distribution detection. Gao et al. build a more challenging dataset FewRel 2.0 and use sufficient experimental results on FewRel 2.0 to show that the state-of-the-art few-shot methods struggle in these two scenarios, and the commonly used techniques for domain adaptation and out-of-distribution detection cannot handle the two challenges well. These findings call for more attention and further efforts to few-shot RE, which is still a challenging open problem.

9.5.5   Open-Domain Relation Extraction

Most RE systems regard the task as relation classification and can only deal with pre-defined relation types. However, relation types in the real-world corpora are typically in rapid growth. For example, the number of relation types in Wikidata grows to over 10,000 in 6 years from 2017 to 2022 [202]. Therefore, handling emerging relations in the open-domain scenario is a challenging problem for RE. Existing methods for open-domain RE can be divided into three categories: extracting open relation phrases, clustering open relation types, and learning with increasing relation types.

Extracting Open Relation Phrases   Open information extraction (OpenIE) aims to extract semi-structured relation phrases [39, 40]. Since the relations are treated as free-form text from the sentences, OpenIE can deal with relations that are not pre-defined. For OpenIE, the traditional statistical methods typically design heuristic rules (e.g., syntactic and lexical constraints) to identify relation phrase candidates and filter out noisy ones via a relation discriminator [41, 169, 189]. Neural OpenIE methods typically learn to generate relation phrases in an encoder-decoder architecture [26, 77], or identify relation phrases in the sentence via sequence labeling [145]. The supervision for neural OpenIE models typically comes from the high-confidence results from the statistical methods. The advantage of OpenIE is that minimal human efforts are required in both relation type design and relation instance annotation. The relational phrases also exhibit good readability to humans. However, due to the diversity of natural language, the same relation type can have different surface forms in different sentences. Therefore, linking various surface forms of relation phrases to the standardized relation types could be difficult.

Clustering Open Relation Types   Open relation extraction (OpenRE) aims to discover new relation types by clustering relational instances into groups. OpenRE methods typically learn discriminative representations for relational instances and cluster these open-domain instances into groups. Compared with OpenIE, OpenRE aims at clustering new types that are out of existing relation types, yet OpenIE only focuses on representing relations with language phrases to get rid of pre-defined types. Generally, the results of OpenIE can be used to support the clustering of OpenRE. Elsahar et al. [38] make an initial attempt to obtain relational instance representations through rich features, including entity types and re-weighted word embeddings, and cluster these handcrafted representations to discover new relation types. Then, some works [67, 103] propose to improve the learning and clustering of relational instance representations by using effective self-supervised signals. Notably, Wu et al. [170] propose to transfer relational knowledge from the supervised data of existing relations to the unsupervised data of open relations. Given labeled relational instances of existing relations, Wu et al. use a relational Siamese network to learn a metric space. Then, the metric space is transferred to measure the similarities of unlabeled sentences, based on which the clustering is performed. Inspired by Wu et al. [170], Zhang et al. [202] further leverage relation hierarchies to learn more discriminative metric space for the clustering of relational instance representations, where the instances of the nearby relations on hierarchies are encouraged to share similar representations. Moreover, since relational instance representations contain rich hierarchical information, the newly discovered relations can be directly appended to the existing relation hierarchy. OpenRE can deal with the diversity of relation surface forms by clustering. However, the specific semantics of relation clusters still needs to be summarized through human efforts.

Learning with Increasing Relation Types After discovering novel relations from the open corpora, the relation classifier needs to be updated to deal with both existing and new relations. A straightforward approach is to re-train the relation classifier using all the instances of existing and new relations together from scratch whenever new relations emerge. However, the approach is not feasible due to the high computational cost. Continual relation learning aims to utilize the instances of novel relations to update a relation classifier continually. A significant challenge of continual relation learning is the catastrophic forgetting [159], where the performance on the existing relations can degrade significantly after training with new relations. To address this problem, some works propose saving several instances for existing classes and re-training the classifier with these memorized instances and new data together [30, 159]. This learning process based on memorized instances is named *memory replay*. However, repeatedly updating the classifier with several memorized instances may cause overfitting of existing relations. Drawing inspirations from the study of mammalian memory in neuroscience [10], EMAR [56] proposes episodic memory activation and reconsolidation mechanism to prevent the overfitting problem. The key idea is that the prototypes of the existing relations should remain discriminative after each time of replaying and activating memorized relation instances. In this way, EMAR can flexibly handle new relations without forgetting or overfitting existing relations.

9.5.6 Contextualized Relation Extraction

As mentioned above, RE systems have been significantly improved in supervised and distantly supervised scenarios. To further improve RE performance, many researchers are working on the contextualized RE by integrating multisource information. In this section, we will describe some typical approaches to contextualized RE in detail.

Utilizing External Information Most existing RE systems stated above only concentrate on the text, regardless of the rich external text-related heterogeneous information, like world knowledge in KGs, visual knowledge in images, and structured or semi-structured knowledge on the Web. Text-related heterogeneous information could provide rich additional context. As mentioned in Sect. 9.4.2, Han et al. [58] propose a joint learning framework for RE, the key idea of which is to jointly learn knowledge and text representations within a unified semantic space via KG-text alignments. In Han's work, for the text part, word and sentence representations are learned via a CNN encoder. For the KG part, entity and relation representations are learned via translation-based methods mentioned in Sect. 9.3.2. The learned representations of the KG and text parts are aligned during the training process, by using entity anchors to share word and entity representations as well as adopting mutual attention to make sentence representations and knowledge representations enhance each other. Apart from this preliminary attempt, many efforts have been devoted to this direction [72, 132, 164, 166].

Incorporating Relational Paths Although existing RE systems have achieved promising results, they still suffer from a major problem: the models can only directly learn from sentences containing both target entities. However, those sentences containing only one of the target entities could also provide helpful information and help build inference chains. For example, if we know that *Alexandre Dumas fils* is the son of *Alexandre Dumas* and *Alexandre Dumas* is the son of *Thomas-Alexandre Dumas*, we can infer that *Alexandre Dumas fils* is the grandson of *Thomas-Alexandre Dumas*. Zeng et al. [199] introduce a path-based RE model incorporating textual relational paths so as to utilize the information of both direct and indirect sentences. The model employs an encoder to represent the semantics of multiple sentences and then builds a relation path encoder to measure the probability distribution of relations given the inference path in text. Finally, the model combines information from both sentences and relational paths and predicts each relation's confidence. This work is the preliminary effort to consider the knowledge of relation paths in text RE. There are also several methods later to consider the reasoning paths of sentence semantic meanings for RE, such as using effective neural models like RNNs to learn relation paths [29], and using distant supervision to annotate implicit relation paths [49] automatically.

9.5.7 Summary

In this section, we elaborate on the approaches to acquiring knowledge. Typically, we focus on RE, and classify RE methods into six groups according to their application scenarios: (1) *sentence-level RE*, which focuses on extracting relations from sentences, (2) *bag-level RE*, which focuses on extracting relations from the bags of sentences annotated by distant supervision, (3) *document-level RE*, which focuses on extracting relations from documents, (4) *few-shot RE*, which focuses on low-resource scenarios, (5) *open-domain RE*, which focuses on continually extracting open-domain relations that are not pre-defined, and (6) *contextualized RE*, which focuses on integrating multisource information for RE.

Note that knowledge acquisition does not just mean RE and includes many other methods, such as KGC, event extraction, etc. Moreover, not all human knowledge is represented in a textual form, and there is also a large amount of knowledge in images, audio, and other knowledge carriers. How to obtain knowledge from these carriers to empower models is also a problem worthy of further consideration by researchers.

9.6 Summary and Further Readings

We have now overviewed the current progress of using knowledge for NLP tasks, including knowledge representation learning, knowledge-guided NLP, and knowledge acquisition. In this last section, we will summarize the contents of this chapter and then provide more readings for reference.

Knowledge representation learning is a critical component of using knowledge since it bridges the gap between knowledge systems that store knowledge and applications that require knowledge. We systemically describe existing methods for knowledge representation learning. Further, we discuss several advanced approaches that deal with the current challenges of knowledge representation learning. For further understanding of knowledge graph representation learning, more related papers can be found in this paper list.[5] There are also some recommended surveys and books [6, 73, 102, 116, 161].

After introducing knowledge representation learning, we introduce the framework of knowledge-guided learning, aiming to improve NLP models with knowledge representations. The framework includes four important directions: knowledge augmentation, knowledge reformulation, knowledge regularization, and knowledge transfer. All these four directions of knowledge-guided learning have been widely advanced in the past few years. Following these four directions, we review typical cases to clarify this knowledge-guided framework, covering information extraction, information retrieval, language modeling, and text generation. Considering the breakthroughs of PTMs, we also use prompts as an example to show the recent trend of knowledge transfer in the era of PTMs. We suggest readers to find further insights from the recent surveys and books about knowledge and PTMs [9, 60, 193].

Based on knowledge representation learning and knowledge-guided learning, we introduce how to acquire more knowledge from plain text to enrich existing knowledge systems. We systematically review knowledge acquisition methods in various textual scenarios, including sentence-level, bag-level, document-level, few-shot, and open-domain acquisition. In this field, we refer further readings to the paper list[6] and the typical surveys [57, 73].

Acknowledgments The contributions of all authors for the second edition are the following: Zhiyuan Liu, Yankai Lin, and Maosong Sun designed the overall architecture of this chapter; Xu Han and Weize Chen drafted the chapter; Zhengyan Zhang and Yuan Yao participated in and proofread some parts of Sects. 9.4 and 9.5. Zhiyuan Liu and Yankai Lin proofread and revised this chapter.

---

[5] https://github.com/thunlp/KRLPapers.

[6] https://github.com/thunlp/NREPapers.

We thank Ruobing Xie for providing initial materials in the first edition and also thank Ning Ding, Chaojun Xiao, Shengding Hu, Xinrong Zhang, Qimin Zhan, Bowen Li, and Shihao Liang for proofreading this chapter.

This chapter is the knowledge representation learning chapter of the second edition of the book *Representation Learning for Natural Language Processing*, with its first edition published in 2020 [101]. As compared to the first edition of this chapter, the main changes include the following: (1) we added the section Knowledge-guided NLP and the section Knowledge Acquisition, and (2) we comprehensively supplemented and updated the information, discussions, examples, and figures in other existing sections.

References

1. Sungjin Ahn, Heeyoul Choi, Tanel Pärnamaa, and Yoshua Bengio. A neural knowledge language model. *arXiv preprint arXiv:1608.00318*, 2016.
2. Barr Avron and Edward A Feigenbaum. *The handbook of artificial intelligence*. Addison-Wesley, 1981.
3. Dzmitry Bahdanau, Kyunghyun Cho, and Yoshua Bengio. Neural machine translation by jointly learning to align and translate. In *Proceedings of ICLR*, 2015.
4. Ivana Balažević, Carl Allen, and Timothy Hospedales. Multi-relational poincaré graph embeddings. In *Proceedings of NeurIPS*, 2019.
5. Islam Beltagy and Raymond J Mooney. Efficient markov logic inference for natural language semantics. In *Proceedings of AAAI Workshop*, 2014.
6. Yoshua Bengio, Aaron Courville, and Pascal Vincent. Representation learning: A review and new perspectives. *IEEE Transactions on Pattern Analysis and Machine Intelligence*, 2013.
7. Tim Berners-Lee, James Hendler, and Ora Lassila. The semantic web. *Scientific american*, 284(5):34–43, 2001.
8. Kurt Bollacker, Colin Evans, Praveen Paritosh, Tim Sturge, and Jamie Taylor. Freebase: a collaboratively created graph database for structuring human knowledge. In *Proceedings of KDD*, 2008.
9. Rishi Bommasani, Drew A Hudson, Ehsan Adeli, Russ Altman, Simran Arora, Sydney von Arx, Michael S Bernstein, Jeannette Bohg, Antoine Bosselut, Emma Brunskill, et al. On the opportunities and risks of foundation models. *arXiv preprint arXiv:2108.07258*, 2021.
10. Bruno Bontempi, Catherine Laurent-Demir, Claude Destrade, and Robert Jaffard. Time-dependent reorganization of brain circuitry underlying long-term memory storage. *Nature*, 400(6745):671–675, 1999.
11. Antoine Bordes, Xavier Glorot, Jason Weston, and Yoshua Bengio. Joint learning of words and meaning representations for open-text semantic parsing. In *Proceedings of AISTATS*, 2012.
12. Antoine Bordes, Nicolas Usunier, Alberto Garcia-Duran, Jason Weston, and Oksana Yakhnenko. Translating embeddings for modeling multi-relational data. In *Proceedings of NeurIPS*, 2013.
13. Antoine Bordes, Jason Weston, Ronan Collobert, and Yoshua Bengio. Learning structured embeddings of knowledge bases. In *Proceedings of AAAI*, 2011.
14. Bernhard E Boser, Isabelle M Guyon, and Vladimir N Vapnik. A training algorithm for optimal margin classifiers. In *Proceedings of COLT*, 1992.
15. David Bostock. *Plato's theaetetus*. Oxford University Press, 1988.
16. Leo Breiman, Jerome Friedman, Charles J Stone, and Richard A Olshen. *Classification and regression trees*. CRC Press, 1984.
17. Tom Brown, Benjamin Mann, Nick Ryder, Melanie Subbiah, Jared D Kaplan, Prafulla Dhariwal, Arvind Neelakantan, Pranav Shyam, Girish Sastry, Amanda Askell, et al. Language models are few-shot learners. In *Proceedings of NeurIPS*, 2020.

18. Razvan C Bunescu and Raymond J Mooney. A shortest path dependency kernel for relation extraction. In *Proceedings of EMNLP*, 2005.
19. Rui Cai, Xiaodong Zhang, and Houfeng Wang. Bidirectional recurrent convolutional neural network for relation classification. In *Proceedings of ACL*, 2016.
20. Andrew Carlson, Justin Betteridge, Richard C Wang, Estevam R Hruschka Jr, and Tom M Mitchell. Coupled semi-supervised learning for information extraction. In *Proceedings of WSDM*, 2010.
21. Mohamed Chabchoub, Michel Gagnon, and Amal Zouaq. Collective disambiguation and semantic annotation for entity linking and typing. In *Proceedings of SWEC*, 2016.
22. Gang Chen, Maosong Sun, and Yang Liu. Towards a universal continuous knowledge base. *AI Open*, 2:197–204, 2021.
23. Weize Chen, Xu Han, Yankai Lin, Hexu Zhao, Zhiyuan Liu, Peng Li, Maosong Sun, and Jie Zhou. Fully hyperbolic neural networks. In *Proceedings of ACL*, 2022.
24. Noam Chomsky. *Syntactic structures*. De Gruyter, 1957.
25. Noam Chomsky. *Aspects of the Theory of Syntax*. MIT Press, 1965.
26. Lei Cui, Furu Wei, and Ming Zhou. Neural open information extraction. In *Proceedings of ACL*, 2018.
27. Aron Culotta and Jeffrey Sorensen. Dependency tree kernels for relation extraction. In *Proceedings of ACL*, 2004.
28. Zhuyun Dai, Chenyan Xiong, Jamie Callan, and Zhiyuan Liu. Convolutional neural networks for soft-matching n-grams in ad-hoc search. In *Proceedings of WSDM*, 2018.
29. Rajarshi Das, Arvind Neelakantan, David Belanger, and Andrew McCallum. Chains of reasoning over entities, relations, and text using recurrent neural networks. In *Proceedings of EACL*, 2017.
30. Cyprien de Masson D'Autume, Sebastian Ruder, Lingpeng Kong, and Dani Yogatama. Episodic memory in lifelong language learning. In *Proceedings of NeurIPS*, 2019.
31. Jia Deng, Wei Dong, Richard Socher, Li-Jia Li, Kai Li, and Li Fei-Fei. ImageNet: A large-scale hierarchical image database. In *Proceedings of CVPR*, 2009.
32. Tim Dettmers, Pasquale Minervini, Pontus Stenetorp, and Sebastian Riedel. Convolutional 2d knowledge graph embeddings. In *Proceedings of AAAI*, 2018.
33. Jacob Devlin, Ming-Wei Chang, Kenton Lee, and Kristina Toutanova. BERT: pre-training of deep bidirectional transformers for language understanding. In *Proceedings of NAACL-HLT*, 2019.
34. Ning Ding, Yujia Qin, Guang Yang, Fuchao Wei, Zonghan Yang, Yusheng Su, Shengding Hu, Yulin Chen, Chi-Min Chan, Weize Chen, et al. Delta tuning: A comprehensive study of parameter efficient methods for pre-trained language models. *arXiv preprint arXiv:2203.06904*, 2022.
35. Bowen Dong, Yuan Yao, Ruobing Xie, Tianyu Gao, Xu Han, Zhiyuan Liu, Fen Lin, Leyu Lin, and Maosong Sun. Meta-information guided meta-learning for few-shot relation classification. In *Proceedings of COLING*, 2020.
36. Li Dong, Furu Wei, Hong Sun, Ming Zhou, and Ke Xu. A hybrid neural model for type classification of entity mentions. In *Proceedings of IJCAI*, 2015.
37. Xin Dong, Evgeniy Gabrilovich, Geremy Heitz, Wilko Horn, Ni Lao, Kevin Murphy, Thomas Strohmann, Shaohua Sun, and Wei Zhang. Knowledge vault: A web-scale approach to probabilistic knowledge fusion. In *Proceedings of KDD*, 2014.
38. Hady Elsahar, Elena Demidova, Simon Gottschalk, Christophe Gravier, and Frederique Laforest. Unsupervised open relation extraction. In *Proceedings of ESWC*, 2017.
39. Oren Etzioni, Michele Banko, Stephen Soderland, and Daniel S Weld. Open information extraction from the web. *Communications of the ACM*, 51(12):68–74, 2008.
40. Oren Etzioni, Anthony Fader, Janara Christensen, Stephen Soderland, et al. Open information extraction: The second generation. In *Proceedings of IJCAI*, 2011.
41. Anthony Fader, Stephen Soderland, and Oren Etzioni. Identifying relations for open information extraction. In *Proceedings of EMNLP*, 2011.

42. Edward A Feigenbaum. Knowledge engineering: The applied side of artificial intelligence. Technical report, Computer Science Department of Stanford University, 1980.
43. Christiane Fellbaum. Wordnet. *The encyclopedia of applied linguistics*, 2012.
44. Jun Feng, Minlie Huang, Li Zhao, Yang Yang, and Xiaoyan Zhu. Reinforcement learning for relation classification from noisy data. In *Proceedings of AAAI*, 2018.
45. Tianyu Gao, Xu Han, Zhiyuan Liu, and Maosong Sun. Hybrid attention-based prototypical networks for noisy few-shot relation classification. In *Proceedings of AAAI*, 2019.
46. Tianyu Gao, Xu Han, Hao Zhu, Zhiyuan Liu, Peng Li, Maosong Sun, and Jie Zhou. FewRel 2.0: Towards more challenging few-shot relation classification. In *Proceedings of EMNLP-IJCNLP*, 2019.
47. Alberto García-Durán, Antoine Bordes, and Nicolas Usunier. Composing relationships with translations. In *Proceedings of EMNLP*, 2015.
48. Robert Geirhos, Jörn-Henrik Jacobsen, Claudio Michaelis, Richard Zemel, Wieland Brendel, Matthias Bethge, and Felix A Wichmann. Shortcut learning in deep neural networks. *Nature Machine Intelligence*, 2(11):665–673, 2020.
49. Michael Glass, Alfio Gliozzo, Oktie Hassanzadeh, Nandana Mihindukulasooriya, and Gaetano Rossiello. Inducing implicit relations from text using distantly supervised deep nets. In *Proceedings of ISWC*, 2018.
50. Kelvin Gu, John Miller, and Percy Liang. Traversing knowledge graphs in vector space. In *Proceedings of EMNLP*, 2015.
51. Yihong Gu, Jun Yan, Hao Zhu, Zhiyuan Liu, Ruobing Xie, Maosong Sun, Fen Lin, and Leyu Lin. Language modeling with sparse product of sememe experts. In *Proceedings of EMNLP*, 2018.
52. Shu Guo, Quan Wang, Lihong Wang, Bin Wang, and Li Guo. Jointly embedding knowledge graphs and logical rules. In *Proceedings of EMNLP*, 2016.
53. Kelvin Guu, Kenton Lee, Zora Tung, Panupong Pasupat, and Mingwei Chang. Retrieval augmented language model pre-training. In *Proceedings of ICML*, 2020.
54. Michael Haenlein and Andreas Kaplan. A brief history of artificial intelligence: On the past, present, and future of artificial intelligence. *California management review*, 61(4):5–14, 2019.
55. Petr Hájek. *Metamathematics of fuzzy logic*. Springer Science & Business Media, 1998.
56. Xu Han, Yi Dai, Tianyu Gao, Yankai Lin, Zhiyuan Liu, Peng Li, Maosong Sun, and Jie Zhou. Continual relation learning via episodic memory activation and reconsolidation. In *Proceedings of ACL*, 2020.
57. Xu Han, Tianyu Gao, Yankai Lin, Hao Peng, Yaoliang Yang, Chaojun Xiao, Zhiyuan Liu, Peng Li, Jie Zhou, and Maosong Sun. More data, more relations, more context and more openness: A review and outlook for relation extraction. In *Proceedings of AACL-IJCNLP*, 2020.
58. Xu Han, Zhiyuan Liu, and Maosong Sun. Neural knowledge acquisition via mutual attention between knowledge graph and text. In *Proceedings of AAAI*, 2018.
59. Xu Han, Zhengyan Zhang, Ning Ding, Yuxian Gu, Xiao Liu, Yuqi Huo, Jiezhong Qiu, Yuan Yao, Ao Zhang, Liang Zhang, et al. Pre-trained models: Past, present and future. *AI Open*, 2021.
60. Xu Han, Zhengyan Zhang, and Zhiyuan Liu. Knowledgeable machine learning for natural language processing. *Communications of the ACM*, 64(11):50–51, 2021.
61. Xu Han, Zhengyan Zhang, and Zhiyuan Liu. Knowledge-guided pre-trained language models. *ZTE CommunicationsM*, 28(2):10–15, 2022.
62. Xu Han, Hao Zhu, Pengfei Yu, Ziyun Wang, Yuan Yao, Zhiyuan Liu, and Maosong Sun. FewRel: A large-scale supervised few-shot relation classification dataset with state-of-the-art evaluation. In *Proceedings of EMNLP*, 2018.
63. Trevor Hastie, Robert Tibshirani, Jerome H Friedman, and Jerome H Friedman. *The elements of statistical learning: data mining, inference, and prediction*, volume 2. Springer, 2009.
64. Frederick Hayes-Roth, Donald A Waterman, and Douglas B Lenat. *Building expert system*. Addison-Wesley, 1983.

65. Shizhu He, Kang Liu, Guoliang Ji, and Jun Zhao. Learning to represent knowledge graphs with gaussian embedding. In *Proceedings of CIKM*, 2015.
66. John Hewitt and Christopher D. Manning. A structural probe for finding syntax in word representations. In *Proceedings of NAACL-HLT*, 2019.
67. Xuming Hu, Lijie Wen, Yusong Xu, Chenwei Zhang, and S Yu Philip. Selfore: Self-supervised relational feature learning for open relation extraction. In *Proceedings of EMNLP*, 2020.
68. Peter Hylton. *Russell, idealism, and the emergence of analytic philosophy*. Oxford University Press, 1990.
69. Ganesh Jawahar, Benoît Sagot, and Djamé Seddah. What does BERT learn about the structure of language? In *Proceedings of ACL*, 2019.
70. Rodolphe Jenatton, Nicolas L Roux, Antoine Bordes, and Guillaume R Obozinski. A latent factor model for highly multi-relational data. In *Proceedings of NeurIPS*, 2012.
71. Guoliang Ji, Kang Liu, Shizhu He, and Jun Zhao. Knowledge graph completion with adaptive sparse transfer matrix. In *Proceedings of AAAI*, 2016.
72. Guoliang Ji, Kang Liu, Shizhu He, and Jun Zhao. Distant supervision for relation extraction with sentence-level attention and entity descriptions. In *Proceedings of AAAI*, 2017.
73. Shaoxiong Ji, Shirui Pan, Erik Cambria, Pekka Marttinen, and S Yu Philip. A survey on knowledge graphs: Representation, acquisition, and applications. *IEEE Transactions on Neural Networks and Learning Systems*, 33(2):494–514, 2021.
74. Zhengbao Jiang, Frank F Xu, Jun Araki, and Graham Neubig. How can we know what language models know? *Transactions of the Association for Computational Linguistics*, 8:423–438, 2020.
75. Nanda Kambhatla. Combining lexical, syntactic, and semantic features with maximum entropy models for extracting relations. In *Proceedings of ACL*, 2004.
76. Vladimir Karpukhin, Barlas Oguz, Sewon Min, Patrick Lewis, Ledell Wu, Sergey Edunov, Danqi Chen, and Wen-tau Yih. Dense passage retrieval for open-domain question answering. In *Proceedings of EMNLP*, 2020.
77. Keshav Kolluru, Samarth Aggarwal, Vipul Rathore, Soumen Chakrabarti, et al. Imojie: Iterative memory-based joint open information extraction. In *Proceedings of ACL*, 2020.
78. Denis Krompaß, Stephan Baier, and Volker Tresp. Type-constrained representation learning in knowledge graphs. In *Proceedings of ISWC*, 2015.
79. John Lafferty, Andrew McCallum, Fernando Pereira, et al. Conditional random fields: Probabilistic models for segmenting and labeling sequence data. In *Proceedings of ICML*, 2001.
80. Thomas K Landauer and Susan T Dumais. A solution to Plato's problem: The latent semantic analysis theory of acquisition, induction, and representation of knowledge. *Psychological Review*, 1997.
81. Ora Lassila, Ralph R. Swick, World Wide, and Web Consortium. Resource description framework (rdf) model and syntax specification, 1998.
82. Quoc V Le and Tomas Mikolov. Distributed representations of sentences and documents. In *Proceedings of ICML*, 2014.
83. Yann LeCun, Yoshua Bengio, and Geoffrey Hinton. Deep learning. *Nature*, 521(7553):436, 2015.
84. Douglas B Lenat. CYC: A large-scale investment in knowledge infrastructure. *Communications of the ACM*, 38(11):33–38, 1995.
85. Omer Levy, Minjoon Seo, Eunsol Choi, and Luke Zettlemoyer. Zero-shot relation extraction via reading comprehension. In *Proceedings of CoNLL*, 2017.
86. Patrick Lewis, Ethan Perez, Aleksandra Piktus, Fabio Petroni, Vladimir Karpukhin, Naman Goyal, Heinrich Küttler, Mike Lewis, Wen-tau Yih, Tim Rocktäschel, et al. Retrieval-augmented generation for knowledge-intensive nlp tasks. In *Proceedings of NeurIPS*, 2020.
87. Diya Li, Lifu Huang, Heng Ji, and Jiawei Han. Biomedical event extraction based on knowledge-driven tree-LSTM. In *Proceedings of NAACL-HLT*, 2019.

88. Jiao Li, Yueping Sun, Robin J. Johnson, Daniela Sciaky, Chih-Hsuan Wei, Robert Leaman, Allan Peter Davis, Carolyn J. Mattingly, Thomas C. Wiegers, and Zhiyong Lu. BioCreative V CDR task corpus: a resource for chemical disease relation extraction. *Database*, 2016, 05 2016.

89. Yankai Lin, Zhiyuan Liu, Huanbo Luan, Maosong Sun, Siwei Rao, and Song Liu. Modeling relation paths for representation learning of knowledge bases. In *Proceedings of EMNLP*, 2015.

90. Yankai Lin, Zhiyuan Liu, Maosong Sun, Yang Liu, and Xuan Zhu. Learning entity and relation embeddings for knowledge graph completion. In *Proceedings of AAAI*, 2015.

91. Yankai Lin, Shiqi Shen, Zhiyuan Liu, Huanbo Luan, and Maosong Sun. Neural relation extraction with selective attention over instances. In *Proceedings of ACL*, 2016.

92. Chunyang Liu, Wenbo Sun, Wenhan Chao, and Wanxiang Che. Convolution neural network for relation extraction. In *Proceedings of ICDM*, 2013.

93. Pengfei Liu, Weizhe Yuan, Jinlan Fu, Zhengbao Jiang, Hiroaki Hayashi, and Graham Neubig. Pre-train, prompt, and predict: A systematic survey of prompting methods in natural language processing. *arXiv preprint arXiv:2107.13586*, 2021.

94. Quan Liu, Hui Jiang, Andrew Evdokimov, Zhen-Hua Ling, Xiaodan Zhu, Si Wei, and Yu Hu. Probabilistic reasoning via deep learning: Neural association models. *arXiv preprint arXiv:1603.07704*, 2016.

95. Tianyu Liu, Kexiang Wang, Baobao Chang, and Zhifang Sui. A soft-label method for noise-tolerant distantly supervised relation extraction. In *Proceedings of EMNLP*, 2017.

96. Weijie Liu, Peng Zhou, Zhe Zhao, Zhiruo Wang, Qi Ju, Haotang Deng, and Ping Wang. K-BERT: Enabling language representation with knowledge graph. In *Proceedings of AAAI*, 2020.

97. Xiao Liu, Fanjin Zhang, Zhenyu Hou, Li Mian, Zhaoyu Wang, Jing Zhang, and Jie Tang. Self-supervised learning: Generative or contrastive. *IEEE Transactions on Knowledge and Data Engineering*, 2021.

98. Yang Liu, Kang Liu, Liheng Xu, Jun Zhao, et al. Exploring fine-grained entity type constraints for distantly supervised relation extraction. In *Proceedings of COLING*, 2014.

99. Yinhan Liu, Myle Ott, Naman Goyal, Jingfei Du, Mandar Joshi, Danqi Chen, Omer Levy, Mike Lewis, Luke Zettlemoyer, and Veselin Stoyanov. RoBERTa: A robustly optimized BERT pretraining approach. *arXiv preprint arXiv:1907.11692*, 2019.

100. Zhenhao Liu, Chenyan Xiong, Maosong Sun, and Zhiyuan Liu. Entity-duet neural ranking: Understanding the role of knowledge graph semantics in neural information retrieval. In *Proceedings of ACL*, 2018.

101. Zhiyuan Liu, Yankai Lin, and Maosong Sun. *Representation Learning for Natural Language Processing*. Springer, 2020.

102. Zhiyuan Liu, Maosong Sun, Yankai Lin, and Ruobing Xie. Knowledge representation learning: A review. *JCRD*, 53(2):247–261, 2016.

103. Diego Marcheggiani and Ivan Titov. Discrete-state variational autoencoders for joint discovery and factorization of relations. *Transactions of the Association for Computational Linguistics*, 4:231–244, 2016.

104. J McCarthy, ML Minsky, and N Rochester. A proposal for the dartmouth summer research project on artificial intelligence. 1955.

105. John McCarthy. History of lisp. *ACM SIGPLAN Notices*, 13(8):217–223, 1978.

106. Pablo N Mendes, Max Jakob, Andrés García-Silva, and Christian Bizer. DBpedia spotlight: shedding light on the web of documents. In *Proceedings of ICSS*, 2011.

107. T Mikolov and J Dean. Distributed representations of words and phrases and their compositionality. In *Proceedings of NeurIPS*, 2013.

108. Tomas Mikolov, Kai Chen, Greg Corrado, and Jeffrey Dean. Efficient estimation of word representations in vector space. In *Proceedings of ICLR*, 2013.

109. Mike Mintz, Steven Bills, Rion Snow, and Dan Jurafsky. Distant supervision for relation extraction without labeled data. In *Proceedings of ACL-IJCNLP*, 2009.

110. Makoto Miwa and Mohit Bansal. End-to-end relation extraction using LSTMs on sequences and tree structures. In *Proceedings of ACL*, 2016.
111. Reiichiro Nakano, Jacob Hilton, Suchir Balaji, Jeff Wu, Long Ouyang, Christina Kim, Christopher Hesse, Shantanu Jain, Vineet Kosaraju, William Saunders, et al. WebGPT: Browser-assisted question-answering with human feedback. *arXiv preprint arXiv:2112.09332*, 2021.
112. Guoshun Nan, Zhijiang Guo, Ivan Sekulić, and Wei Lu. Reasoning with latent structure refinement for document-level relation extraction. In *Proceedings of ACL*, 2020.
113. Allen Newell, John Clifford Shaw, and Herbert A Simon. Empirical explorations of the logic theory machine: a case study in heuristic. In *Proceedings of Western Computer*, 1957.
114. Allen Newell and Fred M Tonge. An introduction to information processing language v. *Communications of the ACM*, 3(4):205–211, 1960.
115. Thien Huu Nguyen and Ralph Grishman. Combining neural networks and log-linear models to improve relation extraction. *arXiv preprint arXiv:1511.05926*, 2015.
116. Maximilian Nickel, Kevin Murphy, Volker Tresp, and Evgeniy Gabrilovich. A review of relational machine learning for knowledge graphs. In *Proceedings of the IEEE*, 2015.
117. Maximilian Nickel, Lorenzo Rosasco, and Tomaso Poggio. Holographic embeddings of knowledge graphs. In *Proceedings of AAAI*, 2016.
118. Maximilian Nickel, Volker Tresp, and Hans-Peter Kriegel. A three-way model for collective learning on multi-relational data. In *Proceedings of ICML*, 2011.
119. Maximilian Nickel, Volker Tresp, and Hans-Peter Kriegel. Factorizing YAGO: scalable machine learning for linked data. In *Proceedings of WWW*, 2012.
120. Sinno Jialin Pan and Qiang Yang. A survey on transfer learning. *TKDE*, 22(10):1345–1359, 2009.
121. Sachin Pawar, Girish K Palshikar, and Pushpak Bhattacharyya. Relation extraction: A survey. *arXiv preprint arXiv:1712.05191*, 2017.
122. Nanyun Peng, Hoifung Poon, Chris Quirk, Kristina Toutanova, and Wen-tau Yih. Cross-sentence n-ary relation extraction with graph LSTMs. *Transactions of the Association for Computational Linguistics*, 5:101–115, 2017.
123. Jeffrey Pennington, Richard Socher, and Christopher Manning. GloVe: Global vectors for word representation. In *Proceedings of EMNLP*, 2014.
124. Matthew E Peters, Mark Neumann, Robert Logan, Roy Schwartz, Vidur Joshi, Sameer Singh, and Noah A Smith. Knowledge enhanced contextual word representations. In *Proceedings of EMNLP-IJCNLP*, 2019.
125. Fabio Petroni, Patrick Lewis, Aleksandra Piktus, Tim Rocktäschel, Yuxiang Wu, Alexander H Miller, and Sebastian Riedel. How context affects language models' factual predictions. In *Proceedings of AKBC*, 2020.
126. Fabio Petroni, Tim Rocktäschel, Sebastian Riedel, Patrick Lewis, Anton Bakhtin, Yuxiang Wu, and Alexander Miller. Language models as knowledge bases? In *Proceedings of EMNLP-IJCNLP*, 2019.
127. Jay Pujara, Hui Miao, Lise Getoor, and William W Cohen. Knowledge graph identification. In *Proceedings of ISWC*, 2013.
128. Yujia Qin, Yankai Lin, Ryuichi Takanobu, Zhiyuan Liu, Peng Li, Heng Ji, Minlie Huang, Maosong Sun, and Jie Zhou. Erica: Improving entity and relation understanding for pre-trained language models via contrastive learning. In *Proceedings of ACL-IJCNLP*, 2021.
129. Chris Quirk and Hoifung Poon. Distant supervision for relation extraction beyond the sentence boundary. In *Proceedings of EACL*, 2017.
130. Arlan Ramsay and Robert D Richtmyer. *Introduction to hyperbolic geometry*. Springer Science & Business Media, 1995.
131. Erich H Reck. *From Frege to Wittgenstein: Perspectives on Early Analytic Philosophy*. Oxford University Press, 2001.
132. Sebastian Riedel, Limin Yao, Andrew McCallum, and Benjamin M Marlin. Relation extraction with matrix factorization and universal schemas. In *Proceedings of NAACL-HLT*, 2013.

133. Dan Roth and Wen-tau Yih. Probabilistic reasoning for entity & relation recognition. In *Proceedings of COLING*, 2002.
134. Dan Roth and Wen-tau Yih. A linear programming formulation for global inference in natural language tasks. In *Proceedings of CoNLL*, 2004.
135. Cicero Nogueira dos Santos, Bing Xiang, and Bowen Zhou. Classifying relations by ranking with convolutional neural networks. In *Proceedings of ACL-IJCNLP*, 2015.
136. Michael Schlichtkrull, Thomas N Kipf, Peter Bloem, Rianne van den Berg, Ivan Titov, and Max Welling. Modeling relational data with graph convolutional networks. In *Proceedings of ESWC*, 2018.
137. Mike Schuster and Kuldip K Paliwal. Bidirectional recurrent neural networks. *IEEE Transactions on Signal Processing*, 45(11):2673–2681, 1997.
138. Sonse Shimaoka, Pontus Stenetorp, Kentaro Inui, and Sebastian Riedel. An attentive neural architecture for fine-grained entity type classification. In *Proceedings of AKBC Workshop*, 2016.
139. Kurt Shuster, Spencer Poff, Moya Chen, Douwe Kiela, and Jason Weston. Retrieval augmentation reduces hallucination in conversation. In *Findings of EMNLP*, 2021.
140. Barry Smith. Ontology. In *The furniture of the world*, pages 47–68. Brill, 2012.
141. Jake Snell, Kevin Swersky, and Richard Zemel. Prototypical networks for few-shot learning. In *Proceedings of NeurIPS*, 2017.
142. Livio Baldini Soares, Nicholas FitzGerald, Jeffrey Ling, and Tom Kwiatkowski. Matching the Blanks: Distributional similarity for relation learning. In *Proceedings of ACL*, pages 2895–2905, 2019.
143. Richard Socher, Danqi Chen, Christopher D Manning, and Andrew Ng. Reasoning with neural tensor networks for knowledge base completion. In *Proceedings of NeurIPS*, 2013.
144. Richard Socher, Brody Huval, Christopher D Manning, and Andrew Y Ng. Semantic compositionality through recursive matrix-vector spaces. In *Proceedings of EMNLP*, 2012.
145. Gabriel Stanovsky, Julian Michael, Luke Zettlemoyer, and Ido Dagan. Supervised open information extraction. In *Proceedings of NAACL-HLT*, 2018.
146. Matthias Steup and Ram Neta. Epistemology. 2005.
147. Fabian M Suchanek, Gjergji Kasneci, and Gerhard Weikum. YAGO: a core of semantic knowledge. In *Proceedings of WWW*, 2007.
148. Tony Sun, Andrew Gaut, Shirlyn Tang, Yuxin Huang, Mai ElSherief, Jieyu Zhao, Diba Mirza, Elizabeth Belding, Kai-Wei Chang, and William Yang Wang. Mitigating gender bias in natural language processing: Literature review. In *Proceedings of ACL*, 2019.
149. Zhiqing Sun, Zhi-Hong Deng, Jian-Yun Nie, and Jian Tang. RotatE: Knowledge graph embedding by relational rotation in complex space. In *Proceedings of ICLR*, 2019.
150. Kumutha Swampillai and Mark Stevenson. Inter-sentential relations in information extraction corpora. In *Proceedings of LREC*, 2010.
151. Kai Sheng Tai, Richard Socher, and Christopher D. Manning. Improved semantic representations from tree-structured long short-term memory networks. In *Proceedings of ACL*, 2015.
152. Jian Tang, Meng Qu, Mingzhe Wang, Ming Zhang, Jun Yan, and Qiaozhu Mei. LINE: Large-scale information network embedding. In *Proceedings of WWW*, 2015.
153. Théo Trouillon, Johannes Welbl, Sebastian Riedel, Éric Gaussier, and Guillaume Bouchard. Complex embeddings for simple link prediction. In *Proceedings of ICML*, 2016.
154. Ashish Vaswani, Noam Shazeer, Niki Parmar, Llion Jones, Jakob Uszkoreit, Aidan N Gomez, and Lukasz Kaiser. Attention is all you need. In *Proceedings of NeurIPS*, 2017.
155. Patrick Verga, Emma Strubell, and Andrew McCallum. Simultaneously self-attending to all mentions for full-abstract biological relation extraction. In *Proceedings of NAACL-HLT*, 2018.
156. Denny Vrandečić and Markus Krötzsch. Wikidata: A free collaborative knowledge base. *Communications of the ACM*, 57(10):78–85, 2014.
157. Eric Wallace, Shi Feng, Nikhil Kandpal, Matt Gardner, and Sameer Singh. Universal adversarial triggers for attacking and analyzing nlp. In *Proceedings of EMNLP-IJCNLP*, 2019.

158. Hong Wang, Christfried Focke, Rob Sylvester, Nilesh Mishra, and William Wang. Fine-tune BERT for DocRED with two-step process. *arXiv preprint arXiv:1909.11898*, 2019.
159. Hong Wang, Wenhan Xiong, Mo Yu, Xiaoxiao Guo, Shiyu Chang, and William Yang Wang. Sentence embedding alignment for lifelong relation extraction. In *Proceedings of NAACL-HLT*, 2019.
160. Qingyun Wang, Lifu Huang, Zhiying Jiang, Kevin Knight, Heng Ji, Mohit Bansal, and Yi Luan. PaperRobot: Incremental draft generation of scientific ideas. In *Proceedings of ACL*, 2019.
161. Quan Wang, Zhendong Mao, Bin Wang, and Li Guo. Knowledge graph embedding: A survey of approaches and applications. *IEEE Transactions on Knowledge and Data Engineering*, 29(12):2724–2743, 2017.
162. Quan Wang, Bin Wang, and Li Guo. Knowledge base completion using embeddings and rules. In *Proceedings of IJCAI*, 2015.
163. Xiaozhi Wang, Tianyu Gao, Zhaocheng Zhu, Zhengyan Zhang, Zhiyuan Liu, Juanzi Li, and Jian Tang. KEPLER: A unified model for knowledge embedding and pre-trained language representation. *Transactions of the Association for Computational Linguistics*, 9:176–194, 2021.
164. Zhen Wang, Jianwen Zhang, Jianlin Feng, and Zheng Chen. Knowledge graph and text jointly embedding. In *Proceedings of EMNLP*, 2014.
165. Zhen Wang, Jianwen Zhang, Jianlin Feng, and Zheng Chen. Knowledge graph embedding by translating on hyperplanes. In *Proceedings of AAAI*, 2014.
166. Zhigang Wang and Juan-Zi Li. Text-enhanced representation learning for knowledge graph. In *Proceedings of IJCAI*, 2016.
167. Jason Weston, Sumit Chopra, and Antoine Bordes. Memory networks. *arXiv preprint arXiv:1410.3916*, 2014.
168. Jason Weston, Emily Dinan, and Alexander H Miller. Retrieve and refine: Improved sequence generation models for dialogue. In *Proceedings of EMNLP*, 2018.
169. Fei Wu and Daniel S Weld. Open information extraction using wikipedia. In *Proceedings of ACL*, 2010.
170. Ruidong Wu, Yuan Yao, Xu Han, Ruobing Xie, Zhiyuan Liu, Fen Lin, Leyu Lin, and Maosong Sun. Open relation extraction: Relational knowledge transfer from supervised data to unsupervised data. In *Proceedings of EMNLP-IJCNLP*, 2019.
171. Yi Wu, David Bamman, and Stuart Russell. Adversarial training for relation extraction. In *Proceedings of EMNLP*, 2017.
172. Chaojun Xiao, Yuan Yao, Ruobing Xie, Xu Han, Zhiyuan Liu, Maosong Sun, Fen Lin, and Leyu Lin. Denoising relation extraction from document-level distant supervision. In *Proceedings of EMNLP*, 2020.
173. Han Xiao, Minlie Huang, and Xiaoyan Zhu. From one point to a manifold: Knowledge graph embedding for precise link prediction. In *Proceedings of IJCAI*, 2016.
174. Han Xiao, Minlie Huang, and Xiaoyan Zhu. TransG: A generative model for knowledge graph embedding. In *Proceedings of ACL*, 2016.
175. Ruobing Xie, Zhiyuan Liu, Tat-seng Chua, Huanbo Luan, and Maosong Sun. Image-embodied knowledge representation learning. In *Proceedings of IJCAI*, 2016.
176. Ruobing Xie, Zhiyuan Liu, Jia Jia, Huanbo Luan, and Maosong Sun. Representation learning of knowledge graphs with entity descriptions. In *Proceedings of AAAI*, 2016.
177. Ruobing Xie, Zhiyuan Liu, and Maosong Sun. Representation learning of knowledge graphs with hierarchical types. In *Proceedings of IJCAI*, 2016.
178. Ji Xin, Yankai Lin, Zhiyuan Liu, and Maosong Sun. Improving neural fine-grained entity typing with knowledge attention. In *Proceedings of AAAI*, 2018.
179. Chenyan Xiong, Jamie Callan, and Tie-Yan Liu. Word-entity duet representations for document ranking. In *Proceedings of SIGIR*, 2017.
180. Chenyan Xiong, Zhuyun Dai, Jamie Callan, Zhiyuan Liu, and Russell Power. End-to-end neural ad-hoc ranking with kernel pooling. In *Proceedings of SIGIR*, 2017.

181. Wenhan Xiong, Jingfei Du, William Yang Wang, and Veselin Stoyanov. Pretrained encyclopedia: Weakly supervised knowledge-pretrained language model. In *Proceedings of ICLR*, 2020.

182. Kun Xu, Yansong Feng, Songfang Huang, and Dongyan Zhao. Semantic relation classification via convolutional neural networks with simple negative sampling. In *Proceedings of EMNLP*, 2015.

183. Yan Xu, Ran Jia, Lili Mou, Ge Li, Yunchuan Chen, Yangyang Lu, and Zhi Jin. Improved relation classification by deep recurrent neural networks with data augmentation. In *Proceedings of COLING*, 2016.

184. Mohamed Yahya, Klaus Berberich, Shady Elbassuoni, and Gerhard Weikum. Robust question answering over the web of linked data. In *Proceedings of CIKM*, 2013.

185. Ikuya Yamada, Akari Asai, Hiroyuki Shindo, Hideaki Takeda, and Yuji Matsumoto. Luke: Deep contextualized entity representations with entity-aware self-attention. In *Proceedings of EMNLP*, 2020.

186. Bishan Yang, Wen-tau Yih, Xiaodong He, Jianfeng Gao, and Li Deng. Embedding entities and relations for learning and inference in knowledge bases. In *Proceedings of ICLR*, 2015.

187. Yuan Yao, Jiaju Du, Yankai Lin, Peng Li, Zhiyuan Liu, Jie Zhou, and Maosong Sun. Codred: A cross-document relation extraction dataset for acquiring knowledge in the wild. In *Proceedings of EMNLP*, 2021.

188. Yuan Yao, Deming Ye, Peng Li, Xu Han, Yankai Lin, Zhenghao Liu, Zhiyuan Liu, Lixin Huang, Jie Zhou, and Maosong Sun. DocRED: A large-scale document-level relation extraction dataset. In *Proceedings of ACL*, 2019.

189. Alexander Yates, Michele Banko, Matthew Broadhead, Michael J Cafarella, Oren Etzioni, and Stephen Soderland. Textrunner: open information extraction on the web. In *Proceedings of NAACL-HLT*, 2007.

190. Deming Ye, Yankai Lin, Jiaju Du, Zhenghao Liu, Peng Li, Maosong Sun, and Zhiyuan Liu. Coreferential reasoning learning for language representation. In *Proceedings of EMNLP*, 2020.

191. Zhi-Xiu Ye and Zhen-Hua Ling. Multi-level matching and aggregation network for few-shot relation classification. In *Proceedings of ACL*, 2019.

192. David Yenicelik, Florian Schmidt, and Yannic Kilcher. How does BERT capture semantics? a closer look at polysemous words. In *Proceedings of BlackboxNLP*, 2020.

193. Wenhao Yu, Chenguang Zhu, Zaitang Li, Zhiting Hu, Qingyun Wang, Heng Ji, and Meng Jiang. A survey of knowledge-enhanced text generation. *ACM Computing Surveys (CSUR)*, 54(11):1–38, 2022.

194. Yuan Zang, Fanchao Qi, Chenghao Yang, Zhiyuan Liu, Meng Zhang, Qun Liu, and Maosong Sun. Word-level textual adversarial attacking as combinatorial optimization. In *Proceedings of ACL*, 2020.

195. Dmitry Zelenko, Chinatsu Aone, and Anthony Richardella. Kernel methods for relation extraction. *JMLR*, 3:1083–1106, 2003.

196. Daojian Zeng, Kang Liu, Yubo Chen, and Jun Zhao. Distant supervision for relation extraction via piecewise convolutional neural networks. In *Proceedings of EMNLP*, 2015.

197. Daojian Zeng, Kang Liu, Siwei Lai, Guangyou Zhou, and Jun Zhao. Relation classification via convolutional deep neural network. In *Proceedings of COLING*, 2014.

198. Shuang Zeng, Runxin Xu, Baobao Chang, and Lei Li. Double graph based reasoning for document-level relation extraction. In *Proceedings of EMNLP*, 2020.

199. Wenyuan Zeng, Yankai Lin, Zhiyuan Liu, and Maosong Sun. Incorporating relation paths in neural relation extraction. In *Proceedings of EMNLP*, 2017.

200. Xiangrong Zeng, Shizhu He, Kang Liu, and Jun Zhao. Large scaled relation extraction with reinforcement learning. In *Proceedings of AAAI*, 2018.

201. Dongxu Zhang and Dong Wang. Relation classification via recurrent neural network. *arXiv preprint arXiv:1508.01006*, 2015.

202. Kai Zhang, Yuan Yao, Ruobing Xie, Xu Han, Zhiyuan Liu, Fen Lin, Leyu Lin, and Maosong Sun. Open hierarchical relation extraction. In *Proceedings of NAACL-HLT*, 2021.

203. Shuai Zhang, Yi Tay, Lina Yao, and Qi Liu. Quaternion knowledge graph embeddings. In *Proceedings of NeurIPS*, 2019.
204. Yuhao Zhang, Peng Qi, and Christopher D Manning. Graph convolution over pruned dependency trees improves relation extraction. In *Proceedings of EMNLP*, 2018.
205. Zhengyan Zhang, Xu Han, Zhiyuan Liu, Xin Jiang, Maosong Sun, and Qun Liu. ERNIE: Enhanced language representation with informative entities. In *Proceedings of ACL*, 2019.
206. Zhengyan Zhang, Xu Han, Hao Zhou, Pei Ke, Yuxian Gu, Deming Ye, Yujia Qin, Yusheng Su, Haozhe Ji, Jian Guan, et al. CPM: A large-scale generative chinese pre-trained language model. *AI Open*, 2:93–99, 2021.
207. Huaping Zhong, Jianwen Zhang, Zhen Wang, Hai Wan, and Zheng Chen. Aligning knowledge and text embeddings by entity descriptions. In *Proceedings of EMNLP*, 2015.
208. Guodong Zhou, Jian Su, Jie Zhang, and Min Zhang. Exploring various knowledge in relation extraction. In *Proceedings of ACL*, 2005.
209. Peng Zhou, Wei Shi, Jun Tian, Zhenyu Qi, Bingchen Li, Hongwei Hao, and Bo Xu. Attention-based bidirectional long short-term memory networks for relation classification. In *Proceedings of ACL*, 2016.
210. Hao Zhu, Yankai Lin, Zhiyuan Liu, Jie Fu, Tat-Seng Chua, and Maosong Sun. Graph neural networks with generated parameters for relation extraction. In *Proceedings of ACL*, 2019.

Chapter 10
Sememe-Based Lexical Knowledge Representation Learning

Yujia Qin, Zhiyuan Liu, Yankai Lin, and Maosong Sun

Abstract Linguistic and commonsense knowledge bases describe knowledge in formal and structural languages. Such knowledge can be easily leveraged in modern natural language processing systems. In this chapter, we introduce one typical kind of linguistic knowledge (sememe knowledge) and a sememe knowledge base named HowNet. In linguistics, sememes are defined as the minimum indivisible units of meaning. We first briefly introduce the basic concepts of sememe and HowNet. Next, we introduce how to model the sememe knowledge using neural networks. Taking a step further, we introduce sememe-guided knowledge applications, including incorporating sememe knowledge into compositionality modeling, language modeling, and recurrent neural networks. Finally, we discuss sememe knowledge acquisition for automatically constructing sememe knowledge bases and representative real-world applications of HowNet.

10.1 Introduction

In the field of NLP, a series of meaningful linguistic units are generally studied, including words, phrases, sentences, discourses, and documents. Specifically, words are typically treated as the smallest usage units since they are deemed as the smallest meaningful objects which can stand by themselves. In fact, word meanings can be further split into smaller components. For instance, the meaning of the word *teacher* comprises the meanings of *education*, *occupation*, and *teach*. From the perspective of linguistics, the minimum indivisible units of meaning are defined as sememes [10].

Y. Qin · Z. Liu (✉) · M. Sun
Department of Computer Science and Technology, Tsinghua University, Beijing, China
e-mail: qyj20@mails.tsinghua.edu.cn; liuzy@tsinghua.edu.cn; sms@tsinghua.edu.cn

Y. Lin
Gaoling School of Artificial Intelligence, Renmin University of China, Beijing, China
e-mail: yankailin@ruc.edu.cn

© The Author(s) 2023
Z. Liu et al. (eds.), *Representation Learning for Natural Language Processing*,
https://doi.org/10.1007/978-981-99-1600-9_10

Linguists contend that the meanings of all the words comprise a closed set of sememes, and the semantic meanings of these sememes are orthogonal to each other. Compared with words, sememes are fairly implicit, and the full set of sememes is difficult to define, not to mention how to decide which sememes a word has. To understand the human language from a finer-grained perspective, it is necessary to fathom the nature of sememe and its connection with words.

To this end, some linguists spend many years identifying sememes from both linguistic knowledge bases (KBs) and dictionaries and labeling each word with sememes, in order to construct sememe-based linguistic KBs. HowNet [23] is the most representative sememe-based linguistic KB. Besides being a linguistic KB, HowNet also contains commonsense knowledge and can be used to unveil the relationships among concepts [23].

In the following sections, we first briefly introduce backgrounds of typical linguistic KB (WordNet) and commonsense KB (ConceptNet), and then we detailedly introduce basic concepts and construction principles of HowNet, as well as the linguistic knowledge and commonsense knowledge in HowNet. Then we discuss how to represent sememe knowledge using neural networks. After that, we introduce sememe-guided NLP techniques, including how to incorporate sememe knowledge into compositionality modeling, language modeling, and recurrent neural networks. Finally, we discuss automatic knowledge acquisition for HowNet and the application of HowNet. Since most of the research (especially the research related to deep learning) in this area is conducted by our group, we will mainly take our works as examples to elaborate on the powerful capability of sememe knowledge. To provide a more comprehensive description, our discussion of the specific methods in this chapter will be more detailed.

10.2   Linguistic and Commonsense Knowledge Bases

Over the years, many human-annotated KBs have been proposed, among which linguistic KBs and commonsense KBs are the most representative ones. Serving as important lexical resources, all of these KBs have pushed forward the understanding of human language and achieved many successes in NLP applications. In this section, we first elaborate on a representative linguistic KB, WordNet, and a commonsense KB, ConceptNet, and then we discuss the unique characteristics of HowNet compared with them.

10.2.1   WordNet and ConceptNet

WordNet and ConceptNet are the most representative KBs aiming at organizing linguistic knowledge and commonsense knowledge, respectively. Both of them have shown importance in various NLP applications. We briefly give an introduction to the construction of both KBs as follows.

WordNet WordNet [48] is a large lexical database and can also be viewed as a KB containing multi-relational data. It was first created in 1985 by George Armitage Miller, a psychology professor in the Cognitive Science Laboratory of Princeton University. Up till now, WordNet has become the most popular lexicon dictionary in the world and has been widely applied in various NLP tasks.

Based on meanings, WordNet groups English nouns, verbs, adjectives, and adverbs into synsets (i.e., sets of cognitive synonyms), which represent unique concepts. Each synset is accompanied by a brief description. In most cases, there are several short sentences illustrating the usage of words in this synset. The synsets and words are linked by conceptual-semantic and lexical relations, covering all WordNet's 117,000 synsets. For example, (1) the words in the same synset are linked with the synonymy relation, which indicates that these words share similar meanings and could be replaced by each other in some contexts; (2) the hypernymy/hyponymy links a general synset and a specific synset, which indicates that the specific synset is a sub-class of the general one, and (3) the antonymy describes the relation among adjectives with opposite meanings.

ConceptNet Besides linguistic knowledge, commonsense knowledge (generic facts about social and physical environments) is also important for general artificial intelligence. ConceptNet [72] is one of the largest freely available commonsense knowledge bases. ConceptNet was first constructed in 2002 in the project of Open Mind Common Sense and was frequently updated in the following years. Like other commonsense knowledge bases, ConceptNet describes the conceptual relations among words. Nodes in ConceptNet are represented as free-text descriptions, and edges stand for symmetric relations like SimilarTo or asymmetric relations like MadeOf.

Thanks to diverse building sources and continual updating, ConceptNet has grown as the largest free commonsense KB that includes more than 21 million edges and 8 million nodes [72]. In addition, ConceptNet supports a variety of languages. Similar to other KBs, ConceptNet can be used to enhance the ability of neural networks on downstream tasks that require commonsense reasoning. Especially with the novel relation ExternalURL, ConceptNet nodes can be easily linked with nodes in other knowledge bases such as WordNet. This capability makes it more convenient to integrate commonsense knowledge and linguistic knowledge for researchers.

10.2.2 HowNet

The above-mentioned KBs take words (WordNet) or concepts (ConceptNet) as basic elements and consist of word-level or concept-level relations. Different from both KBs, HowNet treats sememes as the smallest linguistic objects and additionally focuses on the relation between sememes and words. This is one of the core differences between the design philosophy of HowNet and other KBs.

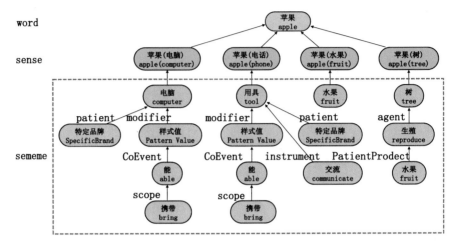

Fig. 10.1 An example of a word annotated with sememes in HowNet

Construction of HowNet We introduce three components for HowNet construction: (1) *Sememe set construction*: the sememe set is determined by analyzing, merging, and sifting the semantics of a great number of Chinese characters and words. Each sememe in HowNet is expressed by a term or a phrase in both English and Chinese to avoid ambiguity. For instance, (*human* | {人}) and (*ProperName* | {专}). All the sememes can be categorized into *seven* types, including part, attribute, attribute value, thing, space, time, and event; (2) *Sememe-sense-word structure definition*: considering the polysemy, HowNet annotates different sets of sememes for different senses of a word, with every sense described in both Chinese and English. Every sense is defined as the root of a "sememe tree." For the sememes belonging to a specific sense, HowNet annotates the relations among these sememes (dubbed as "dynamic roles"). Such relations are the edges of the "sememe tree." We illustrate an example for a word *apple* in Fig. 10.1, where the word *apple* has four senses including *apple(tree)*, *apple(phone)*, *apple(computer)*, and *apple(fruit)*; (3) *Annotation process*: HowNet is constructed by manual annotation of human experts. HowNet was originally built by Zhendong Dong and Qiang Dong in the 1990s and has been frequently updated ever since then, with the latest version of HowNet published in January 2019.

Uniqueness of HowNet The aforementioned KBs share several similarities, for instance, all of them are (1) structured with relational semantic networks, (2) based on the form of natural language, (3) constructed with extensive human labeling, etc. In spite of all these similarities, HowNet owns unique characteristics that differ from WordNet and ConceptNet in construction principles, design philosophy, and foci, which are discussed in the following paragraphs.

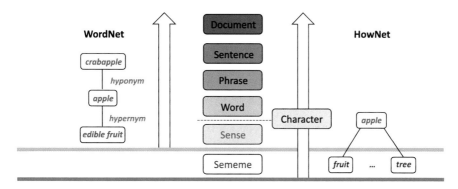

Fig. 10.2 Comparison between WordNet and HowNet. WordNet is based on synsets and their semantic relations, while HowNet investigates sememes and focuses on their relations to word senses. Here we discriminate characters (字) and words (词) in Chinese, with the former being smaller components for the latter

Comparison Between HowNet and WordNet As shown in Fig. 10.2, compared with WordNet, HowNet is unique reflected in the following facets:

1. Basic unit and design philosophy. In human language, words are generally considered as the smallest *usage units*, while sememes are viewed as the smallest *semantic units*. Adhering to reductionism, HowNet considers sememes as the smallest objects and focuses on the relation between words and sememes; instead, the basic block of WordNet is a synset consisting of all the words expressing a specific concept; thus the design of WordNet resembles that of a thesaurus. This is the core difference between the design philosophy of HowNet and WordNet. In addition, according to Kim et al. [39], WordNet is *differential* by nature: instead of explicitly expressing the meaning of a word, WordNet differentiates word senses by placing them into different synsets and further assigning them to different positions in its ontology. Conversely, HowNet is *constructive*, i.e., exploiting sememes from a taxonomy to represent the meaning of each word sense. It is based on the hypothesis that all concepts can be reduced to relevant sememes.

2. Construction principle. WordNet is organized according to *semantic relations* among word meanings. Since word meanings can be represented by synsets, semantic relations can be treated as pointers among synsets. The taxonomy of WordNet is designed not to capture common causality or function, but to show the relations among existing lexemes [39]. Differently, the basic construction principle of HowNet is to form a networked knowledge system of the relations among concepts and the relation between attributes and concepts. Besides, HowNet is constructed using top-down induction: the ultimate sememe set is established by observing and analyzing all the possible basic sememes. After that, human experts evaluate whether every concept can be composed of the subsets of the sememe set.

3. Application scope. Initially designed as a thesaurus, WordNet gradually evolved into a self-contained machine-readable dictionary of semantics. In contrast, HowNet is established towards building a computer-oriented semantic network [22]. In addition, one advantage of WordNet is that it supports multiple languages. For example, since many countries have established lexical databases based on WordNet, it can be easily applied to cross-lingual scenarios. However, since HowNet mainly supports English and Chinese, most of its applications are bilingual. In fact, we have also proposed methods to automatically build sememe KBs for other languages, which will be discussed later.

Comparison Between HowNet and ConceptNet HowNet and ConceptNet differ from each other in the following aspects:

1. Coverage of commonsense knowledge. The commonsense knowledge contained in ConceptNet is relatively explicit, partly because the nodes in ConceptNet are represented as free-text descriptions. In contrast, the notions of nodes in HowNet are purely lexical items (e.g., word senses and sememes with atomic meanings), which correspond to more rock-bottom commonsense knowledge. Therefore, the commonsense knowledge of HowNet is more implicit. For instance, from ConceptNet, we know directly that the concept *buy book* is related to the concept *a bookstore* because the former is a subevent of the latter, while in HowNet, we learn such information by simple induction and reasoning: the word *bookstore* is associated with sememes *publication* and *buy*, and the word *book* consists of the sememe *publication*. Similarly, numerous generic facts about the world can be derived from HowNet. We contend that despite the implicit nature of commonsense knowledge, HowNet actually covers more diverse facets of the world facts than ConceptNet.
2. Foci and construction principle. ConceptNet focuses on everyday episodic concepts and the semantic relations among compound concepts, which are organized hierarchically. These high-level concepts can be contributed by every ordinary person. In contrast, HowNet focuses on rock-bottom linguistic and conceptual knowledge of the human language, and its annotation requires the basic understanding of sememe hierarchy. The above distinction leads to different construction methods for both KBs. HowNet is constructed solely by human handcrafting of linguistic experts, whereas the construction of ConceptNet involves the general public without much background knowledge. In consequence, the annotation of HowNet has higher quality than ConceptNet by nature.
3. Application scope. HowNet annotates sememes for each word sense and thus differentiates different word meanings. Nevertheless, the concepts and relations annotated in ConceptNet may be ambiguous. The ambiguous nature of ConceptNet could hinder it from being directly leveraged in NLP applications, such as word sense disambiguation. In contrast, the sememe knowledge of HowNet can be more easily incorporated into modern neural networks since HowNet overcomes the problem of word ambiguity.

苹果 (IPHONE)

ID: 000000244398

词性: 名词 (noun)

基于义原的定义

{tool|用具:modifier={PatternValue|样式值:CoEvent={able|能:scope={bring|携带:patient={$}}}}{SpeBrand|特定牌子},{communicate|交流:instrument={~}}}

义原树演示

Fig. 10.3 A snapshot of the OpenHowNet website

OpenHowNet To help researchers get access to HowNet data in an easier way, encouraged and approved by the inventors of HowNet, Zhendong Dong and Qiang Dong, we have created OpenHowNet[1] (Fig. 10.3). OpenHowNet is a free open-source sememe KB, which comprises the core data of HowNet. There are two core components of OpenHowNet, i.e., OpenHowNet Web and OpenHowNet API:

1. OpenHowNet Web gives a comprehensive description of HowNet, including statistics of OpenHowNet dataset, research articles relevant to sememe knowledge, history of HowNet, etc. With OpenHowNet, users can easily understand the basic idea of sememe and get familiar with advanced research topics of HowNet. Besides, OpenHowNet supports the visualization of the sememe tree for each sense in HowNet. Together with the tree structure, OpenHowNet Web provides additional information, such as the POS tags, the plain text form of the sememe tree, and semantically related senses. This capability makes it easier for users

---

[1] https://github.com/thunlp/OpenHowNet.

to understand the core linguistic information of each word sense. We also link OpenHowNet to representative KBs such as BabelNet and ConceptNet, which makes it easier for users to get access to information of each word sense outside HowNet.

2. OpenHowNet API supports some important functionalities, e.g., visualizing the sememe tree of a sense, searching senses or sememes, computing word similarity based on sememe tree annotation, etc. We believe such a toolkit can help researchers leverage the sememe annotation in HowNet more easily.

In summary, armed with OpenHowNet, it will be more convenient for beginners to get familiar with the design philosophy of HowNet, easier for senior researchers to utilize the sememe knowledge, and handier for industrial practitioners to deploy their HowNet applications. You can also read our research paper about OpenHowNet [64] for more details.

10.2.3 HowNet and Deep Learning

Back in the early era when statistical learning dominates mainstream NLP techniques, linguistic and commonsense KBs are generally leveraged to provide shallow and primitive information. Typical applications include word similarity calculation [43, 55], word sense disambiguation [3, 85], etc. Ever since the emergence of deep learning, HowNet has renewed a surge of interest in both academic and industrial communities, reflected in the significant proliferation of related research papers. Assisted by the powerful representation capability of deep learning, HowNet is endowed with more imaginative usage to fully exploit its knowledge. Before delving into the usage of sememe knowledge, we introduce several advantages of HowNet.

Advantages of HowNet The sememe knowledge of HowNet owns unique advantages over other linguistic KBs in the era of deep learning, reflected in the following characteristics:

1. In terms of natural language understanding, sememe knowledge is closer to the characteristics of natural language. The sememe annotation breaks the lexical barrier and offers an in-depth understanding of the rich semantic information behind the vocabulary. Compared with other KBs that can only be applied to the word level or the sense level, HowNet provides finer-grained linguistic and commonsense information.

2. Sememe knowledge turns out to be a natural fit for deep learning techniques. By accurately depicting semantic information through a unified sememe labeling system, the meaning of each sememe is clear and fixed and thus can be naturally incorporated into the deep learning model as informative labels/tags of words. As we will show later, most of the modern NLP models are built on word sequences. It is natural and convenient to directly extract the information for each word from HowNet to leverage its knowledge.

3. Sememe knowledge can mitigate poor model performance in low-resource scenarios. Since the sememe set is carefully pre-defined and the total number of sememes is limited, even when there only exists limited supervision, the representations of sememes can still be fully optimized. In contrast, considering the massive word representations needed to be learned, it is generally hard to learn excellent word embeddings, especially for those infrequent words. Thus the well-trained sememe representations can alleviate the problem of insufficient training and enrich the semantic meanings of words in low-resource settings.

How to Incorporate Sememe Knowledge After showing the uniqueness and advantages of HowNet, we briefly introduce several ways (categorized according to Chap. 9) of leveraging sememe knowledge for deep learning techniques:

1. Knowledge augmentation. The first way targets at adding sememe knowledge into the input of neural networks or designing special neural modules that can be inserted into the original networks. In this way, the sememe knowledge can be incorporated explicitly without changing the neural architectures. Since sememes are smaller units of word senses, they always appear together with words. For instance, we can first learn sememe embeddings and directly leverage them to enrich the semantic information of word embeddings.
2. Knowledge reformulation. The second method for incorporating sememe knowledge is to change the original word-based model structures into sememe-based ones. A possible solution is to assign sememe experts in the neural networks. The introduction of sememe experts could properly guide neural models to produce inner hidden representations with rich semantics in a more linguistically informative way.
3. Knowledge regularization. The third way is to design a new training objective function based on sememe knowledge or to use knowledge as extra predictive targets. For instance, we can first extract linguistic information (e.g., the overlap of annotated sememes of different words) from HowNet and then treat it as auxiliary regularization supervision. This approach does not require modifying the specific model architecture but only introduces an additional training objective to regularize the original optimization trajectory.

In the above paragraphs, we only present the high-level ideas for sememe knowledge incorporation. In the next few sections, we will elaborate on these ideas with specific examples to showcase the powerful capabilities of HowNet and its comprehensive applications in NLP.

10.3 Sememe Knowledge Representation

To leverage sememe knowledge, we should first learn to represent it. Sememes do not exist naturally but are labeled by human experts on word senses. We can represent them using techniques similar to word representation learning (WRL). In this section, we first elaborate on how to learn sememe embeddings by representing words as a combination of sememes, and then we introduce how to incorporate sememe knowledge to better learn word representations. The introduction of this section is based on our research works [53, 62].

10.3.1 Sememe-Encoded Word Representation

WRL is a fundamental technique in many NLP tasks such as neural machine translation [75] and language modeling [6]. Many works have been proposed for learning better word representations, among which word2vec [46] strikes an excellent balance between effectiveness and efficiency. Later works propose to leverage existing KBs (such as WordNet [17] and HowNet [74]) to improve word representation.

We first introduce our sememe-encoded word representation learning (SE-WRL) [53]. SE-WRL assumes each word sense is composed of sememes and conducts word sense disambiguation according to the contexts. In this way, we could learn representations of sememes, senses, and words simultaneously. Moreover, SE-WRL proposes an attention-based method to choose an appropriate word sense according to contexts automatically. In the following paragraphs, we introduce three different variants for SE-WRL. For a word w, we denote $S^{(w)}$ as its sense set. $S^{(w)} = \{s_1^{(w)}, \cdots, s_{|S^{(w)}|}^{(w)}\}$ may contain multiple senses. For each sense $s_i^{(w)}$, we denote $X_i^{(w)} = \{x_1^{(s_i)}, \cdots, x_{|X_i^{(w)}|}^{(s_i)}\}$ as the sememe set for this sense, with $\mathbf{x}_j^{(s_i)}$ being the embedding for the correspond sememe $x_j^{(s_i)}$.

Skip-Gram Model Since SE-WRL extends from the skip-gram of word2vec [46], we first give a brief introduction to the skip-gram model. For a series of words $\{w_1, \cdots, w_N\}$, the model targets at maximizing the probability of contextual words based on a centered word w_c. Specifically, we minimize the following loss (more details could be found in Chap. 2):

$$\mathcal{L} = -\sum_{c=l+1}^{n-l} \sum_{-l \leq k \leq l, k \neq 0} \log P(w_{c+k}|w_c), \quad (10.1)$$

where l is the size of the sliding window and $P(w_{c+k}|w_c)$ stands for the predictive probability of the context word w_{c+k} conditioned on the centered word w_c. Denoting V as the vocabulary, the probability is formalized as follows:

$$P(w_{c+k}|w_c) = \frac{\exp(\mathbf{w}_{c+k} \cdot \mathbf{w}_c)}{\sum_{w_s \in V} \exp(\mathbf{w}_s \cdot \mathbf{w}_c)}, \quad (10.2)$$

Simple Sememe Aggregation Model (SSA) SSA is built upon the skip-gram model. It considers the sememes in all senses of a word w and learns the word embedding $\mathbf{w}$ by averaging the embeddings of all its sememes:

$$\mathbf{w} = \frac{1}{m} \sum_{s_i^{(w)} \in S^{(w)}} \sum_{x_j^{(s_i)} \in X_i^{(w)}} \mathbf{x}_j^{(s_i)}, \tag{10.3}$$

where m stands for the total number of the sememes of word w. SSA assumes that the word meaning is composed of smaller semantic units. Since sememes are shared by different words, SSA could utilize sememe knowledge to model semantic correlations among words, and words sharing similar sememes may have close representations.

Sememe Attention over Context Model (SAC) SSA modifies the word embedding to incorporate sememe knowledge. Nevertheless, each word in SSA is still bound to an individual representation, which cannot deal with polysemy in different contexts. Intuitively, we should have distinct embeddings for a word given different contexts. To implement this, we leverage the word sense annotation in HowNet and propose the sememe attention over context model (SAC), with its structure illustrated in Fig. 10.4. For a brief introduction, SAC leverages the attention mechanism

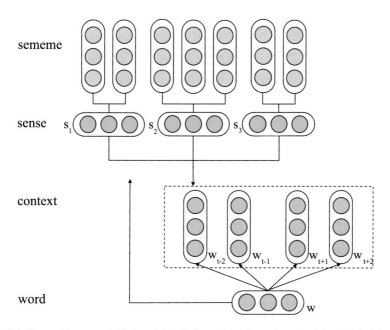

Fig. 10.4 The architecture of SAC model. This figure is re-drawn based on Fig. 10.2 in Niu et al. [53]

to select a proper sense for a word based on its context. More specifically, SAC conducts word sense disambiguation based on contexts to represent the word.

More specifically, SAC utilizes the original embedding of the word w and uses sememe embeddings to represent context word w_c. The word embedding is then employed to choose the proper senses to represent the context word. The context word embedding $\mathbf{w}_c$ can be formalized as follows:

$$\mathbf{w}_c = \sum_{j=1}^{|S^{(w_c)}|} \text{ATT}(s_j^{(w_c)}) \mathbf{s}_j^{(w_c)}, \tag{10.4}$$

where $\mathbf{s}_j^{(w_c)}$ is the j-th sense embedding of w_c and $\text{ATT}(s_j^{(w_c)})$ denotes the attention score of the j-th sense of the word w. The attention score is calculated as:

$$\text{ATT}(s_j^{(w_c)}) = \frac{\exp(\mathbf{w} \cdot \hat{\mathbf{s}}_j^{(w_c)})}{\sum_{k=1}^{|S^{(w_c)}|} \exp(\mathbf{w} \cdot \hat{\mathbf{s}}_k^{(w_c)})}, \tag{10.5}$$

Note $\hat{\mathbf{s}}_j^{(w_c)}$ is different from $\mathbf{s}_j^{(w_c)}$ and is obtained with the average of sememe embeddings (in this way, we could incorporate the sememe knowledge):

$$\hat{\mathbf{s}}_j^{(w_c)} = \frac{1}{|X_j^{(w_c)}|} \sum_{k=1}^{|X_j^{(w_c)}|} \mathbf{x}_k^{(s_j)}. \tag{10.6}$$

The attention technique is based on the assumption that if a context word sense embedding is more relevant to $\mathbf{w}$, then this sense should contribute more to the context word embeddings. Based on the attention mechanism, we represent the context word as a weighted summation of sense embeddings.

Sememe Attention over Target Model (SAT) The aforementioned SAC model selects proper senses and sememes for context words. Intuitively, we could use similar methods to choose the proper senses for the target word by considering the context words as attention. This is implemented by the sememe attention over target model (SAT), which is shown in Fig. 10.5.

Conversely, SAT learns sememe embeddings for target words and original word embeddings for context words. SAT applies context words to compute attention over the senses of w and learn w's embedding. Formally, we have:

$$\mathbf{w} = \sum_{j=1}^{|S^{(w)}|} \text{ATT}(s_j^{(w)}) \mathbf{s}_j^{(w)}, \tag{10.7}$$

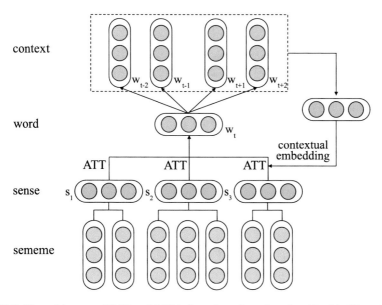

Fig. 10.5 The architecture of SAT model. This figure is re-drawn based on Fig. 3 in Niu et al. [53]

and we can calculate the context-based attention as follows:

$$\text{ATT}(s_j^{(w)}) = \frac{\exp(\mathbf{w}_c' \cdot \hat{\mathbf{s}}_j^{(w)})}{\sum_{k=1}^{|S^{(w)}|} \exp(\mathbf{w}_c' \cdot \hat{\mathbf{s}}_k^{(w)})}, \tag{10.8}$$

where the average of sememe embeddings $\hat{\mathbf{s}}_j^{(w)}$ is also used to learn the embeddings for each sense $s_j^{(w)}$. Here, $\mathbf{w}_c'$ denotes the context embedding, consisting of the embeddings of the contextual words of w_i:

$$\mathbf{w}_c' = \frac{1}{2K} \sum_{k=i-K}^{k=i+K} \mathbf{w}_k, \quad k \neq i, \tag{10.9}$$

where K denotes the window size. SAC merely leverages one target word as attention to choose the context words' senses, whereas SAT resorts to multiple context words as attention to choose the proper senses of target words. Therefore, SAT is better at WSD and results in more accurate and reliable word representations. In general, all the above methods could successfully incorporate sememe knowledge into word representations and achieve better performance.

10.3.2 Sememe-Regularized Word Representation

Besides learning embeddings for sememes, we explore how to incorporate sememe knowledge to improve word representations. We propose two variants [62] for sememe-based word representation: relation-based and embedding-based word representation. By introducing the information of sememe-based linguistic KBs into each word embedding, sememe-guided word representation could improve the performance in downstream applications like sememe prediction.

Sememe Relation-Based Word Representation Relation-based word representation is a simple and intuitive method, which aims to make words with similar sememe annotations have similar embeddings. First, a synonym list is constructed from HowNet, with words sharing a certain number (e.g., 3) of sememes regarded as synonyms. Next, the word embeddings of synonyms are optimized to be closer. Formally, let $\mathbf{w}_i$ be the original word embedding of w_i and $\hat{\mathbf{w}}_i$ be its adjusted word embedding. Denote $\mathrm{Syn}(w_i)$ as the synonym set of word w_i; the loss function is formulated as follows:

$$\mathcal{L}_{\text{sememe}} = \sum_{w_i \in V} \left(\alpha_i \| \mathbf{w}_i - \hat{\mathbf{w}}_i \|^2 + \sum_{w_j \in \mathrm{Syn}(w_i)} \beta_{ij} \| \hat{\mathbf{w}}_i - \hat{\mathbf{w}}_j \|^2 \right), \tag{10.10}$$

where α_i and β_{ij} balance the contribution of the two loss terms and V denotes the vocabulary.

Sememe Embedding-Based Word Representation Despite the simplicity of the relation-based method, it cannot take good advantage of the information of HowNet because it disregards the complicated relations among sememes and words, as well as relations among various sememes. Regarding this limitation, we propose the sememe embedding-based method.

Specifically, sememes are represented using distributed embeddings and placed into the same semantic space as words. This method utilizes sememe embeddings as additional regularizers to learn better word embeddings. Both word embeddings and sememe embeddings are jointly learned.

Formally, a word-sememe matrix $\mathbf{M}$ is built from HowNet, where $\mathbf{M}_{ij} = 1$ indicates that the word w_i is annotated with the sememe x_j; otherwise $\mathbf{M}_{ij} = 0$. The loss function can be defined by factorizing $\mathbf{M}$ as follows:

$$\mathcal{L}_{\text{sememe}} = \sum_{w_i \in V, x_j \in X} (\mathbf{w}_i \cdot \mathbf{x}_j + \mathbf{b}_i + \mathbf{b}'_j - \mathbf{M}_{ij})^2, \tag{10.11}$$

where $\mathbf{b}_i$ and $\mathbf{b}'_j$ are the bias terms of w_i and x_j and X denotes the full sememe set. $\mathbf{w}_i$ and $\mathbf{x}_j$ denote the embeddings of the word w_i and the sememe x_j.

In this method, word embeddings and sememe embeddings are learned in a unified semantic space. The information about the relations among words and sememes is implicitly injected into word embeddings. In this way, the word embeddings are expected to be more suitable for sememe prediction. In summary, either sememe relation-based methods or sememe embedding-based methods could successfully incorporate sememe knowledge into word representations and benefit the performance in specific applications.

10.4 Sememe-Guided Natural Language Processing

In the last section, we introduce how to represent the sememe knowledge annotated in HowNet, with a focus on word representation learning. In fact, linguistic KBs such as HowNet contain rich knowledge, which could also be incorporated into modern neural networks to effectively assist various downstream NLP tasks. In this section, we elaborate on several representative NLP techniques combined with sememe knowledge, including semantic compositionality modeling, language modeling, and sememe-incorporated recurrent neural networks (RNNs). The introduction of this part is based on our research works [29, 61, 66].

10.4.1 Sememe-Guided Semantic Compositionality Modeling

Semantic compositionality (SC) means the semantic meaning of a syntactically complicated unit is influenced by the meanings of the combination rule and the unit's constituents [56]. SC has shown importance in many NLP tasks including language modeling [50], sentiment analysis [45, 70], syntactic parsing [70], etc. For more details of SC, please refer to Chap. 3.

To explore the SC task, we need to represent multiword expressions (MWEs) (embeddings of phrases and compounds). A prior work [49] formulates the SC task with a general framework as follows:

$$\mathbf{p} = f(\mathbf{w}_1, \mathbf{w}_2, \mathcal{R}, \mathcal{K}), \tag{10.12}$$

where $\mathbf{p}$ denotes the MWE embedding, $\mathbf{w}_1$ and $\mathbf{w}_2$ represent the embeddings of two constituents that belong to the MWE, $\mathcal{R}$ is the combination rule, $\mathcal{K}$ means the extra knowledge needed for learning the MWE's semantics, and f denotes the compositionality function.

Most of the existing methods focus on reforming compositionality function f [5, 27, 70, 71], ignoring both $\mathcal{R}$ and $\mathcal{K}$. Some researchers try to integrate combination rule $\mathcal{R}$ to build better SC models [9, 40, 76, 86]. However, few works consider additional knowledge $\mathcal{K}$, except that Zhu et al. [87] incorporate task-specific knowledge into an RNN to solve sentence-level SC.

We argue that the sememe knowledge conduces to modeling SC and propose a novel sememe-based method to model semantic compositionality [61]. To begin with, we conduct an SC degree (SCD) measurement experiment and observe that the

SCD obtained by the sememe formulae is correlated with manually annotated SCDs. Then we present two SC models based on sememe knowledge for representing MWEs, which are dubbed semantic compositionality with aggregated sememe (SCAS) and semantic compositionality with mutual sememe attention (SCMSA). We demonstrate that both models achieve superior performance in the MWE similarity computation task and sememe prediction task. In the following, we first introduce sememe-based SC degree (SCD) computation formulae and then discuss our sememe-incorporated SC models.

Sememe-Based SCD Computation Formulae Despite the fact that SC is a common phenomenon of MWEs, there exist some MWEs that are not fully semantically compositional. As a matter of fact, distinct MWEs have distinct SCDs. We propose to leverage sememes for SCD measurement [61]. We assume that a word's sememes precisely reflect the meaning of a word. Based on this assumption, we propose 4 SCD computation formulae (0, 1, 2, and 3). A smaller number means lower SCD. X_p represents the sememe sets of an MWE. X_{w_1} and X_{w_2} denote the sememe set of MWE's first and second constituent. We briefly introduce these four SCDs as follows:

1. For SCD 0, an MWE is entirely non-compositional, with the corresponding SCD being the lowest. The sememes of the MWE are different from those of its constituents. This implies that the constituents of the MWE cannot compose the MWE's meaning.
2. For SCD 1, the sememes of an MWE and its constituents have some overlap. However, the MWE owns unique sememes that are not shared by its constituents.
3. For SCD 2, an MWE's sememe set is a subset of the sememe sets of constituents. This implies the constituents' meanings cannot accurately infer the meaning of the MWE.
4. For SCD 3, an MWE is entirely semantically compositional and has the highest SCD. The MWE's sememe set is identical to the sememe sets of two constituents. This implies that MWE has the same meaning as the combination of its constituents' meanings.

We show an example for each SCD in Table 10.1, including a Chinese MWE, its two constituents, and their sememes.

SCD Computation Formulae Evaluation In order to test the effectiveness of the proposed formulae, we annotate an SCD dataset [61]. A total number of 500 Chinese MWEs are manually labeled with SCDs. Then we test the correlation between SCDs of the MWEs labeled by humans and those obtained by sememe-based rules. The Spearman's correlation coefficient is 0.74. The high correlation demonstrates the powerful capability of sememes in computing MWEs' SCDs.

Sememe-Incorporated SC Models Next, we discuss the aforementioned sememe-incorporated SC models, covering (1) semantic compositionality with aggregated sememe (SCAS) and (2) semantic compositionality with mutual sememe attention (SCMSA)

Table 10.1 Sememe-based semantic compositionality degree computation formulae and examples. The content of this table is from the original paper [61]

SCD	Our computation formulae	Examples MWEs and constituents	Sememes
3	$X_p = X_{w_1} \cup X_{w_2}$	农民起义(peasant uprising) 农民 (peasant) 起义(uprising)	事情|fact,职位|occupation,政|politics,暴动|uprise,人|human,农|agricultural 职位|**occupation**,人|**human**,农|**agricultural** 暴动|**uprise**,事情|**fact**,政|**politics**
2	$X_p \subseteq (X_{w_1} \cup X_{w_2})$	几何图形(geometric figure) 几何 (geometry; how much) 图形(figure)	数学|math,图像|image 数学|**math**,知识|knowledge,疑问|question,功能词|funcword 图像|**image**
1	$X_p \cap (X_{w_1} \cup X_{w_2}) \neq \emptyset \wedge X_p \not\subset (X_{w_1} \cup X_{w_2})$	应考(engage a test) 应 (deal with; echo; agree) 考(quiz; check)	考试|exam,从事|engage 处理|handle,回应|respond,同意|agree,遵循|obey,功能词|funcword,姓|surname 考试|**exam**,查|check
0	$X_p \cap (X_{w_1} \cup X_{w_2}) = \emptyset$	画句号(finish) 画 (draw) 句号(period)	完毕|finish 画|draw,部件|part,图像|image,文字|character,表示|express 符号|symbol,语文|text

(SCMSA). From now on, we introduce how to integrate combination rules into these models.

We first consider the case when sememe knowledge is incorporated in MWE modeling without combination rules. Following Eq. (10.12), for an MWE $p = \{w_1, w_2\}$, we represent its embedding as:

$$\mathbf{p} = f(\mathbf{w}_1, \mathbf{w}_2, \mathcal{K}), \tag{10.13}$$

where $\mathbf{p} \in \mathbb{R}^d$, $\mathbf{w}_1 \in \mathbb{R}^d$, and $\mathbf{w}_2 \in \mathbb{R}^d$ denote the embeddings of the MWE p, word w_1, and word w_2, d is the embedding dimension, and $\mathcal{K}$ denotes the sememe knowledge. Since an MWE is generally not present in the KB, hence we merely have access to the sememes of w_1 and w_2. Denote X as the set of all the sememes, $X^{(w)} = \{x_1, \cdots, x_{|X^{(w)}|}\} \subset X$ as the sememe set of w, and $\mathbf{x} \in \mathbb{R}^d$ as sememe x's embedding.

1. As illustrated in Fig. 10.6, SCAS concatenates a constituent's embedding and its sememes' embeddings:

$$\mathbf{w}_1' = \sum_{x_i \in X^{(w_1)}} \mathbf{x}_i, \quad \mathbf{w}_2' = \sum_{x_j \in X^{(w_2)}} \mathbf{x}_j, \tag{10.14}$$

where $\mathbf{w}_1'$ and $\mathbf{w}_2'$ denote the aggregated sememe embeddings of w_1 and w_2. We calculate $\mathbf{p}$ as:

$$\mathbf{p} = \tanh(\mathbf{W}_c \, \text{concat}(\mathbf{w}_1 + \mathbf{w}_2; \mathbf{w}_1' + \mathbf{w}_2') + \mathbf{b}_c), \tag{10.15}$$

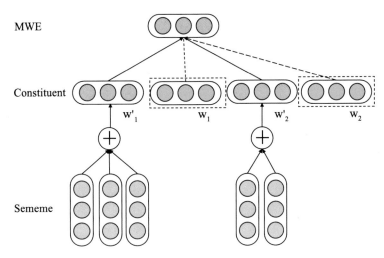

Fig. 10.6 The architecture of SCAS model. This figure is re-drawn based on Fig. 1 in Qi et al. [61]

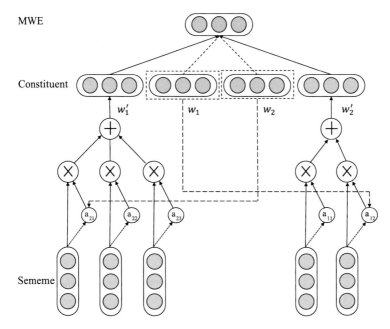

Fig. 10.7 The architecture of the SCMSA model that is introduced. This figure is re-drawn based on Fig. 2 in Qi et al. [61]

where $\mathbf{b}_c \in \mathbb{R}^d$ denotes a bias term and $\mathbf{W}_c \in \mathbb{R}^{d \times 2d}$ denotes a composition matrix.

2. SCAS simply adds up all the sememe embeddings of a constituent. Intuitively, a constituent's sememes may own distinct weights when they are composed of other constituents. To this end, SCMSA (Fig. 10.7) is introduced, which utilizes the attention mechanism to assign weights to sememes (here we take an example to show how to use w_1 to calculate the attention score for w_2):

$$\mathbf{e}_1 = \tanh(\mathbf{W}_a \mathbf{w}_1 + \mathbf{b}_a),$$

$$\alpha_{2,i} = \frac{\exp(\mathbf{x}_i \cdot \mathbf{e}_1)}{\sum_{x_j \in X^{(w_2)}} \exp(\mathbf{x}_j \cdot \mathbf{e}_1)}, \quad (10.16)$$

$$\mathbf{w}_2' = \sum_{x_j \in X^{(w_2)}} \alpha_{2,j} \mathbf{x}_j,$$

where $\mathbf{W}_a \in \mathbb{R}^{d \times d}$ and $\mathbf{b}_a \in \mathbb{R}^d$ are tunable parameters. $\mathbf{w}_1'$ can be calculated in a similar way. $\mathbf{p}$ is obtained the same as Eq. (10.15).

Integrating Combination Rules We can further incorporate combination rules to the sememe-incorporated SC models [61] as follows:

$$\mathbf{p} = f(\mathbf{w}_1, \mathbf{w}_2, \mathcal{K}, \mathcal{R}). \qquad (10.17)$$

MWEs with different combination rules are assigned with totally different composition matrices $\mathbf{W}_c^r \in \mathbb{R}^{d \times 2d}$, where $r \in \mathcal{R}_s$ and $\mathcal{R}_s$ refer to a combination syntax rule set. The combination rules include adjective-noun (Adj-N), noun-noun (NN), verb-noun (V-N), etc. Considering that there exist various combination rules, and some composition matrices are sparse, therefore, the composition matrices may not be well-trained. Regarding this issue, we represent a composition matrix $\mathbf{W}_c$ as the summation of a low-rank matrix containing combination rule information and a matrix containing compositionality information:

$$\mathbf{W}_c = \mathbf{U}_1^r \mathbf{U}_2^r + \mathbf{W}_c^c, \qquad (10.18)$$

where $\mathbf{U}_1^r \in \mathbb{R}^{d \times d_r}$, $\mathbf{U}_2^r \in \mathbb{R}^{d_r \times 2d}$, $d_r \in \mathbb{N}_+$, and $\mathbf{W}_c^c \in \mathbb{R}^{d \times 2d}$. In experiments, the sememe-incorporated models achieve better performance on the MWE similarity computation task and sememe prediction task. These results reveal the benefits of sememe knowledge in compositionality modeling.

10.4.2 Sememe-Guided Language Modeling

Language modeling (LM) targets at measuring the joint probability of a sequence of words. The joint probability reflects the sequence's fluency. LM is a critical component in various NLP tasks, e.g., machine translation [12, 13], speech recognition [38], information retrieval [7, 30, 47, 59], document summarization [4, 67], etc.

Trained with large-scale text corpora, probabilistic language models calculate the conditional probability of the next word based on its contextual words. Traditional language models follow the assumption that words are atomic symbols and thus represent a sequence at the word level. Nevertheless, this does not necessarily hold true. Consider the following example:

The US trade deficit last year is initially estimated to be 40 billion _____ .

Our goal is to predict the word for the blank. At first glance, people may think of a *unit* to fill; after deep consideration, they may realize that the blank should be filled with a *currency unit*. Based on the country (*The US*) the sentence mentions, we can finally know it is an *American currency unit*. Then we can predict the word *dollars*. The *American*, *currency*, and *unit*, which are basic semantic units of the word *dollars*, are also the sememes of the word *dollars*. However, the above process is not explicitly modeled by traditional word-level language models. Hence, explicitly introducing sememes could conduce to language modeling.

In fact, it is non-trivial to incorporate discrete sememe knowledge into neural language models, because it does not fit with the continuous representations of neural networks. To address the above issue, we propose a sememe-driven language model (SDLM) to utilize sememe knowledge [29]. When predicting the next word, (1) SDLM estimates sememes' distribution based on the context; (2) after that, treating those sememes as experts, SDLM employs a sparse expert product to choose the possible senses; (3) then SDLM calculates the word distribution by marginalizing the distribution of senses.

Accordingly, SDLM comprises three components: a sememe predictor, a sense predictor, and a word predictor. The sememe predictor considers the contextual information and assigns a weight for every sememe. In the sense predictor, we regard each sememe as an expert and predict the probability over a set of senses. Lastly, the word predictor calculates the probability of every word. Next, we briefly introduce the design of the three modules.

Sememe Predictor A context vector $\mathbf{g} \in \mathbb{R}^{d_1}$ is considered in the sememe predictor, and the predictor computes a weight for each sememe. Given the context $\{w_1, w_2, \cdots, w_{t-1}\}$, the probability $P(x_k|\mathbf{g})$ whether the next word w_t has the sememe x_k is calculated by:

$$P(x_k|\mathbf{g}) = \text{Sigmoid}(\mathbf{g} \cdot \mathbf{v}_k + b_k), \tag{10.19}$$

where $\mathbf{v}_k \in \mathbb{R}^{d_1}$, $b_k \in \mathbb{R}$ are tunable parameters.

Sense Predictor Motivated by product of experts (PoE) [31], each sememe is regarded as an expert who only predicts the senses connected with it. Given the sense embedding $\mathbf{s} \in \mathbb{R}^{d_2}$ and the context vector $\mathbf{g} \in \mathbb{R}^{d_1}$, the sense predictor calculates $\phi^{(k)}(\mathbf{g}, \mathbf{s})$, which means the score of sense s provided by sememe expert x_k. A bilinear layer parameterized using a matrix $\mathbf{U}_k \in \mathbb{R}^{d_1 \times d_2}$ is chosen to compute $\phi^{(k)}(\cdot, \cdot)$:

$$\phi^{(k)}(\mathbf{g}, \mathbf{s}) = \mathbf{g}^\top \mathbf{U}_k \mathbf{s}. \tag{10.20}$$

The probability $P^{(x_k)}(s|\mathbf{g})$ of sense s given by expert x_k can be formulated as:

$$P^{(x_k)}(s|\mathbf{g}) = \frac{\exp(q_k C_{k,s} \phi^{(k)}(\mathbf{g}, \mathbf{s}))}{\sum_{s' \in S^{(x_k)}} \exp(q_k C_{k,s'} \phi^{(k)}(\mathbf{g}, \mathbf{s}'))}, \tag{10.21}$$

where $C_{k,s}$ is a constant and $S^{(x_k)}$ denotes the set of senses that contain sememe x_k. q_k controls the magnitude of the term $C_{k,s} \phi^{(k)}(\mathbf{g}, \mathbf{s})$. Hence it decides the flatness of the sense distribution output by x_k. Lastly, the predictions can be summarized on sense s by leveraging the probability products computed based on related experts. In other words, the sense s's probability is defined as:

$$P(s|\mathbf{g}) \sim \prod_{x_k \in X^{(s)}} P^{(x_k)}(s|\mathbf{g}), \tag{10.22}$$

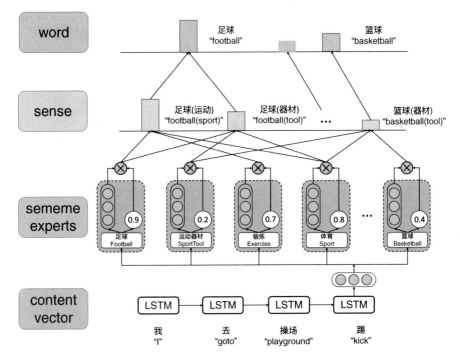

Fig. 10.8 The architecture of SDLM model. This figure is re-drawn based on Fig. 2 in Gu et al. [29]

where $\sim$ indicates that $P(s|\mathbf{g})$ is proportional to $\prod_{x_k \in X^{(s)}} P^{(x_k)}(s|\mathbf{g})$. $X^{(s)}$ denotes the set of sememes of the sense s.

Word Predictor As illustrated in Fig. 10.8, in the word predictor, the probability $P(w|\mathbf{g})$ is calculated through adding up probabilities of s:

$$P(w|\mathbf{g}) = \sum_{s \in S^{(w)}} P(s|\mathbf{g}), \tag{10.23}$$

where $S^{(w)}$ denotes the senses belonging to the word w. When experimenting with both the task of language modeling and headline generation, SDLM achieves remarkable performance, which is due to the benefits of incorporating sememe knowledge. In-depth case studies further reveal that SDLM could improve both the robustness and interpretability of language models.

10.4.3 Sememe-Guided Recurrent Neural Networks

Up until now, we have introduced how to incorporate sememe knowledge into word representation, compositionality modeling, and language modeling. Most of the existing works exploit sememes for limited NLP tasks, and few works have explored leveraging sememes in a general way, e.g., employing sememes for better sequence modeling to achieve better performance in various downstream tasks. In the following paragraphs, we introduce how to incorporate sememes into recurrent neural networks, with the aim of enhancing the ability of sequence modeling [66].

In fact, previous works have tried to incorporate other linguistic KBs into RNNs [1, 54, 78, 81]. The utilized KBs are generally word-level KBs (e.g., WordNet and ConceptNet). Differently, HowNet utilizes sememes to compositionally explain the meanings of words. Consequently, directly adopting existing algorithms to incorporate sememes into RNNs is hard. We propose three algorithms to incorporate sememe knowledge into RNNs [66]. Two representative RNN variants, i.e., LSTM and GRU, are considered.

Preliminaries for RNN Architecture First, let us review some basics about the architectures of LSTM [33]. An LSTM comprises a series of cells, each corresponding to a token. At each step t, the word embedding $\mathbf{w}_t$ is input into the LSTM to produce the cell state $\mathbf{c}_t$ and the hidden state $\mathbf{h}_t$. Based on the previous cell state $\mathbf{c}_{t-1}$ and hidden state $\mathbf{h}_{t-1}$, $\mathbf{c}_t$ and $\mathbf{h}_t$ are calculated as follows:

$$
\begin{aligned}
\mathbf{f}_t &= \text{Sigmoid}(\mathbf{W}_f \, \text{concat}(\mathbf{w}_t; \mathbf{h}_{t-1}) + \mathbf{b}_f), \\
\mathbf{i}_t &= \text{Sigmoid}(\mathbf{W}_I \, \text{concat}(\mathbf{w}_t; \mathbf{h}_{t-1}) + \mathbf{b}_I), \\
\tilde{\mathbf{c}}_t &= \tanh(\mathbf{W}_c \, \text{concat}(\mathbf{w}_t; \mathbf{h}_{t-1}) + \mathbf{b}_c), \\
\mathbf{c}_t &= \mathbf{f}_t \odot \mathbf{c}_{t-1} + \mathbf{i}_t \odot \tilde{\mathbf{c}}_t, \\
\mathbf{o}_t &= \text{Sigmoid}(\mathbf{W}_o \, \text{concat}(\mathbf{w}_t; \mathbf{h}_{t-1}) + \mathbf{b}_o), \\
\mathbf{h}_t &= \mathbf{o}_t \odot \tanh(\mathbf{c}_t),
\end{aligned}
\tag{10.24}
$$

where $\mathbf{f}_t$, $\mathbf{i}_t$, and $\mathbf{o}_t$ denote the output embeddings of the forget gate, input gate, and output gate, respectively. $\mathbf{W}_f$, $\mathbf{W}_I$, $\mathbf{W}_c$, and $\mathbf{W}_o$ are weight matrices and $\mathbf{b}_f$, $\mathbf{b}_I$, $\mathbf{b}_c$, and $\mathbf{b}_o$ are bias terms.

GRU [21] has fewer gates than LSTM and can be viewed as a simplification for LSTM. Given the hidden state $\mathbf{h}_{t-1}$ and the input $\mathbf{w}_t$, GRU has a reset gate $\mathbf{r}_t$ and an update gate $\mathbf{z}_t$ and computes the output $\mathbf{h}_t$ as:

$$
\begin{aligned}
\mathbf{z}_t &= \text{Sigmoid}(\mathbf{W}_z \, \text{concat}(\mathbf{w}_t; \mathbf{h}_{t-1}) + \mathbf{b}_z), \\
\mathbf{r}_t &= \text{Sigmoid}(\mathbf{W}_r \, \text{concat}(\mathbf{w}_t; \mathbf{h}_{t-1}) + \mathbf{b}_r), \\
\tilde{\mathbf{h}}_t &= \tanh(\mathbf{W}_h \, \text{concat}(\mathbf{w}_t; \mathbf{r}_t \odot \mathbf{h}_{t-1}) + \mathbf{b}_h), \\
\mathbf{h}_t &= (\mathbf{1} - \mathbf{z}_t) \odot \mathbf{h}_{t-1} + \mathbf{z}_t \odot \tilde{\mathbf{h}}_t,
\end{aligned}
\tag{10.25}
$$

where $\mathbf{W}_z$, $\mathbf{W}_r$, $\mathbf{W}_h$, $\mathbf{b}_z$, $\mathbf{b}_r$, and $\mathbf{b}_h$ are tunable parameters.

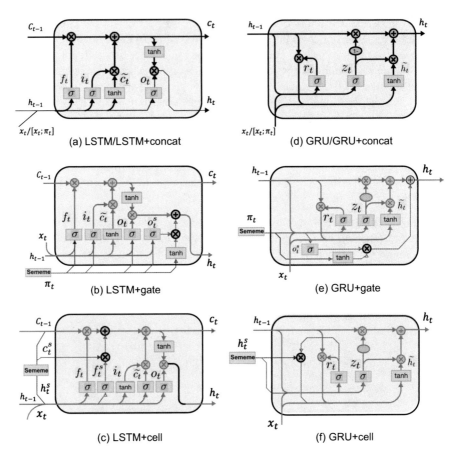

Fig. 10.9 The architectures of three methods for incorporating sememe knowledge into RNNs. This figure is re-drawn based on Fig. 2 in Qin et al. [66]

Next, we elaborate on the three proposed methods of incorporating sememes into RNNs, including simple concatenation (+concat), adding sememe output gate (+gate), and introducing sememe-RNN cell (+cell). We illustrate them in Fig. 10.9.

Simple Concatenation The first method focuses on the input and directly concatenates the summation of the sememe embeddings and the word embedding. Specifically, we have:

$$\boldsymbol{\pi}_t = \frac{1}{|X^{(w_t)}|} \sum_{x \in X^{(w_t)}} \mathbf{x},$$

$$\tilde{\mathbf{w}}_t = \text{concat}(\mathbf{w}_t; \boldsymbol{\pi}_t),$$

(10.26)

where $\mathbf{x}$ is the sememe embedding of x and $\tilde{\mathbf{w}}_t$ denotes the modified word embedding that contains sememe knowledge.

Sememe Output Gate Simple concatenation incorporates sememe knowledge in a shallow way and enhances only the word embeddings. To leverage sememe knowledge in a deeper way, we present the second method by adding a sememe output gate $\mathbf{o}_t^s$. This architecture explicitly models the knowledge flow of sememes. Note that the sememe output gate is designed especially for LSTM and GRU. This output gate controls the flow of sememe knowledge in the whole model. Formally, we have (the modified parts of the model structures are underlined):

$$
\begin{aligned}
\mathbf{f}_t &= \text{Sigmoid}(\mathbf{W}_f \operatorname{concat}(\mathbf{x}_t; \mathbf{h}_{t-1}; \underline{\pi_t}) + \mathbf{b}_f), \\
\mathbf{i}_t &= \text{Sigmoid}(\mathbf{W}_I \operatorname{concat}(\mathbf{x}_t; \mathbf{h}_{t-1}; \underline{\pi_t}) + \mathbf{b}_i), \\
\tilde{\mathbf{c}}_t &= \tanh(\mathbf{W}_c \operatorname{concat}(\mathbf{x}_t; \mathbf{h}_{t-1}) + \mathbf{b}_c), \\
\mathbf{c}_t &= \mathbf{f}_t \odot \mathbf{c}_{t-1} + \mathbf{i}_t \odot \tilde{\mathbf{c}}_t, \\
\mathbf{o}_t &= \text{Sigmoid}(\mathbf{W}_o \operatorname{concat}(\mathbf{x}_t; \mathbf{h}_{t-1}; \underline{\pi_t}) + \mathbf{b}_o), \\
\underline{\mathbf{o}_t^s} &= \underline{\text{Sigmoid}(\mathbf{W}_{o^s} \operatorname{concat}(\mathbf{x}_t; \mathbf{h}_{t-1}; \pi_t) + \mathbf{b}_{o^s})}, \\
\mathbf{h}_t &= \mathbf{o}_t \odot \tanh(\mathbf{c}_t) + \underline{\mathbf{o}_t^s \odot \tanh(\mathbf{W}_c \pi_t)},
\end{aligned}
\tag{10.27}
$$

where $\mathbf{W}_{o^s}$ and $\mathbf{b}_{o^s}$ are tunable parameters.

Similarly, we can rewrite the formulation of a GRU cell as:

$$
\begin{aligned}
\mathbf{z}_t &= \text{Sigmoid}(\mathbf{W}_z \operatorname{concat}(\mathbf{x}_t; \mathbf{h}_{t-1}; \underline{\pi_t}) + \mathbf{b}_z), \\
\mathbf{r}_t &= \text{Sigmoid}(\mathbf{W}_r \operatorname{concat}(\mathbf{x}_t; \mathbf{h}_{t-1}; \underline{\pi_t}) + \mathbf{b}_r), \\
\underline{\mathbf{o}_t^s} &= \underline{\text{Sigmoid}(\mathbf{W}_o \operatorname{concat}(\mathbf{x}_t; \mathbf{h}_{t-1}; \pi_t) + \mathbf{b}_o)}, \\
\tilde{\mathbf{h}}_t &= \tanh(\mathbf{W}_h \operatorname{concat}(\mathbf{x}_t; \mathbf{r}_t \odot \mathbf{h}_{t-1}) + \mathbf{b}_h), \\
\mathbf{h}_t &= (\mathbf{1} - \mathbf{z}_t) \odot \mathbf{h}_{t-1} + \mathbf{z}_t \odot \tilde{\mathbf{h}}_t + \underline{\mathbf{o}_t^s \tanh(\pi_t)},
\end{aligned}
\tag{10.28}
$$

where $\mathbf{b}_o$ is a bias vector, $\mathbf{o}_t^s$ denotes the sememe output gate, and $\mathbf{W}_o$ is a weight matrix.

Sememe-RNN Cell When adding the sememe output gate, despite the fact that sememe knowledge is deeply integrated into the model, the knowledge is still not fully utilized. Taking Eq. (10.27) as an example, $\mathbf{h}_t$ consists of two components: the information in $\mathbf{o}_t \odot \tanh(\mathbf{c}_t)$ has been processed by the forget gate, while the information in $\mathbf{o}_t^s \odot \tanh(\mathbf{W}_c \pi_t)$ is not processed. Thus these two components are incompatible.

To this end, we introduce an additional RNN cell to encode the sememe knowledge. The sememe embedding is fed into a sememe-LSTM cell. Another forget gate processes the sememe-LSTM cell's cell state. After that, the updated state is added to the original state. Moreover, the hidden state of the sememe-LSTM cell is incorporated in both the input gate and the output gate:

$$\mathbf{c}_t^s, \mathbf{h}_t^s = \underline{\text{LSTM}(\pi_t)},$$

$$\mathbf{f}_t = \text{Sigmoid}(\mathbf{W}_f \text{ concat}(\mathbf{x}_t; \mathbf{h}_{t-1}) + \mathbf{b}_f),$$

$$\underline{\mathbf{f}_t^s = \text{Sigmoid}(\mathbf{W}_f^s \text{ concat}(\mathbf{x}_t; \mathbf{h}_t^s) + \mathbf{b}_f^s)},$$

$$\mathbf{i}_t = \text{Sigmoid}(\mathbf{W}_I \text{ concat}(\mathbf{x}_t; \mathbf{h}_{t-1}; \underline{\mathbf{h}_t^s}) + \mathbf{b}_i),$$

$$\tilde{\mathbf{c}}_t = \tanh(\mathbf{W}_c \text{ concat}(\mathbf{x}_t; \mathbf{h}_{t-1}; \underline{\mathbf{h}_t^s}) + \mathbf{b}_c), \qquad (10.29)$$

$$\mathbf{o}_t = \text{Sigmoid}(\mathbf{W}_o \text{ concat}(\mathbf{x}_t; \mathbf{h}_{t-1}; \underline{\mathbf{h}_t^s}) + \mathbf{b}_o),$$

$$\mathbf{c}_t = \mathbf{f}_t \odot \mathbf{c}_{t-1} + \underline{\mathbf{f}_t^s \odot \mathbf{c}_t^s} + \mathbf{i}_t \odot \tilde{\mathbf{c}}_t,$$

$$\mathbf{h}_t = \mathbf{o}_t \odot \tanh(\mathbf{c}_t),$$

where $\mathbf{f}_t^s$ denotes the sememe forget gate and $\mathbf{c}_t^s$ and $\mathbf{h}_t^s$ denote the sememe cell state and sememe hidden state.

For GRU, the transition equation can be modified as:

$$\underline{\mathbf{h}_t^s = \text{GRU}(\pi_t)},$$

$$\mathbf{z}_t = \text{Sigmoid}(\mathbf{W}_z \text{ concat}(\mathbf{x}_t; \mathbf{h}_{t-1}; \underline{\mathbf{h}_t^s}) + \mathbf{b}_z),$$

$$\mathbf{r}_t = \text{Sigmoid}(\mathbf{W}_r \text{ concat}(\mathbf{x}_t; \mathbf{h}_{t-1}; \underline{\mathbf{h}_t^s}) + \mathbf{b}_r), \qquad (10.30)$$

$$\tilde{\mathbf{h}}_t = \tanh(\mathbf{W}_h \text{ concat}(\mathbf{x}_t; \mathbf{r}_t \odot (\mathbf{h}_{t-1} + \underline{\mathbf{h}_t^s})) + \mathbf{b}_h),$$

$$\mathbf{h}_t = (1 - \mathbf{z}_t) \odot \mathbf{h}_{t-1} + \mathbf{z}_t \odot \tilde{\mathbf{h}}_t,$$

where $\mathbf{h}_t^s$ denotes the sememe hidden state.

In experiments of language modeling, sentiment analysis, natural language inference, and paraphrase detection, the sememe-incorporated RNN surpasses the vanilla model, showing the usefulness of sememe knowledge in sequence modeling. These results demonstrate that, by incorporating sememe knowledge into general sequence modeling neural structures, we could enhance the performance on a variety of NLP tasks. Although we focus on RNNs, we contend that similar ideas could also be applied to other neural structures, which is promising to explore in the future.

10.5   Automatic Sememe Knowledge Acquisition

HowNet is built by several linguistic experts for more than 10 years. Apparently, manually constructing HowNet is time-consuming and labor-intensive. Meanwhile, new words or phrases are continually emerging, and the existing words' meanings are always changing as well. In this regard, manual inspection and updates for sememe annotation are becoming more and more overwhelming. Besides, it is also challenging to ensure annotation consistency among experts.

To address these issues, the sememe prediction task is defined to predict the sememes for word senses unannotated in a sememe KB. Ideally, a reliable sememe prediction tool could relieve the annotation burden of human experts. In the following, we first discuss the embedding-based methods for sememe prediction, which serve as the foundation for sememe prediction. After that, we introduce how to leverage internal information for sememe prediction. Finally, we extend the sememe prediction task to a cross-lingual setting. The introduction of this part is based on our research works [35, 60, 62, 77].

10.5.1 Embedding-Based Sememe Prediction

Intuitively, the words with similar meanings have overlapping sememes. Therefore, we strive to represent the semantics of sememes and words and model their semantic relations. To begin with, we introduce our representative sememe prediction algorithms [77], which are based on distributed representation learning [32].

Specifically, two methods are proposed: the first method is sememe prediction with word embeddings (SPWE). For a target word, we look for its relevant words in HowNet based on their embeddings. After that, we assign these relevant words' sememes to the target word. The algorithm is similar to collaborative filtering [68] in recommendation systems. The second method is sememe prediction with (aggregated) sememe embeddings (SPSE/SPASE). We learn sememe embeddings by factorizing the word-sememe matrix extracted from HowNet. Hence, the relation between words and sememes can be measured directly using the dot product of their embeddings, and we can assign relevant sememes to an unlabeled word.

Sememe Prediction with Word Embeddings Inspired by collaborative filtering in the personalized recommendation, words could be seen as users, and sememes can be viewed as products to be recommended. Given an unlabeled word, SPWE recommends sememes according to the word's most related words, assuming that similar words should have similar sememes. Formally, the probability $P(x_j|w)$ of sememe x_j given a word w is defined as:

$$P(x_j|w) = \sum_{w_i \in V} \cos(\mathbf{w}, \mathbf{w}_i) \mathbf{M}_{ij} c^{r_i}. \tag{10.31}$$

$\mathbf{M}$ contains the information of sememe annotation, where $\mathbf{M}_{ij} = 1$ means that the word w_i is annotated with the sememe x_j. V denotes the vocabulary, and $\cos(\cdot, \cdot)$ means the cosine similarity. A high probability $P(x_j|w)$ means the word w should probably be recommended with sememe x_j. A declined confidence factor c^{r_i} is set up for w_i, and r_i denotes the descending rank of $\cos(\mathbf{w}, \mathbf{w_i})$, and $c \in (0, 1)$ denotes a hyper-parameter.

Simple as it may sound, SPWE only leverages word embeddings for computing the similarities of words. In experiments, SPWE is demonstrated to have superior performance in sememe prediction. This is because different from the noisy user-item matrix in recommender systems, HowNet is manually designed by experts, and the word-sememe information can be reliably applied to recommend sememes.

Sememe Prediction with Sememe Embeddings Directly viewing sememes as discrete labels in SPWE could overlook the latent relations among sememes. To consider such latent relations, a sememe prediction with sememe embeddings (SPSE) model is proposed, which learns both word embeddings and sememe embeddings in a unified semantic space.

Inspired by GloVe [58], we optimize sememe embeddings by factorizing the sememe-sememe matrix and the word-sememe matrix. Both matrices can be derived from the annotation in HowNet. SPSE uses word embeddings pre-trained from an unlabeled corpus and freezes them during matrix factorization. After that, both sememe embeddings and word embeddings are encoded in the same semantic space. Then we could use the dot product between them to predict the sememes.

Similar to $\mathbf{M}$, a sememe-sememe matrix $\mathbf{C}$ is extracted, where $\mathbf{C}_{jk}$ is defined as the point-wise mutual information between sememes x_j and x_k. By factorizing $\mathbf{C}$, we finally get two different embeddings ($\mathbf{x}$ and $\bar{\mathbf{x}}$) for each sememe x. Then we optimize the following loss function to get sememe embeddings:

$$
\mathcal{L} = \sum_{w_i \in V, x_j \in X} \left(\mathbf{w}_i \cdot (\mathbf{x}_j + \bar{\mathbf{x}}_j) + \mathbf{b}_i + \mathbf{b}'_j - \mathbf{M}_{ij} \right)^2 + \lambda \sum_{x_j, x_k \in X} \left(\mathbf{x}_j \cdot \bar{\mathbf{x}}_k - \mathbf{C}_{jk} \right)^2,
$$

(10.32)

where $\mathbf{b}_i$ and $\mathbf{b}'_j$ are the bias terms. V and X denote the word vocabulary and the full sememe set. The above loss function consists of two parts, i.e., factorizing $\mathbf{M}$ and $\mathbf{C}$. Two parts are balanced by a hyper-parameter λ.

Considering that every word is generally labeled with 2 to 5 sememes in HowNet, the word-sememe matrix is very sparse, with most of the elements being zero. It is found empirically that, if both "zero elements" and "non-zero elements" are treated in the same way, the performance would degrade. Therefore, we choose distinct factorization strategies for zero and non-zero elements. For the former, the model factorizes them with a small probability (e.g., 0.5%), while for "non-zero elements," the model always chooses to factorize them. Armed with this strategy, the model can pay more attention to those "non-zero elements" (i.e., annotated word-sememe pairs).

Sememe Prediction with Aggregated Sememe Embeddings Based on the property of sememes, we can assume that the words are semantically comprised of sememes. A simple way to model such semantic compositionality is to represent word embeddings as a weighted summation of all its sememes' embeddings.

Based on this intuition, we propose sememe prediction with aggregated sememe embeddings (SPASE). SPASE is also built upon matrix factorization:

$$\mathbf{w}_i = \sum_{x_j \in X^{(w_i)}} \mathbf{M}'_{ij} \mathbf{x}_j, \qquad (10.33)$$

where $X^{(w_i)}$ denotes the sememe set of the word w_i and $\mathbf{M}'_{ij}$ represents the weight of sememe x_j for word w_i. To learn sememe embeddings, we can decompose the word embedding matrix $\mathbf{V}$ into the product of $\mathbf{M}'$ and the sememe embedding matrix $\mathbf{X}$, i.e., $\mathbf{V} = \mathbf{M}'\mathbf{X}$. During training, the pre-trained word embeddings are kept frozen.

Apparently, SPASE follows the assumption that sememes are the semantic units of words. In SPASE, each sememe can be treated as a small semantic component, and each word can be represented with the composition of several semantic units. However, the representation capability of SPASE is limited, especially when modeling the complex semantic relation between sememes and words.

10.5.2 Sememe Prediction with Internal Information

In the previous section, we introduce the automatic lexical sememe prediction proposed in our work [77]. Effective as they are, these methods do not consider the internal information in words, such as the characters of Chinese words. This is important for understanding those uncommon words. In this section, we introduce another work [35], which considers both internal and external information of words to predict sememes.

Specifically, we take the Chinese language as an example. In Chinese, each word typically comprises one or multiple characters, most of which have specific semantic meanings. A previous work [80] contends that over 90% Chinese characters are morphemes. There are two kinds of words in Chinese: single-morpheme words and compound words, where the latter takes up a dominant percentage. As shown in Fig. 10.10, a compound word's meanings are highly related to its internal characters. For instance, the compound word 铁匠(ironsmith) has two characters, 铁(iron) and 匠(craftsman), and 铁匠's semantic meaning could be derived by combining two characters (iron + craftsman → ironsmith).

We present character-enhanced sememe prediction (CSP). Beyond external context, CSP can also utilize character information to improve the performance of sememe prediction [35]. It conducts sememe prediction using the embeddings of a target word and its corresponding characters. Two methods of CSP are proposed to utilize character information, namely, sememe prediction with word-to-character filtering (SPWCF) and sememe prediction with character and sememe embeddings (SPCSE).

Sememe Prediction with Word-to-Character Filtering As mentioned before, sememe prediction can be conducted using similar techniques of collaborative

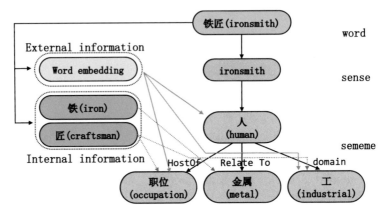

Fig. 10.10 Sememes of the word 铁匠*(ironsmith)* in HowNet. In this figure, we can see that *occupation, human,* and *industrial* can be derived by both internal (characters) and external (contexts) information. However, *metal* can be inferred only using the internal information in the character 铁*(iron)*. This figure is re-drawn based on Fig. 1 in the work of Jin et al. [35]

filtering. If two words have the same characters at the same positions, then these two words should be considered to be similar.

A Chinese character may have different meanings when it appears at different positions in a word [18]. Here we define three positions: *Begin, Middle,* and *End*. For a word $w = \{c_1, c_2, \cdots, c_{|w|}\}$, we define the characters at the *Begin* position as $\pi_B(w) = \{c_1\}$, the characters at the *Middle* position as $\pi_M(w) = \{c_2, \cdots, c_{|w|-1}\}$, and the characters at the *End* position as $\pi_E(w) = \{c_{|w|}\}$. The probability of a sememe x_j given a character c and a position p is defined as follows:

$$P_p(x_j|c) \sim \frac{\sum_{w_i \in V \wedge c \in \pi_p(w_i)} \mathbf{M}_{ij}}{\sum_{w_i \in V \wedge c \in \pi_p(w_i)} |X^{(w_i)}|}, \tag{10.34}$$

where $\mathbf{M}$ denotes the same matrix that is leveraged in SPWE and π_p may be π_B, π_M, or π_E. $\sim$ indicates that the left part is proportional to the right part. Finally, the probability $P(x_j|w)$ of x_j given w is computed as follows:

$$P(x_j|w) \sim \sum_{p \in \{B,M,E\}} \sum_{c \in \pi_p(w)} P_p(x_j|c). \tag{10.35}$$

Simple and efficient as it may seem, SPWCF performs well empirically, and the reason might be that compositional semantics are very common in Chinese compound words, and it is very intuitive to search similar words based on characters.

Sememe Prediction with Character and Sememe Embeddings To further consider the connections among sememes, sememe prediction with character and sememe embeddings (SPCSE) is proposed. Based on internal character information,

SPCSE learns sememe embeddings. Then SPCSE computes the semantic related-ness between words and sememes. When learning character embeddings, we need to consider that characters can be more ambiguous than words. Therefore, we borrow the idea from Chen et al. [18] and learn multiple embeddings for each character. When modeling the meaning of a word, the most representative character (together with its embedding) is selected.

Assume each character c has N_e embeddings: $\mathbf{c}^1, \cdots, \mathbf{c}^{N_e}$. Given a word w and a sememe x, by enumerating all w's character embeddings, we find the embedding that is the closest to the x's embedding. The distance is measured by cosine similarity. The closest character embedding is selected as the representation of the word w. Given $w = \{c_1, \cdots, c_{|w|}\}$ and x_j, we calculate:

$$k^*, r^* = \text{argmin}_{k,r} \left[1 - \cos(\mathbf{c}_k^r, \mathbf{x}_j' + \bar{\mathbf{x}}_j') \right], \tag{10.36}$$

where $\mathbf{x}_j'$ and $\bar{\mathbf{x}}_j'$ are the same as those defined in SPSE (the sememe embeddings of x_j). Using the same $\mathbf{M}$ and $\mathbf{C}$ in SPSE, the sememe embeddings can be obtained by optimizing the following loss:

$$\mathcal{L} = \sum_{w_i \in V, x_j \in X} \left(\mathbf{c}_{k*}^{r*} \cdot \left(\mathbf{x}_j' + \bar{\mathbf{x}}_j' \right) + \mathbf{b}_{k*}^c + \mathbf{b}_j'' - \mathbf{M}_{ij} \right)^2 + \lambda' \sum_{x_j, x_q \in X} \left(\mathbf{x}_j' \cdot \bar{\mathbf{x}}_q' - \mathbf{C}_{jq} \right)^2, \tag{10.37}$$

where $\mathbf{c}_{k*}^{r*}$ is the character embedding of w_i that is the closest to x_j. As the characters and the words do not lie in a unified space, new sememe embeddings are learned with different notations from those in Sect. 10.5.1. $\mathbf{b}_k^c$ and $\mathbf{b}_j''$ denote the biases of c_k and x_j, and λ' is the hyper-parameter balancing two parts. The score function of word $w = \{c_1, \cdots, c_{|w|}\}$ is computed as follows:

$$P(x_j|w) \sim \mathbf{c}_{k*}^{r*} \cdot \left(\mathbf{x}_j' + \bar{\mathbf{x}}_j' \right). \tag{10.38}$$

SPWCF originates from collaborative filtering, but SPCSE is based on factor-izing matrices. Both methods have one thing that is the same: they recommend the sememes of similar words but are different their definition of similarity. SPWCF/SPCSE uses internal information, but SPWE/SPSE utilizes external infor-mation. In consequence, combining the above models through ensembling could lead to better prediction performance.

10.5.3 Cross-lingual Sememe Prediction

Most languages lack sememe-based KBs such as HowNet, which prevents com-puters from better understanding and utilizing human language to some extent.

Therefore, it is necessary to build sememe-based KBs for these languages. In addition, as mentioned before, manually building a sememe-based KB is time-consuming and labor-intensive. To this end, we explore a new task [62], i.e., cross-lingual lexical sememe prediction (CLSP), which aims at automatically predicting lexical sememes for words in other languages.

CLSP encounters unique challenges. On the one hand, there does not exist a consistent one-to-one matching between words from two different languages. For example, the English word "beautiful" can be translated into Chinese words of either 美丽 or 漂亮. Hence, we cannot simply translate the annotation of words in HowNet into another language. On the other hand, since sememe prediction is based on understanding the semantic meanings of words, how to recognize the meanings of a word in other languages is also a critical problem.

To tackle these challenges, we propose a novel model for CLSP [62] to translate sememe-based KBs from a source language to a target language. Our model mainly contains two modules: (1) monolingual word embedding learning, which jointly learns semantic representations of words for the source and the target languages, and (2) cross-lingual word embedding alignment, which bridges the gap between the semantic representations of words in two languages. Learning these word embeddings could conduce to CLSP. Correspondingly, the overall objective function mainly consists of two parts:

$$\mathcal{L} = \mathcal{L}_{\text{mono}} + \mathcal{L}_{\text{cross}}. \tag{10.39}$$

Here, the monolingual term $\mathcal{L}_{\text{mono}}$ is designed to learn monolingual word embeddings for source and target languages, respectively. The cross-lingual term $\mathcal{L}_{\text{cross}}$ aims to align cross-lingual word embeddings in a unified semantic space. In the following, we will introduce the two parts in detail.

Monolingual Word Representation Monolingual word representation is learned using monolingual corpora of source and target languages. Since these two corpora are non-parallel, $\mathcal{L}_{\text{mono}}$ comprises two monolingual sub-models that are independent of each other:

$$\mathcal{L}_{\text{mono}} = \mathcal{L}_{\text{mono}}^{\text{S}} + \mathcal{L}_{\text{mono}}^{\text{T}}, \tag{10.40}$$

where the superscripts S and T denote source and target languages, respectively. To learn monolingual word embeddings, we choose the skip-gram model, which maximizes the predictive probability of context words conditioned on the centered word. Formally, taking the source side for example, given a sequence $\{w_1^S, \cdots, w_n^S\}$, we minimize the following loss:

$$\mathcal{L}_{\text{mono}}^{\text{S}} = -\sum_{c=l+1}^{n-l} \sum_{-l \leq k \leq l, k \neq 0} \log P(w_{c+k}^S | w_c^S), \tag{10.41}$$

where l is the size of the sliding window. $P(w_{c+k}^S | w_c^S)$ stands for the predictive probability of one of the context words conditioned on the centered word w_c^S. It is formalized as follows:

$$P(w_{c+k}^S | w_c^S) = \frac{\exp(\mathbf{w}_{c+k}^S \cdot \mathbf{w}_c^S)}{\sum_{w_s^S \in V^S} \exp(\mathbf{w}_s^S \cdot \mathbf{w}_c^S)}, \quad (10.42)$$

in which V^S denotes the vocabulary of the source language. $\mathcal{L}_{mono}^T$ can be formulated in a similar way.

Cross-Lingual Word Embedding Alignment Cross-lingual word embedding alignment aims to build a unified semantic space for both source and target languages. Inspired by Zhang et al. [84], the cross-lingual word embeddings are aligned with supervision from a seed lexicon. Specifically, $\mathcal{L}_{cross}$ includes two parts: (1) alignment by seed lexicon ($\mathcal{L}_{seed}$) and (2) alignment by matching ($\mathcal{L}_{match}$):

$$\mathcal{L}_{cross} = \lambda_s \mathcal{L}_{seed} + \lambda_m \mathcal{L}_{match}, \quad (10.43)$$

where λ_s and λ_m are hyper-parameters balancing both terms. The seed lexicon term $\mathcal{L}_{seed}$ pulls word embeddings of parallel pairs to be close, which can be achieved as follows:

$$\mathcal{L}_{seed} = \sum_{w_s^S, w_t^T \in \mathcal{D}} (\mathbf{w}_s^S - \mathbf{w}_t^T)^2, \quad (10.44)$$

where $\mathcal{D}$ denotes a seed lexicon w_s^S and w_t^T indicate the words in source and target languages in $\mathcal{D}$.

$\mathcal{L}_{match}$ is designed by assuming that each target word should be matched with a single source word or a special empty word and vice versa. The matching process is defined as follows:

$$\mathcal{L}_{match} = \mathcal{L}_{match}^{T2S} + \mathcal{L}_{match}^{S2T}, \quad (10.45)$$

where $\mathcal{L}_{match}^{T2S}$ and $\mathcal{L}_{match}^{S2T}$ denote target-to-source matching and source-to-target matching.

From now on, we explain the details of target-to-source matching, and source-to-target matching can be derived in a similar way. A latent variable $m_t \in \{0, 1, \cdots, |V^S|\}$ ($t = 1, 2, \cdots, |V^T|$) is first introduced for each target word w_t^T, where $|V^S|$ and $|V^T|$ indicate the vocabulary sizes of the source and target languages, respectively. Here, m_t specifies the index of the source word that w_t^T matches and $m_t = 0$ signifies that the empty word is matched. Then we have $\mathbf{m} = \{m_1, m_2, \cdots, m_{|V^T|}\}$ and can formalize the target-to-source matching term as follows:

$$\mathcal{L}_{match}^{T2S} = -\log P(\mathcal{C}^T | \mathcal{C}^S) = -\log \sum_{\mathbf{m}} P(\mathcal{C}^T, \mathbf{m} | \mathcal{C}^S), \quad (10.46)$$

where $\mathcal{C}^T$ and $\mathcal{C}^S$ denote the target and source corpus. Then we have:

$$P(\mathcal{C}^T, \mathbf{m}|\mathcal{C}^S) = \prod_{w^T \in \mathcal{C}^T} P(w^T, \mathbf{m}|\mathcal{C}^S) = \prod_{t=1}^{|V^T|} P(w_t^T|w_{m_t}^S)^{c(w_t^T)}, \quad (10.47)$$

where $w_{m_t}^S$ is the source word that is matched by w_t^T and $c(w_t^T)$ denotes how many times w_t^T occurs in the target corpus. Here $P(w_t^T|w_{m_t}^S)$ is calculated similar to Eq. (10.42). In fact, the original CLSP model contains another loss function that conducts sememe-based word embedding learning. This loss incorporates sememe information into word representations and conduces to better word embeddings for sememe prediction. The corresponding learning process has been introduced in Sect. 10.3.2.

Sememe Prediction Based on the assumption that relevant words have similar sememes, we propose to predict sememes for a target word in the target language based on its most similar source words. Using the same word-sememe matrix $\mathbf{M}$ in SPWE, the probability of a sememe x_j given a target word w^T is defined as:

$$P(x_j|w^T) = \sum_{w_s^S \in V^S} \cos(\mathbf{w}_s^S, \mathbf{w}^T)\mathbf{M}_{sj}c^{r_s}, \quad (10.48)$$

where $\mathbf{w}_s^S$ and $\mathbf{w}^T$ are the word embeddings for a source word w_s^S and the target word w^T. r_s denotes the descending rank of the word similarity $\cos(\mathbf{w}_s^S, \mathbf{w}^T)$, and c means a hyper-parameter.

Up to now, we have introduced the details for the proposed framework CLSP. In experiments, we take Chinese as the source language with sememe annotations and English as the target language to showcase CLSP. The results show that the model could effectively predict lexical sememes for words with different frequencies in other languages. Besides, the model achieves consistent improvements in two auxiliary experiments including bilingual lexicon induction and monolingual word similarity computation.

10.5.4 Connecting HowNet with BabelNet

Although the aforementioned method demonstrates superiority in cross-lingual sememe prediction, the method can only predict sememes for one language at a time. This means that we need to predict sememes repeatedly for multiple languages, which requires additional efforts to define the lexicon and correct the possible errors. That is why we turn to link HowNet with BabelNet [51].

BabelNet is a multilingual KB that merges Wikipedia and representative linguistic KBs (e.g., WordNet). The node in BabelNet is named BabelNet *Synset*, which contains a definition and multiple words in different languages that share the same meaning, together with some additional information. The edges in BabelNet stand for relations between synsets like *antonym* and *superior*. BabelNet has over 15 million synsets and 364k relations, covering 284 commonly used languages and more than 11 million figures. Words in one synset should be annotated with the same sememe annotations because they have the same meaning. If we connect BabelNet with HowNet, then we can directly predict sememes for multiple languages since each synset in BabelNet supports various languages.

Based on the above motivation, we make the first effort to connect BabelNet and HowNet [60]. We create a "BabelSememe" dataset, which contains BabelNet synsets annotated with sememes. The candidate sememes are the union of all the sememes of the Chinese synonyms in the synset, which are then carefully sifted by over 100 annotators.

Sememe Prediction for BabelNet Synsets (SPBS) BabelSememe dataset is still much smaller than the original BabelNet, and manually annotating synsets is time-consuming. Therefore, we propose the task of sememe prediction for babelNet synsets (SPBS) [60]. The setting of SPBS mostly follows existing sememe prediction frameworks. For a synset $b \in B$, where B denotes all synsets of BabelNet, we calculate the probability $P(x|b)$ of a sememe x and decide its sememe set $X^{(b)}$ as follows:

$$X^{(b)} = \{x \in X | P(x|b) > \delta\}, \tag{10.49}$$

where δ denotes a threshold and X means the full sememe set. To precisely calculate the probability, we need to first obtain the representation of synsets. Intuitively, we introduce two methods to learn the synset representation [60]: SPBS with semantic representation and SPBS with relational representation. By leveraging the edges in BabelNet Synset and sememe relations in HowNet, we can obtain the relational representation. In the following paragraphs, we introduce the above two methods in detail.

SPBS with Semantic Representation (SPBS-SR) Similar to the aforementioned method for sememe prediction, we can force synsets with similar semantic representation to have similar sememe annotations:

$$P(x|b) \sim \sum_{b' \in B} \cos(\mathbf{b}, \mathbf{b}')' I_{X^{(b')}}(x)' c^{r_{b'}}, \tag{10.50}$$

where b' denotes a synset in BabelNet and $\mathbf{b}, \mathbf{b}'$ mean the semantic representation of b and b'. $I(\cdot)$ is a function that indicates whether x lies within the sememe set of b'. c is a hyper-parameter, and $r_{b'}$ is the descending rank of cosine similarities, which makes the model focus more on similar synsets. To obtain the semantic representation of the synsets, we resort to NASARI representations [15], which utilize the Wikipedia pages related to these synsets to learn their representations.

SPBS with Relational Representation (SPBS-RR) Some of the synsets in Babel-Net are annotated with relations, and most of the relations originate from WordNet. In addition, there are four types of relations of sememes in HowNet: *hypernym*, *hyponym*, *antonym*, and *converse*.

If we define a new relation *have_sememe*, which is denoted as r_h, to represent that a synset consists of one specific sememe, then we can assign such a relation (edge) pointing from some synset nodes to some sememe nodes. In this way, we define triplets $\langle h, r, t \rangle$, where $h, t \in X \cup B$. $r \in R_X \cup R_B \cup \{r_h\}$ stands for the relation. R_X and R_B denote the originally defined relations in HowNet and BabelNet. Borrowing the idea from TransE [11] on knowledge representation learning, we can jointly learn the representation of all the nodes and relations as follows:

$$\mathcal{L}_1 = \sum_{(h,r,t) \in G} \max[0, \gamma + d(\mathbf{h} + \mathbf{r}, \mathbf{t}) - d(\mathbf{h} + \mathbf{r}, \mathbf{t}')], \tag{10.51}$$

where t' denotes another node that is different from t and $\mathbf{t}'$ is the embedding of t'. γ denotes a margin and d means the Euclidean distance.

Following the definition of BabelNet synset that the representation of a synset is related to the summation of all its sememes' representations, we have:

$$\mathcal{L}_2 = \sum_{b \in B} \|\mathbf{b} + \mathbf{r}_b - \sum_{x \in X^{(b)}} \mathbf{x}\|^2, \tag{10.52}$$

where $\mathbf{r}_b$ is a special semantic equivalence relation standing for the difference between one synset representation $\mathbf{b}$ and the summation of all its sememes' representations. To sum up, the overall loss function is defined as follows:

$$\mathcal{L} = \lambda_1 \mathcal{L}_1 + \lambda_2 \mathcal{L}_2, \tag{10.53}$$

where λ_1 and λ_2 are the hyper-parameters balancing both losses. Now we can formulate the probability $P(x|b)$ of a sememe given a synset using the difference between the representations:

$$P(x|b) \sim \frac{1}{d(\mathbf{b} + \mathbf{r}_h, \mathbf{x})}. \tag{10.54}$$

Since SPBS-SR and SPBS-SR follow different assertions, combining them can take both semantics and relations into consideration. In this way, we could achieve better performance.

10.5.5 Summary and Discussion

In this section, we introduce the task of sememe prediction, which is designed for reducing human labor in creating sememe-based KBs. Prior efforts in this direction are spent on defining the sememe prediction task [77]. Their methods are based on collaborative filtering and matrix factorization. Others take the internal information of words into account when predicting sememes [35]. Beyond sememe prediction within one language, the task of cross-lingual lexical sememe prediction is proposed, together with a bilingual word representation learning and an alignment model [62]. Researchers have also tried to connect existing sememe-based KBs with multilingual KBs, e.g., BabelNet.

There also exist important research works that we do not elaborate on in this chapter. For instance, some researchers propose to automatically predict sememes using the word descriptions in the Wikipedia websites [41]; others resort to leveraging dictionary definitions for better performance and robustness [25]. In the above works, efforts are mainly spent on annotating the sememe set for each word sense. In fact, we have also explored how to predict the hierarchical structure of sememe annotations [79]. Considering the significant importance and powerful capability of sememe knowledge, we believe it is essential to design better algorithms for automatically building sememe-based KBs.

10.6 Applications

In the previous sections, we have introduced how to leverage the sememe knowledge to enhance advanced neural networks, including word representation, language modeling, and recurrent neural networks. Benefiting from the rich semantic knowledge, HowNet has been successfully applied to various NLP tasks and achieved significant performance improvements. A typical application is word similarity computation [43], in which the similarity of two given words is computed by measuring the degree of resemblance of their sememe trees. Other applications include word sense disambiguation [85], question classification [73], and sentiment analysis [20, 26]. In this section, we introduce another two practical applications of HowNet, i.e., Chinese lexicon expansion and reverse dictionary. The introduction of this part is based on our research works [82, 83].

10.6.1 Chinese LIWC Lexicon Expansion

Linguistic inquiry and word count (LIWC) [57] has been widely used for computational text analysis in social science. LIWC computes the percentages of words in a given text that fall into over 80 linguistic, psychological, and topical categories.[2] Not only can LIWC be used for text classification, but it can also be utilized to examine the underlying psychological states of a writer or a speaker. LIWC was initially developed to address content analytic issues in experimental psychology. Nowadays, it has been widely applied to various fields such as computational linguistics [28], demographics [52], health diagnostics [14], social relationship [36], etc.

Despite the fact that Chinese is the most spoken language in the world, the original LIWC does not support Chinese. Fortunately, Chinese LIWC [34] has been released to fill the vacancy. In the following, we mainly focus on Chinese LIWC and would use the term "LIWC" to stand for "Chinese LIWC" if not otherwise specified. While LIWC has been used in a variety of fields, its lexicon contains fewer than 7,000 words. This is insufficient because according to a previous work [42], there are at least 56,008 common words in Chinese. Moreover, LIWC lexicon does not consider emerging words or phrases from the Internet. Therefore, it is reasonable and necessary to expand the LIWC lexicon so that it can cover more scientific research purposes. Apparently, manual annotation is labor-intensive. To this end, automatic LIWC lexicon expansion is proposed.

In LIWC lexicon, words are labeled with different categories, which form a special hierarchy. Formally, LIWC lexicon expansion is a hierarchical multi-label classification task, which predicts the joint probability of a series of labels $P(y_1, y_2, \cdots, y_L | w)$ given a word w. Hierarchical classification algorithms can be naturally applied to LIWC lexicon. For instance, Chen et al. [19] propose hierarchical SVM (support vector machine), which is a modified version of SVM based on the hierarchical problem decomposition approach. Another line of work attempts to use neural networks in the hierarchical classification [16, 37]. In addition, researchers [8] have presented a novel algorithm that can be used on both tree-structured and DAG (directed acyclic graph)-structured hierarchies. However, these methods are too generic without considering the special properties of words and LIWC lexicon. In fact, many words and phrases have multiple meanings (i.e., polysemy) and can thus be classified into multiple leaf categories. Additionally, many categories in LIWC are fine-grained, thus making it more difficult to distinguish them. To address these issues, we propose to incorporate sememe information when expanding the lexicon [82], which will be discussed after a brief introduction to the basic model.

Basic Decoder for Hierarchical Classification The basic model exploits the well-known seq2seq decoder [75] for hierarchical classification. The original seq2seq decoder is often trained to predict the next word w_t conditioned on all the previously predicted words $\{w_1, \cdots, w_{t-1}\}$. To leverage the seq2seq decoder, we can first transform hierarchical labels into a sequence. Note here the encoder of the seq2seq model is used to encode the information of the target word and the decoder of the seq2seq model is used for label prediction.

---

[2] https://www2.fgw.vu.nl/werkbanken/dighum/tools/tool_list/liwc.php.

Specifically, denote Y as the label set and $\pi: Y \to Y$ as the parent relationship, where $\pi(y)$ is the parent node of $y \in Y$. Given a word w, its labels form a tree-structure hierarchy. We enumerate every path starting from the root node to each leaf node and transform the path into a sequence $\{y_1, y_2, \cdots, y_L\}$ where $\pi(y_i) = y_{i-1}$, $\forall i \in [2, L]$. Here, L means the number of levels in the hierarchy. In this way, when the model predicts a label y_i, it takes into consideration the probability of parent label sequence $\{y_1, \cdots, y_{i-1}\}$. Formally, we define a probability over the label sequence:

$$P(y_1, y_2, \cdots, y_L | w) = \prod_{i=1}^{L} P(y_i | y_1, \cdots, y_{i-1}, w). \tag{10.55}$$

The decoder is modeled using an LSTM. At the i-th step, the decoder takes the label embedding $\mathbf{y}_{i-1}$ and the previous hidden state $\mathbf{h}_{i-1}$ as input and then predicts the current label. Denote $\mathbf{h}_i$ and $\mathbf{o}_i$ as the hidden state and output state of the i-th step; the conditional probability is computed as:

$$P(y_i | y_1, \cdots, y_{i-1}, w) = \mathbf{o}_i \odot \tanh(\mathbf{h}_i), \tag{10.56}$$

where $\odot$ is an element-wise multiplication. To consider the information from w, the initial state $\mathbf{h}_0$ is chosen to be the word embedding $\mathbf{w}$.

Hierarchical Decoder with Sememe Attention   As mentioned above, the basic decoder uses word embeddings as the initial state, and each word in the basic decoder model only has one representation. Considering that many words are polysemous and many categories are fine-grained, it is difficult to handle these properties using a single real-valued vector.

As illustrated in Fig. 10.11, we utilize the attention mechanism [2] to incorporate sememe information when predicting the word label sequence.

Similar to the basic decoder approach, word embeddings are applied as the initial state of the decoder. The primary difference is that at the i-th step, concat($\mathbf{y}_{i-1}$; $\mathbf{c}_i$) instead of $\mathbf{y}_{i-1}$ is input into the decoder, where $\mathbf{c}_i$ is the context vector. $\mathbf{c}_i$ depends on a set of sememe embeddings $\{\mathbf{x}_1, \cdots, \mathbf{x}_N\}$, where N denotes the total number of sememes of all the senses of the word w. More specifically, the context vector $\mathbf{c}_i$ is computed as a weighted summation of the sememe embeddings as follows:

$$\mathbf{c}_i = \sum_{j=1}^{N} \alpha_{ij} \mathbf{x}_j. \tag{10.57}$$

The weight α_{ij} of each sememe embedding $\mathbf{x}_j$ is calculated as follows:

$$\alpha_{ij} = \frac{\exp(\mathbf{v} \cdot \tanh(\mathbf{W}_1 \mathbf{y}_{i-1} + \mathbf{W}_2 \mathbf{x}_j))}{\sum_{k=1}^{N} \exp(\mathbf{v} \cdot \tanh(\mathbf{W}_1 \mathbf{y}_{i-1} + \mathbf{W}_2 \mathbf{x}_k))}, \tag{10.58}$$

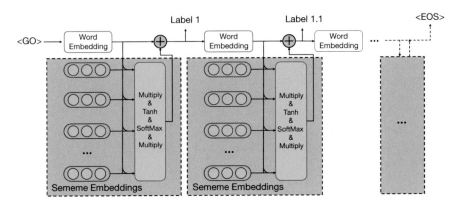

Fig. 10.11 The architecture of the sememe attention decoder with word embeddings as the initial state. This figure is re-drawn based on Fig. 3 in Zeng et al. [82]

where $\mathbf{v}$ is a trainable vector and $\mathbf{W}_1$ and $\mathbf{W}_2$ are weight matrices. At each time step, the decoder chooses which sememes to pay attention to when predicting the current word label. With the support of sememe attention, the decoder can differentiate multiple meanings in a word and the fine-grained categories and thus can expand a more accurate and comprehensive lexicon.

10.6.2 Reverse Dictionary

The task of reverse dictionary [69] is defined as the dual task of the normal dictionary: it takes the definition as input and outputs the target words or phrases that match the semantic meaning of the definition. In real-world scenarios, reverse dictionaries not only assist the public in writing articles but can also help anomia patients, who cannot organize words due to neurological disorders. In addition, reverse dictionaries conduce to NLP tasks such as sentence representations and text-to-entity mapping.

Some commercial reverse dictionary systems (e.g., OneLook) are satisfying in performance but are closed source. Existing reverse dictionary algorithms face the following problems: (1) Human-written inputs differ a lot from word definitions, and models trained on the latter have poor generalization abilities on user inputs. (2) It is hard to predict those low-frequency target words due to limited training data for them. They may actually appear frequently according to Zipf's law.

Multi-channel Reverse Dictionary Model To address the aforementioned problems, we propose the multi-channel reverse dictionary (MCRD) [83], which utilizes POS tag, morpheme, word category, and sememe information of candidate words. MCRD embeds the queried definition into hidden states and computes similarity scores with all the candidates and the query embeddings. As shown in Fig. 10.12,

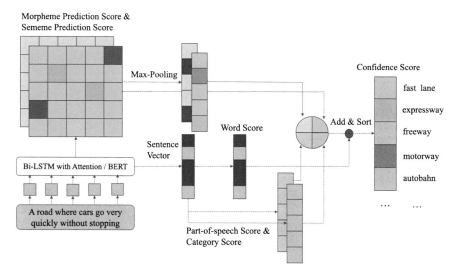

Fig. 10.12 Architecture of multi-channel reverse dictionary model. This figure is re-drawn based on Fig. 2 in Zhang et al. [83]

inspired by the inference process of humans, the model further considers particular characteristics of words, i.e., POS tag, word category, morpheme, and sememe.

Basic Framework We first introduce the basic framework, which embeds the queried definition into representations, i.e., $\mathbf{Q} = \{\mathbf{q}_1, \cdots, \mathbf{q}_{|Q|}\}$. The model feeds $\mathbf{Q}$ into a BiLSTM model and obtains the hidden states as follows (here we can also use more advanced neural structures to obtain the hidden states, and we take the BiLSTM as an example):

$$\{\overrightarrow{\mathbf{h}_1}, \cdots, \overrightarrow{\mathbf{h}_{|Q|}}\}, \{\overleftarrow{\mathbf{h}_1}, \cdots, \overleftarrow{\mathbf{h}_{|Q|}}\} = \text{BiLSTM}(\mathbf{q}_1, \cdots, \mathbf{q}_{|Q|}),$$
$$\mathbf{h}_i = \text{concat}(\overrightarrow{\mathbf{h}_i}; \overleftarrow{\mathbf{h}_i}). \tag{10.59}$$

Then the hidden states are passed into a weighted summation module, and we have the definition embedding $\mathbf{v}$:

$$\mathbf{v} = \sum_{i=1}^{|Q|} \alpha_i \mathbf{h}_i,$$
$$\alpha_i = \mathbf{h}_t^\top \mathbf{h}_i, \tag{10.60}$$
$$\mathbf{h}_t = \text{concat}(\overrightarrow{\mathbf{h}_{|Q|}}; \overleftarrow{\mathbf{h}_1}).$$

Finally, the definition embedding is mapped into the same semantic space as words, and dot products are used to represent word-word confidence score $sc_{w,\text{word}}$:

$$\mathbf{v}_{\text{word}} = \mathbf{W}_{\text{word}}\mathbf{v} + \mathbf{b}_{\text{word}},$$
$$sc_{w,\text{word}} = \mathbf{v}_{\text{word}}^{\top}\mathbf{w}, \tag{10.61}$$

where $\mathbf{W}_{\text{word}}$ and $\mathbf{b}_{\text{word}}$ are trainable weights and $\mathbf{w}$ denotes the word embedding.

Internal Channels: POS Tag Predictor To return words with POS tags relevant to the input query, we predict the POS tag of the target word. The intuition is that human-written queries can usually be easily mapped into one of the POS tags.

Denote the union of the POS tags of all the senses of a word w as P_w. We can compute the POS score of the word w with the sentence embedding $\mathbf{v}$:

$$\mathbf{sc}_{\text{pos}} = \mathbf{W}_{\text{pos}}\mathbf{v} + \mathbf{b}_{\text{pos}},$$
$$sc_{w,\text{pos}} = \sum_{p \in P_w} [\mathbf{sc}_{\text{pos}}]_{\text{index}^{\text{pos}}(p)}, \tag{10.62}$$

where $\text{index}^{\text{pos}}(p)$ means the id of POS tag p and operator $[x]_i$ denotes the i-th element of x. In this way, candidates with qualified POS tags are assigned a higher score.

Internal Channels: Word Category Predictor Semantically related words often belong to distinct categories, despite the fact that they could have similar word embeddings (for instance, "bed" and "sleep"). Word category information can help us eliminate these semantically related but not similar words. Following the same equation, we can get the word category score of candidate word w as follows:

$$\mathbf{sc}_{\text{cat},k} = \mathbf{W}_{\text{cat},k}\mathbf{v} + \mathbf{b}_{\text{cat},k},$$
$$sc_{w,\text{cat}} = \sum_{k=1}^{K} [\mathbf{sc}_{\text{cat},k}]_{\text{index}_k^{\text{cat}}(w)}, \tag{10.63}$$

where $\mathbf{W}_{\text{cat},k}$ and $\mathbf{b}_{\text{cat},k}$ are trainable weights. K denotes the number of the word hierarchy of w and $\text{index}_k^{\text{cat}}(w)$ denotes the id of the k-th word hierarchy.

Internal Channels: Morpheme Predictor Similarly, all the words have different morphemes, and each morpheme may share similarities with some words in the word definition. Therefore, we can conduct the morpheme prediction of query Q at the word level:

$$\mathbf{sc}_{\text{mor}}^i = \mathbf{W}_{\text{mor}}\mathbf{h}_i + \mathbf{b}_{\text{mor}}, \tag{10.64}$$

where $\mathbf{W}_{\text{mor}}$ and $\mathbf{b}_{\text{mor}}$ are trainable weights. The final score of whether the query Q has the morpheme j can be viewed as the maximum score of all positions:

$$[\mathbf{sc}_{\text{mor}}]_j = \max_{i \leq |Q|} [\mathbf{sc}^i_{\text{mor}}]_j, \tag{10.65}$$

where the operator $[x]_j$ means the j-th element of x. Denote the union of the morphemes of all the senses of a word w as M_w, we can then compute the morpheme score of the word w and query Q as follows:

$$sc_{w,\text{mor}} = \sum_{m \in M_w} [\mathbf{sc}_{\text{mor}}]_{\text{index}^{\text{mor}}(m)}, \tag{10.66}$$

where $\text{index}^{\text{mor}}(m)$ means the id of the morpheme m.

Internal Channels: Sememe Predictor Similar to the morpheme predictor, we can also use sememe annotations of words and the sememe predictions of the query at the word level and then compute the sememe score of all the candidate words:

$$\begin{aligned} \mathbf{sc}^i_{\text{sem}} &= \mathbf{W}_{\text{sem}}\mathbf{h}_i + \mathbf{b}_{\text{sem}}, \\ [\mathbf{sc}_{\text{sem}}]_j &= \max_{i \leq |Q|} [\mathbf{sc}^i_{\text{sem}}]_j, \\ sc_{w,\text{sem}} &= \sum_{x \in X_w} [\mathbf{sc}_{\text{sem}}]_{\text{index}^{\text{sem}}(x)}, \end{aligned} \tag{10.67}$$

where X_w is the sememe set of all w's sememes and $\text{index}^{\text{sem}}(x)$ denotes the id of the sememe x. With all the internal channel scores of candidate words, we can finally get the confidence scores by combining them as follows:

$$sc_w = \lambda_{\text{word}} sc_{w,\text{word}} + \sum_{c \in \mathbb{C}} \lambda_c sc_{w,c}, \tag{10.68}$$

where $\mathbb{C}$ is the aforementioned channels: *POS tag*, *morpheme*, *word category*, and *sememes*. A series of λ are assigned to balance different terms. With the sememe annotations as additional knowledge, the model could achieve better performance, even outperforming commercial systems.

WantWords WantWords[3] [65] is an open-source online reverse dictionary system that is based on multi-channel methods. WantWords employs BERT as the sentence encoder and thus performs more stably and flexibly. WantWords supports both monolingual and cross-lingual modes. For the monolingual mode, when the query only contains one word, WantWords compares the query word embedding with candidate word embeddings and doubles the score of a candidate word if it is a

---

[3] https://wantwords.net/.

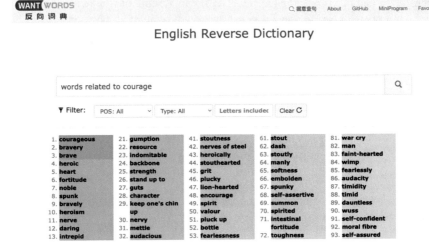

Fig. 10.13 A snapshot of WantWords. We show an example of an English reverse dictionary

synonym of the query word. To support the cross-lingual mode, WantWords uses Baidu Translation API[4] to translate queries into the target language. Up till now, WantWords has handled more than 25 million queries from 2 million users, with 120 thousand daily active users. The success of WantWords again demonstrates the usefulness of sememe knowledge in real-world NLP applications. We give an example in Fig. 10.13.

10.7   Summary and Further Readings

In this chapter, we first give an introduction to the most well-known sememe knowledge base, HowNet, which uses about 2, 000 predefined sememes to annotate over 100, 000 Chinese/English words and phrases. Different from other linguistic KBs (e.g., WordNet) or commonsense KBs (e.g., ConceptNet), HowNet focuses on the minimum semantic units (sememes) and captures the compositional relations between sememes and words.

To model the sememe knowledge, we elaborate on three models, namely, the simple sememe aggregation model (SSA), sememe attention over context model (SAC), and sememe attention over target model (SAT). After that, we introduce how to exploit the sememe knowledge for NLP. Specifically, we show that sememe knowledge can be well incorporated into word representation, semantic composition, language modeling, and sequence modeling. To further enrich the annotation of

---

[4] https://fanyi.baidu.com.

HowNet, we detail how to automatically predict sememes for both monolingual and cross-lingual unannotated words and how to connect HowNet with a representative multilingual KB, i.e., BabelNet. Finally, we introduce two applications of sememe knowledge, including Chinese LIWC lexicon expansion and reverse dictionary.

Further Reading and Future Work For further learning of sememe knowledge-based NLP, you can read the book written by the authors of HowNet [24], which detailedly introduces the basic information about HowNet. You can also find more related papers in this paper list[5] to easily get familiar with this interesting research field. We also recommend you to read our review on sememe knowledge computation [63], where we discuss recent advances in application and expansion of sememe knowledge bases. There are also some research directions worth exploring in the future:

1. *Building Sememe KBs for Other Languages.* The original annotations in HowNet only support two languages: Chinese and English. As far as we know, there are no sememe-based KBs in other languages. Since HowNet and its sememe knowledge have been verified as helpful for better understanding human language, it will be of great significance to annotate sememes for words and phrases in other languages. As we have mentioned above, the cross-lingual sememe prediction task can be leveraged to automatically create sememe-based KBs, and we think it is promising to make efforts in this direction. It should also be mentioned that compared to words, sememes may cover less textual knowledge to some extent.

2. *Utilizing Structures of Sememe Annotations.* The sememe annotations in HowNet are hierarchical, and sememes associated with a word are actually organized as a tree structure. However, existing attempts still do not fully exploit the structural information of sememes. Instead, in current methods, sememes are simply regarded as semantic labels. In fact, the structures of sememes also contain abundant semantic information and may conduce to a deeper understanding of lexical semantics. Besides, existing sememe prediction studies predict unstructured sememes only, and it is an interesting task to predict sememes' structures.

3. *Leveraging Sememe Knowledge in Low-Resource Scenarios.* One of the most important and typical characteristics of sememes is that limited sememes can represent unlimited semantics, which can play an important and positive role in tackling low-resource scenarios. In word representation learning, the representations of low-frequency words can be improved by their sememes, which have been well learned with the high-frequency words annotated with sememes. We believe sememe knowledge will be beneficial to other low-resource scenarios, e.g., low-resource language NLP tasks. We also encourage future work to apply sememe knowledge to more NLP applications.

---

[5] https://github.com/thunlp/SCPapers.

Acknowledgments The contributions of all authors for the second edition are as follows: Zhiyuan Liu, Yankai Lin, and Maosong Sun designed the overall architecture of this chapter; Yujia Qin drafted this chapter. Zhiyuan Liu and Yankai Lin proofread and revised this chapter.

We also thank Fanchao Qi and Chenghao Yang for preparing the initial draft materials of the first edition, thank Yining Ye for drawing figures and preparing the initial draft materials for Sect. 10.5.2, 10.5.4, and 10.7, and thank Biru Zhu, Yining Ye, Zihan Cai, Zhe Wang, Xu Han, Yankai Lin, Chaojun Xiao, and Zhiyuan Liu for proofreading the chapter.

This is the sememe knowledge representation learning chapter of the second edition of the book *Representation Learning for Natural Language Processing*, with its first edition published in 2020 [44]. As compared to the first edition of this chapter, the main changes include the following: (1) we additionally added basic backgrounds for human knowledge bases, one sememe knowledge acquisition method, and several applications of sememe knowledge, and (2) we rewrote the original chapter and changed the narrative logic of the chapter.

References

1. Sungjin Ahn, Heeyoul Choi, Tanel Pärnamaa, and Yoshua Bengio. A neural knowledge language model. *arXiv preprint arXiv:1608.00318*, 2016.
2. Dzmitry Bahdanau, Kyunghyun Cho, and Yoshua Bengio. Neural machine translation by jointly learning to align and translate. In *Proceedings of ICLR*, 2015.
3. Satanjeev Banerjee and Ted Pedersen. An adapted lesk algorithm for word sense disambiguation using wordnet. In *Proceedings of CICLing*, pages 136–145. Springer, 2002.
4. Michele Banko, Vibhu O Mittal, and Michael J Witbrock. Headline generation based on statistical translation. In *Proceedings of ACL*, 2000.
5. Marco Baroni and Roberto Zamparelli. Nouns are vectors, adjectives are matrices: Representing adjective-noun constructions in semantic space. In *Proceedings of EMNLP*, 2010.
6. Yoshua Bengio, Réjean Ducharme, Pascal Vincent, and Christian Jauvin. A neural probabilistic language model. *Journal of Machine Learning Research*, 3, 2003.
7. Adam Berger and John Lafferty. Information retrieval as statistical translation. In *Proceedings of SIGIR*, 1999.
8. Wei Bi and James T Kwok. Multi-label classification on tree-and dag-structured hierarchies. In *Proceedings of ICML*, 2011.
9. William Blacoe and Mirella Lapata. A comparison of vector-based representations for semantic composition. In *Proceedings of EMNLP-CoNLL*, 2012.
10. Leonard Bloomfield. A set of postulates for the science of language. *Language*, 2(3):153–164, 1926.
11. Antoine Bordes, Nicolas Usunier, Alberto Garcia-Duran, Jason Weston, and Oksana Yakhnenko. Translating embeddings for modeling multi-relational data. In *Proceedings of NeurIPS*, 2013.
12. Thorsten Brants, Ashok C Popat, Peng Xu, Franz J Och, and Jeffrey Dean. Large language models in machine translation. In *Proceedings of EMNLP*, 2007.
13. Peter F Brown, John Cocke, Stephen A Della Pietra, Vincent J Della Pietra, Fredrick Jelinek, John D Lafferty, Robert L Mercer, and Paul S Roossin. A statistical approach to machine translation. *Computational Linguistics*, 16(2):79–85, 1990.
14. Wilma Bucci and Bernard Maskit. Building a weighted dictionary for referential activity. *Computing Attitude and Affect in Text*, pages 49–60, 2005.
15. José Camacho-Collados, Mohammad Taher Pilehvar, and Roberto Navigli. Nasari: Integrating explicit knowledge and corpus statistics for a multilingual representation of concepts and entities. *Artificial Intelligence*, 240:36–64, 2016.
16. Ricardo Cerri, Rodrigo C Barros, and André CPLF De Carvalho. Hierarchical multi-label classification using local neural networks. *Journal of Computer and System Sciences*, 80(1):39–56, 2014.

17. Xinxiong Chen, Zhiyuan Liu, and Maosong Sun. A unified model for word sense representation and disambiguation. In *Proceedings of EMNLP*, 2014.
18. Xinxiong Chen, Lei Xu, Zhiyuan Liu, Maosong Sun, and Huanbo Luan. Joint learning of character and word embeddings. In *Proceedings of IJCAI*, 2015.
19. Yangchi Chen, Melba M Crawford, and Joydeep Ghosh. Integrating support vector machines in a hierarchical output space decomposition framework. In *Proceedings of IGARSS*, 2004.
20. Lei Dang and Lei Zhang. Method of discriminant for chinese sentence sentiment orientation based on hownet. *Application Research of Computers*, 4:43, 2010.
21. Rahul Dey and Fathi M Salem. Gate-variants of gated recurrent unit (gru) neural networks. In *Proceedings of MWSCAS*, 2017.
22. Zhendong Dong. Conceptual difference between wordnet and hownet. In *Proceedings of NLPKE*. IEEE, 2007.
23. Zhendong Dong and Qiang Dong. Hownet-a hybrid language and knowledge resource. In *Proceedings of NLP-KE*, 2003.
24. Zhendong Dong and Qiang Dong. *HowNet and the Computation of Meaning (With CD-Rom)*. World Scientific, 2006.
25. Jiaju Du, Fanchao Qi, Maosong Sun, and Zhiyuan Liu. Lexical sememe prediction using dictionary definitions by capturing local semantic correspondence. *arXiv preprint arXiv:2001.05954*, 2020.
26. Xianghua Fu, Guo Liu, Yanyan Guo, and Zhiqiang Wang. Multi-aspect sentiment analysis for chinese online social reviews based on topic modeling and hownet lexicon. *Knowledge-Based Systems*, 37:186–195, 2013.
27. Edward Grefenstette and Mehrnoosh Sadrzadeh. Experimental support for a categorical compositional distributional model of meaning. In *Proceedings of EMNLP*, 2011.
28. Justin Grimmer and Brandon M Stewart. Text as data: The promise and pitfalls of automatic content analysis methods for political texts. *Political analysis*, 21(3):267–297, 2013.
29. Yihong Gu, Jun Yan, Hao Zhu, Zhiyuan Liu, Ruobing Xie, Maosong Sun, Fen Lin, and Leyu Lin. Language modeling with sparse product of sememe experts. In *Proceedings of EMNLP*, 2018.
30. Djoerd Hiemstra. A linguistically motivated probabilistic model of information retrieval. In *Proceedings of TPDL*, 1998.
31. G. E Hinton. Products of experts. In *Proceedings of ICANN*, 1999.
32. Geoffrey E Hinton. Learning distributed representations of concepts. In *Proceedings of CogSci*, 1986.
33. Sepp Hochreiter and Jürgen Schmidhuber. Long short-term memory. *Neural Computation*, 1997.
34. Chin-Lan Huang, CK Chung, Natalie Hui, Yi-Cheng Lin, Yi-Tai Seih, WC Chen, and JW Pennebaker. The development of the chinese linguistic inquiry and word count dictionary. *Chinese Journal of Psychology*, 54(2):185–201, 2012.
35. Huiming Jin, Hao Zhu, Zhiyuan Liu, Ruobing Xie, Maosong Sun, Fen Lin, and Leyu Lin. Incorporating chinese characters of words for lexical sememe prediction. In *Proceedings of ACL*, 2018.
36. Ewa Kacewicz, James W Pennebaker, Matthew Davis, Moongee Jeon, and Arthur C Graesser. Pronoun use reflects standings in social hierarchies. *Journal of Language and Social Psychology*, 33(2):125–143, 2014.
37. Sanjeev Kumar Karn, Ulli Waltinger, and Hinrich Schütze. End-to-end trainable attentive decoder for hierarchical entity classification. In *Proceedings of EACL*, 2017.
38. Slava Katz. Estimation of probabilities from sparse data for the language model component of a speech recognizer. *IEEE Transactions on Acoustics, Speech, and Signal Processing*, 35(3):400–401, 1987.
39. Hyesook Kim, Shanshan Chen, and Tony Veale. Analogical reasoning with a synergy of wordnet and hownet, 2010.
40. Thomas Kober, Julie Weeds, Jeremy Reffin, and David Weir. Improving sparse word representations with distributional inference for semantic composition. In *Proceedings of EMNLP*, 2016.

41. Wei Li, Xuancheng Ren, Damai Dai, Yunfang Wu, Houfeng Wang, and Xu Sun. Sememe prediction: Learning semantic knowledge from unstructured textual wiki descriptions. *arXiv preprint arXiv:1808.05437*, 2018.
42. Xingjian Li et al. Lexicon of common words in contemporary chinese, 2008.
43. Qun Liu. Word similarity computing based on hownet. *Computational linguistics and Chinese language processing*, 7(2):59–76, 2002.
44. Zhiyuan Liu, Yankai Lin, and Maosong Sun. *Representation Learning for Natural Language Processing*. Springer, 2020.
45. Andrew L Maas, Raymond E Daly, Peter T Pham, Dan Huang, Andrew Y Ng, and Christopher Potts. Learning word vectors for sentiment analysis. In *Proceedings of ACL*, 2011.
46. Tomas Mikolov, Kai Chen, Greg Corrado, and Jeffrey Dean. Efficient estimation of word representations in vector space. In *Proceedings of ICLR*, 2013.
47. David RH Miller, Tim Leek, and Richard M Schwartz. A hidden markov model information retrieval system. In *Proceedings of SIGIR*, 1999.
48. George A Miller. WordNet: a lexical database for English. *Communications of the ACM*, 1995.
49. Jeff Mitchell and Mirella Lapata. Vector-based models of semantic composition. In *Proceedings of ACL*, 2008.
50. Jeff Mitchell and Mirella Lapata. Language models based on semantic composition. In *Proceedings of EMNLP*, 2009.
51. Roberto Navigli and Simone Paolo Ponzetto. Babelnet: The automatic construction, evaluation and application of a wide-coverage multilingual semantic network. *Artificial Intelligence*, 193:217–250, 2012.
52. Matthew L Newman, Carla J Groom, Lori D Handelman, and James W Pennebaker. Gender differences in language use: An analysis of 14,000 text samples. *Discourse Processes*, 45(3):211–236, 2008.
53. Yilin Niu, Ruobing Xie, Zhiyuan Liu, and Maosong Sun. Improved word representation learning with sememes. In *Proceedings of ACL*, 2017.
54. Prasanna Parthasarathi and Joelle Pineau. Extending neural generative conversational model using external knowledge sources. In *Proceedings of EMNLP*, 2018.
55. Ted Pedersen, Siddharth Patwardhan, and Jason Michelizzi. WordNet: Similarity: measuring the relatedness of concepts. In *Proceedings of NAACL-HLT*, 2004.
56. Francis Jeffry Pelletier. The principle of semantic compositionality. *Topoi*, 13(1):11–24, 1994.
57. James W Pennebaker, Roger J Booth, and Martha E Francis. Linguistic inquiry and word count: Liwc [computer software]. *Austin, TX: liwc. net*, 2007.
58. Jeffrey Pennington, Richard Socher, and Christopher Manning. GloVe: Global vectors for word representation. In *Proceedings of EMNLP*, 2014.
59. Jay M Ponte and W Bruce Croft. A language modeling approach to information retrieval. In *Proceedings of SIGIR*, 1998.
60. Fanchao Qi, Liang Chang, Maosong Sun, Sicong Ouyang, and Zhiyuan Liu. Towards building a multilingual sememe knowledge base: predicting sememes for babelnet synsets. In *Proceedings of AAAI*, 2020.
61. Fanchao Qi, Junjie Huang, Chenghao Yang, Zhiyuan Liu, Xiao Chen, Qun Liu, and Maosong Sun. Modeling semantic compositionality with sememe knowledge. In *Proceedings of ACL*, 2019.
62. Fanchao Qi, Yankai Lin, Maosong Sun, Hao Zhu, Ruobing Xie, and Zhiyuan Liu. Cross-lingual lexical sememe prediction. In *Proceedings of EMNLP*, 2018.
63. Fanchao Qi, Ruobing Xie, Yuan Zang, Zhiyuan Liu, and Maosong Sun. Sememe knowledge computation: a review of recent advances in application and expansion of sememe knowledge bases. *Frontiers of Computer Science*, 15(5):1–11, 2021.
64. Fanchao Qi, Chenghao Yang, Zhiyuan Liu, Qiang Dong, Maosong Sun, and Zhendong Dong. Openhownet: An open sememe-based lexical knowledge base. *arXiv preprint arXiv:1901.09957*, 2019.

65. Fanchao Qi, Lei Zhang, Yanhui Yang, Zhiyuan Liu, and Maosong Sun. Wantwords: An open-source online reverse dictionary system. In *Proceedings of EMNLP: System Demonstrations*, 2020.
66. Yujia Qin, Fanchao Qi, Sicong Ouyang, Zhiyuan Liu, Cheng Yang, Yasheng Wang, Qun Liu, and Maosong Sun. Improving sequence modeling ability of recurrent neural networks via sememes. *IEEE/ACM Transactions on Audio, Speech, and Language Processing*, 28:2364–2373, 2020.
67. Alexander M Rush, Sumit Chopra, and Jason Weston. A neural attention model for abstractive sentence summarization. In *Proceedings of EMNLP*, 2015.
68. Badrul Sarwar, George Karypis, Joseph Konstan, and John Riedl. Item-based collaborative filtering recommendation algorithms. In *Proceedings of WWW*, 2001.
69. Gerardo Sierra. The onomasiological dictionary: a gap in lexicography. In *Proceedings of the Ninth Euralex International Congress*, 2000.
70. Richard Socher, John Bauer, Christopher D Manning, and Andrew Y Ng. Parsing with compositional vector grammars. In *Proceedings of ACL*, 2013.
71. Richard Socher, Brody Huval, Christopher D Manning, and Andrew Y Ng. Semantic compositionality through recursive matrix-vector spaces. In *Proceedings of EMNLP*, 2012.
72. Robyn Speer, Joshua Chin, and Catherine Havasi. Conceptnet 5.5: An open multilingual graph of general knowledge. In *Proceedings of AAAI*, 2017.
73. Jingguang Sun, Dongfeng Cai, Dexin Lv, and Yanju Dong. Hownet based chinese question automatic classification. *Journal of Chinese Information Processing*, 21(1):90–95, 2007.
74. Maosong Sun and Xinxiong Chen. Embedding for words and word senses based on human annotated knowledge base: A case study on hownet. *Journal of Chinese Information Processing*, 30:1–6, 2016.
75. Ilya Sutskever, Oriol Vinyals, and Quoc V Le. Sequence to sequence learning with neural networks. In *Proceedings of NeurIPS*, 2014.
76. David Weir, Julie Weeds, Jeremy Reffin, and Thomas Kober. Aligning packed dependency trees: a theory of composition for distributional semantics. *Computational Linguistics*, 42(4):727–761, December 2016.
77. Ruobing Xie, Xingchi Yuan, Zhiyuan Liu, and Maosong Sun. Lexical sememe prediction via word embeddings and matrix factorization. In *Proceedings of IJCAI*, 2017.
78. Bishan Yang and Tom Mitchell. Leveraging knowledge bases in LSTMs for improving machine reading. In *Proceedings of ACL*, 2017.
79. Yining Ye, Fanchao Qi, Zhiyuan Liu, and Maosong Sun. Going deeper: Structured sememe prediction via transformer with tree attention. In *Findings of ACL*, 2022.
80. Binyong Yin. Quantitative research on Chinese morphemes. *Studies of the Chinese Language*, 5:338–347, 1984.
81. Tom Young, Erik Cambria, Iti Chaturvedi, Hao Zhou, Subham Biswas, and Minlie Huang. Augmenting end-to-end dialogue systems with commonsense knowledge. In *Proceedings of AAAI*, 2018.
82. Xiangkai Zeng, Cheng Yang, Cunchao Tu, Zhiyuan Liu, and Maosong Sun. Chinese liwc lexicon expansion via hierarchical classification of word embeddings with sememe attention. In *Proceedings of AAAI*, 2018.
83. Lei Zhang, Fanchao Qi, Zhiyuan Liu, Yasheng Wang, Qun Liu, and Maosong Sun. Multi-channel reverse dictionary model. In *Proceedings of AAAI*, 2020.
84. Meng Zhang, Haoruo Peng, Yang Liu, Huan-Bo Luan, and Maosong Sun. Bilingual lexicon induction from non-parallel data with minimal supervision. In *Proceedings of AAAI*, 2017.
85. Yuntao Zhang, Ling Gong, and Yongcheng Wang. Chinese word sense disambiguation using hownet. In *Proceedings of ICNC*, 2005.
86. Yu Zhao, Zhiyuan Liu, and Maosong Sun. Phrase type sensitive tensor indexing model for semantic composition. In *Proceedings of AAAI*, 2015.
87. Xiaodan Zhu, Parinaz Sobhani, and Hongyu Guo. DAG-Structured long short-term memory for semantic compositionality. In *Proceedings of NAACL-HLT*, 2016.

Chapter 11
Legal Knowledge Representation Learning

Chaojun Xiao, Zhiyuan Liu, Yankai Lin, and Maosong Sun

Abstract The law guarantees the regular functioning of the nation and society. In recent years, legal artificial intelligence (legal AI), which aims to apply artificial intelligence techniques to perform legal tasks, has received significant attention. Legal AI can provide a handy reference and convenient legal services for legal professionals and non-specialists, thus benefiting real-world legal practice. Different from general open-domain tasks, legal tasks have a high demand for understanding and applying expert knowledge. Therefore, enhancing models with various legal knowledge is a key issue of legal AI. In this chapter, we summarize the existing knowledge-intensive legal AI approaches regarding knowledge representation, acquisition, and application. Besides, future directions and ethical considerations are also discussed to promote the development of legal AI.

11.1 Introduction

The law is the cornerstone of human civilization, and it guarantees the regular functioning of our state and society. The development of law has always been an important symbol of the development of human civilization. The practice of law can be traced back to 3000 BC. Ancient Egyptian law regulated social norms and encouraged people to be at peace with each other [97]. Over the millennia of development and progress, the law can reach every aspect of human activities nowadays. It regulates and mediates the relations and interactions between people, institutions, and state authorities. For example, international law concerns relations between sovereign nations; administrative law regulates the behavior of state authorities; criminal and civil law guarantees the fundamental rights of citizens.

C. Xiao · Z. Liu (✉) · M. Sun
Department of Computer Science and Technology, Tsinghua University, Beijing, China
e-mail: xiaocj20@mails.tsinghua.edu.cn; liuzy@tsinghua.edu.cn; sms@tsinghua.edu.cn

Y. Lin
Gaoling School of Artificial Intelligence, Renmin University of China, Beijing, China
e-mail: yankailin@ruc.edu.cn

Z. Liu et al. (eds.), *Representation Learning for Natural Language Processing*,
https://doi.org/10.1007/978-981-99-1600-9_11

The legal domain is knowledge-intensive and requires a high level of knowledge for its practitioners. It takes years of study and work experience for legal practitioners, including judges and lawyers, to be qualified for their jobs, which results in the scarcity and irreplaceability of legal practitioners. Besides, as society develops and technology advances, public communication and interactions become more frequent, leading to more legal disputes and increasing demand for legal services. Therefore, the following two urgent challenges in real-world judicial practice are increasingly prominent: (1) High caseload for public authorities. Take China, the world's most populous country, as an example; according to the statistics, the grassroots courts in China need to hear more than 30 million cases per year, with each judge hearing an average of 238 cases a year [1]. This means that each judge has high work pressure. The same situation also exists in other countries, such as the United States [30]. (2) Scarcity of legal services for the public. The scarcity of lawyers leads to the high cost of legal services, and in the United States, roughly 86% low-income people who encounter legal problems cannot obtain adequate and timely legal assistance [23].

With the development of artificial intelligence, the interdisciplinary discipline, legal artificial intelligence (Legal AI), has received increasing attention in recent years [3–5, 10, 32, 35, 122]. Legal AI aims to empower legal tasks with AI techniques and help legal professionals to deal with repetitive and cumbersome mental work. Thus, legal AI can assist legal professionals in improving their work efficiency. Besides, legal AI can also provide convenient legal services to people unfamiliar with legal knowledge. Notably, as most legal data are presented in the textual form, many legal AI methods focus on applying natural language processing (NLP) techniques, i.e., legal NLP, which is the focus of this chapter.

The core of legal NLP tasks lies in automatic case analysis and case understanding, which requires the models to understand the legal facts and the corresponding knowledge. Due to the knowledge-intensive feature of case analysis, straightforwardly applying general methods is suboptimal. Enhancing legal NLP systems with legal knowledge is crucial for effective case analysis. Figure 11.1 presents a real-world case document consisting of several parts, including the case fact description, the court's views, and the judgment results. In this case, the judges are required to apply the corresponding law to the specific circumstances of the defendant's case. After the judge determines the applicable law based on the attributes of the "domain name" and the purpose of the defendant's behavior, the judge then makes the final decision on the crime and the prison term. This example shows that the flexible application of legal knowledge is crucial in case analysis.

Unlike the widely used relational triple knowledge in the open domain, the structure of legal knowledge is complex and diverse. In daily work and communication, legal knowledge mainly presents in textual form. Take the two most representative legal systems as an example. In civil law systems, legal knowledge is mainly contained in laws and regulations, and judges need to analyze and decide cases according to the principles in the laws and regulations. In common law systems,

---

Fact Description: Alice bid for the domain name "..." and handed it over to Company A for maintenance. Bob premeditatedly stole the domain name "...", he first cracked the password of the mailbox bound by the domain name, and then bound the domain name to his own mailbox. Bob sold the domain name "..." for RMB 125,000 on the online platform.

Court's Views: Internet domain name has exclusivity and uniqueness ... Internet domain name is a scarce resource ... Internet domain names have market exchange value ... In this case, the perpetrator used technical means to realize the illegal possession of the domain name ...

Results: According to *Article 264 of the Criminal Law of the People's Republic of China*, Bob committed the crime of theft and was sentenced to four years and seven months in prison and a fine of RMB 50,000.

Fig. 11.1 An example of a legal case document, which consists of the fact description, court's views, and judgment results. The case document is translated from the cases published by the Supreme People's Court of the People's Republic of China

legal knowledge is mainly contained in opinions and decisions in previous cases, and judges need to summarize the principles from the past decisions of relevant courts to make judgments for current cases. The textual laws and case documents together form essential legal knowledge sources. Furthermore, many researchers formalize textual legal knowledge into various structured knowledge, such as legal elements, legal events, and legal logical rules, to facilitate the efficiency and fairness of real-world legal case analysis. Structured legal knowledge can decompose a case analysis task into several simplified subtasks and thus reduce the analysis complexity. Both textual and structured knowledge are essential and beneficial for legal NLP systems.

The complexity of structures of legal knowledge poses challenges to legal NLP models. Early legal NLP systems mainly utilize the legal knowledge following the mix of symbolism and rationalism paradigm [83, 95]. These works usually suffer from poor transferability and can only utilize a single type of knowledge. Inspired by the connectionism and empiricism approaches, many efforts have been devoted to designing neural models to integrate the knowledge for legal tasks [15, 31, 41, 59, 66, 105, 110, 120]. We term these methods as legal knowledge representation learning, which attempts to encode legal knowledge with different structures into a unified distributed representation space. Legal representation learning can help transform human-friendly textual and structured knowledge into machine-friendly model knowledge (modeledge) and gradually becomes a common paradigm for legal NLP.

In this chapter, we summarize the state of the arts of legal NLP from the perspective of the definition, acquisition, and application of legal knowledge. Notably, the Supreme People's Court of the People's Republic of China has published more than

100 million case documents,[1] greatly contributing to the development of legal AI. According to the statistics, China has the highest number of research papers on legal AI in recent years [80]. Therefore, in this chapter, the examples and models mainly come from legal AI research for Chinese cases.

In the following sections, we first present the typical tasks and real-world applications in legal NLP in Sect. 11.2, where all these tasks are knowledge-intensive and require multiple types of knowledge for reasoning. Then, we introduce several widely used legal knowledge nowadays, including textual and structured legal knowledge, in Sect. 11.3. Moreover, in Sect. 11.4, we focus on knowledge-guided legal NLP methods, which attempt to learn machine-friendly model knowledge from human-friendly legal knowledge. Based on the discussion in Chap. 9, we divide the existing knowledge-guided methods into four groups according to which model components are fused with knowledge. To promote future research, we discuss some directions in Sect. 11.5 and potential ethical risks of existing legal NLP methods in Sect. 11.6.

11.2 Typical Tasks and Real-World Applications

Recently, many legal tasks have been formally defined from the computational perspective to promote the application of AI techniques in the legal domain. To facilitate the introduction of subsequent sections, we will briefly describe the definition and challenges of several typical legal tasks, including legal judgment prediction, legal information retrieval, and legal question answering. Though many tasks have been intensively studied in recent years, not all of them have been widely used in real-world systems due to unsatisfactory performance and ethical considerations. In this section, we also briefly introduce some real-world applications of existing legal NLP methods.

Legal Judgment Prediction (LJP) LJP aims to predict the judgment results when giving the fact description and claims. LJP is one of the most practical tasks in legal NLP. Take widely studied Chinese cases as an example; the cases can be classified into three categories: administrative cases, civil cases, and criminal cases. The claims in administrative cases and civil cases are usually diverse, which introduces challenges for task formalization and evaluation. For example, in divorce dispute cases, the claims often include the distribution of property, child custody issues, etc. In contrast, the claims in criminal cases usually are homogeneous and request for courts to impose a certain punishment on the defendants, such as a fine, a prison term, or a death penalty. The homogeneity of claims in criminal cases brings convenience to the evaluation and formalization of LJP. Therefore, existing LJP research mainly focuses on criminal cases, and only limited research has been

---

[1] https://wenshu.court.gov.cn/

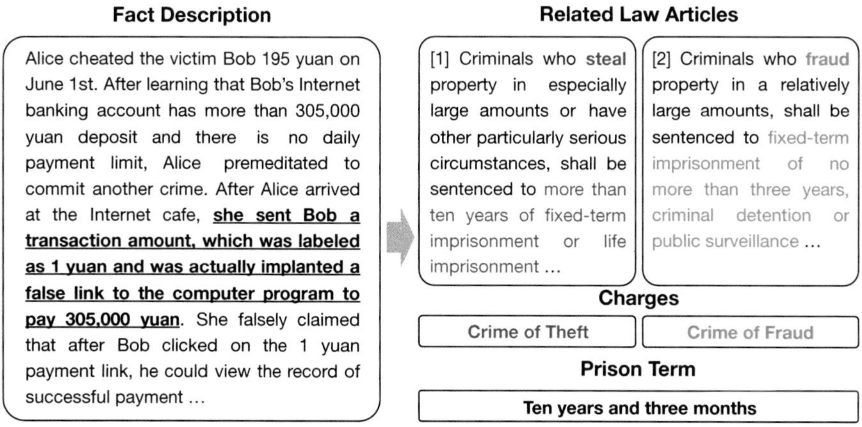

Fact Description

Alice cheated the victim Bob 195 yuan on June 1st. After learning that Bob's Internet banking account has more than 305,000 yuan deposit and there is no daily payment limit, Alice premeditated to commit another crime. After Alice arrived at the Internet cafe, **she sent Bob a transaction amount, which was labeled as 1 yuan and was actually implanted a false link to the computer program to pay 305,000 yuan**. She falsely claimed that after Bob clicked on the 1 yuan payment link, he could view the record of successful payment …

Related Law Articles

[1] Criminals who steal property in especially large amounts or have other particularly serious circumstances, shall be sentenced to more than ten years of fixed-term imprisonment or life imprisonment …

[2] Criminals who fraud property in a relatively large amounts, shall be sentenced to fixed-term imprisonment of no more than three years, criminal detention or public surveillance …

Charges

| Crime of Theft | Crime of Fraud |

Prison Term

| Ten years and three months |

Fig. 11.2 An example of legal judgment prediction. Given the fact description, the model is required to predict the judgment results, including the relevant law articles, the charges, and the prison term. The case document is translated from the cases published by the Supreme People's Court of the People's Republic of China

conducted on civil cases and administrative cases [33, 64]. In this subsection, we will mainly introduce the LJP task for criminal cases. As shown in Fig. 11.2, when given the textual fact description, the model is required to predict the judgment results, including the related law articles, charges, and the prison term, in turn.

With the rapid progress of end-to-end distributed representation learning, judgment prediction tasks are formalized as text classification or regression tasks. LJP mainly faces the following challenges: (1) Long-tail problem. The number of law articles and charges is large, and the number of cases for each category is imbalanced. And existing data-hungry methods cannot perform well for low-frequency categories. (2) Interpretability. Real-world applications are required to provide not only accurate predictions but also meaningful explanations for the results.

LJP has been studied since the 1950s and has been of interest to researchers from various countries, including China [66, 120], the United States [47, 52], Europe [13], and Korea [44]. Early works explore predicting the court decision with mathematical and statistical approaches from hand-crafted features [49, 52, 74]. Recent years witness the progress of neural judgment prediction models [13, 18, 78, 122]. To alleviate the long-tail problem and improve the interpretability of the LJP models, structured legal knowledge is often utilized to guide the model learning [41, 120, 121], which we will discuss in the following sections. LJP can provide potential judgment suggestions for judges and thus reduce their work stress. Automatic LJP models can also offer basic consult advice for the public. However, due to poor interpretability and unsatisfactory performance, LJP models usually face potential ethical risks and cannot be directly applied to real-world legal systems. The details of ethical issues are discussed in Sect. 11.6.

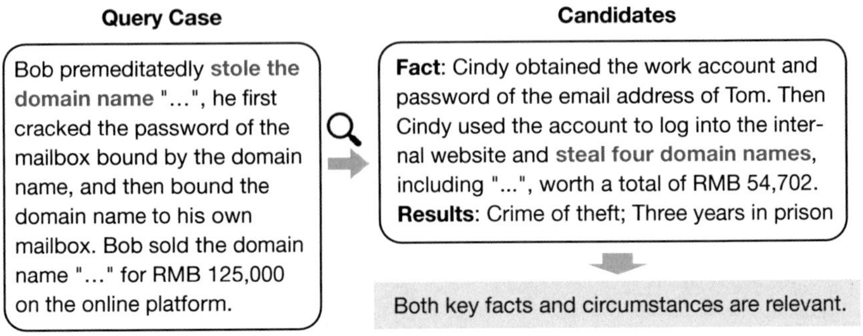

Query Case

Bob premeditatedly stole the domain name "...", he first cracked the password of the mailbox bound by the domain name, and then bound the domain name to his own mailbox. Bob sold the domain name "..." for RMB 125,000 on the online platform.

Candidates

Fact: Cindy obtained the work account and password of the email address of Tom. Then Cindy used the account to log into the internal website and steal four domain names, including "...", worth a total of RMB 54,702. **Results**: Crime of theft; Three years in prison

Both key facts and circumstances are relevant.

Fig. 11.3 An example of legal IR. The case document is translated from the cases published by the Supreme People's Court of the People's Republic of China

Legal Information Retrieval (Legal IR) Legal IR aims to retrieve similar cases, laws, regulations, and other information for supporting legal case analysis. Legal IR is essential for both civil and common law systems, where judges need to first retrieve relevant knowledge from the amounts of laws and cases and then make decisions based on the relevant knowledge. Manual retrieval from large-scale knowledge sources is very time-consuming and labor-intensive. Therefore, automatic legal IR based on factual descriptions is an essential task. As shown in Fig. 11.3, given a query case and a candidate set with several cases or law articles, legal IR models are required to calculate relevant scores between the query and candidates and then rank the candidates according to the relevant scores for final retrieval outputs.

Legal IR faces the following challenges: (1) Long text matching. Due to involving complex facts, the case documents usually contain thousands of tokens. The models are supposed to locate the key information in the long text and generate expressive case representation in the semantic space [89, 105]. (2) Diverse definitions of similarity. In open-domain IR, similarity mainly refers to topical similarity. But legal IR aims to find supporting evidence for case analysis, and the definition of similarity may be diverse for different requirements, including similarity from the aspects of related laws, occurring events, and focuses of dispute [90, 91]. For example, the cases shown in Fig. 11.3 are similar in terms of occurring events, but the related laws of the two cases are different, as the value of the stolen property in the query case is much higher than that of the candidate cases.

Conventional statistical legal IR relies heavily on laborious hand-crafted rules and expert knowledge [4], via legal issue decomposition [115] or ontological framework enhancing [87]. Statistical methods mainly focus on lexical similarity and suffer from the token mismatch problem. Recent neural-based methods have been proven effective in capturing semantic similarity between cases [68, 79]. According to the data structures, we can divide neural legal IR models into text-based and network-based methods. Text-based models compute the relevant scores

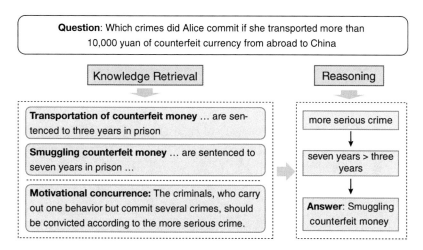

Fig. 11.4 An example of legal question answering. (The figure is re-drawn according to Fig. 1 in [123])

based on the textual content of cases [56, 57, 67, 113]. And network-based models utilize the citation network of cases to learn the representation and then recommend relevant cases and laws for queries [8, 42, 50, 109]. These works achieve good performance and make legal IR a widely used technique for various real-world applications.

Legal Question Answering (Legal QA) Legal QA aims to answer questions in the legal domain automatically. Figure 11.4 shows an example of legal QA, which needs to find the relevant legal knowledge given the question and then perform reasoning to finally get the answer to the question. An important task of lawyers and legal professionals is to provide legal advice, i.e., answering questions from people unfamiliar with legal knowledge. As mentioned before, the scarcity of legal professionals and the high cost of legal services prevent most low-income people from receiving timely and effective legal assistance. In such a situation, legal QA can be an effective way to achieve convenient and inexpensive legal consultation services.

Legal QA involves complex reasoning steps. As shown in Fig. 11.4, the model needs to perform multiple steps of reasoning based on the retrieved legal knowledge. According to the observation and statistics from a real-world legal question dataset [123], there are five challenging reasoning types for legal QA. (1) Lexical matching. It is a basic reasoning type for legal QA, which requires locating the relevant information and answers based on lexical matching. (2) Concept understanding. Legal questions usually involve abstract legal concepts. As shown in Fig. 11.4, after finding the relevant laws for facts, we can find that *Alice* commits two crimes with only one behavior. Then the models are supposed to associate the fact with the abstract concept of "Motivational concurrence" for the final decision.

(3) Numerical analysis. Case analysis sometimes requires performing numerical calculations. For example, the question in Fig. 11.4 requires comparing two prison terms to select the more serious charges. (4) Multi-paragraph reading. To answer the questions, the models must read and synthesize information from multiple paragraphs. (5) Multi-hop reasoning. It means that we need to conduct multiple steps of logical reasoning. These reasoning requirements make legal QA a challenging task.

Some researchers construct large-scale legal QA datasets and verify the performance of existing open-domain QA methods [29, 123]. They find that existing methods are suboptimal for legal QA, as legal QA usually involves complicated fact and knowledge reasoning. Recently, owing to the powerful large-scale pre-trained models (PTMs), many researchers begin exploring QA with complex reasoning using chain-of-thought prompting [102, 103] and behavior learning [75, 77]. Besides, to build knowledge-intensive models, some researchers enhance PTMs with knowledge graphs [98, 117] and knowledge retrieval [38, 54]. We argue that powerful PTMs also bring great potential for improving the performance of legal QA systems.

Real-World Applications In previous paragraphs, we introduce the legal AI tasks that have received much attention in academic research. Due to the unsatisfactory model performance, not all tasks have been applied in real-world systems. To provide an overview of the current situation of legal AI applications, we focus on widely used legal technology systems in real-world scenarios in this subsection.

With the development of the Internet, human activities and interactions have become more frequent. Meanwhile, people have become more aware of their rights. It leads to an increasing demand for legal services in recent years. According to legal industry reports, annual revenues for legal services have exceeded 150 billion yuan in China and exceeded 300 billion dollars in the United States [70]. The huge market size has given rise to many real-world legal application systems, designed to provide convenient legal services. These applications can be divided into two stages, legal information applications and legal intelligent applications.

Legal Information Applications Legal information applications aim to electronically manage and coordinate information and personnel in complex legal services. For example, in the United States, the Federal Judicial Center launched the COURTRAN project for electronic court records in 1975, and the subsequent PACER system has provided more convenient tools for electronic judicial litigation [71]. In China, the Supreme People's Court reported that China has successfully built legal information systems, where the digitization of case documents and online filing have been realized [19]. In terms of commercial companies, software for case file management and website for online legal consulting services have also received great attention [45]. In summary, legal information applications can help store and transfer information efficiently and reduce the communication cost between legal practitioners and the public.

Legal Intelligent Applications The development of legal informatization has provided basic data support and applicable scenario support for legal intelligent applications. Different from legal information applications, legal intelligent applications focus on the understanding, reasoning, and prediction of legal data to help achieve efficient knowledge acquisition and data analysis [112]. For example, to help judges and lawyers quickly find past similar cases, case retrieval and recommendation technology are now widely used, and systems providing case retrieval services have appeared in various countries around the world, such as WestLaw[2] and LexisNexis[3] in the United States and Faxin[4] in China. Besides, aiming to check the legality of contract terms, automatic contract review is gradually becoming a focus of legal commercial applications. Automatic contract review can provide risk warnings and assessments for public business activities, thus reducing the occurrence of contract disputes. Many startups are established for this application, such as PowerLaw,[5] LegalSifter,[6] etc.

It is worth mentioning that though many legal AI tasks have been widely studied in academic research, the application of these tasks in real-world systems is still limited. There are two main reasons. First, the performance of existing models is not satisfactory, as legal case analysis has a high demand for knowledge understanding and complex reasoning. Second, legal AI tasks, such as LJP, involve the basic rights and interests of citizens, but the potential ethical risks caused by the uninterpretability of legal models are not well studied yet. Therefore, some legal AI tasks are still in the research stage while not being applied. This still requires researchers and developers around the world to work together to develop safe, reliable, and accurate legal AI applications.

11.3 Legal Knowledge Representation and Acquisition

Legal tasks rely heavily on expert knowledge and require the model to associate relevant legal knowledge based on the understanding of facts to conduct complex analysis. In this section, we will introduce two important types of legal knowledge with different forms, including textual and structured knowledge. Legal knowledge is naturally presented in the textual form for daily work and communication. To promote work efficiency, some researchers formalize textual knowledge into structured knowledge. Both two types of knowledge play a crucial role in case analysis.

11.3.1 Legal Textual Knowledge

The majority of resources in the legal domain are presented in text forms, such as laws and regulations, legal cases, etc. These data can provide a rich reference basis for legal case analysis and enable models to effectively mine legal judgment patterns from them. In this subsection, we will introduce in detail the widely used legal textual knowledge.

Laws and Regulations Laws and regulations are a set of rules created by governmental institutions to regulate human and institutional behaviors. They are the basis of real-world legal systems and are the origin of other legal knowledge. Especially in the civil law system, all cases should be judged based on the related law articles, which provide comprehensive and valuable knowledge for case analysis. Notably, laws and regulations usually involve many abstract concepts, which are challenging for models to understand.

Legal Cases Legal cases cannot only provide amounts of training instances for models but also serve as a helpful knowledge base for case analysis. Different from laws and regulations which contain many abstract concepts, legal cases contain records of real-world facts and court views. Especially in common law systems, the judges are required to make a judgment based on the decision of relevant past cases and synthesize the rules of past cases as applicable to the current facts. Even for civil law systems, past cases can also provide a valuable reference for judges in some countries. Therefore, legal cases are an important supplement to the knowledge of laws and regulations.

Figure 11.5 presents an example of the fact description in a case and the corresponding law article. From the example, we can observe that the case document records the concrete real-world scenario, while the law article gives an abstract definition of the crime of position encroachment. The law articles are the basis for the judgment of legal cases, and the legal cases are the concrete manifestation of the law in the real world. Both laws and cases are important legal knowledge sources, and the legal AI models are required to associate the abstract concepts in laws and specific scenarios in facts for effective case analysis during knowledge applications.

Legal Textual Knowledge Representation Learning Legal textual knowledge is a very important knowledge source for legal AI, and how to learn informative representation for laws and cases lies in the core of knowledge-intensive legal AI tasks. In this subsection, we will introduce how to represent textual knowledge for downstream tasks. We denote laws and cases as legal documents. Similar to the development of NLP in other domains, early research mainly represents the legal documents with symbolic representation [2, 47, 72]. Inspired by the progress of distributed representation learning [48, 73, 81, 82], pre-trained embeddings and neural network are widely applied for legal case analysis [41, 66, 120, 122] in recent years.

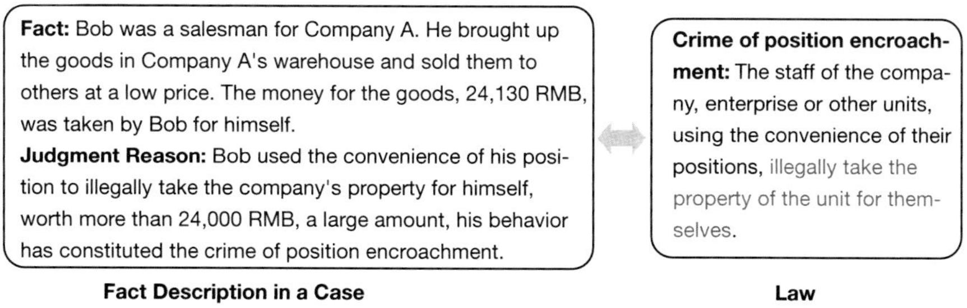

> **Fact:** Bob was a salesman for Company A. He brought up the goods in Company A's warehouse and sold them to others at a low price. The money for the goods, 24,130 RMB, was taken by Bob for himself.
> **Judgment Reason:** Bob used the convenience of his position to illegally take the company's property for himself, worth more than 24,000 RMB, a large amount, his behavior has constituted the crime of position encroachment.

Fact Description in a Case

> **Crime of position encroachment:** The staff of the company, enterprise or other units, using the convenience of their positions, illegally take the property of the unit for themselves.

Law

Fig. 11.5 An example of the fact description in a case and the corresponding law article. The case document is translated from the cases published by the Supreme People's Court of the People's Republic of China. The law is translated from Article 271 of the Criminal Law of the People's Republic of China

Nowadays, large-scale PTMs have been proven effective in capturing knowledge from the unlabeled corpus. In legal NLP, OpenCLaP [124] and Legal-BERT [15] are the earliest PTMs, which show that domain-specific pre-training can lead to performance improvement [36, 118]. Then Lawformer [105] is proposed to reduce the computational complexity of PTMs for long legal documents with a sparse attention mechanism. In addition, there are strict rules and regulations regarding the writing of legal documents. As a result, legal documents are written with extra attention to protecting privacy and avoiding offensive content. Such high-quality text data can also help mitigate the ethical risks of open-domain PTMs. Henderson et al. [39] show that models pre-trained on legal documents can perform well in learning contextual privacy rules and toxicity norms. Though these works can achieve promising results, the applied approaches ignore the specific characteristics of legal documents. For example, cases usually involve multiple semantic labels, such as the causes of action and the relevant laws. Designing legal-specific models that can effectively capture the semantic information of legal documents is still a challenging task.

11.3.2 Legal Structured Knowledge

In recent years, many researchers attempt to represent legal textual knowledge into structured knowledge. On the one hand, legal textual knowledge has its intrinsic structures. For example, the laws and regulations usually can be translated into the structure of "if...then...," which can be further converted into logical rules; the facts of a legal case usually consist of several key events; then it can be represented as a structured event timeline. On the other hand, structured knowledge is conducive to knowledge understanding and retention. Therefore, various types of legal structured knowledge have been widely used in legal AI tasks. In this subsection, we will introduce the definition and acquisition of several legal structured knowledge.

```
   ┌──────────── sell_drug_to ────────────┐
   │                                        ↓
```
Alice sold methamphetamine to Bob the entrance of People's Park for 300 yuan.

Fig. 11.6 An example of legal cases with a relational triple

Legal Relation Knowledge Relational factual knowledge organizes the knowledge in a triplet format and is widely used in knowledge graphs. Similar to relational knowledge in the open domain, legal triples can emphasize the important information in the legal domain. As shown in Fig. 11.6, the triple (*Alice*, sell_drug_to, *Bob*) can be regarded as the summarization of the given case, which will help models to capture the key information and benefit the downstream tasks. The legal relation extraction methods are similar to methods in the open domain. Please refer to Chap. 9 for details about relation extraction methods.

Legal Event Knowledge Recognizing the events and the corresponding causal relations between these events is the basis of legal case analysis. Following event definition in open domain [27, 99], legal events refer to crucial information from cases about what happened and what is involved. Figure 11.7 presents a legal case with the annotated event information and the corresponding judgment results. An event consists of one event trigger and several event arguments, such as time, place, and involved participants. Here, *crashed into* triggers the Bodily_harm event, in which *Alice*, *Bob*, and *Green Avenue* are the event actor, patient, and place. Enhancing case representation with legal event information can benefit various downstream case analysis tasks, such as LJP and legal IR. In this example, *Alice* causes a traffic accident, and the following Desertion and Escaping events lead to the Death event of *Bob*. Based on these occurred events, we can easily find the relevant law articles and give the final judgment results.

Legal events can be regarded as a summarization of legal cases and help models perform accurate case analysis. Therefore, many efforts have been devoted to legal event extraction (LEE). Early works mainly focus on utilizing hand-crafted symbolic features to extract legal events [6, 51]. Inspired by the success of neural networks, Li et al. [55] formalize LEE as a sequence labeling task and employ Bi-LSTM with CRF as the output layer to compute role labels for each token. Furthermore, Shen et al. [92] introduce a hierarchical event schema for LEE, which can capture the connections between different arguments and events. However, these works only attempt to extract events of limited event types and cannot be widely adopted in real-world applications. To this end, a large-scale legal event detection dataset, LEVEN [110] is proposed. LEVEN contains 8, 116 legal documents and 150, 977 annotated event mentions in 108 event types, which can serve as a reliable evaluation benchmark for LEE. It has been verified that many open-domain event detection methods can achieve good results on LEVEN. But we are also looking forward to future works that achieve comprehensive legal event extraction with high coverage of event types and event arguments.

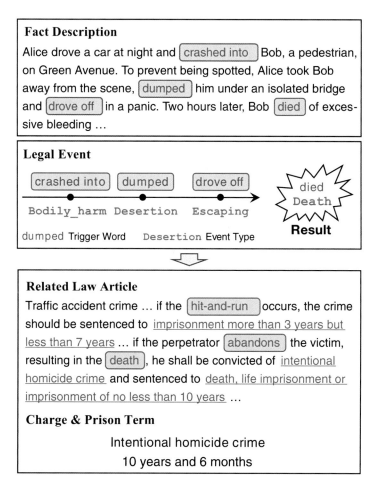

Fig. 11.7 An example of legal events. We can summarize the fact description as the legal event timeline, which can help in making judgments. (The figure is re-drawn according to Fig. 1 in [110])

Legal Element Knowledge Legal elements, also known as legal attributes, refer to properties of cases, based on which we can directly make judgment decisions. Figure 11.8 presents an example of attribute-based charge prediction. Here Theft and Robbery are two confusing charges, whose definitions only differ in a specific action. From the legal attributes, we can observe that due to the violence committed by *Bob*, he should be sentenced to Robbery crime instead of Theft crime. The element of violence or not is the key to distinguishing these two charges. The intellectual origin of legal attributes is the elemental trial [22, 84, 96], an important legal principle that requires judges to conduct a trial solely based on crucial legal elements. The elemental trial can help judges clarify trial procedures and avoid ethical issues.

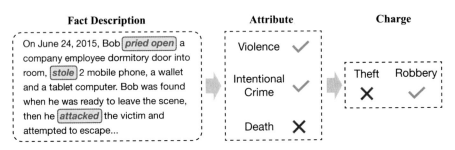

Fig. 11.8 An example of legal attributes. (The figure is re-drawn according to Fig. 1 in [41])

Legal element extraction is usually formalized as a multi-label text classification task for legal cases, where each element is a binary value with yes/no as the answer. Zhong et al. [122] evaluate typical text representation models for element classification and find that existing methods can achieve good results on legal element classification. Thus, element extraction is usually treated as an intermediate auxiliary task to promote the performance of downstream tasks, such as confusing charges discrimination [41] and interpretable judgment prediction [121]. Though legal elements are important for case analysis, existing methods only perform extraction on limited types of elements and have low coverage of charges. There is still a lack of benchmarks for comprehensive training and evaluation of element extraction.

Besides, legal element extraction is also a popular topic for contract analysis, where each element is a basic component of contracts, such as parties and beginning and ending dates [12, 14, 40, 101, 116]. This task can improve the speed of contract reading and facilitate the following contract review.

Legal Logical Knowledge Laws and regulations are in nature logical statements, and legal case analysis is a process of determining whether defendants violate the logical propositions contained in the law. Therefore, many researchers explore representing legal knowledge in a logical form and enhance case analysis with logical reasoning.

Legal logical knowledge can be divided into two categories: (1) Coarse-grained heuristic logic. Inspired by the various legal principles, coarse-grained heuristic logic mainly involve the task-level or charge-level logical rules. For example, task-level logical dependencies [111, 120] require the models to perform multi-task prediction following a specific human-inspired logical order. Elemental trial process [121] designs charge-level key elements for each charge and requires the models to analyze the cases by answering element-oriented questions. (2) Fine-grained first-order logic. As we mentioned before, laws and regulations are a set of rules expressed in natural language. And for better integration of these rules, some researchers attempt to represent laws as a set of fine-grained first-order logic rules [33]. For example, the law article in Fig. 11.5 can be represented as a logic rule:

$$X_{\text{staff}} \wedge X_{\text{property}} \rightarrow Y_{\text{crime}}, \tag{11.1}$$

where X_{staff} refers to if the defendant is the staff of the unit, X_{property} refers to if the defendant illegally takes the unit's property, and Y_{crime} refers to if the defendant should be sentenced to the crime of position encroachment. With the first-order logic rule, the judgment for the crime of position encroachment can be decomposed into two subtasks (i.e., X_{staff} and X_{property}).

Legal logical knowledge can help break down the case analysis into several logical substeps. Thus it can help improve the interpretability and reasoning ability of the models.

11.3.3 Discussion

Legal textual and structured knowledge play a significant role in legal applications, and different types of knowledge possess different characteristics and play different roles. In this subsection, we will discuss the advantages of legal textual and structured knowledge.

Textual Knowledge Textual knowledge possesses the following characteristics: (1) High coverage. Legal textual knowledge, including both laws and cases, is the origin of other forms of legal knowledge. Almost all scenarios can find their counterparts in the textual knowledge. (2) Updating over time. With the continuous refinement of laws and the increasing number of cases, legal textual knowledge is growing over time. Take the statistics from China as an example; there are more than a thousand national laws and more than 100 million legal cases nowadays, and this legal knowledge is updating and growing rapidly every year. Therefore, the two valuable characteristics make textual knowledge indispensable for existing legal NLP models. However, the textual knowledge is diverse in expression, which also makes it hard to retrieve and integrate legal textual knowledge for downstream tasks.

Structured Knowledge Structured knowledge possesses the following characteristics: (1) Concise and condensed. Structured knowledge often contains vital information that allows for a quick grasp of the case's specifics. Hence structured knowledge can benefit downstream models to capture key information from complicated cases. (2) Interpretable. The symbolic case representation derived from structured knowledge can provide intermediate interpretations for the prediction results. For example in Fig. 11.8, the element results can provide explanations for distinguishing confusing charges. However, structured knowledge requires labor-intensive and time-consuming manual annotation. Therefore, it needs further exploration for automatic structured knowledge acquisition.

Towards Model Knowledge In the era of deep learning, there is also an important type of knowledge, named model knowledge or modeledge. Modeledge refers to knowledge implicitly contained in models. Different from textual and structured knowledge which are explicit human-friendly knowledge, implicit modeledge is machine-friendly and can be easily utilized by AI systems. How to transform textual and structured legal knowledge into model knowledge is a popular research topic and will be discussed in the following section.

11.4 Knowledge-Guided Legal NLP

Legal knowledge, including textual and structured knowledge, is essential for case analysis. However, both two types of knowledge are presented in a human-friendly form, and it is not straightforward to enhance legal knowledge into legal NLP models. To this end, many efforts have been devoted to knowledge-guided legal NLP methods, which aim to embed explicit textual and structured knowledge into implicit model knowledge. Following the knowledgeable framework introduced in Chap. 9, in this section, we introduce knowledge-guided legal NLP methods from the perspective of which component is enhanced with legal knowledge.

11.4.1 Input Augmentation

Input augmentation methods integrate knowledge into the inputs of the models. Regarding the integration methods, the approaches can be mainly divided into two categories: text concatenation and embedding augmentation. Figure 11.9 illustrates the two categories of input augmentation methods.

Text Concatenation Text concatenation aims to concatenate the knowledge text with the original text and directly feed the concatenation into the model without the architecture modification. For example, Zhong et al. [123] concatenate relevant

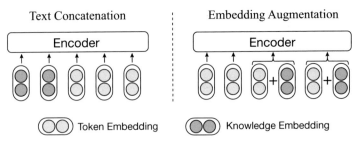

Fig. 11.9 The architecture of input augmentation methods

knowledge for legal question answering. Given the question, the authors retrieve relevant regulations from the knowledge base and concatenate the retrieved knowledge with questions to predict the final answers. Integrating knowledge via direct text concatenation can successfully inject knowledge into models but also may introduce noise if the textual knowledge is irrelevant to the given inputs.

Embedding Augmentation Embedding augmentation aims to integrate knowledge via fuse knowledge embeddings with original text embeddings. For example, Yao et al. [110] enhance PTMs with legal event knowledge by fusing the event embeddings in the input layer. They first extract legal events from fact description and then add additional event type embeddings with origin token embeddings as the inputs of PTMs for further applications.

11.4.2 Architecture Reformulation

Architecture reformulation refers to methods that design model architectures according to heuristic rules in the legal domain. From the perspective of the inspiration source, the works for architecture reformulation can be divided into two categories. One is inspired by the human thought process and the other is inspired by the knowledge structure.

Inspiration from Human Thought Process For human judges, there exist thinking logic patterns for legal case analysis, following which we can design models in line with judicial logic. For example, criminal judgment consists of multiple subtasks, including relevant law prediction, charge prediction, and prison term prediction. Judges usually would make decisions step by step, and the thought steps are closely related to each other. For example, the results of charges rely on the relevant laws, and the results of prison terms rely on the charges and relevant laws. Inspired by this observation, TopJudge [120] proposes to capture the logical dependency between different subtasks of LJP via a topological graph, where each node represents a subtask, and each edge represents the information dependency between subtasks. Figure 11.10 presents the model architecture for TopJudge, where the logical dependency is captured via RNN cells. As the prediction of prison terms relies on the results of laws and charges, the cell for prison term takes fact representation as inputs and the hidden vectors from laws and charges as cell states. Furthermore, inspired by the elemental trial principle, which requires judges to analyze cases from the perspective of legal elements, Zhong et al. [121] propose an iterative model to predict the charges by judging the key elements step by step. In this way, the thought process of the model is open and transparent, which can bring interpretability to LJP.

Inspiration from Knowledge Structure There are many different types of legal knowledge, and different types of knowledge can be utilized with different modules. The attention modules mentioned in the input augmentation section can also

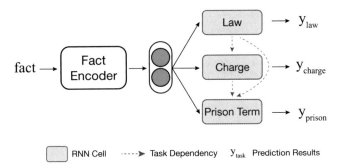

Fig. 11.10 Model architecture for TopJudge. (The figure is re-drawn according to Fig. 2 and Fig. 3 from [120])

be regarded as a specific input architecture for knowledge fusion. In addition, structured knowledge also plays a very important role in the design of network architecture. Some researchers build legal-specific output layers. For instance, as the definition of criminal charges follows a hierarchical structure, where each specific charge is the leaf node and similar charges are in the same subtree, Liu et al. [62] design a hierarchical classifier layer, where the model predicts the final charges following the path from the root node to the leaf charge node.

11.4.3 Objective Regularization

Objective regularization methods integrate legal knowledge into the objective functions. By introducing additional expert prior into the objective functions, the model can better capture key information from the text and improve downstream task performance. There are two mainstream approaches for objective regularization: regularization on new targets and regularization on existing targets. The former aims to design new training tasks, while the latter aims to build new constraints to the existing targets.

Regularization on New Targets Constructing additional supervision signals is a widely used strategy for legal case analysis. Xu et al. [108] improve prison term prediction by requiring the model to predict the seriousness of charges. Feng et al. [31] introduce the event extraction task for judgment prediction. Hu et al. [41] utilize the legal element knowledge via a multi-task framework. As shown in Fig. 11.11, the model is required to predict both the charges and element values. Then the model can generate an element-aware case representation and better distinguish the confusing charges.

Regularization on Existing Targets Legal case analysis usually consists of multiple subtasks, and there is a logical association between different subtasks. To

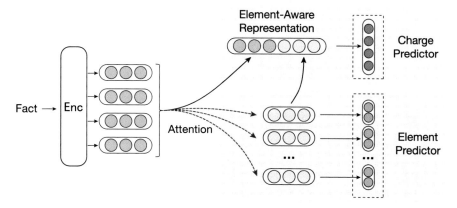

Fig. 11.11 The model architecture of charge prediction with the additional element prediction task as regularization

this end, many researchers attempt to construct extra constraints between different subtasks to improve the consistency across different tasks. For example, Feng et al. [31] add a penalty for legal event-based consistency, which requires that if an event trigger is detected, then all and only its corresponding argument types can be extracted. Chen et al. [20] add regularization for three legal judgment prediction subtasks from the perspective of causal inference. Regularization for multi-task learning can help the model produce consistent case analysis results and improve the reliability of legal AI.

11.4.4 Parameter Transfer

Parameter transfer refers to methods that train models on source tasks and then transfer the parameters to the target tasks to achieve knowledge transfer. As for the source task, existing approaches can be divided into two categories: transferring from self-supervised pre-trained models or transferring from other supervised tasks.

Pre-trained Models A typical paradigm of parameter transfer is pre-trained language models, which transfer parameters trained with self-supervised tasks to downstream applications. Early works only transfer the word embeddings to the target domain [16, 26]. Further, the pre-training-fine-tuning paradigm is widely used to transfer the knowledge learned from large-scale unsupervised data to downstream tasks. Legal PTMs have been proposed for various languages, such as Chinese [105, 124], English [15, 39], and French [28]. Furthermore, as most legal documents consist of thousands of tokens, a sparse attention-based model, Lawformer [105], is proposed for the legal domain. As shown in Fig. 11.12, instead of applying a fully connected self-attention mechanism, Lawformer utilizes the

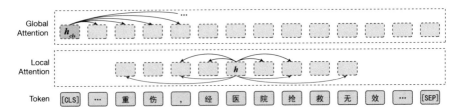

Fig. 11.12 The sparse attention mechanism applied in Lawformer. (The figure is re-drawn according to Fig. 2 from [105])

sparse attention mechanism, where the local attention requires each token to only attend its neighbor tokens, and the global attention only requires limited tokens to attend the whole sequence. Hence, the sparse attention mechanism decreases the computational complexity to linear complexity. These methods mainly adopt existing open-domain methods to the legal domain and do not design legal-specific pre-training tasks and model architectures, which is also important for future research.

Cross-task Transfer In addition to transferring parameters via self-supervised pre-training, some researchers attempt to train models on some source supervised tasks and then transfer the model to target tasks. For example, Shao et al. [89] conduct transfer learning from legal entailment to legal case retrieval. Gupta et al. [37] first train the model with open-domain datasets and then conduct further tuning for legal conference resolution. Cross-task transfer requires source tasks to be similar to the target tasks so that the task knowledge can be successfully transferred across tasks.

11.5 Outlook

Although legal NLP is currently well developed and makes good progress on many tasks, there is still a long way to go for the real-world applications of legal NLP methods. In this section, we list four directions for future research.

More Data As a family of neural networks, existing legal NLP models are data-hungry and require large amounts of high-quality labeled data. However, legal tasks often require complex reasoning about the case facts, which has a high requirement of expertise for annotators. As a result, the annotation of legal datasets is usually time-consuming and costly. For example, the annotation of CUAD, a legal contract review dataset, took dozens of law students a year and over 2 million dollars [40].

PTMs have shown their effectiveness in capturing knowledge from large-scale unlabeled data [25, 85]. Especially, the self-supervised pre-training can effectively improve the ability of few-shot learning [11, 34], which can help alleviate the data-scarce problem. In addition, with the continuous disclosure of legal documents and the accumulation of various legal data on the Internet, we can easily access publicly available legal data, which provides a substantial data basis for legal PTMs.

Some works attempt to train legal PTMs [15, 28, 39, 105]. However, the legal data used in these models are still limited, containing only legal cases, contracts, etc. Most of the pre-training tasks simply follow the tasks in the open domain, and the model size is also still limited. Therefore, we argue that it is very important to use more data and design legal-specific pre-training tasks to train larger legal PTMs with more capabilities.

More Knowledge Legal tasks place high demands on the understanding and application of legal knowledge. As mentioned in previous sections, many knowledge-guided legal NLP approaches have achieved significant progress in recent years [41, 66, 120, 121]. However, more knowledge is still desired for legal NLP methods.

As for textual knowledge, existing applications are limited by the ability of knowledge retrieval due to the gap between abstract knowledge and concrete facts. As for structured knowledge, existing applications focus on a limited number of case types, resulting in low coverage. Therefore, improving the ability to utilize textual knowledge and increasing the coverage of structured knowledge is an important issue to be addressed. Moreover, as for the combination of multiple knowledge, using more types of knowledge and more amount of knowledge in legal NLP models is also a very important research direction.

More Interpretability While existing neural network-based approaches have achieved high accuracy, the black-box characteristics of neural models pose a great ethical risk to real-world applications. For example, if gender bias is introduced in case analysis models, the uninterpretability will make it challenging to detect such bias and harm legal fairness. Moreover, the main goal of legal AI is to use technology to assist in legal tasks, which requires the models to cooperate with human experts for decision making. If legal NLP models can only give results without explanations, it will significantly increase the time cost for human experts to understand the model's results and reduce the credibility of the models.

Thus, while improving the accuracy of legal models, we also need to pay extra attention to the interpretability of legal models. Existing efforts explore improving the interpretability via outputting the intermediate states and results [33, 121], extracting the prediction evidence [59, 113], and generating the corresponding explanations [58, 111]. These methods mainly focus on specific tasks and usually need additional annotation, and it is still challenging to design a general and efficient framework for explainable legal AI.

More Intelligence Legal case analysis often involves complex legal reasoning, including abstract concept understanding, numerical analysis, multi-hop reasoning, and multi-passage information synthesis, which are still open problems in NLP. Therefore, complex case analysis reasoning requires models with more cognitive intelligence capabilities. In the open domain, many studies have demonstrated that large-scale PTMs can manipulate tools to complete complex tasks. For example,

WebGPT learns to use search engines to answer complex questions [75], and CC-Net can manipulate computers to finish some human instructions [43]. This gives the possibility to implement legal models with more intelligence, which is desired for complex case analysis. Enabling models to manipulate legal search engines, numerical calculators, etc. to complete complex case reasoning is also a very important future research direction for legal AI.

11.6 Ethical Consideration

Existing research has shown that legal AI systems can help improve work efficiency and alleviate a considerable workload for legal practitioners. However, legal work involves the essential rights and interests of individuals, and while emphasizing efficiency, we should also pay attention to the fairness and justice of the legal system. At present, legal AI systems are still in the stage of rapid development, and the ethical risks of legal AI systems have not been fully explored. In this section, we discuss the ethical risks of legal AI systems and what principles should be followed in the application of legal AI systems.

Ethical Risks In the application of legal AI systems, both the inevitable model bias and careless use may cause ethical risks.

Model Bias While the neural architecture brings significant performance improvement to legal NLP tasks, its black-box characteristics make it difficult to discover and detect the potential bias of the models. Many existing methods prove that the models may learn the bias from the training data, such as gender bias [94] and racial bias [76]. Besides, recent popular PTMs are usually trained on large-scale open-domain data, which may contain various types of bias [88, 104]. These potential model biases may result in severe unfair treatment of individuals, which is a serious ethical risk. Therefore, it is very important to detect and eliminate the bias of legal AI systems.

Some works attempt to explore fairness evaluation for legal models, especially LJP models. For example, FairLex [17] collects a fairness evaluation benchmark across five attributes (gender, age, region, language, and area) for four jurisdictions (European Council, USA, Switzerland, and China). CaLF [100] is a metric for real-world legal fairness and explores bias elimination with adversarial training. However, these works are still limited to specific tasks and attributes. It still needs further efforts to explore the general fairness evaluation across multiple tasks and attributes.

Misuse The goal of legal AI systems is to assist legal practitioners in their work, and legal AI systems must be used with the guidance of professionals. Due to the inevitable errors of legal models, it is very important to ensure that legal AI systems are used correctly. Specifically, legal AI systems should not be used to make final decisions that may affect the rights and interests of individuals, and legal practitioners should still be responsible for the final decision. It is desired to clarify the boundaries of the applications of legal AI systems and to ensure that the legal AI systems are used correctly.

Besides, the legal AI models are supposed to be well evaluated and used in the appropriate scenario. Each legal AI model is usually trained with specific datasets for a specific scenario, and misuse in an inappropriate scenario can boost the error rate of the models. For example, existing legal case retrieval models are trained with long cases as inputs, while for real-world applications, we may need to retrieve relevant cases with keywords or short sentences as inputs. The gap between model training and application may lead to the misuse of legal AI systems and obtain suboptimal performance.

Application Principles The potential risks of legal AI may cause serious consequences, and it is very important to ensure that legal AI systems are used correctly and ethically. In this section, we discuss the application principles of legal AI systems from the perspective of purpose, methodology, and monitoring [60, 119].

People-Oriented The ultimate goal of legal AI systems is to provide assistance and support for legal practitioners. The design of legal AI methods should be people-oriented, which means legal AI systems are designed to provide explainable references, but not the final decision, for legal practitioners, and the legal practitioners should be responsible for the final results.

Human-in-the-Loop In real-world legal AI applications, humans and AI models should cooperate to form a human-in-the-loop paradigm. Specifically, complex legal tasks should be divided into several subtasks, and the legal AI systems carry out the subtasks suitable for automation, including information storage, information extraction, knowledge retrieval, etc., while the legal practitioners carry out the subtasks that involve important decisions. In this way, the legal AI systems can provide assistance and support for legal practitioners, and the legal practitioners can provide the necessary control and training signals for the legal AI systems. Thus, the human-in-the-loop principle can improve reliability and avoid the misuse of legal AI systems.

Transparency As mentioned in the previous section, the black-box characteristics of legal AI systems make model biases inevitable. Therefore, it is very important to ensure the transparency of algorithms and details of legal AI systems. It means the public and legal practitioners should be able to understand the model's reasoning process, algorithm principles, and prediction results. In this way, transparency can help improve the credibility of legal AI systems and prevent the misuse of legal AI systems.

11.7   Open Competitions and Benchmarks

The formalization and definition of legal tasks are the basis of legal AI and require the efforts of both AI researchers and legal researchers. Besides, the evaluation of legal tasks requires large-scale human-annotated data. To this end, some organizations have formalized many legal tasks and collected large-scale legal datasets for the research community. There are three popular open challenges, including COLIEE,[7] AILA,[8] and CAIL,[9] which provide a large number of legal datasets for the research community.

Competition on Legal Information Extraction/Entailment, COLIEE, is held since 2014 and encourages the competitors to perform automatic legal retrieval and entailment. The data in COLIEE is collected from Japanese and Canadian case documents.

Artificial Intelligence for Legal Assistance, AILA, is held since 2019. AILA focuses on legal retrieval in 2019 and then extends the tasks to rhetorical labeling and legal summarization in 2020 and 2021. The data of AILA is collected from cases published by the Indian Supreme Court.

Challenge of AI in Law, CAIL, is held since 2018. CAIL publishes datasets for various legal NLP tasks. The data of CAIL is collected from cases published by the People's Supreme Court of the People's Republic of China.

We summarize several representative large-scale datasets for legal NLP in Table 11.1. These datasets provide valuable training and evaluation resources for the development of legal AI. We hope more researchers and organizations can collect and release more large-scale legal datasets for the research community to promote the development of legal AI systems.

11.8   Summary and Further Readings

In this chapter, we first introduce three typical legal knowledge-intensive tasks and their challenges, including legal judgment prediction, legal information retrieval, and legal question answering. All three tasks can provide a handy reference for legal services. Next, we describe the textual and structured legal knowledge, which is summarized by legal experts to facilitate case analysis. Further, we introduce knowledge-guided legal NLP methods from the perspective of how to integrate legal knowledge into neural models. We also discuss some advanced topics that aim to further promote the development of legal NLP approaches.

---

[7] https://sites.ualberta.ca/~rabelo/COLIEE2022/

[8] https://sites.google.com/view/aila-2021/track-description

[9] http://cail.cipsc.org.cn

Table 11.1 Large-scale datasets for legal AI tasks

Dataset	Open challenge	Task	Language
CAIL2018 [106]	CAIL2018	Judgment prediction	Chinese
SCM [107]	CAIL2019	Case matching	Chinese
CJRC [29]	CAIL2019-2021	Reading comprehension	Chinese
FE [93]	CAIL2019	Element extraction	Chinese
Argmine [114]	CAIL2020-2022	Argument-pair extraction	Chinese
JecQA [123]	CAIL2020-2022	Question answering	Chinese
Summarization[a]	CAIL2020-2022	Summarization	Chinese
IE [21]	CAIL2021-2022	Relation extraction	Chinese
LeCaRD [68]	CAIL2021-2022	Case retrieval	Chinese
FactLabel[b]	CAIL2021	Element extraction	Chinese
LEVEN [110]	CAIL2022	Event detection	Chinese
ELAM [113]	CAIL2022	Case retrieval	Chinese
Proofread[c]	CAIL2022	Grammar error correction	Chinese
AILA2019 [7]	AILA2019	Case/statute retrieval	Indian
AILA2020 [9]	AILA2020	Case/statute retrieval, rhetorical labeling	Indian
AILA2021 [79]	AILA2021	Summarization	Indian
COLIEE-Task1	COLIEE2014-2017	Case retrieval	Japanese
COLIEE-Task2	COLIEE2014-2017	Case entailment	Japanese
COLIEE-Task1	COLIEE2018-2021	Case retrieval	Japanese, Canadian
COLIEE-Task2	COLIEE2018-2021	Case entailment	Japanese, Canadian
COLIEE-Task3	COLIEE2018-2021	Statute retrieval	Japanese
COLIEE-Task4	COLIEE2018-2021	Statute entailment	Japanese
COLIEE-Task5	COLIEE2021	Question answering	Japanese
ILDC [69]	–	Judgment prediction	Indian
HLDC [46]	–	Bail prediction	Hindi
ECHR [13]	–	Judgment prediction	English
CUAD [40]	–	Contract review	English
EDGAR [53]	–	Contract review	English
FairLex [17]	–	Fairness evaluation	English
BSARD [65]	–	Statute retrieval	French

[a] http://cail.cipsc.org.cn/task_summit.html?raceID=4&cail_tag=2022
[b] http://cail.cipsc.org.cn/task_summit.html?raceID=6&cail_tag=2021
[c] http://cail.cipsc.org.cn/task2.html?raceID=2&cail_tag=2022

As for the introduction to legal tasks, Cui et al. [24] give a comprehensive overview of the datasets, subtasks, and methods of legal judgment prediction. Sansone et al. [86] and Locke et al. [63] review recent progress on legal information retrieval systems.

As for the survey of legal AI, Zhong et al. [122] provide insightful discussion and experiments on how existing deep learning methods perform on legal tasks. Bommasani et al. [10] discuss the opportunities and risks of the application in the legal domain of large-scale PTMs.

Acknowledgments The contributions of all authors of this chapter are as follows: Zhiyuan Liu, Yankai Lin, and Maosong Sun designed the overall architecture of this chapter. Chaojun Xiao drafted this chapter. Zhiyuan Liu and Yankai Lin proofread and revised this chapter.

We also thank Ganqu Cui, Yuzhong Wang, Zheni Zeng, Shengding Hu, Feng Yao, Run Li, and Pengle Zhang for proofreading the chapter.

This chapter about legal knowledge representation learning is the newly complemented content in the second edition of the book *Representation Learning for Natural Language Processing*. The first edition of the book was published in 2020 [61].

References

1. Work report of the Supreme People's Court of the People's Republic of China (in Chinese). 2022.
2. Nikolaos Aletras, Dimitrios Tsarapatsanis, Daniel Preoţiuc-Pietro, and Vasileios Lampos. Predicting judicial decisions of the European Court of Human Rights: A natural language processing perspective. *PeerJ Computer Science*, 2:e93, 2016.
3. Zhenwei An, Yuxuan Lai, and Yansong Feng. Natural language understanding for legal documents (in Chinese). *Journal of Chinese Information Processing*, 36(8):1–11, 2022.
4. Trevor Bench-Capon, Michał Araszkiewicz, Kevin Ashley, Katie Atkinson, Floris Bex, Filipe Borges, Daniele Bourcier, Paul Bourgine, Jack G Conrad, Enrico Francesconi, et al. A history of ai and law in 50 papers: 25 years of the international conference on ai and law. *Artificial Intelligence and Law*, 20(3):215–319, 2012.
5. Trevor Bench-Capon and Giovanni Sartor. A model of legal reasoning with cases incorporating theories and values. *Artificial Intelligence*, 150(1-2):97–143, 2003.
6. Anderson Bertoldi, Rove Chishman, Sandro José Rigo, and Thaís Domênica Minghelli. Cognitive linguistic representation of legal events. In *Proceedings of COGNITIVE*, 2014.
7. Paheli Bhattacharya, Kripabandhu Ghosh, Saptarshi Ghosh, Arindam Pal, Parth Mehta, Arnab Bhattacharya, and Prasenjit Majumder. Fire 2019 aila track: Artificial intelligence for legal assistance. In *Proceedings of FIRE*, 2019.
8. Paheli Bhattacharya, Kripabandhu Ghosh, Arindam Pal, and Saptarshi Ghosh. Hier-spcnet: a legal statute hierarchy-based heterogeneous network for computing legal case document similarity. In *Proceedings of SIGIR*, 2020.
9. Paheli Bhattacharya, Parth Mehta, Kripabandhu Ghosh, Saptarshi Ghosh, Arindam Pal, Arnab Bhattacharya, and Prasenjit Majumder. Overview of the fire 2020 aila track: Artificial intelligence for legal assistance. In *Proceedings of FIRE*, 2020.
10. Rishi Bommasani, Drew A Hudson, Ehsan Adeli, Russ Altman, Simran Arora, Sydney von Arx, Michael S Bernstein, Jeannette Bohg, Antoine Bosselut, Emma Brunskill, et al. On the opportunities and risks of foundation models. *arXiv preprint arXiv:2108.07258*, 2021.
11. Tom Brown, Benjamin Mann, Nick Ryder, Melanie Subbiah, Jared D Kaplan, Prafulla Dhariwal, Arvind Neelakantan, Pranav Shyam, Girish Sastry, Amanda Askell, et al. Language models are few-shot learners. In *Proceedings of NeurIPS*, 2020.

12. Ilias Chalkidis and Ion Androutsopoulos. A deep learning approach to contract element extraction. In *JURIX*, 2017.
13. Ilias Chalkidis, Ion Androutsopoulos, and Nikolaos Aletras. Neural legal judgment prediction in English. In *Proceedings of ACL*, 2019.
14. Ilias Chalkidis, Ion Androutsopoulos, and Achilleas Michos. Extracting contract elements. In *Proceedings of ICAIL*, 2017.
15. Ilias Chalkidis, Manos Fergadiotis, Prodromos Malakasiotis, Nikolaos Aletras, and Ion Androutsopoulos. Legal-BERT: The muppets straight out of law school. In *Proceedings of EMNLP Findings*, 2020.
16. Ilias Chalkidis and Dimitrios Kampas. Deep learning in law: early adaptation and legal word embeddings trained on large corpora. *Artificial Intelligence and Law*, 27(2):171–198, 2019.
17. Ilias Chalkidis, Tommaso Pasini, Sheng Zhang, Letizia Tomada, Sebastian Schwemer, and Anders Søgaard. Fairlex: A multilingual benchmark for evaluating fairness in legal text processing. In *Proceedings of ACL*, 2022.
18. Huajie Chen, Deng Cai, Wei Dai, Zehui Dai, and Yadong Ding. Charge-based prison term prediction with deep gating network. In *Proceedings of EMNLP-IJCNLP*, 2019.
19. Su Chen, He Tian, Yanbin Lyu, and Hu Changming. Annual report on informatization of Chinese courts (in Chinese). Technical report, 2022.
20. Wenqing Chen, Jidong Tian, Liqiang Xiao, Hao He, and Yaohui Jin. Exploring logically dependent multi-task learning with causal inference. In *Proceedings of EMNLP*, 2020.
21. Yanguang Chen, Yuanyuan Sun, Zhihao Yang, and Hongfei Lin. Joint entity and relation extraction for legal documents with legal feature enhancement. In *Proceedings of COLING*, 2020.
22. Jerome Alan Cohen. The criminal procedure law of the People's Republic of China. *The Journal of Criminal Law and Criminology*, 1982.
23. Legal Services Corporation. The justice gap: Measuring the unmet civil legal needs of low-income americans, 2017.
24. Junyun Cui, Xiaoyu Shen, Feiping Nie, Zheng Wang, Jinglong Wang, and Yulong Chen. A survey on legal judgment prediction: Datasets, metrics, models and challenges. *arXiv preprint arXiv:2204.04859*, 2022.
25. Jacob Devlin, Ming-Wei Chang, Kenton Lee, and Kristina Toutanova. BERT: pre-training of deep bidirectional transformers for language understanding. In *Proceedings of NAACL-HLT*, 2019.
26. Jenish Dhanani, Rupa Mehta, and Dipti Rana. Effective and scalable legal judgment recommendation using pre-learned word embedding. *Complex & Intelligent Systems*, pages 1–15, 2022.
27. George R Doddington, Alexis Mitchell, Mark A Przybocki, Lance A Ramshaw, Stephanie M Strassel, and Ralph M Weischedel. The automatic content extraction (ace) program-tasks, data, and evaluation. In *Proceedings of LREC*, 2004.
28. Stella Douka, Hadi Abdine, Michalis Vazirgiannis, Rajaa El Hamdani, and David Restrepo Amariles. JuriBERT: A masked-language model adaptation for French legal text. In *Proceedings of the Natural Legal Language Processing Workshop 2021*, 2021.
29. Xingyi Duan, Baoxin Wang, Ziyue Wang, Wentao Ma, Yiming Cui, Dayong Wu, Shijin Wang, Ting Liu, Tianxiang Huo, Zhen Hu, et al. CJRC: A reliable human-annotated benchmark dataset for Chinese judicial reading comprehension. In *Proceedings of CCL*, 2019.
30. Donald J Farole and Lynn Langton. *County-based and local public defender offices, 2007*. US Department of Justice, Office of Justice Programs, Bureau of Justice . . . , 2010.
31. Yi Feng, Chuanyi Li, and Vincent Ng. Legal judgment prediction via event extraction with constraints. In *Proceedings of ACL*, 2022.
32. Jens Frankenreiter and Michael A Livermore. Computational methods in legal analysis. *Annual Review of Law and Social Science*, 16:39–57, 2020.
33. Leilei Gan, Kun Kuang, Yi Yang, and Fei Wu. Judgment prediction via injecting legal knowledge into neural networks. In *Proceedings of AAAI*, 2021.

34. Tianyu Gao, Adam Fisch, and Danqi Chen. Making pre-trained language models better few-shot learners. In *Proceedings of ACL*, 2021.
35. Anne von der Lieth Gardner. *An artificial intelligence approach to legal reasoning*. MIT press, 1987.
36. Nicolas Garneau, Eve Gaumond, Luc Lamontagne, and Pierre-Luc Déziel. Criminelbart: a French Canadian legal language model specialized in criminal law. In *Proceedings of ICAIL*, 2021.
37. Ajay Gupta, Devendra Verma, Sachin Pawar, Sangameshwar Patil, Swapnil Hingmire, Girish K Palshikar, and Pushpak Bhattacharyya. Identifying participant mentions and resolving their coreferences in legal court judgements. In *Proceedings of TSD*, 2018.
38. Kelvin Guu, Kenton Lee, Zora Tung, Panupong Pasupat, and Mingwei Chang. Retrieval augmented language model pre-training. In *Proceedings of ICML*, 2020.
39. Peter Henderson, Mark S Krass, Lucia Zheng, Neel Guha, Christopher D Manning, Dan Jurafsky, and Daniel E Ho. Pile of law: Learning responsible data filtering from the law and a 256gb open-source legal dataset. *arXiv preprint arXiv:2207.00220*, 2022.
40. Dan Hendrycks, Collin Burns, Anya Chen, and Spencer Ball. CUAD: An expert-annotated NLP dataset for legal contract review. *arXiv preprint arXiv:2103.06268*, 2021.
41. Zikun Hu, Xiang Li, Cunchao Tu, Zhiyuan Liu, and Maosong Sun. Few-shot charge prediction with discriminative legal attributes. In *Proceedings of COLING*, 2018.
42. Zihan Huang, Charles Low, Mengqiu Teng, Hongyi Zhang, Daniel E Ho, Mark S Krass, and Matthias Grabmair. Context-aware legal citation recommendation using deep learning. In *Proceedings of ICAIL*, 2021.
43. Peter C Humphreys, David Raposo, Tobias Pohlen, Gregory Thornton, Rachita Chhaparia, Alistair Muldal, Josh Abramson, Petko Georgiev, Adam Santoro, and Timothy Lillicrap. A data-driven approach for learning to control computers. In *Proceedings of ICML*, 2022.
44. Wonseok Hwang, Dongjun Lee, Kyoungyeon Cho, Hanuhl Lee, and Minjoon Seo. A multi-task benchmark for korean legal language understanding and judgement prediction. *arXiv preprint arXiv:2206.05224*, 2022.
45. Johnathan Jenkins. What can information technology do for law. *Harv. JL & Tech.*, 21:589, 2007.
46. Arnav Kapoor, Mudit Dhawan, Anmol Goel, TH Arjun, Akshala Bhatnagar, Vibhu Agrawal, Amul Agrawal, Arnab Bhattacharya, Ponnurangam Kumaraguru, and Ashutosh Modi. Hldc: Hindi legal documents corpus. *arXiv preprint arXiv:2204.00806*, 2022.
47. Daniel Martin Katz, Michael J Bommarito, and Josh Blackman. A general approach for predicting the behavior of the Supreme Court of the United States. *PloS one*, 12(4):e0174698, 2017.
48. Yoon Kim. Convolutional neural networks for sentence classification. In *Proceedings of EMNLP*, 2014.
49. Fred Kort. Predicting Supreme Court decisions mathematically: A quantitative analysis of the "right to counsel" cases. *American Political Science Review*, 51(1):1–12, 1957.
50. Sushanta Kumar, P Krishna Reddy, V Balakista Reddy, and Aditya Singh. Similarity analysis of legal judgments. In *Proceedings of COMPUTE*, 2011.
51. Nikolaos Lagos, Frederique Segond, Stefania Castellani, and Jacki O'Neill. Event extraction for legal case building and reasoning. In *Proceedings of IIP*, 2010.
52. Benjamin E Lauderdale and Tom S Clark. The Supreme Court's many median justices. *American Political Science Review*, 106(4):847–866, 2012.
53. Spyretta Leivaditi, Julien Rossi, and Evangelos Kanoulas. A benchmark for lease contract review. *arXiv preprint arXiv:2010.10386*, 2020.
54. Patrick Lewis, Ethan Perez, Aleksandra Piktus, Fabio Petroni, Vladimir Karpukhin, Naman Goyal, Heinrich Küttler, Mike Lewis, Wen-tau Yih, Tim Rocktäschel, et al. Retrieval-augmented generation for knowledge-intensive nlp tasks. In *Proceedings of NeurIPS*, 2020.
55. Chuanyi Li, Yu Sheng, Jidong Ge, and Bin Luo. Apply event extraction techniques to the judicial field. In *Proceedings of UbiComp-ISWC*, 2019.

56. Bulou Liu, Yueyue Wu, Yiqun Liu, Fan Zhang, Yunqiu Shao, Chenliang Li, Min Zhang, and Shaoping Ma. Conversational vs traditional: Comparing search behavior and outcome in legal case retrieval. In *Proceedings of SIGIR*, 2021.
57. Bulou Liu, Yueyue Wu, Fan Zhang, Yiqun Liu, Zhihong Wang, Chenliang Li, Min Zhang, and Shaoping Ma. Query generation and buffer mechanism: Towards a better conversational agent for legal case retrieval. *Information Processing & Management*, 59(5):103051, 2022.
58. Liting Liu, Wenzheng Zhang, Jie Liu, Wenxuan Shi, and Yalou Huang. Interpretable charge prediction for legal cases based on interdependent legal information. In *Proceedings of IJCNN*, 2021.
59. Xiao Liu, Da Yin, Yansong Feng, Yuting Wu, and Dongyan Zhao. Everything has a cause: Leveraging causal inference in legal text analysis. In *Proceedings of NAACL*, 2021.
60. Yiqun Liu. Establishing a robust system of rules for the application of legal artificial intelligence to achieve a higher level of digital justice (in Chinese). *China Internet Civilization Conference*, 2022.
61. Zhiyuan Liu, Yankai Lin, and Maosong Sun. *Representation Learning for Natural Language Processing*. Springer, 2020.
62. Zhiyuan Liu, Cunchao Tu, and Maosong Sun. Legal cause prediction with inner descriptions and outer hierarchies. In *Proceedings of CCL*, 2019.
63. Daniel Locke and Guido Zuccon. Case law retrieval: problems, methods, challenges and evaluations in the last 20 years. *arXiv preprint arXiv:2202.07209*, 2022.
64. Shangbang Long, Cunchao Tu, Zhiyuan Liu, and Maosong Sun. Automatic judgment prediction via legal reading comprehension. In *Proceedings of CCL*, 2019.
65. Antoine Louis and Gerasimos Spanakis. A statutory article retrieval dataset in French. In *Proceedings of ACL*, 2022.
66. Bingfeng Luo, Yansong Feng, Jianbo Xu, Xiang Zhang, and Dongyan Zhao. Learning to predict charges for criminal cases with legal basis. In *Proceedings of EMNLP*, 2017.
67. Yixiao Ma, Qingyao Ai, Yueyue Wu, Yunqiu Shao, Yiqun Liu, Min Zhang, and Shaoping Ma. Incorporating retrieval information into the truncation of ranking lists for better legal search. In *Proceedings of SIGIR*, 2022.
68. Yixiao Ma, Yunqiu Shao, Yueyue Wu, Yiqun Liu, Ruizhe Zhang, Min Zhang, and Shaoping Ma. LeCaRD: a legal case retrieval dataset for Chinese law system. In *Proceedings of SIGIR*, 2021.
69. Vijit Malik, Rishabh Sanjay, Shubham Kumar Nigam, Kripabandhu Ghosh, Shouvik Kumar Guha, Arnab Bhattacharya, and Ashutosh Modi. ILDC for CJPE: Indian legal documents corpus for court judgment prediction and explanation. In *Proceedings of ACL-IJCNLP*, 2021.
70. MarketLine. Legal services in the United States. Technical report, 2021.
71. Peter W Martin. Online access to court records-from documents to data, particulars to patterns. *Vill. L. Rev.*, 53:855, 2008.
72. Masha Medvedeva, Michel Vols, and Martijn Wieling. Using machine learning to predict decisions of the European Court of Human Rights. *Artificial Intelligence and Law*, 28(2):237–266, 2020.
73. T Mikolov and J Dean. Distributed representations of words and phrases and their compositionality. In *Proceedings of NeurIPS*, 2013.
74. Stuart S Nagel. Applying correlation analysis to case prediction. *Tex. L. Rev.*, 42:1006, 1963.
75. Reiichiro Nakano, Jacob Hilton, Suchir Balaji, Jeff Wu, Long Ouyang, Christina Kim, Christopher Hesse, Shantanu Jain, Vineet Kosaraju, William Saunders, et al. WebGPT: Browser-assisted question-answering with human feedback. *arXiv preprint arXiv:2112.09332*, 2021.
76. Ziad Obermeyer, Brian Powers, Christine Vogeli, and Sendhil Mullainathan. Dissecting racial bias in an algorithm used to manage the health of populations. *Science*, 366(6464):447–453, 2019.
77. Long Ouyang, Jeff Wu, Xu Jiang, Diogo Almeida, Carroll L Wainwright, Pamela Mishkin, Chong Zhang, Sandhini Agarwal, Katarina Slama, Alex Ray, et al. Training language models to follow instructions with human feedback. *arXiv preprint arXiv:2203.02155*, 2022.

78. Sicheng Pan, Tun Lu, Ning Gu, Huajuan Zhang, and Chunlin Xu. Charge prediction for multi-defendant cases with multi-scale attention. In *Proceedings of ChineseCSWC*, 2019.
79. Vedant Parikh, Upal Bhattacharya, Parth Mehta, Ayan Bandyopadhyay, Paheli Bhattacharya, Kripa Ghosh, Saptarshi Ghosh, Arindam Pal, Arnab Bhattacharya, and Prasenjit Majumder. Aila 2021: Shared task on artificial intelligence for legal assistance. In *Proceedings of FIRE*, 2021.
80. So-Hui Park, Dong-Gu Lee, Jin-Sung Park, and Jun-Woo Kim. A survey of research on data analytics-based legal tech. *Sustainability*, 13(14):8085, 2021.
81. Jeffrey Pennington, Richard Socher, and Christopher Manning. GloVe: Global vectors for word representation. In *Proceedings of EMNLP*, 2014.
82. Matthew Peters, Mark Neumann, Mohit Iyyer, Matt Gardner, Christopher Clark, Kenton Lee, and Luke Zettlemoyer. Deep contextualized word representations. In *Proceedings of NAACL-HLT*, 2018.
83. James Popple. *A pragmatic legal expert system*. Dartmouth (Ashgate), 1996.
84. François Quintard-Morénas. The presumption of innocence in the French and Anglo-American legal traditions. *The American Journal of Comparative Law*, 58(1):107–149, 2010.
85. Colin Raffel, Noam Shazeer, Adam Roberts, Katherine Lee, Sharan Narang, Michael Matena, Yanqi Zhou, Wei Li, and Peter J Liu. Exploring the limits of transfer learning with a unified text-to-text transformer. *Journal of Machine Learning Research*, 21:1–67, 2020.
86. Carlo Sansone and Giancarlo Sperlí. Legal information retrieval systems: State-of-the-art and open issues. *Information Systems*, 106:101967, 2022.
87. Manavalan Saravanan, Balaraman Ravindran, and Shivani Raman. Improving legal information retrieval using an ontological framework. *Artificial Intelligence and Law*, 17(2):101–124, 2009.
88. Patrick Schramowski, Cigdem Turan, Nico Andersen, Constantin A Rothkopf, and Kristian Kersting. Large pre-trained language models contain human-like biases of what is right and wrong to do. *Nature Machine Intelligence*, 4(3):258–268, 2022.
89. Yunqiu Shao, Jiaxin Mao, Yiqun Liu, Weizhi Ma, Ken Satoh, Min Zhang, and Shaoping Ma. BERT-PLI: Modeling paragraph-level interactions for legal case retrieval. In *Proceedings of IJCAI*, 2020.
90. Yunqiu Shao, Yueyue Wu, Yiqun Liu, Jiaxin Mao, and Shaoping Ma. Understanding relevance judgments in legal case retrieval. *ACM Transactions on Information Systems*, 2022.
91. Yunqiu Shao, Yueyue Wu, Yiqun Liu, Jiaxin Mao, Min Zhang, and Shaoping Ma. Investigating user behavior in legal case retrieval. In *Proceedings of SIGIR*, 2021.
92. Shirong Shen, Guilin Qi, Zhen Li, Sheng Bi, and Lusheng Wang. Hierarchical Chinese legal event extraction via pedal attention mechanism. In *Proceedings of COLING*, 2020.
93. Yi Shu, Yao Zhao, Xianghui Zeng, and Qingli Ma. Cail2019-fe. Technical report, 2019.
94. Tony Sun, Andrew Gaut, Shirlyn Tang, Yuxin Huang, Mai ElSherief, Jieyu Zhao, Diba Mirza, Elizabeth Belding, Kai-Wei Chang, and William Yang Wang. Mitigating gender bias in natural language processing: Literature review. In *Proceedings of ACL*, 2019.
95. Richard E Susskind. The latent damage system: A jurisprudential analysis. In *Proceedings of ICAIL*, 1989.
96. Victor Tadros and Stephen Tierney. The presumption of innocence and the human rights act. *The Modern Law Review*, 67(3):402–434, 2004.
97. Russ VerSteeg. *Law in ancient Egypt*. Carolina Academic Press, 2002.
98. Xiaozhi Wang, Tianyu Gao, Zhaocheng Zhu, Zhengyan Zhang, Zhiyuan Liu, Juanzi Li, and Jian Tang. KEPLER: A unified model for knowledge embedding and pre-trained language representation. *Transactions of the Association for Computational Linguistics*, 9:176–194, 2021.
99. Xiaozhi Wang, Ziqi Wang, Xu Han, Wangyi Jiang, Rong Han, Zhiyuan Liu, Juanzi Li, Peng Li, Yankai Lin, and Jie Zhou. MAVEN: A massive general domain event detection dataset. In *Proceedings of EMNLP*, 2020.

100. Yuzhong Wang, Chaojun Xiao, Shirong Ma, Haoxi Zhong, Cunchao Tu, Tianyang Zhang, Zhiyuan Liu, and Maosong Sun. Equality before the law: legal judgment consistency analysis for fairness. *arXiv preprint arXiv:2103.13868*, 2021.
101. Zihan Wang, Hongye Song, Zhaochun Ren, Pengjie Ren, Zhumin Chen, Xiaozhong Liu, Hongsong Li, and Maarten de Rijke. Cross-domain contract element extraction with a bidirectional feedback clause-element relation network. In *Proceedings of SIGIR*, 2021.
102. Jason Wei, Maarten Bosma, Vincent Zhao, Kelvin Guu, Adams Wei Yu, Brian Lester, Nan Du, Andrew M Dai, and Quoc V Le. Finetuned language models are zero-shot learners. In *Proceedings of ICLR*, 2021.
103. Jason Wei, Xuezhi Wang, Dale Schuurmans, Maarten Bosma, Ed Chi, Quoc Le, and Denny Zhou. Chain of thought prompting elicits reasoning in large language models. *arXiv preprint arXiv:2201.11903*, 2022.
104. Laura Weidinger, John Mellor, Maribeth Rauh, Conor Griffin, Jonathan Uesato, Po-Sen Huang, Myra Cheng, Mia Glaese, Borja Balle, Atoosa Kasirzadeh, et al. Ethical and social risks of harm from language models. *arXiv preprint arXiv:2112.04359*, 2021.
105. Chaojun Xiao, Xueyu Hu, Zhiyuan Liu, Cunchao Tu, and Maosong Sun. Lawformer: A pretrained language model for Chinese legal long documents. *AI Open*, 2:79–84, 2021.
106. Chaojun Xiao, Haoxi Zhong, Zhipeng Guo, Cunchao Tu, Zhiyuan Liu, Maosong Sun, Yansong Feng, Xianpei Han, Zhen Hu, Heng Wang, et al. Cail2018: A large-scale legal dataset for judgment prediction. *arXiv preprint arXiv:1807.02478*, 2018.
107. Chaojun Xiao, Haoxi Zhong, Zhipeng Guo, Cunchao Tu, Zhiyuan Liu, Maosong Sun, Tianyang Zhang, Xianpei Han, Zhen Hu, Heng Wang, et al. Cail2019-scm: A dataset of similar case matching in legal domain. *arXiv preprint arXiv:1911.08962*, 2019.
108. Zhuopeng Xu, Xia Li, Yinlin Li, Zihan Wang, Yujie Fanxu, and Xiaoyan Lai. Multi-task legal judgement prediction combining a subtask of the seriousness of charges. In *Proceedings of CCL*, 2020.
109. Jun Yang, Weizhi Ma, Min Zhang, Xin Zhou, Yiqun Liu, and Shaoping Ma. LegalGNN: Legal information enhanced graph neural network for recommendation. *ACM Transactions on Information Systems (TOIS)*, 40(2):1–29, 2021.
110. Feng Yao, Chaojun Xiao, Xiaozhi Wang, Zhiyuan Liu, Lei Hou, Cunchao Tu, Juanzi Li, Yun Liu, Weixing Shen, and Maosong Sun. LEVEN: A large-scale Chinese legal event detection dataset. In *Proceedings of ACL Findings*, 2022.
111. Hai Ye, Xin Jiang, Zhunchen Luo, and Wenhan Chao. Interpretable charge predictions for criminal cases: Learning to generate court views from fact descriptions. In *Proceedings of NAACL*, 2018.
112. Yueyue Wu Yiqun Liu. Informatization and intelligence: discernment in the context of justice (in Chinese). *Chinese Applied Jurisprudence*, 02:15–30, 2021.
113. Weijie Yu, Zhongxiang Sun, Jun Xu, Zhenhua Dong, Xu Chen, Hongteng Xu, and Ji-Rong Wen. Explainable legal case matching via inverse optimal transport-based rationale extraction. In *Proceedings of SIGIR*, 2022.
114. Jian Yuan, Zhongyu Wei, Yixu Gao, Wei Chen, Yun Song, Donghua Zhao, Jinglei Ma, Zhen Hu, Shaokun Zou, Donghai Li, et al. Overview of smp-cail2020-argmine: The interactive argument-pair extraction in judgement document challenge. *Data Intelligence*, 3(2):287–307, 2021.
115. Yiming Zeng, Ruili Wang, John Zeleznikow, and Elizabeth Kemp. Knowledge representation for the intelligent legal case retrieval. In *Proceedings of KES*, 2005.
116. Shuo Zhang, Junzhou Zhao, Pinghui Wang, Nuo Xu, Yang Yang, Yiting Liu, Yi Huang, and Junlan Feng. Learning to check contract inconsistencies. In *Proceedings of AAAI*, 2021.
117. Zhengyan Zhang, Xu Han, Zhiyuan Liu, Xin Jiang, Maosong Sun, and Qun Liu. ERNIE: Enhanced language representation with informative entities. In *Proceedings of ACL*, 2019.
118. Lucia Zheng, Neel Guha, Brandon R Anderson, Peter Henderson, and Daniel E Ho. When does pretraining help? assessing self-supervised learning for law and the casehold dataset of 53,000+ legal holdings. In *Proceedings of ICAIL*, 2021.

119. Xi Zheng. Risks and regulation of judicial use of artificial intelligence(in Chinese). *People's Court Daily*, 6, 2021.
120. Haoxi Zhong, Zhipeng Guo, Cunchao Tu, Chaojun Xiao, Zhiyuan Liu, and Maosong Sun. Legal judgment prediction via topological learning. In *Proceedings of EMNLP*, 2018.
121. Haoxi Zhong, Yuzhong Wang, Cunchao Tu, Tianyang Zhang, Zhiyuan Liu, and Maosong Sun. Iteratively questioning and answering for interpretable legal judgment prediction. In *Proceedings of AAAI*, 2020.
122. Haoxi Zhong, Chaojun Xiao, Cunchao Tu, Tianyang Zhang, Zhiyuan Liu, and Maosong Sun. How does nlp benefit legal system: A summary of legal artificial intelligence. In *Proceedings of ACL*, 2020.
123. Haoxi Zhong, Chaojun Xiao, Cunchao Tu, Tianyang Zhang, Zhiyuan Liu, and Maosong Sun. Jec-qa: a legal-domain question answering dataset. In *Proceedings of AAAI*, 2020.
124. Haoxi Zhong, Zhengyan Zhang, Zhiyuan Liu, and Maosong Sun. Open Chinese language pre-trained model zoo. *Technical report*, 2019.

Chapter 12
Biomedical Knowledge Representation Learning

Zheni Zeng, Zhiyuan Liu, Yankai Lin, and Maosong Sun

Abstract As a subject closely related to our life and understanding of the world, biomedicine keeps drawing much attention from researchers in recent years. To help improve the efficiency of people and accelerate the progress of this subject, AI techniques especially NLP methods are widely adopted in biomedical research. In this chapter, with biomedical knowledge as the core, we launch a discussion on knowledge representation and acquisition as well as biomedical knowledge-guided NLP tasks and explain them in detail with practical scenarios. We also discuss current research progress and several future directions.

12.1 Introduction

There is a widely adopted perspective that the twenty-first century is the age of biology [30]. Actually, biomedicine has always occupied an important position and maintained a relatively rapid development. Researchers devote to explore how the life systems (e.g., cells, organisms, individuals, and populations) work, what the mechanism of genetics (e.g., DNA and RNA) is, how the external environment (e.g., chemicals and drugs) affects the systems, and many other important topics [42]. Recent flourish has been brought by the development of emerging interdisciplinary domains [92], among which biomedical NLP draws much attention as a representative topic in **AI for science**, which aims to apply modern AI tools to various areas of science to achieve efficient scientific knowledge acquisition and applications.

Z. Zeng · Z. Liu (✉) · M. Sun
Department of Computer Science and Technology, Tsinghua University, Beijing, China
e-mail: zzn20@mails.tsinghua.edu.cn; liuzy@tsinghua.edu.cn; sms@tsinghua.edu.cn

Y. Lin
Gaoling School of Artificial Intelligence, Renmin University of China, Beijing, China
e-mail: yankailin@ruc.edu.cn

© The Author(s) 2023
Z. Liu et al. (eds.), *Representation Learning for Natural Language Processing*,
https://doi.org/10.1007/978-981-99-1600-9_12

433

12.1.1 Perspectives for Biomedical NLP

The prospect of biomedical NLP is to improve human experts' efficiency by mining useful information and finding potential implicit laws automatically, and this is closely related to two branches of biology: computational biology and bioinformatics. Computational biology emphasizes solving biological problems with the favor of computer science. Researchers use computer languages and mathematical logics to describe and simulate the biological world. Bioinformatics studies the collection, processing, storage, dissemination, analysis, and interpretation of biological information. Bioinformatics research mainly focuses on the two aspects of genomics and proteomics.[1] The two terms are now generally used interchangeably.

According to the format of the processed data, we can make an inductive analysis of biomedical NLP from two perspectives. The first perspective is **NLP tasks in biomedical domain text**, in which we regard biomedicine as a specific domain of natural language documents; therefore the basic tasks are common with general domain NLP, while the corpus has its own features. Typical tasks [17] include named entity recognition, term linking, relation extraction, information retrieval, document classification, question answering, etc.

The other perspective is **NLP methods for biomedical materials**, in which the NLP techniques are adopted and transferred for modeling non-natural-language data and solving biomedical problems, such as the data mining of genetic and protein sequences [38]. As shown in Fig. 12.1, biomedical materials include natural language documents and other materials. The latter can be expressed in sequences, graphs, and other forms, and therefore the representation learning technique we introduce in the previous chapters can be employed to help model biomedical material. To ensure the effectiveness of general NLP techniques in the new scenario, adjustments are required to better fit with the data characteristics (e.g., fewer tokens for genetic sequences compared with natural language).

Overall, the biomedical natural language documents contain linguistic and commonsense knowledge and also provide explicit and flexible descriptions for biomedical knowledge. Meanwhile, the special materials in biomedical domain contain even more subject knowledge in implicit expressions. We believe that the two perspectives are gradually fusing together to achieve more universal biomedical material processing, and we will go into more detail about this trend later.

12.1.2 Role of Knowledge in Biomedical NLP

A characteristic of biomedical NLP is that expert knowledge is of key importance to get a deep comprehension of the processing materials. This even restricts the

---

[1] https://www.genome.gov/genetics-glossary/Bioinformatics

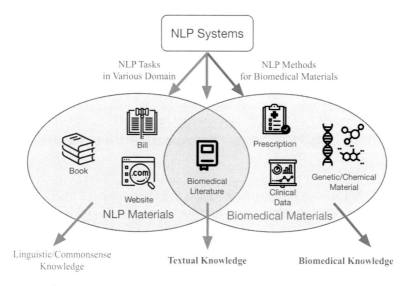

Fig. 12.1 Introduction to biomedical knowledge and biomedical NLP. Icons are bought or freely downloaded from IconFinder (https://www.iconfinder.com/)

scale of golden datasets due to the high cost and difficulty of manual annotation. Therefore, we emphasize the **knowledge representation**, **knowledge acquisition**, and **knowledge-guided NLP** methods for the biomedical domain.

First, biomedical materials have to be expressed properly to fit automatic computing, and this benefits from the development of knowledge representation methods such as distributed representations. Next, echoing the basic goals of AI for science, we expect the biomedical NLP systems to assist us in extracting and summarizing useful information or rules in a mass of unstructured materials, which is an important part of the knowledge acquisition process. Nevertheless, we have mentioned above that the biomedical NLP datasets are hard to reach on a large scale, which is one reason that the data-driven deep learning system performance in the biomedical domain is not always satisfying. To improve the performance of these intelligent systems under limited conditions, the knowledge-guided NLP methods become especially important. With the help of biomedical knowledge, NLP models trained on the general domain can be easily transferred to biomedical tasks with minimal supervision. For instance, the definition and synonyms of terms in biomedical ontologies can guide models to get a deeper comprehension of biomedical terms occurring in the processing texts.

In Sect. 12.2, we first introduce the representation and acquisition of biomedical knowledge, which comes from two types of materials: natural language text and other biomedical data. Later, in Sect. 12.3, we focus on the knowledge-guided biomedical NLP methods which are divided into four groups according to the discussion in Chap. 9. After learning about the basic situation of biomedical knowledge representation learning, we will then explore several typical application scenarios in Sect. 12.4 and discuss some advanced topics that are worth researching in Sect. 12.5.

12.2   Biomedical Knowledge Representation and Acquisition

Going back decades, AI systems for biomedical decision support have already shown the importance of knowledge representation and acquisition. The former is the basis of practical usage and the latter ensures the sustainability of expert systems with growing knowledge. Biomedical knowledge is represented in a structured manner in that period. For instance, DENDRAL [10] is an expert system providing advice for chemical synthesis, and production rules in DENDRAL first recognize the situation and then generate corresponding actions. This two-stage process is similar to human reasoning and has a strong explanation capability. Other systems also represent the knowledge in the form of frame, relations, and so on [37]. Correspondingly, the acquisition of biomedical knowledge mainly relies on manual collection, and the assistant information extraction systems are conducted mainly based on the results of manual feature engineering.

With the development of machine learning, knowledge representation and acquisition have been raised to new heights. Our following discussion is divided into two different sources of knowledge: natural language text materials and other materials, which can correspond to the two perspectives mentioned in the last section.

12.2.1   Biomedical Knowledge from Natural Language

Text Biomedical textual knowledge is scattered in various natural language documents, patents, clinical records, etc. Various knowledge representation learning methods in the general domain are applied to these natural language text materials. What is special about biomedical texts is that we have to achieve a deep comprehension of the key biomedical terms. Therefore, we are going to first discuss various term-oriented biomedical tasks that researchers explore. Further, we turn to pre-trained models (PTMs) to achieve the overall understanding of language descriptions (including sentences, paragraphs, and even document materials around these terms).

Term-Oriented Biomedical Knowledge   Biomedical terms, including the professional concepts and entities in the biomedical domain, are important carriers of domain knowledge. Common biomedical terms that we may process include chemicals/genetics/protein entities, disease/drug/examination/treatment items, cell/tissue/organ parts, and others. To process the biomedical natural language materials better, deeper comprehension of these biomedical terms is necessary. Dictionary-based and rule-based methods are very manpower demanding, and it is difficult to maintain immediacy and hold complicated scenarios [46]. To grasp and analyze the data features automatically, machine learning and statistical learning are adopted to get more generalized term representations and achieve better acquisition performances [87], while still far from satisfaction. Further, deep learning has been rapidly developed and proven its effectiveness in the biomedical domain; therefore we are going to mainly introduce biomedical term process methods in the deep learning approach, which is currently the mainstream solution for biomedical knowledge representation and acquisition.

Biomedical Term Representations   The mainstream term representation methods are in a self-supervised manner, which is to predict the missing parts for the given context, hoping to get general feature representations for various downstream tasks.

Many works in the general domain such as word embeddings are directly used in biomedical scenarios without adaptation. The skip-gram version of word2vec [16], for example, is proven to get satisfying performance on the biomedical term semantic relatedness task [65]. Besides, researchers also try distributed representations especially trained for biomedical terms [22, 70, 102], introducing extra information such as the UMLS [9] ontology and the medical subject headings (MeSH) [54]. Based on the shallow term embeddings, we can go a step further to adopt deep neural networks such as CNNs and BiLSTMs to get the deep distributed representations for biomedical terms [48, 74].

In recent years, PTM is the most popular choice to generate distributed representations as the basis of various downstream tasks. Inspired by the PTMs including BERT that have achieved more and more surprisingly great performances in the general domain, researchers quickly transfer the self-supervised approach to the biomedical domain. SciBERT [7] is one of the earliest PTMs that are specially adapted for the scientific corpus, followed by BioBERT [47] which further refines the corpus field of the model target into biomedicine. The specific operation is very simple: replacing the pre-training corpus of BERT in the general domain (e.g., Wikipedia, books, and news) with biomedical texts (e.g., literature and medical records). Some other biomedical PTMs also follow this strategy, such as SciFive [73] which is adapted from T5.

To sum up, the knowledge representation methods in the general domain are adapted to biomedical terms quite well. Special hints including information from the subject ontologies can provide extra help to generate better representations.

Biomedical Term Knowledge Acquisition The identification of terms involves many subtasks: recognition (NER), classification (typing), mapping (linking), and so on [46]. We introduce several mainstream solutions in chronological order.

Some simple supervised learning methods are applied for term recognition, such as the hidden Markov model (HMM) for term recognition and classification [19, 24] and the support vector machine for biomedical NER [67]. These systems mainly rely on the pre-defined domain-specific word features and perform not well enough for some lack-of-data classes. Neural networks including LSTM are also widely adopted for biomedical term knowledge acquisition [31, 33]. Unsupervised approaches are also explored and proven to be effective [101].

With the help of biomedical PTMs, we can better acquire and organize knowledge from the mass of unstructured text. At the term level, PTMs encode the long text and get a dense representation, which can then be fed into classifiers or softmax layers to finish NER, entity typing, and linking precisely. The tuning methods are sometimes specially designed, such as conducting self-alignment training on the pair tuples of the entity names and categorical labels in several ontologies [40, 55]. Though the methodology of PTMs has been successfully adapted to the biomedical domain, there still exist domain-special problems waiting for solving. For example, compared with the general domain, biomedical texts have more nested entities because of the terminology naming convention. For example, the DNA entity *IL-2 promoter* also contains a shorter protein entity *IL-2*, and *G1123S/D* refers to two separate entities *G1123S* and *G1123D* as shown in Fig. 12.2. We can solve the nested entity problem by separating different types of entities in different output layers [26] or by detecting the boundaries and assembling all the possible combinations [15, 89].

Language-Described Biomedical Knowledge As we can see, the research we have discussed concerns more about the special terms with professional biomedical knowledge. Nevertheless, other words and phrases in the language materials also contain rich information such as commonsense knowledge and can express much more flexible biomedical facts and attributes than isolated terms. It is necessary to

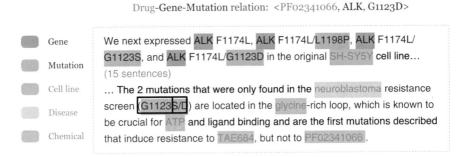

Fig. 12.2 Instance of biomedical textual knowledge acquisition. (Text is taken from [35])

represent the whole language descriptions instead of only biomedical terms, and this can be achieved by domain PTMs quite well. Based on the representations of language materials, the biomedical knowledge scattered in the unstructured text can be acquired and organized into a structured form.

We now introduce the overall development of the language-described biomedical knowledge extraction. The popular datasets are mostly small-scale and focus on specific types of relations, like the BC5CDR chemical-disease relation detection dataset and the ChemProt chemical-protein interaction dataset [49]. These simple tasks can sometimes be finished quite well with the help of distributed representations, even if they are generated by simple neural networks without pre-training [85]. However, in practical scenarios, biomedical knowledge exists in more sophisticated information extraction (e.g., N-ary relations, overlapping relations and events). Since scientific facts usually involve stricter conditions, few of them can be expressed clearly with only a triplet. For example, the effect of drugs on a disease is related to the characteristics of the sample, the course of the disease, etc. As shown in Fig. 12.2, the text mentioning N-ary relations is usually quite long and may cross several paragraphs. PTMs show their effectiveness due to their capability of capturing long-distant dependence for sophisticated relations in long documents, encoding the mentions, and then getting the distributed entity representations for the final prediction [41].

Summary Overall, researchers solve the simple biomedical text processing scenarios quite well by transferring many knowledge representation and acquisition methods in the general domain of NLP, while the challenges still exist from practical perspectives. Knowledge storage structures with stronger expressive ability, plenty of annotated data, and targeted-designed architectures are urgently expected for sophisticated biomedical knowledge representation and acquisition.

12.2.2 Biomedical Knowledge from Biomedical Language Materials

Biomedical materials contain not only textual materials scattered in natural language but also some materials unique to the biomedical field. These materials have their own special structures in which rich knowledge exists, and we collectively refer to them here as biomedical language (e.g., genetic language) materials. Compared with natural language materials, biomedical language materials like genetic sequences are not easy to comprehend and require extensive experience and background information for analysis. Fortunately, modern neural networks can process not only natural language documents but also most of the sequential data including some representations of chemical and genetic substances. Besides, deep learning methods can also be applied to represent and acquire biomedical knowledge in other forms such as graphs. In this section, we consider genetic language, protein language, and

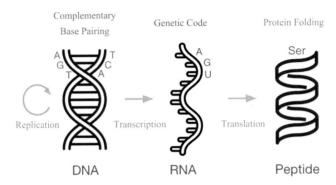

Fig. 12.3 Genetic central dogma. Icons are bought or freely downloaded from IconFinder

chemical language for discussion, and substances expressed by these languages are linked by the genetic central dogma [21]. As shown in Fig. 12.3, genetic sequences are expressed to get proteins, which react with various chemicals to execute their functions.

Genetic Language There are altogether only five different types of nucleic acid, among which A, G, C, and T are in the DNA sequences and A, G, C, and U are in the RNA sequences. Since the coding region of the unwinding DNA is transcribed to generate an mRNA sequence with a very fixed correspondence, i.e., A-T(U) and G-C, the processing methods for DNA and RNA sequences are often similar. We mainly discuss DNA sequences in this section. We first introduce basic tasks for DNA sequence processing and then discuss the similarities and differences between genetic language and natural language. In terms of genetic language, we show related work about tokenization and encoding methods.

Basic Tasks for Genetic Sequence Processing First, let's take a look at various downstream property prediction tasks for genetic sequences. Some of them emphasize the high-level semantic understanding of DNA sequences [5, 81] (long-distance dependency capturing and gene expression prediction), such as transcriptional activity, histone modifications, TF binding, and DNA accessibility in various cell types and tissues for held-out chromatin. Other tasks evaluate the low-level semantic understanding of DNA sequences [68] (precise recognition of basic regulatory elements), such as the prediction of promoters, transcription factor binding sites (TFBSs), and splice sites.

Features of Genetic Language Although both DNA/RNA language and natural language are textual sequences, there still exist differences between them. Firstly, the genetic sequences are quite long and dull, thus not as reader-friendly for human beings as natural language sequences. However, the NLP models are actually good at reading and learning from the mass of data and finding patterns. Secondly, compared with natural language, genetic language has a much smaller vocabulary (only five types of nucleic acid as mentioned above); therefore low-level semantic modeling is important for overall sequence comprehension, about which researchers have launched many explorations as the following introduction.

Genetic Language Tokenization   Early works express the sequences via one-hot coding [82]. The nucleic-acid-level features can be captured by converting the sequences into 2D binary matrices. Based on the tokenized results, convolutional layers and sequence learning modules such as LSTMs are applied to get the final distributed representations [34]. More researchers use the k-mer (substrings of length k monomers contained within a biological sequence) tokenizer to take co-occurrence information into account. In other words, the encoding of each position in the gene sequences will be considered together with the preceding and following positions (a sliding window with a total length of k) [63]. Other methods such as byte pair encoding [80] have also been proven to be useful.

Genetic Sequence Representation   The shallow models can hardly process the long sequences which may have thousands of base pairs, while Transformers [88] can capture the long-distance dependency quite well thanks to its attention module. Further, the self-supervised pre-training for Transformers is proven to be also effective on the genetic language [39]. Besides, improved versions of Transformers are implemented and achieve good performances on DNA tasks. For instance, Enformer [5] is designed to enlarge the receptive field. To be more specific, the ideas from computer vision can be borrowed to use deep convolution layers to expand the region that each neuron can process. Enformer replaces the base Transformer layer with seven convolutional layers to capture the low-level semantic information. The captured features are fed into 11 Transformer layers and processed by the separately trained organism-specific heads. Experimental results show that Enformer improves gene expression prediction, variant effect prediction, and mutation effect prediction.

Protein Language   Protein sequence processing has a lot in common with genetic sequences. There exist altogether 20 types of amino acids in the human body, so protein language is a special language with low readability and a small vocabulary size as well. We also discuss some basic tasks and methods first and then introduce a representative work in protein sequence processing.

Basic Tasks for Protein Sequence Processing   The sequence specificity of DNA- and RNA-binding proteins [2] is a basic task that we are concerned about, because RNA sequences are translated to obtain an amino acid sequence and the two types of sequences are highly related. Moreover, the spatial structure analysis is another unique and important task for protein sequences, since the protein quaternary structure determines the properties and functions.

We have introduced the similarity of genetic and protein language, which allows most genetic sequence processing methods to be adapted to proteins. However, there are also some special methods for protein sequence processing. A significant fact is that structural and functional similarities exist between homologous protein sequences, which can help supervise protein representation learning. By contact prediction and pairwise comparison, we can conduct multi-task training of protein sequence distributed representations [8] and conversely assist spatial structure prediction.

Landmark Work for Protein Spatial Structure Analysis AlphaFold [43] proposed by DeepMind has achieved a breakthrough in highly accurate protein structure prediction and become the champion of the Critical Assessment of protein structure prediction challenge. The system incorporates multiple sequence alignment (MSA) [11] templates and pairwise information for the protein sequence representation. It is built based on a variant of Transformers, which is named as EvoFormer. The column and row attention of MSA sequences and pair representations are fed into EvoFormer blocks. Peptide bond angles and distances are then predicted by the subsequent modules. The interfaces and tools for AlphaFold interaction have been developed quite well, and it is easy for users without an AI background to master. This reflects the essence of interdisciplinary research: division of labor and cooperation to improve efficiency.

Besides, it is worth mentioning that the initial results generated by AlphaFold can be further improved with the help of molecular dynamics knowledge. Incorporating domain knowledge also shows its effectiveness in some other scenarios, such as using chemical reaction templates for retrosynthesis learning [28]. Overall, the combination of professional knowledge and data-driven deep learning is getting better results, which is an important development trend for biomedical NLP.

Chemical Language Apart from biological sequences, chemical substances (especially small molecules) can also be encoded and expressed into molecule representations, which can help finish property prediction and filtering. These representations play similar roles as the molecule fingerprint (a commonly used abstract molecular representation that converts the molecular structure into a series of binary sequences by checking whether some specific substructures exist).

Early Fashions for Chemical Substance Representation In the early days of applying machine learning to assist the prediction of molecular properties, molecule descriptors such as nuclear charges and atomic positions are provided for nonlinear statistical regression [77]. Essentially, people still need to manually select features for the molecule descriptors. To alleviate the labor of manual feature engineering, data-driven deep learning systems have gradually become the main approach for the analysis of molecules.

For current deep learning systems of chemical substance representations, we classify according to the different expressions of chemical substances, for which there are several common methods as shown in Fig. 12.4.

Graph Representations One of the clearest ways is the 2D and 3D topology diagrams [23, 45] describing the inner chemical structure of molecules. This naturally corresponds to the essential elements of **graphs**. In molecular graphs, the nodes represent the atoms, and the edges represent the connections (chemical bond, hydrogen bond, van der Waals force, etc.). Graph representation learning bridges chemical expression and machine learning [95], and we have introduced graph

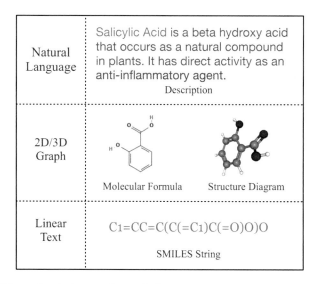

Fig. 12.4 Different chemical expression methods

representation learning in detail in Chap. 6. Graph Transformer [98], for example, is currently one of the most popular approaches in molecular graph representation learning [76]. With the graph representation learning methods, we can achieve two main tasks for molecular processing: molecular graph understanding to capture the topology information of molecular structures and predict properties [45] and molecular graph generation to provide assistance for drug discovery and refinement [59]. Overall, graph representation learning has already been proven to be an effective approach to chemical analysis.

Linear Text and Other Representations There are also some other solutions for expressing chemical substances. For example, **linear text** such as the structural formula, structural abbreviation, and simplified molecular input line entry specification (SMILES) [79] can be adopted for chemical expression. The straightforward advantage of linear text expressions is that they can naturally be fed into any NLP model. Although different from natural language text, the SMILES text expressing molecules and chemical reactions can also be processed by the Transformer-based models, if only with the assistance of specially designed tokenizers [50] and pre-training tasks [90]. Nevertheless, the linear text losses some structural information, and the 2D topologic and 3D spatial hints are still proven to be important. The atom coordinates computed according to SMILES help improve the performance of SMILES processing models [93], and this inspires us that the domain knowledge (e.g., molecule 3D procure) will enhance the NLP models when processing biomedical materials.

Summary Apart from substances related to central dogma, there exist some other types of special materials in the biomedical domain, such as image data and numeric data. The former including molecule images and medical magnetic resonance images [58] can be automatically processed by AI systems to some extent. The latter such as continuous monitoring health data is also processed with NLP methods adapted to the biomedical domain [94]. In summary, the materials waiting for processing are in versatile forms, and deep learning methods have already achieved satisfying performances on many of them. Further, to achieve deep comprehension and precise capture of biomedical knowledge, we believe that adaptive and universal processing of various materials will gradually become the trend in biomedical NLP research.

12.3 Knowledge-Guided Biomedical NLP

We have already discussed the development and basic characteristics of biomedical knowledge representation and acquisition. Conversely, domain knowledge can guide and enhance biomedical NLP systems to better finish those knowledge-intensive tasks. Though the commonsense and facts in the general domain can be learned in a self-supervised manner, the biomedical knowledge we use to guide the systems is more professional and has to be additionally introduced. The guidance from domain knowledge bases can even assist human experts and help improve their performances, let alone the biomedical NLP systems. In this section, we introduce the basic ideas and representative work for knowledge-guided biomedical NLP, according to the four types of methods mentioned in Chap. 9: input augmentation, architecture reformulation, objective regularization, and parameter transfer.

12.3.1 Input Augmentation

To guide neural networks with biomedical knowledge, one simple solution is to directly provide the knowledge as the input augmentation of the systems. There exist different sources of knowledge that can augment the input, as we are going to introduce later. One mainstream source is the biomedical knowledge graph (KG) which contains human knowledge and facts organized in a structured form. Besides, knowledge may also come from linguistic rules, experimental results, and other unstructured records. The problem for input augmentation is to select helpful information, encode, and fuse it with the processing input.

Encoding Knowledge Graph Information from professional KGs is of high quality and suitable for guiding models in downstream tasks. Usually, we rely on basic entity recognition and linking tools to select the subgraphs or triplets from KGs that are related to the current context and further finish more sophisticated tasks such as reading comprehension and information extraction. We now give three instances: (1) *Improving word embeddings with the help of KGs.* Graph representation learning approaches like GCN-based methods can get better-initialized embeddings for the link prediction task based on biomedical KGs [3]. (2) *Augmenting the inputs with knowledge.* Models such as the hybrid Transformers can encode token sequences and triplet sequences at the same time and incorporate the knowledge into the raw text [6]. (3) *Mounting the knowledge by extra modules.* Extra modules are designed to encode the knowledge, such as a graph-based network encoding KG subgraphs to assist biomedical event extraction [36]. As shown in Fig. 12.5, the related terms in the UMLS ontology are parsed and form a subgraph, which is encoded and concatenated into the hidden layer of the SciBERT text encoder to assist event trigger and type classification. There also exist other examples, such as the separate KG encoder providing entity embeddings for the lexical layers in the original Transformer [27] and the KG representations trained by TransE being attached to the attention layers [14].

Encoding Other Information Apart from KG information, there are other types of knowledge that are proven to be helpful. Syntactic information, for example, is a significant part of linguistic knowledge. Though not a part of biomedical expert knowledge, syntactic information can also be provided as augmented input to better analyze sentences, recognize entities, and so on [86]. For non-textual material processing tasks, such as the discovery of the relationship between basal gene expression and drug response, researchers believe that experimentally verified prior knowledge including the protein and genetic interactions is important. The information can be concatenated with the original input substances to get representations and show the effectiveness of input augmentation [25]. Overall, introducing extra knowledge usually shows at least no harm to the performance, while we need to decide whether the knowledge is related and helpful to the specific tasks, through human experience or automatic filtering.

12.3.2   *Architecture Reformulation*

Human prior knowledge is sometimes reflected in the design of model architectures, as we have mentioned in the representation learning of biomedical data. This is especially significant when we try to process domain-specific materials, such as the substances we have introduced in the last section. After all, the backbone models are designed for general materials (e.g., natural language documents, natural images), which may have remarkable differences from biomedical substances. Here we analyze two examples in detail: Enformer [5] and MSA Transformer [75].

Enformer is an adapted version of Transformers framework for DNA sequences, and we provide the model architecture in Fig. 12.6. The general idea of this model has already been introduced when we discuss genetic sequences. Here we take a look at two designs in Enformer that help the model better capture the low-level semantic information in the super-long genetic sequences, and this information is of key importance for the high-level sequence analysis. First, Enformer emphasizes

Fig. 12.5 Encoding UMLS information to assist event extraction. (The figure is re-drawn according to Figs. 1 and 2 from GEANet paper [36])

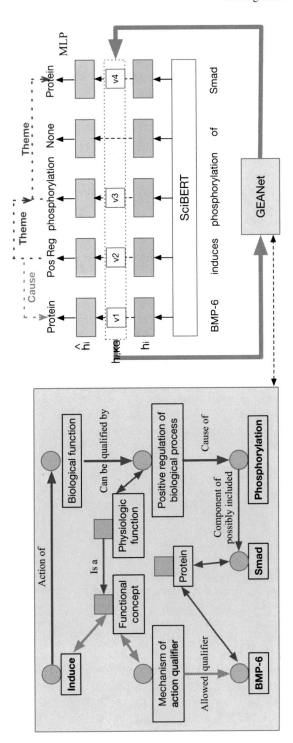

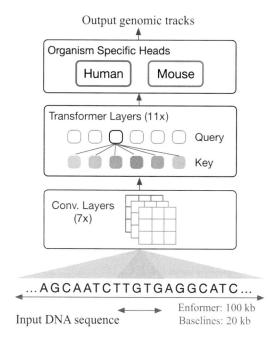

Fig. 12.6 Model architecture for Enformer. (The figure is re-drawn according to Fig. 1a from DeepMind's Enformer paper [5])

the relative position information, selects the relative positional encoding basis function carefully, and uses a concatenation of exponential, gamma, and central mask encodings. Second, convolutional layers are applied to capture the low-level features, enlarging the receptive field and greatly expanding the number of relevant enhancers seen by the model.

When discussing AlphaFold, we have mentioned the significance of MSA information. Inspired by this idea, MSA Transformer is proposed to process multiple protein sequences. The model architecture is shown in Fig. 12.7. The normal Transformers conduct attention calculations separately for each sequence. However, different sequences in the same protein family share information including the co-evolution signal. MSA Transformer introduces the column attention corresponding to the row attention of each sequence and is trained with a variant of the masked language modeling across different protein families. Experimental results show that MSA Transformer gets obviously better performance compared with processing only single sequences, and this becomes the basic paradigm of processing protein sequences.

12.3.3 Objective Regularization

Formalizing new tasks from extra knowledge can change the optimization target of the model and guide the models to finish the target task better. In the biomedical

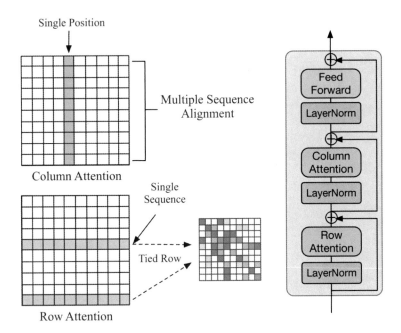

Fig. 12.7 Model architecture for MSA Transformer. (The figure is re-drawn according to Fig. 1 from MSA Transformer paper [75])

domain, there are plenty of ready-made tasks that can be adopted for objective regularization once chosen carefully, and we do not need to specially formalize new tasks. Usually, we conduct multi-task training in the downstream **adaptation** period. Some researchers also explore objective regularization in the **pre-training** period, and PTMs learn the knowledge contained in the multiple pre-training tasks. We will give examples of these two modes and conduct a comparative analysis.

Multi-task Adaptation The introduced multiple tasks can be the same or slightly different from the target task. For the former one, we usually collect several datasets (may be differently distributed or in various language styles) for the same task. For instance, the biomedical NER model has shared parameters while separated output layers for various datasets to deal with the style gap [12, 20]. When we do not have ready-made datasets, KGs can help generate more silver data for training, such as utilizing the KG shortest dependency path for relation extraction augmentation [84]. Further, different tasks can also benefit each other, such as several language understanding tasks (biomedical NER, sentence similarity, and relation extraction) in the BLUE benchmark [71]. Similarly, when dealing with non-textual biomedical materials, we can conduct multi-task adaptation to require the models to understand different properties of the same substances. For example, the molecular encoder reads the SMILES strings and learns comprehensive capability on five different molecule property classification tasks [53], and the knowledge in these tasks assists in improving the performance of each other.

Multi-task Pre-training Pre-training itself is a knowledge-guided method, which we will introduce later in the next subsection. When it comes to multi-task pre-training, with knowledge of KGs or expert ontologies, we can create extra data and conduct knowledgeable pre-training tasks. The domain-specific PTMs we have mentioned such as SciBERT and BioBERT simply keep the masked language modeling training strategy. To introduce more knowledge, biomedical PTMs with specially designed pre-training tasks are proposed. One instance is masked entity prediction, e.g., MC-BERT [100] is trained with the Chinese medical entities and phrases masked instead of randomly picked characters with the assistance of biomedical KGs. The other instance is entity detection and linking, e.g., KeBioLM [97] annotates the large-scale corpus with the SciSpacy [66] tool and introduces the entity-oriented tasks during pre-training, which essentially integrates the entity understanding capabilities of the annotation tool. The PTMs enhanced by extra pre-training tasks usually show much better performance on the corresponding downstream tasks. In short, the multi-task pre-training period implicitly injects knowledge from the KGs/ontologies or the ready-made annotation tools, and this can help improve the capability of the PTMs in related aspects.

Comparing the two approaches above, we can find that multi-task adaptation is a more direct way to change the optimization target for the target task, and therefore the introduced datasets have to be of high quality and highly related to our target data. In contrast, the requirement for multi-task pre-training is less stringent since the pre-training period is conducted on a sufficiently large corpus that is insensitive to small disturbances, while the assistance of pre-training tasks is also not so explicit and remarkable compared with multi-task adaptation.

12.3.4 Parameter Transfer

One of the most common paradigms of transfer learning is the pre-training-fine-tuning paradigm. In this way, the data-driven deep learning systems can be applied to specific domains which may lack annotated data. The knowledge learned from the source domain corpus/tasks can help improve the performance of the target domain tasks. Taking the PTMs as an example, they transfer the commonsense, linguistic knowledge, and other useful information from the large-scale pre-training corpus to the downstream tasks. We now discuss two types of parameter transfer: between different data domains and between tasks.

Cross-Domain Transfer The models pre-trained in the general domains are frequently transferred to the biomedical domain, and two of the most common scenarios are the processing of natural language documents and images. For example, the model pre-trained on ImageNet can better understand medical images and finish melanoma screening [61]. Compared with randomly initialized models, PTMs such as BERT can also achieve satisfying performances when fine-tuned on biomedical text processing datasets.

Nevertheless, with more biomedical corpora obtained, we do not have to rely on general domain pre-training now. Experimental results have shown that domain-specific pre-training has a more obvious improvement than general domain pre-training [64]. In fact, each domain may have its own characteristics, such as some empirical results in the biomedical domain showing that pre-training from scratch gains more over continual pre-training of general-domain PTMs [32], which is contrary to popular belief and waiting for further exploration.

Cross-Task Transfer Models can be tuned on other tasks or styles of data before being transferred to the target task data, and knowledge learned from other tasks is contained in the initialized parameters. Specific to the biomedical domain, the high cost of biomedical data annotation limits the scale of golden samples labeled by human experts. Some methods can generate large-scale silver datasets automatically, such as distant supervision, which assumes that a piece of text/image expresses the already-known relation, if only the head and tail entities appear in it. Sometimes it is too absolute to directly change the optimization target. Instead, we consider using the cross-task transfer method to utilize the knowledge of the introduced task more softly. Pre-training on the silver-standard corpora and then tuning on the golden-standard datasets is proved to be effective [29]. Another example is cross-species biomedical data for transfer learning, in which the underlying biological laws of different species have similarities; therefore the biological data from other species can be used for pre-training before fine-tuning with the data from the target species and achieving higher accuracy for DNA sequence site prediction [52, 56].

Summary To sum up, knowledge-guided NLP methods are widely used in biomedical tasks, such as parameter transfer which can be easily conducted, being proven useful in various scenarios and becoming an essential paradigm. For textual material processing, the structured biomedical expert knowledge in KGs is suitable for providing augmented input and designing better objective functions. For non-textual material processing, architecture reformulation is usually necessary due to the differences in the data characteristics between various forms of raw materials. Some special materials naturally provide clues for objective regularization, such as multiple properties for the given molecule. The satisfying performances achieved by the above methods inspire us to emphasize the significance of knowledge-guided biomedical NLP.

12.4 Typical Applications

In this section, we explain the practical significance of biomedical knowledge representation learning through three specific application scenarios. Literature processing is a typical scenario for biomedical natural language material processing, and retrosynthetic prediction focuses more on biomedical language (chemical language) material processing. Both the two applications belong to AI for science problems, attempting to search from a large space and collect useful information to improve the efficiency of human researchers. We then talk about diagnosis assistance, which is of high practical value in our daily life.

12.4.1 Literature Processing

The size of the biomedical literature is expanding rapidly, and it is hardly possible for researchers to keep pace with every aspect of biomedical knowledge development. We provide an example of a literature processing pipeline in Fig. 12.8 to show how biomedical NLP helps improve our efficiency. We divide the pipeline into four stages: literature screening, information extraction, question answering, and result analysis.

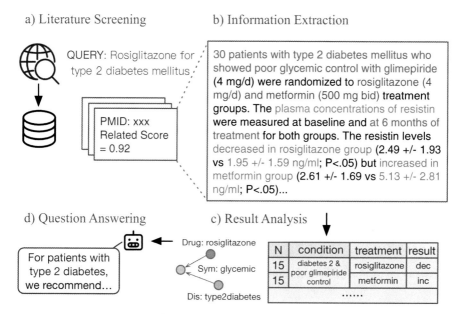

Fig. 12.8 A possible pipeline for biomedical literature processing. (Text in the example is taken from [44])

Literature Screening In our usual academic search process, we first screen the mass of literature returned by our search engine. We require the information retrieval model to return a relevance score ranking according to the query conditions, which may describe the type and age limit of the document, the entities or relation pairs we are concerned about, and other details. Echoing the importance of the biomedical terms we have mentioned, sometimes the document representations in the biomedical information retrieval models emphasize the key biomedical terms in the documents and queries for better matching [1].

Information Extraction We have already introduced some significant tasks for biomedical information extraction, such as term recognition, linking, and relation extraction. After we get the targeted literature by screening, we have to mine the text, extract the useful information, and convert it into a structured form just as we do in those extraction tasks. This stage usually relies on the knowledge-transferred PTMs reading and understanding the long documents.

Result Analysis and Question Answering We may also care about advanced meta-relations between the extracted structured knowledge items or facts. An example is to perform a meta-analysis for clinical randomized controlled trials [4], which is one of the most convincing pieces of evidence in evidence-based medicine. The process of inductively analyzing the results of different trials does not necessarily need to be fully automated, while we expect the AI system to help us do a quality assessment and conclusion highlighting and therefore largely improve our efficiency. Based on the analysis result, we may even get assistance from the conversation systems generating reasonable responses to medical questions and providing effective suggestions for further research.

12.4.2 Retrosynthetic Prediction

Organic synthesis is an essential application for modern organic chemistry and plays an important role in drug discovery, material science, and other fields. To design synthetic routines for the target molecules more efficiently, AI systems are applied for chemical reaction reading, such as the reaction classification task. Further, we expect the systems to achieve deep comprehension of the reactions and can therefore generate single-step reaction predictions. Eventually, the multi-step retrosynthesis task, reasoning the synthetic routes for the given target product, can also be finished automatically with the help of extra information from knowledge bases or ontologies.

Chemical Reaction Classification Machine learning methods can help researchers analyze large-scale reaction records and summarize useful reaction templates, which is a significant form of chemical knowledge [18]. These templates can further guide human researchers or AI systems to design synthetic routines.

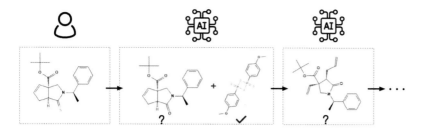

Fig. 12.9 A possible solution for automatic multi-step retrosynthesis. Icons are bought or freely downloaded from IconFinder

Single-Step Reaction Prediction In recent years, models such as the Transformers are pre-trained on the large-scale reaction corpus, and they are proven to be effective when predicting the single-step reactions without the guidance of templates [91].

Multi-step Reaction Prediction For predicting multi-step reactions, most of the current methods search for reasonable routes based on the already-known reaction knowledge in the knowledge bases [13]. With the development of biomedical deep learning models, we may also explore end-to-end generation for multi-step retrosynthesis in the future, as shown in Fig. 12.9. Specifically, the heuristic algorithm for searching routes, the query of knowledge bases, and other operations may all be finished with unified models guided by chemical knowledge.

12.4.3   Diagnosis Assistance

There exists a huge demand for diagnosis assistance. The scarce medical resources in some areas call for AI systems to provide patients with auxiliary knowledge for simple daily situations. This can reduce the pressure on medical resources and improve the work efficiency of hospital systems.

We first take a look at several basic tasks in diagnosis assistance. The most practical application is *automatic triage*. The system is fed with the symptom descriptions from the patients and predicts the suitable clinic. This is essentially a disease classification problem. A similar task is *medicine prescription*, which requires processing more complex diagnostic information (including the text of complaints, quantified findings, and even images) and providing advice with the aid of medical knowledge. Further, the *doctor-patient conversation* is a challenging task due to the gap between the colloquial style of patients and the standard terms and structured items in KGs. The system must first recognize the key information and finish linking and then provide correct and helpful knowledge with good interpretability and readability.

Since safety is significant for issues related to medical care, the assistance systems have to be supported by plenty of knowledge and provide explainable suggestions. Incorporating knowledge representations with text representations achieves significantly better performance on the diagnosis assistance task [51].

12.5 Advanced Topics

We have introduced the current development in biomedical knowledge representation learning. There are several consensuses for biomedical NLP through which we can further discuss and get inspiration about future trends. We have discussed the significance of high-quality training data for the current deep learning biomedical systems, and data scarcity can lead to research in two ways: by guiding the models with the knowledge to adapt with few data or by incorporating different data forms from multiple sources. Besides, the black-box property of deep learning systems brings challenges for domain research since biomedical applications are highly related to human life and emphasizes safety and ethical justification. Next, we will elaborate on the above two solution paths and one main concern.

Knowledgeable Warm Start There is a term in the field of recommendation algorithms called the cold-start [78] problem, which describes impaired performance when lacking user history. Extended to more deep learning applications such as biomedical NLP, we also face the cold-start challenge under scarce-data scenarios and often alleviate the problem with the help of transfer learning or other methods. For biomedical NLP tasks, data annotation is difficult, and we always have few supervision signals for model training. Therefore, it becomes more important to achieve warm start training for biomedical NLP systems.

As we have mentioned above, knowledge can guide deep learning systems in several different ways even when the data is comparably plenty, such as biomedical PTMs transferring linguistic and commonsense knowledge to help achieve a warm start. When it comes to the low-resource scenarios, there have been a few explorations. Knowledge-aware modules such as the self-attention layer introducing external KGs are designed for biomedical few-shot learning [96]. Special tuning strategies such as entity-aware masking are also applied and proved to be effective under low-resource problems [72]. Still, the knowledgeable warm start problem is rarely discussed in a targeted manner or even just clearly raised, although it is prevalent in biomedical NLP tasks. We believe that it deserves more attention and research.

Cross-Modal Knowledge Processing Though the annotated datasets are small-scale, we have various forms of biomedical data that are linked to each other by biomedical knowledge. Apart from the regular cross-modal tasks (about which we can learn more details in Chap. 7) including medical image captioning, other types

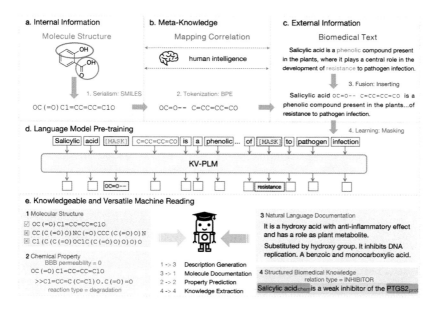

Fig. 12.10 KV-PLM model bridging molecular structure expressions and natural language descriptions. This figure is taken from the original paper [99] with CC BY 4.0 license (https://www.nature.com/articles/s41467-022-28494-3)

of materials can also be versatilely processed. For example, natural language and chemical language can describe the same chemical entities, and they may provide complementary information from different perspectives. KV-PLM [99] has proved that the connections between natural language descriptions and molecular structures can be modeled in an unsupervised manner through pre-training (Fig. 12.10). It can even surpass human professionals in the molecular property comprehension task and reveal its potential in drug discovery. Follow-up works further incorporate other materials such as molecular graphs with the text [83].

Different expressions for biomedical terms have diverse emphases. Bridging them together and capturing the mapping relations between various data forms through a large number of observations, just as humans do, is a form of meta-knowledge learning, enabling a deeper understanding of terms while alleviating data scarcity issues. As long as we can design tokenizers to utilize different structures uniformly, the advantages of data-driven deep learning systems can be carried forward.

Interpretability, Privacy, and Ease of Use   There exist some other concerns about biomedical NLP. The first one is the interpretability problem, which we have discussed in Chap. 8. Most deep learning systems are black boxes that have poor interpretability, and this leads to distrust of automated decision-making, especially under medical scenarios closely related to human lives. Directly predicting the prescription without providing symptom analysis and disease diagnosis makes it hard for users to assess the credibility of the recommendations. This is not only related to safety but also some ethical problems including accident liability determination. There are already some researchers that focus on the interpretability of biomedical NLP due to its importance [60].

The second one is the privacy problem. The ethical controversy of privacy always exists when we talk about AI development. For example, the genetic sequence training data of deep learning models may be leaked by privacy attacks, and the genetic traits and disease information of the system users may be illegally sold. Some methods such as private aggregation of teacher ensembles can alleviate the privacy leakage problem [69], while it still needs more effort to be solved.

Thirdly, as the assistance tool for domain research, biomedical NLP systems are supposed to be designed as easily as possible to use. Some toolkits and online demos are developed [103], while most of them still propose quite high requirements for the users' devices and programming foundation. There is a huge market for user-friendly platforms, and we hope the AI community to implement useful aids as soon as possible.

12.6 Summary and Further Readings

In this chapter, we discuss the representation learning of biomedical NLP. As an emerging interdisciplinary field, biomedical NLP has undergone rapid development in recent years, especially after deep learning methods such as PTMs appeared. We first introduce the knowledge representation and acquisition in biomedical materials, including natural language text materials and other materials, of which the latter adapts the advanced NLP algorithms and models to the biomedicine scenarios. Further, we explain the knowledge-guided methods in the biomedical domain in the four aspects: input augmentation, architecture reformulation, objective regularization, and parameter transfer. Future directions in this field have also been discussed.

For further understanding of biomedical knowledge representation learning, we recommend reading some surveys about the early works [62] and the comprehensive analysis for PTMs [32] which is the recent-year representative results.

Acknowledgments The contributions of all authors are as follows: Zhiyuan Liu, Yankai Lin, and Maosong Sun designed the overall architecture of this chapter; Zheni Zeng drafted this chapter. Zhiyuan Liu and Yankai Lin proofread and revised this chapter.

We thank Ganqu Cui, Yankai Lin, Yuan Yao, Xu Han, Chenyang Song, Zeyu Pan, Kunlun Zhu, and Ruiyi Fang for proofreading the chapter and proposing valuable revisions.

This chapter about biomedical knowledge representation learning is the newly complemented content in the second edition of the book *Representation Learning for Natural Language Processing*. The first edition of the book was published in 2020 [57].

References

1. Maristella Agosti, Stefano Marchesin, and Gianmaria Silvello. Learning unsupervised knowledge-enhanced representations to reduce the semantic gap in information retrieval. *ACM Transactions on Information Systems (TOIS)*, 38(4):1–48, 2020.
2. Babak Alipanahi, Andrew Delong, Matthew T Weirauch, and Brendan J Frey. Predicting the sequence specificities of DNA- and RNA-binding proteins by deep learning. *Nature Biotechnology*, 33(8):831–838, 2015.
3. Mona Alshahrani, Maha A Thafar, and Magbubah Essack. Application and evaluation of knowledge graph embeddings in biomedical data. *PeerJ Computer Science*, 7:e341, 2021.
4. Ashwin Karthik Ambalavanan and Murthy V Devarakonda. Using the contextual language model bert for multi-criteria classification of scientific articles. *Journal of Biomedical Informatics*, 112:103578, 2020.
5. Žiga Avsec, Vikram Agarwal, Daniel Visentin, Joseph R Ledsam, Agnieszka Grabska-Barwinska, Kyle R Taylor, Yannis Assael, John Jumper, Pushmeet Kohli, and David R Kelley. Effective gene expression prediction from sequence by integrating long-range interactions. *Nature Methods*, 18(10):1196–1203, 2021.
6. Helena Balabin, Charles Tapley Hoyt, Colin Birkenbihl, Benjamin M Gyori, John Bachman, Alpha Tom Kodamullil, Paul G Plöger, Martin Hofmann-Apitius, and Daniel Domingo-Fernández. STonKGs: a sophisticated transformer trained on biomedical text and knowledge graphs. *Bioinformatics*, 38(6):1648–1656, 2022.
7. Iz Beltagy, Kyle Lo, and Arman Cohan. SciBERT: A pretrained language model for scientific text. In *Proceedings of EMNLP-IJCNLP*, 2019.
8. Tristan Bepler and Bonnie Berger. Learning protein sequence embeddings using information from structure. In *Proceedings of ICLR*, 2018.
9. Olivier Bodenreider. The unified medical language system (UMLS): integrating biomedical terminology. *Nucleic Acids Research*, 32(suppl_1):D267–D270, 2004.
10. Bruce Buchanan, Georgia Sutherland, and Edward A Feigenbaum. Heuristic DENDRAL: A program for generating explanatory hypotheses. *Organic Chemistry*, 1969.
11. Humberto Carrillo and David Lipman. The multiple sequence alignment problem in biology. *SIAM Journal on Applied Mathematics*, 48(5):1073–1082, 1988.
12. Zhaoying Chai, Han Jin, Shenghui Shi, Siyan Zhan, Lin Zhuo, and Yu Yang. Hierarchical shared transfer learning for biomedical named entity recognition. *BMC Bioinformatics*, 23(1):1–14, 2022.
13. Binghong Chen, Chengtao Li, Hanjun Dai, and Le Song. Retro*: learning retrosynthetic planning with neural guided A* search. In *Proceedings of ICML*, 2020.
14. Jing Chen, Baotian Hu, Weihua Peng, Qingcai Chen, and Buzhou Tang. Biomedical relation extraction via knowledge-enhanced reading comprehension. *BMC Bioinformatics*, 23(1):1–19, 2022.
15. Yanping Chen, Ying Hu, Yijing Li, Ruizhang Huang, Yongbin Qin, Yuefei Wu, Qinghua Zheng, and Ping Chen. A boundary assembling method for nested biomedical named entity recognition. *IEEE Access*, 8:214141–214152, 2020.
16. Kenneth Ward Church. Word2vec. *Natural Language Engineering*, 23(1):155–162, 2017.
17. Kevin Bretonnel Cohen and Dina Demner-Fushman. *Biomedical natural language processing*, volume 11. John Benjamins Publishing Company, 2014.
18. Connor W Coley, William H Green, and Klavs F Jensen. Machine learning in computer-aided synthesis planning. *Accounts of Chemical Research*, 51(5):1281–1289, 2018.
19. Nigel Collier, Chikashi Nobata, and Jun' ichi Tsujii. Extracting the names of genes and gene products with a hidden Markov model. In *Proceedings of COLING*, 2000.
20. Gamal Crichton, Sampo Pyysalo, Billy Chiu, and Anna Korhonen. A neural network multi-task learning approach to biomedical named entity recognition. *BMC Bioinformatics*, 18(1):1–14, 2017.
21. Francis Crick. Central dogma of molecular biology. *Nature*, 227(5258):561–563, 1970.

22. Lance De Vine, Guido Zuccon, Bevan Koopman, Laurianne Sitbon, and Peter Bruza. Medical semantic similarity with a neural language model. In *Proceedings of CIKM*, 2014.
23. David K Duvenaud, Dougal Maclaurin, Jorge Aguileraiparraguirre, Rafael Gomezbombarelli, Timothy D Hirzel, Alan Aspuruguzik, and Ryan P Adams. Convolutional networks on graphs for learning molecular fingerprints. In *Proceedings of NeurIPS*, 2015.
24. Sean R Eddy. What is a hidden Markov model? *Nature Biotechnology*, 22(10):1315–1316, 2004.
25. Amin Emad, Junmei Cairns, Krishna R Kalari, Liewei Wang, and Saurabh Sinha. Knowledge-guided gene prioritization reveals new insights into the mechanisms of chemoresistance. *Genome Biology*, 18(1):1–21, 2017.
26. Hao Fei, Yafeng Ren, and Donghong Ji. Recognizing nested named entity in biomedical texts: A neural network model with multi-task learning. In *Proceedings of BIBM*, 2019.
27. Hao Fei, Yafeng Ren, Yue Zhang, Donghong Ji, and Xiaohui Liang. Enriching contextualized language model from knowledge graph for biomedical information extraction. *Briefings in Bioinformatics*, 22(3):bbaa110, 2021.
28. Michael E Fortunato, Connor W Coley, Brian C Barnes, and Klavs F Jensen. Data augmentation and pretraining for template-based retrosynthetic prediction in computer-aided synthesis planning. *Journal of chemical information and modeling*, 60(7):3398–3407, 2020.
29. John M Giorgi and Gary D Bader. Transfer learning for biomedical named entity recognition with neural networks. *Bioinformatics*, 34(23):4087–4094, 2018.
30. Anne Glover. The 21st century: the age of biology. In *OECD Forum on Global Biotechnology, Paris*, 2012.
31. Mourad Gridach. Character-level neural network for biomedical named entity recognition. *Journal of biomedical informatics*, 70:85–91, 2017.
32. Yu Gu, Robert Tinn, Hao Cheng, Michael Lucas, Naoto Usuyama, Xiaodong Liu, Tristan Naumann, Jianfeng Gao, and Hoifung Poon. Domain-specific language model pretraining for biomedical natural language processing. *ACM Transactions on Computing for Healthcare (HEALTH)*, 3(1):1–23, 2021.
33. Maryam Habibi, Leon Weber, Mariana Neves, David Luis Wiegandt, and Ulf Leser. Deep learning with word embeddings improves biomedical named entity recognition. *Bioinformatics*, 33(14):i37–i48, 2017.
34. Hamid Reza Hassanzadeh and May D Wang. DeeperBind: Enhancing prediction of sequence specificities of DNA binding proteins. In *Proceedings of BIBM*, 2016.
35. Johannes M Heuckmann, Michael Hölzel, Martin L Sos, Stefanie Heynck, Hyatt Balke-Want, Mirjam Koker, Martin Peifer, Jonathan Weiss, Christine M Lovly, Christian Grütter, et al. ALK mutations conferring differential resistance to structurally diverse ALK inhibitors. *Clinical Cancer Research*, 17(23):7394–7401, 2011.
36. Kung-Hsiang Huang, Mu Yang, and Nanyun Peng. Biomedical event extraction with hierarchical knowledge graphs. In *Findings of EMNLP*, 2020.
37. Donna L Hudson and Maurice E Cohen. *Neural networks and artificial intelligence for biomedical engineering*. Wiley Online Library, 2000.
38. Hitoshi Iuchi, Taro Matsutani, Keisuke Yamada, Natsuki Iwano, Shunsuke Sumi, Shion Hosoda, Shitao Zhao, Tsukasa Fukunaga, and Michiaki Hamada. Representation learning applications in biological sequence analysis. *Compuqihnology Journal*, 19:3198–3208, 2021.
39. Yanrong Ji, Zhihan Zhou, Han Liu, and Ramana V Davuluri. DNABERT: pre-trained bidirectional encoder representations from Transformers model for DNA-language in genome. *Bioinformatics*, 37(15):2112–2120, 2021.
40. Zongcheng Ji, Qiang Wei, and Hua Xu. BERT-based ranking for biomedical entity normalization. *AMIA Summits on Translational Science Proceedings*, 2020:269, 2020.
41. Robin Jia, Cliff Wong, and Hoifung Poon. Document-level n-ary relation extraction with multiscale representation learning. In *Proceedings of NAACL-HLT*, 2019.
42. George Brooks Johnson and Peter H Raven. *Biology: Principles & Explorations*. Recording for the Blind & Dyslexic, 2007.

43. John Jumper, Richard Evans, Alexander Pritzel, Tim Green, Michael Figurnov, Olaf Ronneberger, Kathryn Tunyasuvunakool, Russ Bates, Augustin Žídek, Anna Potapenko, et al. Highly accurate protein structure prediction with AlphaFold. *Nature*, 596(7873):583–589, 2021.
44. Hye Seung Jung, Byung-Soo Youn, Young Min Cho, Kang-Yeol Yu, Hong Je Park, Chan Soo Shin, Seong Yeon Kim, Hong Kyu Lee, and Kyong Soo Park. The effects of rosiglitazone and metformin on the plasma concentrations of resistin in patients with type 2 diabetes mellitus. *Metabolism*, 54(3):314–320, 2005.
45. Steven Kearnes, Kevin McCloskey, Marc Berndl, Vijay Pande, and Patrick Riley. Molecular graph convolutions: moving beyond fingerprints. *Journal of Computer-aided Molecular Design*, 30(8):595–608, 2016.
46. Michael Krauthammer and Goran Nenadic. Term identification in the biomedical literature. *Journal of Biomedical Informatics*, 37(6):512–526, 2004.
47. Jinhyuk Lee, Wonjin Yoon, Sungdong Kim, Donghyeon Kim, Sunkyu Kim, Chan Ho So, and Jaewoo Kang. BioBERT: a pre-trained biomedical language representation model for biomedical text mining. *Bioinformatics*, 36(4):1234–1240, 2020.
48. Haodi Li, Qingcai Chen, Buzhou Tang, Xiaolong Wang, Hua Xu, Baohua Wang, and Dong Huang. CNN-based ranking for biomedical entity normalization. *BMC Bioinformatics*, 18(11):79–86, 2017.
49. Jiao Li, Yueping Sun, Robin J. Johnson, Daniela Sciaky, Chih-Hsuan Wei, Robert Leaman, Allan Peter Davis, Carolyn J. Mattingly, Thomas C. Wiegers, and Zhiyong Lu. BioCreative V CDR task corpus: a resource for chemical disease relation extraction. *Database*, 2016, 05 2016.
50. Xinhao Li and Denis Fourches. SMILES pair encoding: a data-driven substructure tokenization algorithm for deep learning. *Journal of Chemical Information and Modeling*, 61(4):1560–1569, 2021.
51. Xuedong Li, Yue Wang, Dongwu Wang, Walter Yuan, Dezhong Peng, and Qiaozhu Mei. Improving rare disease classification using imperfect knowledge graph. *BMC Medical Informatics and Decision Making*, 19(5):1–10, 2019.
52. Zutan Li, Hangjin Jiang, Lingpeng Kong, Yuanyuan Chen, Kun Lang, Xiaodan Fan, Liangyun Zhang, and Cong Pian. Deep6mA: a deep learning framework for exploring similar patterns in DNA N6-methyladenine sites across different species. *PLoS Computational Biology*, 17(2):e1008767, 2021.
53. Sangrak Lim and Yong Oh Lee. Predicting chemical properties using self-attention multi-task learning based on SMILES representation. In *Proceedings of ICPR*, 2021.
54. Carolyn E Lipscomb. Medical subject headings (MeSH). *Bulletin of the Medical Library Association*, 88(3):265, 2000.
55. Fangyu Liu, Ehsan Shareghi, Zaiqiao Meng, Marco Basaldella, and Nigel Collier. Self-alignment pretraining for biomedical entity representations. In *Proceedings of NAACL-HLT*, 2021.
56. Quanzhong Liu, Jinxiang Chen, Yanze Wang, Shuqin Li, Cangzhi Jia, Jiangning Song, and Fuyi Li. DeepTorrent: a deep learning-based approach for predicting DNA N4-methylcytosine sites. *Briefings in Bioinformatics*, 22(3):bbaa124, 2021.
57. Zhiyuan Liu, Yankai Lin, and Maosong Sun. *Representation Learning for Natural Language Processing*. Springer, 2020.
58. Alexander Selvikvåg Lundervold and Arvid Lundervold. An overview of deep learning in medical imaging focusing on MRI. *Zeitschrift für Medizinische Physik*, 29(2):102–127, 2019.
59. Omar Mahmood, Elman Mansimov, Richard Bonneau, and Kyunghyun Cho. Masked graph modeling for molecule generation. *Nature Communications*, 12(1):1–12, 2021.
60. Sherin Mary Mathews. Explainable artificial intelligence applications in NLP, biomedical, and malware classification: a literature review. In *Intelligent Computing-proceedings of the Computing Conference*, pages 1269–1292. Springer, 2019.

61. Afonso Menegola, Michel Fornaciali, Ramon Pires, Flávia Vasques Bittencourt, Sandra Avila, and Eduardo Valle. Knowledge transfer for melanoma screening with deep learning. In *Proceedings of ISBI*, 2017.

62. Seonwoo Min, Byunghan Lee, and Sungroh Yoon. Deep learning in bioinformatics. *Briefings in Bioinformatics*, 18(5):851–869, 2017.

63. Xu Min, Wanwen Zeng, Ning Chen, Ting Chen, and Rui Jiang. Chromatin accessibility prediction via convolutional long short-term memory networks with k-mer embedding. *Bioinformatics*, 33(14):i92–i101, 2017.

64. Milad Moradi, Kathrin Blagec, Florian Haberl, and Matthias Samwald. GPT-3 models are poor few-shot learners in the biomedical domain. *arXiv preprint arXiv:2109.02555*, pages arXiv–2109, 2021.

65. TH Muneeb, Sunil Sahu, and Ashish Anand. Evaluating distributed word representations for capturing semantics of biomedical concepts. In *Proceedings of BioNLP*, 2015.

66. Mark Neumann, Daniel King, Iz Beltagy, and Waleed Ammar. ScispaCy: Fast and robust models for biomedical natural language processing. In *Proceedings of the 18th BioNLP Workshop and Shared Task*, pages 319–327, 2019.

67. William S Noble. What is a support vector machine? *Nature Biotechnology*, 24(12):1565–1567, 2006.

68. Mhaned Oubounyt, Zakaria Louadi, Hilal Tayara, and Kil To Chong. DeePromoter: robust promoter predictor using deep learning. *Frontiers in genetics*, 10:286, 2019.

69. Nicolas Papernot, Martın Abadi, Ulfar Erlingsson, Ian Goodfellow, and Kunal Talwar. Semi-supervised knowledge transfer for deep learning from private training data. *stat*, 1050:7, 2016.

70. Shengwen Peng, Ronghui You, Hongning Wang, Chengxiang Zhai, Hiroshi Mamitsuka, and Shanfeng Zhu. DeepMeSH: deep semantic representation for improving large-scale MeSH indexing. *Bioinformatics*, 32(12):i70–i79, 2016.

71. Yifan Peng, Qingyu Chen, and Zhiyong Lu. An empirical study of multi-task learning on BERT for biomedical text mining. In *Proceedings of the 19th SIGBioMed Workshop on Biomedical Language Processing*, pages 205–214, 2020.

72. Gabriele Pergola, Elena Kochkina, Lin Gui, Maria Liakata, and Yulan He. Boosting low-resource biomedical QA via entity-aware masking strategies. In *Proceedings of EACL*, 2021.

73. Long N Phan, James T Anibal, Hieu Tran, Shaurya Chanana, Erol Bahadroglu, Alec Peltekian, and Grégoire Altan-Bonnet. Scifive: a text-to-text transformer model for biomedical literature. *arXiv preprint arXiv:2106.03598*, 2021.

74. Minh C Phan, Aixin Sun, and Yi Tay. Robust representation learning of biomedical names. In *Proceedings of ACL*, 2019.

75. Roshan M Rao, Jason Liu, Robert Verkuil, Joshua Meier, John Canny, Pieter Abbeel, Tom Sercu, and Alexander Rives. MSA transformer. In *Proceedings of ICML*, pages 8844–8856. PMLR, 2021.

76. Yu Rong, Yatao Bian, Tingyang Xu, Weiyang Xie, Ying Wei, Wenbing Huang, and Junzhou Huang. Self-supervised graph transformer on large-scale molecular data. In *Proceedings of NeurIPS*, 2020.

77. Matthias Rupp, Alexandre Tkatchenko, Klaus-Robert Müller, and O Anatole Von Lilienfeld. Fast and accurate modeling of molecular atomization energies with machine learning. *Physical Review Letters*, 108(5):058301, 2012.

78. Andrew I Schein, Alexandrin Popescul, Lyle H Ungar, and David M Pennock. Methods and metrics for cold-start recommendations. In *Proceedings of ACM SIGIR*, 2002.

79. Philippe Schwaller, Teodoro Laino, Théophile Gaudin, Peter Bolgar, Christopher A Hunter, Costas Bekas, and Alpha A Lee. Molecular Transformer: a model for uncertainty-calibrated chemical reaction prediction. *ACS Central Science*, 5(9):1572–1583, 2019.

80. Yusuxke Shibata, Takuya Kida, Shuichi Fukamachi, Masayuki Takeda, Ayumi Shinohara, Takeshi Shinohara, and Setsuo Arikawa. Byte Pair encoding: A text compression scheme that accelerates pattern matching. 1999.

81. Toshiyuki Shiraki, Shinji Kondo, Shintaro Katayama, Kazunori Waki, Takeya Kasukawa, Hideya Kawaji, Rimantas Kodzius, Akira Watahiki, Mari Nakamura, Takahiro Arakawa, et al. Cap analysis gene expression for high-throughput analysis of transcriptional starting point and identification of promoter usage. In *Proceedings of the NAS*, 2003.
82. Maria Stepanova, Feng Lin, and Valerie C-L Lin. A hopfield neural classifier and its FPGA implementation for identification of symmetrically structured DNA motifs. *The Journal of VLSI Signal Processing Systems for Signal, Image, and Video Technology*, 48(3):239–254, 2007.
83. Bing Su, Dazhao Du, Zhao Yang, Yujie Zhou, Jiangmeng Li, Anyi Rao, Hao Sun, Zhiwu Lu, and Ji-Rong Wen. A molecular multimodal foundation model associating molecule graphs with natural language. *arXiv preprint arXiv:2209.05481*, 2022.
84. Peng Su, Yifan Peng, and K Vijay-Shanker. Improving BERT model using contrastive learning for biomedical relation extraction. In *Proceedings of the 20th Workshop on Biomedical Language Processing*, pages 1–10, 2021.
85. Cong Sun, Zhihao Yang, Leilei Su, Lei Wang, Yin Zhang, Hongfei Lin, and Jian Wang. Chemical–protein interaction extraction via gaussian probability distribution and external biomedical knowledge. *Bioinformatics*, 36(15):4323–4330, 2020.
86. Yuanhe Tian, Wang Shen, Yan Song, Fei Xia, Min He, and Kenli Li. Improving biomedical named entity recognition with syntactic information. *BMC Bioinformatics*, 21(1):1–17, 2020.
87. Yoshimasa Tsuruoka and Jun'ichi Tsujii. Improving the performance of dictionary-based approaches in protein name recognition. *Journal of biomedical informatics*, 37(6):461–470, 2004.
88. Ashish Vaswani, Noam Shazeer, Niki Parmar, Llion Jones, Jakob Uszkoreit, Aidan N Gomez, and Lukasz Kaiser. Attention is all you need. In *Proceedings of NeurIPS*, 2017.
89. Jue Wang, Lidan Shou, Ke Chen, and Gang Chen. Pyramid: A layered model for nested named entity recognition. In *Proceedings of ACL*, 2020.
90. Sheng Wang, Yuzhi Guo, Yuhong Wang, Hongmao Sun, and Junzhou Huang. SMILES-BERT: large scale unsupervised pre-training for molecular property prediction. In *Proceedings of ACM-BCB*, 2019.
91. Xiaorui Wang, Yuquan Li, Jiezhong Qiu, Guangyong Chen, Huanxiang Liu, Benben Liao, Chang-Yu Hsieh, and Xiaojun Yao. Retroprime: A diverse, plausible and transformer-based method for single-step retrosynthesis predictions. *Chemical Engineering Journal*, 420:129845, 2021.
92. Ying Wei, Jun Zhou, Yin Wang, Yinggang Liu, Qingsong Liu, Jiansheng Luo, Chao Wang, Fengbo Ren, and Li Huang. A review of algorithm & hardware design for AI-based biomedical applications. *IEEE Transactions on Biomedical Circuits and Systems*, 14(2):145–163, 2020.
93. Fang Wu, Qiang Zhang, Dragomir Radev, Jiyu Cui, Wen Zhang, Huabin Xing, Ningyu Zhang, and Huajun Chen. Molformer: Motif-based Transformer on 3D heterogeneous molecular graphs. *arXiv preprint arXiv:2110.01191*, 2021.
94. Cao Xiao, Edward Choi, and Jimeng Sun. Opportunities and challenges in developing deep learning models using electronic health records data: a systematic review. *Journal of the American Medical Informatics Association*, 25(10):1419–1428, 2018.
95. Hai-Cheng Yi, Zhu-Hong You, De-Shuang Huang, and Chee Keong Kwoh. Graph representation learning in bioinformatics: trends, methods and applications. *Briefings in Bioinformatics*, 23(1):bbab340, 2022.
96. Shujuan Yin, Weizhong Zhao, Xingpeng Jiang, and Tingting He. Knowledge-aware few-shot learning framework for biomedical event trigger identification. In *Proceedings of BIBM*, 2020.
97. Zheng Yuan, Yijia Liu, Chuanqi Tan, Songfang Huang, and Fei Huang. Improving biomedical pretrained language models with knowledge. In *Proceedings of the 20th Workshop on Biomedical Language Processing*, pages 180–190, 2021.
98. Seongjun Yun, Minbyul Jeong, Raehyun Kim, Jaewoo Kang, and Hyunwoo J Kim. Graph transformer networks. *Advances in neural information processing systems*, 32, 2019.

99. Zheni Zeng, Yuan Yao, Zhiyuan Liu, and Maosong Sun. A deep-learning system bridging molecule structure and biomedical text with comprehension comparable to human professionals. *Nature Communications*, 13(1):1–11, 2022.

100. Ningyu Zhang, Qianghuai Jia, Kangping Yin, Liang Dong, Feng Gao, and Nengwei Hua. Conceptualized representation learning for Chinese biomedical text mining. *arXiv preprint arXiv:2008.10813*, 2020.

101. Shaodian Zhang and Noémie Elhadad. Unsupervised biomedical named entity recognition: Experiments with clinical and biological texts. *Journal of Biomedical Informatics*, 46(6):1088–1098, 2013.

102. Yijia Zhang, Qingyu Chen, Zhihao Yang, Hongfei Lin, and Zhiyong Lu. BioWordVec, improving biomedical word embeddings with subword information and MeSH. *Scientific Data*, 6(1):1–9, 2019.

103. Zhaocheng Zhu, Chence Shi, Zuobai Zhang, Shengchao Liu, Minghao Xu, Xinyu Yuan, Yangtian Zhang, Junkun Chen, Huiyu Cai, Jiarui Lu, et al. Torchdrug: A powerful and flexible machine learning platform for drug discovery. *arXiv preprint arXiv:2202.08320*, 2022.

Chapter 13
OpenBMB: Big Model Systems for Large-Scale Representation Learning

Guoyang Zeng, Xu Han, Zhengyan Zhang, Zhiyuan Liu, Yankai Lin, and Maosong Sun

Abstract Big pre-trained models (PTMs) have received increasing attention in recent years from academia and industry for their excellent performance on downstream tasks. However, huge computing power and sophisticated technical expertise are required to develop big models, discouraging many institutes and researchers. In order to facilitate the popularization of big models, we introduce OpenBMB, an open-source suite of big models, to break the barriers of computation and expertise of big model applications. In this chapter, we will introduce the core toolkits in OpenBMB, including BMTrain for efficient training, OpenPrompt and OpenDelta for efficient tuning, BMCook for efficient compression, and BMInf for efficient inference.

13.1 Introduction

Since the emergence of pre-trained models (PTMs) represented by BERT [7] in 2018 to the subsequent release of GPT-3 [4] with 175 billion parameters in 2020, PTMs have attracted increasing attention from academia and industry. In Chap. 5, we have introduced that these PTMs perform well on various downstream tasks, and model performance can be further improved by increasing model parameter size. From BERT and GPT-3 to the recently proposed big models [11, 18, 39, 40],

G. Zeng
ModelBest Inc. and OpenBMB, Beijing, China
e-mail: zengguoyang@modelbest.cn

X. Han · Z. Zhang · Z. Liu (✉) · M. Sun
Department of Computer Science and Technology, Tsinghua University, Beijing, China
e-mail: hanxu2022@tsinghua.edu.cn; zy-z19@mails.tsinghua.edu.cn; liuzy@tsinghua.edu.cn; sms@tsinghua.edu.cn

Y. Lin
Gaoling School of Artificial Intelligence, Renmin University of China, Beijing, China
e-mail: yankailin@ruc.edu.cn

© The Author(s) 2023
Z. Liu et al. (eds.), *Representation Learning for Natural Language Processing*,
https://doi.org/10.1007/978-981-99-1600-9_13

the parameter sizes of these models have gradually grown from hundreds of millions to trillions. By 2022, the record for the maximum parameter size has been raised to over 1 trillion [11].

Although increasing model parameter size brings better model performance, to let most institutions and individuals enjoy the power of big models still faces several challenges:

Computation Barrier Bigger models inevitably require higher computing costs in their training, tuning, and inference. For example, as shown in the original paper of GPT-3, more than 1, 000 high-performance GPUs are used to train GPT-3. Most small- and medium-sized institutes cannot afford such a colossal computing cluster. Moreover, even after completing the training process, deploying these big models for a specific application still requires to cost thousands of dollars at a time to perform the model inference, indicating that in addition to training big models, using these big models is also expensive.

Expertise Barrier Many deep learning techniques that work well on small models become inefficient or inapplicable on big models. For example, full-parameter fine-tuning has been widely used to tune PTMs for solving specific downstream tasks. If full-parameter fine-tuning is performed to adapt big models for each downstream task, we have to spend terabytes of disk and memory space to store all task-specific fine-tuned models. Moreover, as parameter size grows to billions, some novel characteristics specific to big models emerge, such as in-context learning [26], chain of thoughts [36], and the lottery ticket hypothesis [13]. Hence, novel techniques such as prompt learning [24], delta tuning [9], and MoEfication [42] are specifically developed for big models. These techniques form an expertise barrier to those researchers and practitioners who are interested in big models and do not have extensive experience.

To address these barriers, we introduce OpenBMB, an open-source suite for building big model systems more efficiently, to make big models available to everyone. OpenBMB contains several toolkits for different application scenarios of big models, including *BMTrain* for efficient training (in Sect. 13.2), *OpenPrompt and OpenDelta* for efficient tuning (in Sect. 13.3), *BMCook* for efficient compression (in Sect. 13.4), and *BMInf* for efficient inference (in Sect. 13.5). Figure 13.1 shows how these toolkits work together to form an effective and efficient big model system. In this chapter, we will introduce these toolkits in detail, covering the critical technologies involved in these toolkits and how to use code to run these toolkits. More details of advanced big model techniques, such as prompt learning and delta tuning, can be found in Chap. 5.

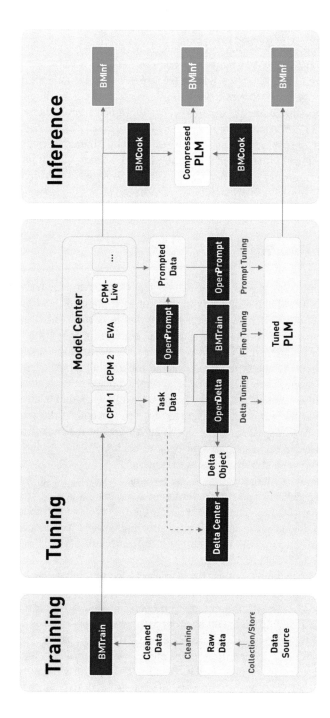

Fig. 13.1 An overview of the whole OpenBMB architecture. (The figure comes from the website of OpenBMB https://www.openbmb.cn)

13.2 BMTrain: Efficient Training Toolkit for Big Models

The success of big models is due to large-scale data and huge parameter size. The large-scale data facilitates big models to acquire versatile knowledge during the pre-training stage, while the huge parameter size enables big models to store the acquired knowledge well. However, utilizing large-scale data and huge parameter sizes is a double-edged sword, simultaneously bringing significant performance enhancement and a severe computation barrier. The key to solving the computational barrier of big model training is to adopt more efficient distributed learning strategies to take full advantage of distributed computing clusters. Next, we will introduce the distributed learning strategies used in BMTrain and how to use BMTrain to set these strategies to train big models efficiently.

13.2.1 Data Parallelism

Intuitively, training a big model on a GPU cluster is similar to organizing an event with many partners. Imagine that you have a lot of tasks to do and you want to get them done as quickly as possible; a good solution is to distribute these tasks equally to your partners so that everyone is busy and no one is idle. This is the principle of data parallelism. As shown in Fig. 13.2, the training data is divided into several parts, and each GPU is responsible for training a part of the data. For the model parameters, each GPU stores the whole model. In each iteration, each GPU calculates the gradient for its own part. After all the gradients are computed, the gradients of different GPUs are averaged together as the overall gradient. The overall gradient is then passed to the optimizer to update the model parameters. Finally, the updated parameters are sent back to each GPU for the next iteration. This process is repeated until the training is finished.

Data parallelism is largely similar to training a model on a single GPU, except that data parallelism adds two additional stages of gradient aggregation and parameter synchronization. Since the training data is evenly partitioned to each GPU, each GPU performs the same amount of computations. As a result, all GPUs can work concurrently to achieve the highest utilization, rather than having some nodes idle while others are under high load.

Data parallelism can effectively utilize large-scale computing power clusters, but some limitations are gradually exposed as the model size increases. Since data parallelism requires keeping the complete model parameters on each GPU, this method is not competent when the number of model parameters exceeds the capacity of one single GPU. For example, for a model with 10 billion parameters, it would take more than 40 GB of memory to store the model parameters and gradients, which is far beyond the capacity of most GPUs. Besides, GPUs with a capacity of 40 GB are expensive and cost inefficient. This parallel strategy needs to be further improved to enable big models to be trained with limited resources.

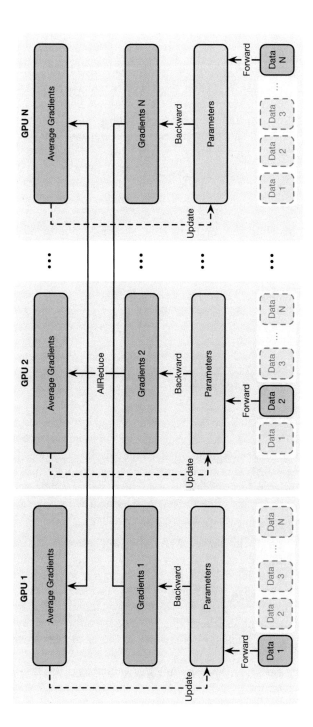

Fig. 13.2 Data parallelism partitions the input data evenly to each GPU before the forward propagation and aggregates the gradients from all GPUs to update the model parameters after the backward propagation

13.2.2 ZeRO Optimization

ZeRO (zero redundancy optimizer) [29] is a strategy that allows efficient training for models with a parameter size far exceeding the capacity of one single GPU. As mentioned in the data parallelism section, the complete model parameters, gradients, or optimizer states cannot be stored on a single GPU to train big models. As shown in Fig. 13.3, the ZeRO algorithm is proposed to further optimize data parallelism by partitioning the model parameters, gradients, and optimizer states evenly to each GPU and only temporarily aggregating them when needed. During the forward and backward propagations, the model parameters are aggregated twice, and the model gradients are aggregated only once while updating the optimizer states. Therefore, ZeRO needs to communicate three times in each iteration, one more time than data parallelism.

Although bringing additional communication, the ZeRO algorithm can train big models on a GPU cluster with limited capacity. For example, with ZeRO, it is possible to train a model with 175 billion parameters on 64× A100 GPUs, while data parallelism can only train a model with no more than 3 billion parameters. Empirically, if the number of GPUs is increased indefinitely, the number of model parameters that can be trained using the ZeRO algorithm can also be increased indefinitely.

13.2.3 Quickstart of BMTrain

BMTrain is an open-source toolkit for big model training. It provides easy-to-use interfaces based on PyTorch to help users accelerate the training process using the data parallelism and ZeRO algorithms. For commonly used model architectures such as Transformers, BMTrain implements a series of customized optimizations and makes writing distributed training code just like writing single-node training code. With BMTrain, only four steps are required to modify a single-node training code to drive a distributed cluster for training acceleration.

Step 1: Initializing BMTrain Since BMTrain is built based on PyTorch, similar to PyTorch calling the function *init_process_group* at the beginning of code to initialize distributed training, BMTrain requires to use *init_distributed* at the beginning of code to initialize BMTrain (Fig. 13.4). The function *init_distributed* also supports users to set random seeds. By setting random seeds, BMTrain can provide users with a reliable and reproducible mechanism to control all random processes, ensuring that multiple runs of the same code can yield stable results.

Step 2: Enabling ZeRO Optimization After initializing BMTrain, the data parallel and ZeRO algorithms can be applied by making some simple modifications to the single-node training code. As shown in Fig. 13.5, users should modify the model construction code to replace *Module* and *Parameter* in PyTorch with

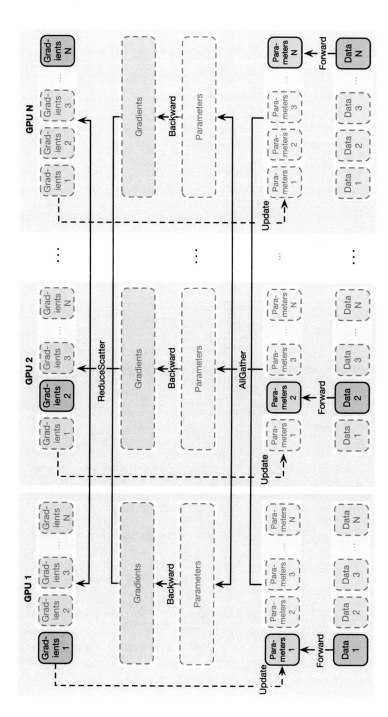

Fig. 13.3 ZeRO partitions the model parameters, gradients, and optimizer states evenly to each GPU and only gathers them together temporarily when they are needed for the computation

```
1
2  import bmtrain as bmt
3
4  bmt.init_distributed(
5      seed=0,
6      zero_level=3,   # Support both ZeRO-2 and ZeRO-3.
7  )
```

Fig. 13.4 An example of initializing BMTrain

```
1
2  import torch
3
4  class MyModule(torch.nn.Module):
5      def __init__(self):
6          super().__init__()
7          self.param = torch.nn.Parameter(torch.empty(1024))
8          self.module_list = torch.nn.ModuleList([
9              SomeTransformerBlock(),
10             SomeTransformerBlock()
11         ])
```

(a)

```
1
2  import torch
3  import bmtrain as bmt
4
5  class MyModule(bmt.DistributedModule):
6      def __init__(self):
7          super().__init__()
8          self.param = bmt.DistributedParameter(torch.empty(1024))
9          self.module_list = torch.nn.ModuleList([
10             bmt.CheckpointBlock(SomeTransformerBlock()),
11             bmt.CheckpointBlock(SomeTransformerBlock())
12         ])
```

(b)

Fig. 13.5 An example of enabling the data parallel and ZeRO algorithms. (**a**) An example of the single-node training code built on PyTorch. (**b**) An example of modifying the single-node training code to the BMTrain version

DistributedModule and *DistributedParameter* implemented in BMTrain. To further alleviate memory overhead, wrapping some model layers with *CheckpointBlock* enables the activation checkpointing mechanism, details of which can be found in the work [6].

Step 3: Communication Optimization (Optional) For those model architectures with multiple layers, such as Transformers, BMTrain provides a communication optimization module to reduce the time overhead by overlapping communication and computation. As shown in Fig. 13.6, to enable this optimization, users need to replace *ModuleList* in PyTorch with *TransformerBlockList* implemented in BMTrain.

```
1
2  import torch
3
4  class MyModule(torch.nn.Module):
5      def __init__(self):
6          super().__init__()
7          self.param = torch.nn.Parameter(torch.empty(1024))
8          self.module_list = torch.nn.ModuleList([
9              SomeTransformerBlock(),
10             SomeTransformerBlock()
11         ])
12
13     def forward(self):
14         x = self.param
15         for module in self.module_list:
16             x = module(x, 1, 2, 3)
17         return x
```

(a)

```
1
2  import torch
3  import bmtrain as bmt
4
5  class MyModule(bmt.DistributedModule):
6      def __init__(self):
7          super().__init__()
8          self.param = bmt.DistributedParameter(torch.empty(1024))
9          self.module_list = bmt.TransformerBlockList([
10             bmt.CheckpointBlock(SomeTransformerBlock()),
11             bmt.CheckpointBlock(SomeTransformerBlock()),
12         ])
13
14     def forward(self):
15         x = self.param
16         x = self.module_list(x, 1, 2, 3)
17         return x
```

(b)

Fig. 13.6 An example of enabling the communication optimization algorithm. (**a**) An example of using *ModuleList* to define a multi-layer model in PyTorch. (**b**) An example of using *TransformerBlockList* to replace *ModuleList*

Step 4: Launching the Distributed Training The code optimized by BMTrain can be run using the same launcher as the PyTorch distributed module. Figure 13.7 shows the distributed training command depending on specific PyTorch versions, and this command should be executed on all nodes in the cluster at the same time. In the launching command, users should specify the communication protocol, including the IP address and port number.

```
 1
 2  # PyTorch version < 1.9
 3  python3 -m torch.distributed.launch --nproc_per_node ${GPU_PER_NODE} \
 4  --nnodes ${NNODES} --node_rank ${NODE_RANK} \
 5  --master_addr ${MASTER_ADDR} --master_port ${MASTER_PORT} train.py
 6
 7  # PyTorch version >= 1.9
 8  torchrun --nnodes=${NNODES} --nproc_per_node=${GPU_PER_NODE}
 9  --rdzv_id=1 --rdzv_backend=c10d \
10  --rdzv_endpoint=${MASTER_ADDR}:${MASTER_PORT} train.py
```

Fig. 13.7 An example of launching the training code *train.py* in BMTrain

13.3 OpenPrompt and OpenDelta: Efficient Tuning Toolkit for Big Models

In Chap. 5, we have introduced the general capabilities of big models and their excellent performance on a wide range of downstream tasks. In this section, we focus on showing the critical role that prompt learning and delta tuning play in applications and how to adapt big models to downstream tasks using OpenPrompt and OpenDelta. More details of prompt learning and delta tuning can be found in Chap. 5.

13.3.1 Serving Multiple Tasks with a Unified Big Model

Tuning models is an essential part of the practical application of big models, which can adapt the generic capabilities of big models to specific tasks. Due to big models' high capability and generality, a big model can be adapted to dozens or hundreds of different downstream tasks. Before prompt learning and delta tuning, full-parameter fine-tuning was the mainstream method to adapt big models, where a complete model was fine-tuned and stored for each downstream task. As the parameter size of models increases, the overhead of the full-parameter fine-tuning approach on the memory storage becomes more and more heavy. Moreover, as the number of downstream tasks continues to increase, so does the memory requirement to deploy these models. These issues associated with full-parameter fine-tuning seriously increase the cost of using and deploying big models.

As compared with full-parameter fine-tuning, prompt learning and delta tuning are more friendly to big models, because these two special tuning approaches can adapt big models to different downstream tasks by changing only a small number of parameters or even without changing model parameters, as shown in Fig. 13.8. Such a feature solves the storage problem of big models caused by full-parameter fine-tuning and further reduces the problem of heavy resource consumption when deploying multiple models for different tasks. In addition, both prompt learning and delta tuning usually design task instructions in the form of natural languages

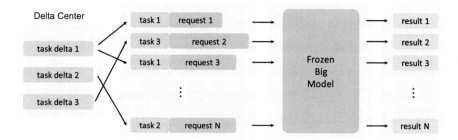

Fig. 13.8 Big models are fixed on the cluster. Different delta objects are dynamically loaded for different tasks with almost no overhead per task

to stimulate the knowledge of big models to solve specific problems. Compared with using a programming language to control big models, using task instructions to control big models is more human-friendly.

Based on prompt learning and delta tuning, we can deploy a big model on a GPU cluster and dynamically task-specific *task instructions* or *delta objects* rather than task-specific models. These task instructions and delta objects can cooperate with the big model to handle multiple downstream tasks. In this way, the resources consumed during deployment will not grow with the increase of downstream tasks. Aggregating all tasks into the same cluster also brings higher utilization, avoiding the cluster being partially idle due to unbalanced task requests.

13.3.2   Quickstart of OpenPrompt

OpenPrompt is an open-source prompt learning framework with high extensibility. OpenPrompt supports a variety of mainstream prompt learning methods [21, 30] and can help users more easily apply prompt learning on existing models or develop new models. As shown in Fig. 13.9, a *PromptModel* consists of a PTM as the backbone, one or more *Templates* to wrap the raw input with task instructions, and one or more *Verbalizers* to map the task labels to the vocabulary of PTMs. Developing a prompt learning pipeline using OpenPrompt requires the following steps to build a *PromptModel*.

Step 1: Defining the Task The first step in using OpenPrompt is defining the details of the task. Taking sentiment classification as an example, which aims to judge whether the input sentence is positive or negative, Fig. 13.10 shows how to define the dataset used for the prompt learning of the sentiment classification task.

Step 2: Loading the PTM A PTM is the backbone of *PromptModel*. OpenPrompt supports directly loading models obtained from some online model hubs, such

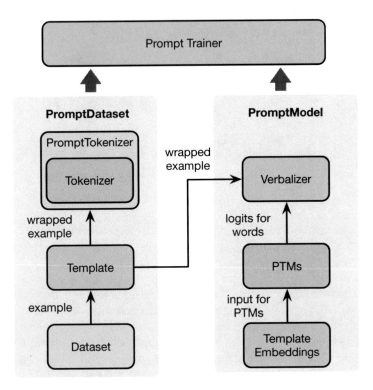

Fig. 13.9 The framework of OpenPrompt. (The figure is re-drawn according to Fig. 1 from OpenPrompt paper [8])

as ModelCenter[3] built on BMTrain and Huggingface Transformers[4] built on PyTorch (Fig. 13.11).

Step 3: Defining the Task Instruction At least one task instruction is required to make a *PromptModel*. The *Template* is used to modify the original input with the defined task instruction. Figure 13.12 shows how to join the task instruction *It was* and the field *text_a* of the dataset defined in Step 1.

Step 4: Defining the Verbalizer At least one verbalizer is required for *Prompt-Model*. Verbalizer is an important part of prompt learning that maps the output words to the task labels. Figure 13.13 maps the word *bad* to the *negative* label and maps *good*, *wonderful*, and *great* to the *positive* label.

---

[3] https://github.com/OpenBMB/ModelCenter/

[4] https://github.com/huggingface/transformers/

```
1
2  import openprompt
3  from openprompt.data_utils import InputExample
4
5  # There are two classes in the sentiment classification task, one for negative and one for
        positive.
6  classes = [
7      "negative",
8      "positive"
9  ]
10
11 # For simplicity, Here only uses two examples.
12 # text_a is the input text of the data, some other datasets may have multiple input
        sentences in one example (e.g., natural language inference).
13 dataset = [
14     InputExample(
15         guid = 0,
16         text_a = "Albert Einstein was one of the greatest intellects of his time.",
17     ),
18     InputExample(
19         guid = 1,
20         text_a = "The film was badly made.",
21     ),
22 ]
```

Fig. 13.10 An example of defining the task for prompt learning in OpenPrompt

```
1
2  import openprompt
3  from openprompt.plms import load_plm
4
5  # Here we use BERT as an example to show how to load the PTM for adaption.
6  plm, tokenizer, model_config, WrapperClass = load_plm("bert", "bert-base-cased")
```

Fig. 13.11 An example of loading the PTM for prompt learning in OpenPrompt

```
1
2  import openprompt
3  from openprompt.prompts import ManualTemplate
4
5  promptTemplate = ManualTemplate(
6      text = '{"placeholder": "text_a"} It was {"mask"}',
7      tokenizer = tokenizer,
8  )
```

Fig. 13.12 An example of defining the task instruction for prompt learning in OpenPrompt

```
1
2  import openprompt
3  from openprompt.prompts import ManualVerbalizer
4
5  promptVerbalizer = ManualVerbalizer(
6      classes = classes,
7      label_words = {
8          "negative": ["bad"],
9          "positive": ["good", "wonderful", "great"],
10     },
11     tokenizer = tokenizer,
12 )
```

Fig. 13.13 An example of defining the verbalizer for prompt learning in OpenPrompt

```
1
2  import openprompt
3  from openprompt import PromptForClassification
4
5  promptModel = PromptForClassification(
6      template = promptTemplate,
7      plm = plm,
8      verbalizer = promptVerbalizer,
9  )
```

Fig. 13.14 An example of building the *PromptModel* for prompt learning in OpenPrompt

```
1
2  import openprompt
3  from openprompt import PromptDataLoader
4
5  data_loader = PromptDataLoader(
6      dataset = dataset,
7      tokenizer = tokenizer,
8      template = promptTemplate,
9      tokenizer_wrapper_class = WrapperClass,
10 )
```

Fig. 13.15 The example of building the dataloader for prompt learning in OpenPrompt

Step 5: Combining Different Modules into a PromptModel Based on the PTM, task instruction, and verbalizer obtained earlier, we can define a *PromptModel* using the code in Fig. 13.14.

Step 6: Defining the PromptDataLoader In order to learn the defined prompt, we also need a dataloader to sample data from the dataset. The *PromptDataLoader* in Fig. 13.15 is an extension of the PyTorch dataloader used to sample data for prompt learning.

Step 7: Training and Inference After defining the *PromptModel* and *PromptDataLoader*, we can train and infer the defined prompt. All the code for training and inference can be implemented with PyTorch. Figure 13.16 represents an inference example based on OpenPrompt.

```
1  import torch
2
3  promptModel.eval()
4
5  # The predictions would be 0 and 1 for the classes 'negative' and 'positive', respectively.
6  with torch.no_grad():
7      for batch in data_loader:
8          logits = promptModel(batch)
9          preds = torch.argmax(logits, dim = -1)
10         print(classes[preds])
```

Fig. 13.16 An example of model inference based on prompt learning in OpenPrompt

```
1
2  from transformers import AutoModelForSequenceClassification
3
4  # Load BART as the backbone.
5  model = AutoModelForSequenceClassification.from_pretrained("facebook/bart-base")
```

Fig. 13.17 The example of loading the PTM for delta tuning in OpenDelta

```
1
2  import opendelta
3  from opendelta import AdapterModel
4
5  # This will apply adapter tuning to the self-attn and feed-forward layers.
6  delta_model = AdapterModel(model)
```

Fig. 13.18 The example of specifying the delta object for delta tuning in OpenDelta

13.3.3 QuickStart of OpenDelta

OpenDelta is an open-source delta tuning toolkit that can perform delta tuning without modifying the code of the backbone PTM. By using OpenDelta, users could easily implement prefix tuning [23], adapter tuning [20], LoRA [17], or any other types of delta tuning. OpenDelta also supports sharing delta objects so that users can load the delta objects trained by others or save and publish users' own delta objects. To adapt PTMs using OpenDelta only needs the following three steps.

Step 1: Loading the PTM Similar to OpenPrompt, OpenDelta requires a PTM to be loaded first as the backbone for subsequent tuning. The code in Fig. 13.17 shows how to load BART [22].

Step 2: Adding the Delta Object After loading the PTM, OpenDelta requires users to specify the parameters to be tuned (i.e., delta object). The code in Fig. 13.18 shows how to use the built-in modifications (i.e., adapter layers) in OpenDelta to specify the parameters that require to be tuned.

```
 1
 2 import opendelta
 3
 4 # Freeze the parameters of the backbone.
 5 delta_model.freeze_module(exclude=["deltas"], set_state_dict=True)
```

Fig. 13.19 The example of freezing the backbone for delta tuning in OpenDelta

Step 3: Freezing the Backbone After adding the delta object, the parameters of the backbone model are not automatically frozen. Users need to use the *freeze_module* method provided by OpenDelta to freeze the backbone model (Fig. 13.19).

After completing the above three steps, the model can be trained with the training script. After the training is completed, users can also use the *state_dict* interface to obtain the delta object's parameters, similar to obtaining parameters in PyTorch. The delta object obtained using OpenDelta usually only needs very little space to store, which is very space-efficient and can be easily stored and shared with others.

13.4   BMCook: Efficient Compression Toolkit for Big Models

The research on model compression started long ago, and in recent years several vital directions, such as quantization [2, 3, 31, 32], distillation [16, 19, 28, 33], and pruning [5, 10, 14, 35, 37, 38, 43], have been widely explored. Before the emergence of big models, model compression techniques were mainly applied to adapt models to various low-resource end devices, such as cell phones and cameras, or to some real-time applications that require low-latency inference. After the popularity of big models, the inference of these big models requires more expensive high-end devices than conventional deep models, making the compression process more critical.

In order to make big models run on common devices like various consumer GPUs, we have to combine different compression techniques to minimize the parameter size of these models without degrading the model performance too much. Therefore, as shown in Figure 13.20, BMCook systematically implements model

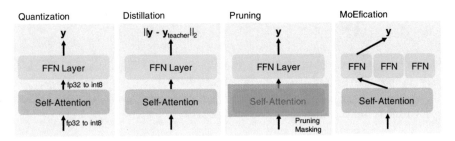

Fig. 13.20 The unified compression framework of BMCook [41]. (The figure is re-drawn according to Fig. 1 from BMCook paper [41])

quantization, model distillation, model pruning, and model MoEfication. To take full advantage of these compression techniques, BMCook builds a unified compression framework to support the combination of these compression techniques. Under this unified framework, BMCook can support arbitrary combinations of different compression techniques. Next, we will introduce typical model compression techniques and how to use these techniques to compress big models with the help of BMCook. And more details about BMInf can be found in the paper [41].

13.4.1 Model Quantization

Model quantization aims to represent the parameters of big models with those lower-bit data types rather than the commonly used 32-bit floating-point types. By using lower-bit data types to represent model parameters, both the memory and computation costs can be significantly reduced. For example, representing a model with an 8-bit fixed-point type is 4 times faster than representing the same model with a 32-bit floating-point type.

The widely used model quantization methods mainly follow two paradigms to quantize model parameters: post-training quantization (PTQ) and quantization-aware training QAT. PTQ [3] aims to directly quantize model parameters after the model learning is completed. PTQ is simple, but it may bring significant performance degradation since lower-bit data types simultaneously bring efficiency improvement and accuracy degradation. QAT [32] is proposed to alleviate the degradation caused by quantization. Specifically, QAT simulates the quantization process during learning models so that model parameters can be quantized with the guide of training data. With linear layers as an example, QAT replaces all linear layers of models with quantized linear layers. In quantized linear layers, the matrix multiplication operation is replaced by the quantized matrix multiplication.

Existing popular deep learning frameworks, such as PyTorch and TensorFlow, have already supported PTQ. Considering this and toward better performance, BMCook mainly adopts QAT to compress models.

13.4.2 Model Distillation

Model distillation aims to transfer model knowledge from larger teacher models to smaller student models. Conventional distillation methods mainly focus on adding the KL divergence between the output results of teacher models and those of student models as an additional training objective [16], so that student models perform similarly to teacher models.

After the emergence of PTMs, making the inner computation between student models and teacher students closer facilitates distilling PTMs [19, 28, 33].

Specifically, these distillation methods add the MSE loss between student and teacher models' hidden states. Note that distillation methods only provide additional training objectives rather than directly reducing the parameter size. Owing to this, any other compression methods can be combined with distillation methods to further improve the performance of compressed models.

Both the distillation methods based on the KL-divergence and MSE loss functions are implemented in BMCook.

13.4.3 Model Pruning

Model pruning is widely used to prune redundant model parameters. The existing model pruning methods mainly follow two paradigms to prune model parameters: structured pruning and unstructured pruning. Structured pruning aims to remove complete redundant modules such as model layers [10, 35, 37, 43], yet unstructured pruning focuses on removing individual parameters [5, 14, 38]. Since the pruned parameters do not need to be stored in memory and are also not involved in model computation, model pruning can thus reduce the requirements of memory and computing power.

In BMCook, both structured pruning and unstructured pruning are supported. However, unstructured pruning cannot guarantee efficiency gain since existing parallel processing devices (e.g., GPUs) do not sufficiently support arbitrary sparse computation operations [44]. Considering that 2:4 sparsity is well supported by Sparse Tensor Core [45], BMCook thus implements the unstructured pruning with 2:4 sparsity, i.e., BMCook forces every four continuous parameters to have two zeros. In this way, BMCook can guarantee that its sparse operations can be at least twice as fast as dense ones.

13.4.4 Model MoEfication

Since Transformers adopt ReLU [27] as the activation function of the feedforward networks (FFNs), bringing a sparse activation phenomenon, we can only use the part of FFNs for a specific input without affecting the model performance. To this end, model MoEfication [42] is proposed to transform Transformers to the mixture-of-expert (MoE) versions [11], which can significantly reduce the computation costs of Transformers. Model MoEfication only selects parts of model parameters for computation rather than changing or removing model parameters. To this end, model MoEfication can be viewed as a post-processing technique that can be applied to an already compressed model to further improve efficiency.

13.4.5   QuickStart of BMCook

Compressing a model with BMCook requires to set concrete compression strategies using a configuration file. The configuration file includes hyper-parameter settings, model configurations, and which compression methods to use for model compression. Figure 13.21 is an example of the configuration file. In the configuration file, we can set whether to use each compression method or not and determine which modules of the model each compression method is used for.

After setting up the configuration file, some code needs to be modified to drive BMCook to compress the model. Next, we will introduce how to use BMCook to run each compression method.

Model Quantization   As shown in Fig. 13.22, to perform QAT in BMCook, we only need to use the function *BMQuant.quantize* to operate the given model. The function will automatically replace all linear modules in the model with QAT linear modules.

```
 1
 2  {
 3      "distillation": {
 4          "ce_scale": 0,
 5          "mse_hidn_scale": 1,
 6          "mse_hidn_proj": false
 7      },
 8      "pruning": {
 9          "is_pruning": true,
10          "pruning_mask_path": "prune_mask.bin",
11          "pruned_module": ["ffn.ffn.w_in.w.weight", "ffn.ffn.w_out.weight"],
12          "mask_method": "m4n2_1d"
13      },
14      "quantization": {
15          "is_quant": true
16      },
17      "MoEfication": {
18          "is_moefy": false,
19          "first_FFN_module": ["ffn.layernorm_before_ffn"]
20      }
21  }
```

Fig. 13.21   An example of the configuration file of BMCook

```
1
2  import bmcook
3  from bmcook.quant import BMQuant
4
5  # config is the the configuration file of BMCook.
6  BMQuant.quantize(model, config)
```

Fig. 13.22   An example of performing model quantization in BMCook

```
 1
 2  import bmcook
 3  from bmcook.distilling import BMDistill
 4
 5  # config is the the configuration file of BMCook.
 6  # foward_fn is the distillation loss function.
 7  BMDistill.set_forward(model, teacher_model, foward_fn, config)
```

Fig. 13.23 An example of performing model distillation in BMCook

```
 1
 2  import bmcook
 3  from bmcook.pruning import BMPrune
 4
 5  # config is the the configuration file of BMCook.
 6  # Set the pruning masking matrix.
 7  BMPrune.compute_mask(model, config)
 8
 9  # Modify the optimizer to freeze the masked parameters.
10  BMPrune.set_optim_for_pruning(optimizer)
```

Fig. 13.24 An example of performing model pruning in BMCook

```
 1
 2  import bmcook
 3  from bmcook.moe import BMMoE
 4
 5  # config is the the configuration file of BMCook.
 6  # Retain the hidden states during the forward propagation
 7  BMMoE.get_hidden(model, config, Trainer.forward)
```

Fig. 13.25 An example of performing model MoEfication in BMCook

Model Distillation As shown in Fig. 13.23, to perform model distillation in BMCook, we need to use the function *BMDistill.set_forward* to combine the distillation loss function with the original loss function of the student model.

Model Pruning In BMCook, model pruning is implemented based on the pruning masking mechanism, which requires to modify the optimizer to freeze those masked parameters. As shown in Fig. 13.24, model pruning in BMCook can be performed by using *BMPrune.compute_mask* and *BMPrune.set_optim_for_pruning*, which can help the model compute the pruning masking matrix and modify the optimizer.

Model MoEfication Model MoEfication is to transform a dense model into a sparse one. It requires the hidden states during the forward propagation to learn how to select experts. As shown in Fig. 13.25, to perform MoEfication it is a need to use *BMMoE.get_hidden*.

13.5   BMInf: Efficient Inference Toolkit for Big Models

How to achieve efficient model inference has long been of great interest to the industry. ONNX,[5] proposed in 2017, is a general model representation format compatible with various mainstream deep learning frameworks such as PyTorch and TensorFlow. Owing to the excellent compatibility of the ONNX format, many tools for accelerating model inference based on the ONNX format have sprung up. The most well-known toolkits are TensorRT[6] and onnxruntime.[7] Up to now, transforming big models to the ONNX format and then using TensorRT or onnxruntime for model inference have become a widely used paradigm in the industry. FasterTransformer[8] is released by NVIDIA in 2019, which is a toolkit for the efficient inference of Transformer-based models on CUDA devices, considering that recently proposed big models are mainly based on Transformers.

Although these toolkits for accelerating model inference have achieved promising results, they cannot meet inference requirements for those models with more than 10 billion parameters. The main bottleneck of big model inference is GPUs' computing power and capacity. For a model with 10 billion parameters, at least 20 GB of memory is required to store the model, and a throughput of over 10^{13} FLOPs is also required to reach a usable inference speed. Such requirements are beyond the computing power and capacity of most GPUs. To make it possible to run big models on consumer GPUs, we need to leverage more heterogeneous devices such as CPUs and RAMs. Moreover, we also need to apply quantization and memory scheduling techniques to reduce the computing power and memory requirements of big models. These techniques, which need to be better supported, are well integrated into BMInf. Next, we will introduce these inference techniques and how to use BMInf for model inference. And more details about BMInf can be found in the paper [15].

13.5.1   Accelerating Big Model Inference

Optimizing the efficiency of model inference is usually closely related to specific hardware platforms. Since CUDA-based GPUs are now widely used in academia and industry, we focus on introducing inference acceleration techniques around the GPUs based on the CUDA architecture.

One of the most common optimizations is kernel fusion [1, 12, 34]. In the CUDA architecture, the basic operation unit is the kernel, and each kernel needs

---

[5] https://onnx.ai/supported-tools.html/

[6] https://developer.nvidia.com/tensorrt/

[7] https://onnxruntime.ai/

[8] https://github.com/NVIDIA/FasterTransformer/

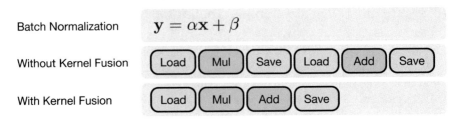

Fig. 13.26 The example of batch normalization. As shown in this figure, kernel fusion saves time in accessing global memory and improves efficiency

to read data from and write results back to global memory. Complex operators of neural networks usually require multiple kernels to cooperate, leading to redundant computation and storage. For example, for batch normalization, if it is implemented as the combination of multiplication and addition kernels, the intermediate variables need to be written back to global memory. However, using kernel fusion to integrate the multiplication and addition kernel into one kernel, the intermediate variables can be stored in registers rather than global memory, which would be much faster (Fig. 13.26).

Another common optimization is model quantization. In Sect. 13.4, we have described how to use BMCook for model quantization, and we thus will not introduce more quantization details here. For CUDA-based devices, using shorter data types (i.e., lower-bit data types) is usually much faster than using longer ones. Therefore, model quantization can significantly improve the efficiency of model inference. Using different quantization strategies may result in different running efficiencies for different models. For example, for a model based on Transformers, the main time overhead is in the computation of its linear layers, and quantizing the linear layers of this model will lead to efficiency improvements. Such a quantization strategy is implemented in BMInf.

13.5.2 Reducing the Memory Footprint of Big Models

Due to the huge parameter size, only using GPUs for big model inference requires each GPU to have a huge memory capacity. This requirement raises the threshold for applying big models to specific tasks. In order to run big models on consumer GPUs with limited memory capacities, it is critical to take advantage of CPUs to alleviate the requirement for GPU storage capacity. Specifically, when performing big model inference, we can put the part of parameters that will not be used in a short time on CPUs instead of storing all parameters on GPUs all the time. These parameters stored on CPUs will be transferred to GPUs when they need to participate in the computation.

Table 13.1 The inference performance of the big model with 10 billion parameters on different GPUs using PyTorch with BMInf

Toolkit	GPU	Memory capacity (GB)	Encoding speed (tokens/s)	Decoding speed (tokens/s)
PyTorch	V100	32	–	3
	A100	40	–	7
BMInf	GTX 1060	6	718	5
	GTX 1080Ti	11	1,200	12
	RTX 2080Ti	11	2,275	19
	V100	32	2,966	20
	A100	40	4,365	26

Note that such storage optimization comes with a price. The frequent passing of parameters between CPUs and GPUs requires non-negligible additional communication time. To this end, along with using both GPUs and CPUs, CPU-GPU scheduling is also applied to overlap the time of passing parameters and model computation. By using the CPU-GPU scheduling, the parameters that GPUs will use soon can be transferred from CPUs to GPUs beforehand.

Based on accelerating big model inference and reducing the memory footprint of big models, it is possible to run extremely big models on a single small-capacity GPU with limited computation overhead. BMInf implements kernel fusion, model quantization, and automatic CPU-GPU scheduling at the layer level. Table 13.1 shows the performance of BMInf running a big model with 10 billion parameters on different GPUs. From the table, we can see that combining all the above techniques enables us to run a big model even on a single GPU with only 6 GB capacity.

13.5.3   QuickStart of BMInf

As shown in Fig. 13.27, BMInf is an out-of-the-box toolkit that requires only minor modifications to the original model inference code. After implementing the model requiring to perform inference with PyTorch, users only need to load the parameters of the model on the CPU and then use the *wrapper* function of BMInf to wrap the model. The wrapped model can automatically apply optimization techniques according to specific hardware conditions during the inference process without manual intervention. As shown in Table 13.1, just with a few lines of code, BMInf can help a big model achieve inference speeds far exceeding its PyTorch implementation on consumer GPUs.

```
 1
 2  import bminf
 3
 4  # Use PyTorch to implement your model and initialize the model on the CPU.
 5  model = MyModel()
 6
 7  # Load the parameters of your model before using the wrapper function.
 8  model.load_state_dict(model_checkpoint)
 9
10  # Apply the wrapper function to optimize the inference process of your model with BMInf.
11  with torch.cuda.device(CUDA_DEVICE_INDEX):
12      model = bminf.wrapper(model)
```

Fig. 13.27 Using the *wrapper* function to wrap the model can drive BMInf to optimize model inference

13.6 Summary and Further Readings

In this chapter, we review the progress of big models in NLP and highlight the difficulties and corresponding solutions to run big models for practical applications. Then, we introduce how to use the toolkits of OpenBMB to achieve efficient training, tuning, compression, and inference. To learn more about the latest progress of big models, please follow BMList,[9] which contains information about the latest release of big models and is very helpful for readers to understand the trends of big models. We also recommend finding more details on efficient and efficient big models in Chap. 5.

Acknowledgments The contributions of all authors for the second edition are as follows: Zhiyuan Liu, Yankai Lin, and Maosong Sun designed the overall architecture of this chapter; Guoyang Zeng, Xu Han, and Zhengyan Zhang drafted the chapter; Xu Han, Zhiyuan Liu, and Yankai Lin proofread the chapter.

The initial materials in the first edition were provided by Xu Han, Zhengyan Zhang, and Cheng Yang.

This chapter is the resource chapter of the second edition of the book *Representation Learning for Natural Language Processing*, with its first edition published in 2020 [25]. As compared to the first edition of this chapter, the main changes include the following: (1) this edition focuses more on the introduction of open-source toolkits of big models, and (2) this edition adds example code to introduce how to use these toolkits.

References

1. Reza Yazdani Aminabadi, Samyam Rajbhandari, Ammar Ahmad Awan, Cheng Li, Du Li, Elton Zheng, Olatunji Ruwase, Shaden Smith, Minjia Zhang, Jeff Rasley, et al. DeepSpeed-inference: enabling efficient inference of transformer models at unprecedented scale. In *Proceedings of SC*, 2022.
2. Haoli Bai, Wei Zhang, Lu Hou, Lifeng Shang, Jin Jin, Xin Jiang, Qun Liu, Michael Lyu, and Irwin King. Binarybert: Pushing the limit of bert quantization. In *Proceedings of ACL*, 2021.

---

[9] https://openbmb.github.io/BMList/

3. Ron Banner, Yury Nahshan, and Daniel Soudry. Post training 4-bit quantization of convolutional networks for rapid-deployment. *Proceedings of NeurIPS*, 2019.
4. Tom Brown, Benjamin Mann, Nick Ryder, Melanie Subbiah, Jared D Kaplan, Prafulla Dhariwal, Arvind Neelakantan, Pranav Shyam, Girish Sastry, Amanda Askell, et al. Language models are few-shot learners. In *Proceedings of NeurIPS*, 2020.
5. Tianlong Chen, Jonathan Frankle, Shiyu Chang, Sijia Liu, Yang Zhang, Zhangyang Wang, and Michael Carbin. The lottery ticket hypothesis for pre-trained bert networks. In *Proceedings of NeurIPS*, 2020.
6. Tianqi Chen, Bing Xu, Chiyuan Zhang, and Carlos Guestrin. Training deep nets with sublinear memory cost. *arXiv preprint arXiv:1604.06174*, 2016.
7. Jacob Devlin, Ming-Wei Chang, Kenton Lee, and Kristina Toutanova. BERT: pre-training of deep bidirectional transformers for language understanding. In *Proceedings of NAACL-HLT*, 2019.
8. Ning Ding, Shengding Hu, Weilin Zhao, Yulin Chen, Zhiyuan Liu, Haitao Zheng, and Maosong Sun. Openprompt: An open-source framework for prompt-learning. In *Proceedings of ACL: System Demonstrations*, 2022.
9. Ning Ding, Yujia Qin, Guang Yang, Fuchao Wei, Zonghan Yang, Yusheng Su, Shengding Hu, Yulin Chen, Chi-Min Chan, Weize Chen, et al. Delta tuning: A comprehensive study of parameter efficient methods for pre-trained language models. *arXiv preprint arXiv:2203.06904*, 2022.
10. Angela Fan, Edouard Grave, and Armand Joulin. Reducing transformer depth on demand with structured dropout. In *Proceedings of ICLR*, 2019.
11. William Fedus, Barret Zoph, and Noam Shazeer. Switch transformers: Scaling to trillion parameter models with simple and efficient sparsity. *Journal of Machine Learning Research*, 23(120):1–39, 2022.
12. Jiří Filipovič, Matúš Madzin, Jan Fousek, and Luděk Matyska. Optimizing cuda code by kernel fusion: application on blas. *The Journal of Supercomputing*, 71(10):3934–3957, 2015.
13. Jonathan Frankle and Michael Carbin. The lottery ticket hypothesis: Finding sparse, trainable neural networks. In *Proceedings of ICLR*, 2018.
14. Song Han, Jeff Pool, John Tran, and William Dally. Learning both weights and connections for efficient neural network. In *Proceedings of NeurIPS*, 2015.
15. Xu Han, Guoyang Zeng, Weilin Zhao, Zhiyuan Liu, Zhengyan Zhang, Jie Zhou, Jun Zhang, Jia Chao, and Maosong Sun. BMInf: An efficient toolkit for big model inference and tuning. In *Proceedings of ACL: System Demonstrations*, 2022.
16. Geoffrey Hinton, Oriol Vinyals, Jeff Dean, et al. Distilling the knowledge in a neural network. *arXiv preprint arXiv:1503.02531*, 2(7), 2015.
17. Edward J Hu, Phillip Wallis, Zeyuan Allen-Zhu, Yuanzhi Li, Shean Wang, Lu Wang, Weizhu Chen, et al. LoRA: Low-rank adaptation of large language models. In *Proceedings of ICLR*, 2021.
18. Sumit Kumar Jha, Rickard Ewetz, Alvaro Velasquez, Laura Pullum, and Susmit Jha. BLOOM large language models and the chomsky hierarchy. 2022.
19. Xiaoqi Jiao, Yichun Yin, Lifeng Shang, Xin Jiang, Xiao Chen, Linlin Li, Fang Wang, and Qun Liu. Tinybert: Distilling bert for natural language understanding. In *Findings of EMNLP*, 2020.
20. Rabeeh Karimi Mahabadi, James Henderson, and Sebastian Ruder. Compacter: Efficient low-rank hypercomplex adapter layers. In *Proceedings of NeurIPS*, 2021.
21. Brian Lester, Rami Al-Rfou, and Noah Constant. The power of scale for parameter-efficient prompt tuning. In *Proceedings of EMNLP*, 2021.
22. Mike Lewis, Yinhan Liu, Naman Goyal, Marjan Ghazvininejad, Abdelrahman Mohamed, Omer Levy, Veselin Stoyanov, and Luke Zettlemoyer. Bart: Denoising sequence-to-sequence pre-training for natural language generation, translation, and comprehension. In *Proceedings of ACL*, 2020.
23. Xiang Lisa Li and Percy Liang. Prefix-tuning: Optimizing continuous prompts for generation. In *Proceedings of ACL-IJCNLP*, 2021.
24. Pengfei Liu, Weizhe Yuan, Jinlan Fu, Zhengbao Jiang, Hiroaki Hayashi, and Graham Neubig. Pre-train, prompt, and predict: A systematic survey of prompting methods in natural language processing. *arXiv preprint arXiv:2107.13586*, 2021.

25. Zhiyuan Liu, Yankai Lin, and Maosong Sun. *Representation Learning for Natural Language Processing*. Springer, 2020.
26. Sewon Min, Xinxi Lyu, Ari Holtzman, Mikel Artetxe, Mike Lewis, Hannaneh Hajishirzi, and Luke Zettlemoyer. Rethinking the role of demonstrations: What makes in-context learning work? *arXiv preprint arXiv:2202.12837*, 2022.
27. Vinod Nair and Geoffrey E Hinton. Rectified linear units improve restricted boltzmann machines. In *Proceedings of ICML*, 2010.
28. Geondo Park, Gyeongman Kim, and Eunho Yang. Distilling linguistic context for language model compression. In *Proceedings of EMNLP*, 2021.
29. Samyam Rajbhandari, Jeff Rasley, Olatunji Ruwase, and Yuxiong He. ZeRO: Memory optimizations toward training trillion parameter models. In *Proceedings of SC*, 2020.
30. Timo Schick, Helmut Schmid, and Hinrich Schütze. Automatically identifying words that can serve as labels for few-shot text classification. In *Proceedings of COLING*, 2020.
31. Sheng Shen, Zhen Dong, Jiayu Ye, Linjian Ma, Zhewei Yao, Amir Gholami, Michael W Mahoney, and Kurt Keutzer. Q-bert: Hessian based ultra low precision quantization of bert. In *Proceedings of AAAI*, 2020.
32. Pierre Stock, Angela Fan, Benjamin Graham, Edouard Grave, Rémi Gribonval, Herve Jegou, and Armand Joulin. Training with quantization noise for extreme model compression. In *Proceedings of ICLR*, 2020.
33. Siqi Sun, Yu Cheng, Zhe Gan, and Jingjing Liu. Patient knowledge distillation for BERT model compression. In *Proceedings of EMNLP-IJCNLP*, 2019.
34. Xiaohui Wang, Ying Xiong, Xian Qian, Yang Wei, Lei Li, and Mingxuan Wang. Lightseq2: Accelerated training for transformer-based models on gpus. *arXiv preprint arXiv:2110.05722*, 2021.
35. Ziheng Wang, Jeremy Wohlwend, and Tao Lei. Structured pruning of large language models. In *Proceedings of EMNLP*, pages 6151–6162, 2020.
36. Jason Wei, Xuezhi Wang, Dale Schuurmans, Maarten Bosma, Ed Chi, Quoc Le, and Denny Zhou. Chain of thought prompting elicits reasoning in large language models. *arXiv preprint arXiv:2201.11903*, 2022.
37. Mengzhou Xia, Zexuan Zhong, and Danqi Chen. Structured pruning learns compact and accurate models. In *Proceedings of ACL*, 2022.
38. Dongkuan Xu, Ian En-Hsu Yen, Jinxi Zhao, and Zhibin Xiao. Rethinking network pruning–under the pre-train and fine-tune paradigm. In *Proceedings of NAACL-HLT*, 2021.
39. Aohan Zeng, Xiao Liu, Zhengxiao Du, Zihan Wang, Hanyu Lai, Ming Ding, Zhuoyi Yang, Yifan Xu, Wendi Zheng, Xiao Xia, et al. GLM-130B: An open bilingual pre-trained model. *arXiv preprint arXiv:2210.02414*, 2022.
40. Susan Zhang, Stephen Roller, Naman Goyal, Mikel Artetxe, Moya Chen, Shuohui Chen, Christopher Dewan, Mona Diab, Xian Li, Xi Victoria Lin, et al. OPT: Open pre-trained transformer language models. *arXiv preprint arXiv:2205.01068*, 2022.
41. Zhengyan Zhang, Baitao Gong, Yingfa Chen, Xu Han, Guoyang Zeng, Weilin Zhao, Yanxu Chen, Zhiyuan Liu, and Maosong Sun. BMCook: A task-agnostic compression toolkit for big models. In *Proceedings of EMNLP: System Demonstrations*, 2022.
42. Zhengyan Zhang, Yankai Lin, Zhiyuan Liu, Peng Li, Maosong Sun, and Jie Zhou. Moefication: Conditional computation of transformer models for efficient inference. *arXiv preprint arXiv:2110.01786*, 2021.
43. Zhengyan Zhang, Fanchao Qi, Zhiyuan Liu, Qun Liu, and Maosong Sun. Know what you don't need: Single-shot meta-pruning for attention heads. *AI Open*, 2:36–42, 2021.
44. Ningxin Zheng, Bin Lin, Quanlu Zhang, Lingxiao Ma, Yuqing Yang, Fan Yang, Yang Wang, Mao Yang, and Lidong Zhou. SparTA: Deep-learning model sparsity via tensor-with-sparsity-attribute. In *Proceedings of OSDI*, 2022.
45. Aojun Zhou, Yukun Ma, Junnan Zhu, Jianbo Liu, Zhijie Zhang, Kun Yuan, Wenxiu Sun, and Hongsheng Li. Learning n:m fine-grained structured sparse neural networks from scratch. In *Proceedings of ICLR*, 2021.

Chapter 14
Ten Key Problems of Pre-trained Models: An Outlook of Representation Learning

Ning Ding, Weize Chen, Zhengyan Zhang, Shengding Hu, Ganqu Cui, Yuan Yao, Yujia Qin, Zheni Zeng, Xu Han, Zhiyuan Liu, Yankai Lin, and Maosong Sun

Abstract The aforementioned representation learning methods have shown their effectiveness in various NLP scenarios and tasks. Large-scale pre-trained language models (i.e., big models) are the state of the art of representation learning for NLP and beyond. With the rapid growth of data scale and the development of computation devices, big models bring us to a new era of AI and NLP. Standing on the new giants of big models, there are many new challenges and opportunities for representation learning. In the last chapter, we will provide a 2023 outlook for the future directions of representation learning techniques for NLP by summarizing ten key open problems for pre-trained models.

14.1 Pre-trained Models: New Era of Representation Learning

It is about 2 years after the publication of this book's first edition. The 2 years have witnessed the astonishing rise of large-scale pre-trained models, also well known as big models or foundation models. With the development of pre-trained modeling techniques, representation learning is exhibiting the following remarkable trends.

The order of the first nine authors is determined according to the order of their corresponding sections.

N. Ding · W. Chen · Z. Zhang · S. Hu · G. Cui · Y. Yao · Y. Qin · Z. Zeng · X. Han · Z. Liu (✉) · M. Sun
Department of Computer Science and Technology, Tsinghua University, Beijing, China
e-mail: liuzy@tsinghua.edu.cn; sms@tsinghua.edu.cn

Y. Lin
Gaoling School of Artificial Intelligence, Renmin University of China, Beijing, China
e-mail: yankailin@ruc.edu.cn

Unified Architecture of Representation Learning Ever since the initiative of parallel distributed processing (PDP) in 1980s, hundreds of neural network architectures have been proposed to address various objects, tasks, and domains, with some landmark architectures such as Hopfield networks [48], Boltzmann machines [2], self-organizing map (SOM) [63], recurrent neural networks (RNN), convolutional neural networks (CNN) [70], long short-term memory (LSTM) [47], ResNet [45], and Transformers [118].

As the evolution of neural network techniques, some architectures step down and others emerge. At the early stage of the deep learning era, there were still many architectures specifically designed with different characteristics. For example, at that time, we have to conduct experiments to find which architectures are more suitable for the given NLP task, among CNNs, GRUs, and LSTMs or their variants. After the neural architecture Transformers was proposed in 2017, especially after the pre-trained language models BERT and GPT built on Transformers as the backbone, we can find optimal neural architectures across various domains and tasks are becoming more and more unified from diverse schemes as shown in Chap. 5. The neural architecture Transformers has been the most widely used backbone across almost all NLP tasks, ranging from natural language understanding to generation.

The unifying trend also happens across multiple modalities, and the neural architecture Transformers has shown its power beyond NLP, to CV as shown in Chap. 7, and to other data such as biomedical structures as shown in Chap. 12. The unified architecture across multiple modalities will help to model rich knowledge of cross-modal interaction and further facilitate learning from heterogeneous data.

Of course, the unification process has not been completed, and there is no evidence showing Transformers will be the ultimate neural architecture for representation learning. It will be an important research topic in the future.

Unified Model Capability for Multiple Tasks With the "pre-training-fine-tuning" pipeline, pre-trained language models also build unified model capability from large-scale unlabeled corpora for multiple downstream tasks. The unified capability of pre-trained models becomes more significant as model parameters grow into a billion scale with more data and more computation power.

The evidence is their unprecedented power in zero-shot learning and few-shot learning as shown in Chap. 5. For example, we can use no more than 1% additional parameters by parameter-efficient delta tuning to adapt big models to specific complicated tasks. It makes us conjecture that big models may have learned all essential knowledge by pre-training from large-scale corpora, and the function of delta tuning is only to inform big models which internal knowledge should be stimulated for the specific downstream tasks.

The unified model capability revealed by big pre-trained models makes them completely different from conventional machine learning approaches including statistical learning and deep learning. It requires the exploration of a new theoretical foundation and efficient optimization methods conditioned on pre-trained big models.

Moreover, with the above-mentioned characteristics of unified model architecture and model capability, we believe pre-trained models to some extent indicate the maturity of distributed representation learning for AI, with great potential for extensive use in each area requiring AI for assistance. It will open a new era of AI and NLP from research to application. Standing on the new giants of big pre-trained models, there are also many new challenges and opportunities for representation learning. Here we summarize ten key open problems for pre-trained models and hope more efforts will be devoted to these problems and promote wide applications of big model techniques.

14.2   Ten Key Problems of Pre-trained Models

In this section, we summarize ten key open problems of pre-trained models, including theoretical foundation, next-generation architecture, high-performance computing, parameter-efficient delta tuning, controllable generation, safety and ethics, cross-modality, cognitive learning, innovative applications, and big model systems.

Note that these open problems are raised based on our research experiences on pre-trained models and deep learning. It does not indicate other problems beyond these ten are not or less important.

14.2.1   P1: Theoretical Foundation of Pre-trained Models

As pre-trained models (PTMs) [9, 42] become the infrastructure of modern NLP, the theoretical principles behind them become exceedingly intriguing to the community. Self-contained and rigorous mathematical theories could efficaciously guide the ameliorations of neural structures, pre-training objectives, and adaptations of PTMs and even pave the road to more powerful artificial intelligence. However, the sad truth is that we are still far from a complete understanding of PTMs. Their mechanism intersects with deep neural networks, transfer learning, and self-supervised learning in an intricate way, and moreover, considerable empirical evidence suggests that the potential of PTMs has not been fully explored.

The specialty of PTMs comes from the universal generalization capability expressed by adaptations to various tasks. Constructed on the basis of deep neural networks (typically deep Transformers [118]), PTMs are firstly pre-trained on massive unsupervised corpora and then adapted to particular downstream tasks. After optimizing a general language modeling objective in the pre-training phase, PTMs are able to yield tremendous generalization capability on a wide range of NLP tasks that involve language data, even with a few examples and a small amount of optimization [30, 50, 71].

In this subsection, we hold the mindset of seekers and discuss the theoretical foundation of the miraculous generalization capability of PTMs by decomposing it into several sub-questions.

What Is the Appropriate Mathematical Description of the Generalization Capability? When dealing with machine learning and deep learning models, calculus, linear algebra, and probability theory are among the most common choices as tools, while more advanced (and complicated) mathematics are almost untouched at the current stage. This may limit our understanding because real linear and nonlinear operations in the representation space are difficult to inscribe with these tools. Some argue that the probability theory framework that is widely used to describe generative models is intractable in the situation of capturing the correlations of high-dimensional variables [96]. Under this circumstance, other mathematical tools need to be adopted and evaluated to interpret the utilities of neural networks and even PTMs [43, 127]. For example, recent progress in geometric deep learning [11] elaborates different types of neural networks through the lens of symmetry and invariance, bringing new inspiration to the community. There are also works that attempt to provide mathematical frameworks for the revolutionary trigger point, i.e., the Transformer model [34]. Nevertheless, merely attempting to elucidate the neural network architecture may still be insufficient to understand PTMs, and grasping the relationship between pre-training and adaptation is crucial as well.

Why Does Pre-training Bring the Generalization to Downstream Tasks? Compared to traditional deep learning, the most obvious difference, and the key to success, is the far-flung pre-training phase over numerous data. The simplicity of the pre-training task and the effortlessness of the adaptation to complex tasks urge us to wonder about the principles of how pre-training and adaptation are related. From a vague point of view, PTMs' colossal capacity makes it possible to induce a type of *general knowledge*, while adaptation is a process to expose such knowledge [142]. This is, of course, an incomplete and unverifiable explanation, but a series of delta tuning [30] efforts implicitly guided by this insight has yielded remarkable results in a parameter-efficient manner. Taking a closer and simplified look, such knowledge can be modeled as coherence structures in a latent space with the Bayesian framework [128, 134]. Switching to another pragmatic perspective, analyzing the loss landscape may bring new insights into the relationship between pre-training and adaptation [73], where the pre-training phase produces a readily optimizable initialization landscape for PTMs surrounded by local optimums. Modern supervised learning theory aims to explore the bounds of theoretical adaptation loss via empirical adaptation loss and generalization errors. And studies of self-supervised learning borrow from progress in supervised learning theory to bound from the adaptation loss with pre-training loss in certain preconditions [4, 43, 116, 134]. Although analysis of pre-training and adaptation could move our understanding of PTMs one step forward, the special capabilities that come with model scaling take the ultimate goal even further.

How Are the Model Capacity and Capabilities Related? One of the most fascinating empirical observations of PTMs is the power expressed by merely scaling the size. It is not just a matter of accuracy on standard classification or generation tasks; large PTMs will counter-intuitively emerge with unprecedented capabilities as the number of parameters increases. Models with tens of billions of parameters would give surprising adaptation performance a small number of trainable parameters prepended to the input layer [71]. GPT-3 [13], a model with 175 billion parameters, shows an extraordinary capability of in-context learning, which uses several examples to stimulate the model to imitatively make predictions without tuning a single parameter. Large-scale models could even directly learn from tokenized behaviors of humans to carry out complex tasks such as using search engines [89] and playing sandbox games [7]. Experimental studies indicate that special capabilities of large models do not accumulate linearly but emerge at a certain point [129]. Although such power of scale is verified under different scenarios, it is still hardly framed theoretically.

The success of PTMs could simultaneously attribute to data, objectives, and neural architectures, and it seems to be difficult to separate modules of the process and study them independently without interfering with each other. Overall, the exploration of the theoretical foundation of PTMs is a necessarily arduous journey, whereas any promising conclusions could have profound influences. We encourage the readers of our book to keep an open mind and attempt to apply theoretical tools beyond NLP, machine learning, and even computer science to analyze the behaviors of PTMs and develop corresponding frameworks.

14.2.2 P2: Next-Generation Model Architecture

It has already been 5 years since Transformer was first released. The high capability and ease of parallelism have enabled Transformer-based models to efficiently scale up and achieve near-human or even beyond-human performance on numerous tasks. During these 5 years, we have witnessed the boom of Transformer-based PTMs and the realization of more and more previously unimaginable goals one by one [13, 89]. We have also been a witness to the spreading of Transformer's territory from NLP to other fields such as computer vision [32, 80], robotics [39, 106], etc. Undoubtedly, Transformer must be one of the most revolutionary model architectures in the history of deep learning.

Despite the power of Transformer-based models, as we have introduced in the first key problem, there is still not yet a sound theory that is able to elucidate the mechanism of Transformer. Besides, Transformer is a data-hungry and resource-intensive architecture, and the problem is further exacerbated as the model size increases [1]. Though Transformer is an epoch-making architecture, we still believe that it will not be the ultimate of neural networks. A natural question we would like to ask is *what could be the next-generation architecture for neural networks?*

From a historical perspective, we find that many of the earlier breakthroughs in neural networks were inspired by other disciplines. For example, the convolution in CNNs is borrowed from the research on the receptive field in cats' visual cortex [53], and the memory in LSTMs is also designed to mimic the mechanisms of the human brain. Therefore in this subsection, we will stand at the intersection of different disciplines and focus on neural network architectures that are inspired by different fields. Specifically, we will introduce some architectures inspired by dynamical systems, geometry, and neuroscience. While, at this time, these structures may not be able to outperform Transformer significantly, they all have their own potential and their own strengths that are still worth paying attention to.

Dynamical Systems Inspired Architectures A dynamical system is a system whose state is evolving over time, e.g., the random motion of particles, where the location of each particle changes over time. Looking at the propagation of hidden states between different layers of a deep neural network, it is intuitive to associate it with a discrete dynamical system by interpreting the layer depth as *time step*. Indeed, many works have drawn the connection between deep neural networks and discrete dynamical systems described by ordinary differential equations (ODEs) [82, 132]. The hidden state propagation in ResNet [45] exactly resembles the forward Euler discretization of an ODE. Therefore, the computation in ResNet can be seen as implicitly solving an ODE defined by the model parameters. Apart from the dynamical systems described by ODEs, the dynamical systems described by controlled differential equations (CDEs) [17, 101] and stochastic differential equations (SDEs) [61, 76] are also shown to be closely related to neural networks.

A number of advantages stem from the dynamical system perspective of neural networks. Examples are as follows:

(1) *GPU memory efficiency.* By introducing the adjoint state method [15] in the numerical optimization problem, the GPU memory consumption can be reduced from $\mathcal{O}(L)$ for ResNet (L denotes the number of layers) to $\mathcal{O}(1)$ [20, 76].

(2) *Adaptive computational time.* Ideally, models should spend less time on simple samples and more time on complex ones. However, current architectures treat the instances with different complexity equally. By leveraging the adaptive step-size solvers in numerical optimization literature, models can have adaptive time costs for different instances [18, 37].

Through the perspective of dynamical systems, neural networks can be naturally generalized to continuous systems, and plentiful theories in the dynamical systems can step in to inspire new designs for neural networks. We believe it is a promising area to explore.

Geometry Inspired Architectures Humans live in a Euclidean world. Therefore, we naturally accept the assumption that the geometry of the neural networks should also be Euclidean. However, this is not the case, as the data that the neural networks handle differs from what we are exposed to. Many complex data, such as graph data, have been shown to exhibit non-Euclidean properties [12]. Intuitively, when the neural networks are also non-Euclidean, they should be able to handle the data better due to the matching of the geometry.

Considering the non-Euclidean geometries in the neural networks brings several benefits: (1) *Greater capability in modeling structured features both theoretically and empirically.* Many real-life graphs are known to be tree-like. However, even when the dimension of the Euclidean space is unbounded, tree structures still cannot be embedded with arbitrarily low distortion, i.e., some information will always be lost. However, it can be easily achieved in a two-dimensional hyperbolic space, which is a non-Euclidean space [102]. And in practice, there have also been a lot of graph-related works demonstrating the effectiveness of low-dimensional hyperbolic models [16, 22, 91]. (2) *Combinability with the dynamical system.* Geometry can also collaborate with the dynamical system we mentioned above. From the perspective of geometry, the layers in neural networks can be seen as transformations on the coordinate representation of the data manifold. From the perspective of the dynamical system, the depth of neural networks can be continuous. When combined, it is possible to give a continuous transformation process from the data manifold to the final linearly separable manifold for different classes [12, 84]. It has the potential to provide a more intuitive understanding of how neural networks gradually transform the data from input features to features that can eventually be used for classification.

In all, non-Euclidean geometry offers a prominent direction for neural networks. It is a promising approach to handle the structured data and to combine with other perspectives to offer better insight into neural networks.

Neuroscience Inspired Architectures When thinking, unlike neural networks, we don't need to consume large amounts of energy, nor do we spike our brain temperature to near 100 °C. Although still called *neural* networks, today's artificial neural networks (ANNs) have already become much more energy-hungry and resource-demanding than the human nervous system. Compared to ANNs, the sparsity of human brains allows them to consume much less energy than ANNs. Therefore, inspired by the sparsity of neuronal interconnections in the human brain, researchers have experimented with designing neural networks with sparsity from two dimensions: spatial sparsity and temporal sparsity.

The human brain has sparse neuronal connections and relatively distinct functional partitions. That is, neuronal connections in the human brain are spatially sparse. This allows us to accomplish a simple task without using neurons from the whole brain. Inspired by the spatial sparsity, the mixture of experts (MoE) structure is proposed [33, 54]. Unlike conventional neural networks which are densely connected, MoE divides each layer into several experts and additionally includes a router to route every input to only a few experts. Since not all the experts are involved in the computation, the inference can be faster than densely connected networks. The advantage of MoE models in terms of computational cost allows them to scale up very efficiently. In addition, because different inputs will be processed by different experts, ideally, different experts can learn to handle different aspects of a task (or even multiple tasks), making it suitable for artificial general intelligence. Indeed, MoE models have been shown to reach state of the art on several benchmarks with fewer computational cost [88].

In addition to spatial sparsity, the human brain also exhibits temporal sparsity, i.e., neurons do not transmit signals every time step. Spike neural networks (SNNs) [38] mimic the behavior of information propagation between neurons interconnected by synapses. When the pre-synaptic neuron is activated, it sends a signal in the form of synaptic current to the post-synaptic neuron, and the current strength is proportional to the weight of the synapse. The incoming synaptic currents change the membrane potential of the post-synaptic neuron, and when the membrane potential reaches a certain threshold, the post-synaptic neuron emits a spike, and its membrane potential is reset to its resting potential. The biggest advantage brought by the SNNs is the extremely low energy consumption. Because SNNs only consume energy when emitting spikes, the energy consumption of SNNs can be extremely low compared to mainstream neural networks [60, 87, 112]. Also, *neuromorphic*, which is the specialized hardware for SNN, allows both computation and parameter storage on the same chip, further boosting the efficiency [26]. Although the performances of SNNs are often slightly lower than the mainstream neural networks on datasets such as MNIST [70] and CIFAR-10 [66], the low energy characteristic of SNN makes it promising for the future.

Looking back at history, in 2012 AlexNet [67] was proposed, and since then deep neural networks such as CNNs and RNNs take the lead in machine learning. Five years later in 2017, Transformer was introduced and gradually replaced the models such as RNNs. Now, in 2022, we are celebrating another 5-year period, wondering what could be the next-generation neural network. We believe Transformer will not be the ultimate of the neural networks, and we are eager to see more researchers think about and explore the next-generation neural network architecture and propose more economical, more efficient, and more effective models.

14.2.3   P3: High-Performance Computing of Big Models

Numerous parameters of big models come with exceedingly expensive computation and storage costs, imposing substantial challenges on both training and inference. In fact, improving the computational efficiency of big models is a complicated process in which many fundamental aspects should be considered. In particular, meliorations across the computational infrastructure, algorithms, and specific applications can be simultaneously conducted. In this subsection, we discuss high-performance computing of big models from these three perspectives.

High-Performance Computational Infrastructure We collectively refer to the hardware and software as the computational infrastructure, which is the foundation for both the training and inference of big models and even deep neural networks. In general, high-performance computational infrastructure can be further exploited from the following directions: (1) Parallel computing methods, including data parallelism [113], tensor parallelism [52, 90], pipeline parallelism [104], and hybrid parallelism [95], could fully utilize distributed computing capabilities to accelerate the computation of big models. (2) We should take advantage of heterogeneous computing devices [56], including multi-level computing devices consisting of GPUs and CPUs, and multi-level storage devices consisting of VRAMs, RAMs, and disks, to reduce the computing cost while ensuring the computing efficiency. (3) Considering big models have large-scale parameters, we should investigate techniques to reduce the memory overhead, including tensor offloading [100, 107] and tensor rematerialization [21, 62], facilitating us to compute bigger models using fewer computing devices. (4) Moreover, high-performance tensor programs [122] are also critical for making deploying big models efficient, especially sparse tensor programs [149] considering the sparsity of neural networks.

High-Performance Algorithms Existing work on big models enjoys the emergent ability that comes with increasing parameters while ignoring the efficiency of the parameter utilization. If we draw an analogy between big models and the brain, we will find that the brain consumes much less cost for the similar billions of parameters (neurons) due to some enigmatic mechanism brought by evolution. Recently, two Turing Award Winners, Yoshua Bengio and Yann LeCun, also highlight the importance of neuroscience for AI [140], and they believe that the next generation of AI will be largely driven by neuroscience. Hence, it is a promising way to design new algorithms by utilizing knowledge of neuroscience. We will discuss several important brain-inspired mechanisms as examples and hope these methods can inspire more explorations. (1) Learning from memory mechanisms of human brains [115]. We should build an explicit memory system to store the information and retrieve relevant pieces for a given input instead of computing all parameters [40, 72]. (2) Learning from System 1 and System 2 of human brains [25]. We should design a system that can automatically switch between the fast and the accurate modes for inputs with different levels of complexity [135]. (3) Inspired by recent work highlighting the importance of cooperation between brain regions [114], we should also explore how to compose multiple big models to achieve better performance [3], which is more efficient than training a bigger model from scratch.

High-Performance Application When dealing with limited resources of edge devices such as mobile phones, our approach should shift from squeezing the performance out of computing devices to compressing the big models themselves for efficient deployment. As introduced in Chap. 5, there are many compression techniques, such as knowledge distillation [46] and parameter pruning [41], that could compress big models to acceptable scales. Overall, in terms of high-performance applications, we believe the following future directions show considerable potential. (1) Computing hardware sets boundaries for our compression techniques. To this end, properties of application hardware must be considered to find the best compression architecture with minimal latency [121] or energy costs [125] rather than FLOPs. (2) Different downstream tasks may exhibit different characteristics, thereby requiring compression strategies with disparate focuses. We should explore task-aware compression to utilize the specific patterns of different tasks, such as vocabulary reconstruction [136] for tasks of a specific language and decoder-oriented compression for generation tasks [77]. (3) Many compression approaches could achieve similar results but are orthogonal in technical aspects. To this end, we could take advantage of multiple compression techniques to achieve higher compression ratios. Some preliminary works have begun to investigate combinational compression and have already achieved some promising results [148]. However, how to combine all existing methods to achieve optimal inference acceleration within an acceptable performance degradation still remains an open problem.

The development of high-performance computing is an important driving force for deep learning, especially for big pre-trained models. In the past, the performance gains have mainly come from the growth of computing power. In the future, we need to devote more efforts on how to improve the utilization of computing power. On the one hand, it can lower both the bar of using big models for anyone who is interested in AI and the carbon footprint of computing big models. On the other hand, in the post-Moore era, there is limited room for further improvement in computing power, and new methods should shift from relying on the growth of computing power to improving efficiency.

14.2.4 P4: Effective and Efficient Adaptation

Before the arrival of the era of PTMs, empirical improvements of NLP applications are primarily achieved by considerations across aspects of models, algorithms, task-specific characteristics, etc. After PTMs take the stage, researchers find that prominent advancements in almost all NLP tasks can be delivered by merely scaling up PTMs. Such a success of scaling, despite elusive, has fueled a surge of development of big models with billions [93] and even hundreds of billions of parameters [13]. Accordingly, the emergence of big models triggers provoking explorations of advanced model adaptations, which suggests that the full-parameter fine-tuning approach used in early PTMs is not the optimal solution for model adaptation. It is neither effective across all forms of datasets nor economically efficient on common computation devices. That is to say, the inherent characteristics of the big model itself must be taken into account, and innovative strategies for model adaptations should be established. To this end, how to effectively and efficiently adapt big models becomes a pivotal research issue. The problem is threefold in this subsection, including computationally practical adaptation, task-wise effective adaptation, and advanced adaptation with complex reasoning.

Computationally Practical Adaptation The huge size of big models is a blessing in terms of experimental performance, whereas a curse in terms of the adaptation process. Deploying and adapting these models to assorted tasks require considerable computational and storage resources that are prohibitive to common researchers. Instead of updating all the parameters of big models, recent studies of delta tuning [30, 49, 50, 75] find that only a tiny portion of parameters could yield comparable or even better performance of full-parameter fine-tuning. These trainable parameters can be represented as different structures or in different positions in big models. But a consistent empirical characteristic is that the larger the model, the better the performance of this paradigm. Delta tuning reifies conceptual capabilities to solve particular tasks in a concrete and lightweight manner. The resulting lightweight delta objects are easy to store and share across tasks and users, imposing considerable maneuverability on big models and unleashing the imagination of the industrialized use of these behemoths. Despite the efficiency, there are dark clouds still hanging over this topic. For example, it is difficult to assess the optimal amount of tunable parameters for different tasks, and the convergence of delta tuning is relatively slower than full-parameter fine-tuning. In addition, the theoretical principles behind the success of delta tuning can also help the community further understand big models. The revolution in terms of model adaptations does not only occur at the parameter optimization level but also at the level of data and tasks. Next, we take prompt learning as a landing point to discuss the task-wise effective adaptation of big models.

Task-Wise Effective Adaptation Taking BERT [29] as an example, PTMs in the early stage first produce representations for current inputs and adopt extra classifiers to carry out adaptations to downstream tasks. This seemingly established approach may actually be counter-intuitive since there is a considerable chasm between pre-training and adaptations. Empirical evidence shows that inserting additional contexts, i.e., prompts, and transferring downstream tasks to pre-training tasks could substantially shrink the gap and yield promising performance, especially in the low-data regimes. Prompts could be generated and constructed by different means and forms, but fundamentally, this technique implies a trend of unification of NLP tasks, which includes the unification of pre-training tasks and downstream tasks, as well as the unification between different downstream tasks. Prompt learning has shown intriguing attributes such as zero- and few-shot learning, task generalization, and structural unification of datasets. Besides, the flexibility of prompts makes it possible to smooth the logic chain of big models and stimulate complex reasoning capabilities.

Advanced Adaptation with Complex Reasoning The reasoning capability of big models has been a long-standing debate that no one can perfectly arbitrate, where the existence, representation, and stimulation methods have been suspending research questions for years. Intuitively for human beings, solving more complex questions is almost equivalent to more comprehensive reasoning ability. When it comes to big models or, more generally, neural networks, continuous studies about *shortcuts* and record-breaking performance of complex tasks create a confrontational situation. With no intention of philosophizing the argument, we look at this only from the perspective of performing complex tasks, where big models could produce striking logical processes in numerical and commonsense reasoning tasks [130]. Consistent with the aforementioned two points of computationally practical and task-wise effective adaptations, such reasoning capabilities emerge at a certain point of model scaling, which implies that models should have sufficient capacity and be trained on sufficient data in pre-training to elicit complex reasoning. However, such reasoning abilities to perform complex tasks are not stable in practice, where they show different variances for different data and are extremely demanding in terms of stimulation manners. This puts researchers in the awkward position that we are all vaguely aware of the enormous potential of big models, but have few clues about how to hit that upper limit.

In summary, research considerations of big model adaptations could be encapsulated in three points according to the above statements: First, big models should be computationally practical so that they can fully replace previous approaches when their training and storage are no longer an unattainable goal for the community. Delta tuning is a highly prospective attempt at the algorithmic level, and perhaps the community also needs to make efforts on computational systems and hardware. Second, the predictive power of big models could be realized by new types of data and task organization, and prompt learning is the product brought by the development of big models, which also pushes us to adopt a more unified perspective when looking at the tasks. Finally, to further tap the potential of big models, complex reasoning must be explored, and this is a key step for artificial intelligence to enter the cognitive level instead of making simple predictions.

14.2.5 P5: Controllable Generation with Pre-trained Models

Generating data distribution is a long-standing challenge for the machine learning community due to its inherent high dimensionality and intractability. Fortunately, the unprecedented capabilities accompanied by PTMs have brought this goal within reach and thus sparked a new surge of research. In empirical inspections of large-scale PTMs, researchers have discovered their impressive ability to generate high-quality text [13], images [94], videos [108], or programming codes [19]. However, PTMs are black boxes, which make us passively accept the generated results rather than actively controlling the model to produce contents that match a specific requirement. How to precisely introduce conditional constraints to control the generated results poses a major challenge for PTMs. Specifically, the challenge of controllable generation comes from three facets: a unified framework for diverse controls, the compositionality of controls, as well as a well-recognized evaluation benchmark.

A Unified Framework for Diverse Controls The primary objective of controlled generation is to meet the diverse practical desires of users concerning content, features, and styles. Diverse controls result in dispersed research efforts. For example, depending on the category of the input, separate models are trained for generation from paragraphs [36], dialogues [145], tables [109], etc. Regarding the properties of the generated text, requirements for sentiment orientation [51] or keyword satisfaction [147] are accomplished by distributional change or insertion-based methods, respectively. In spite of the proliferation of works on diverse controls, we would prefer to use a unified framework to accomplish all these controls rather than designing specific methods to meet each requirement. A unified framework can not only encourage research to be iterated rapidly and convergently but also enable the investigation of the relatedness and combinatoriality of diverse controls. Recently, there have been several research works in this direction: (1) *Prompt-based methods.* Either by injecting a control code [58] or continuous parameters [75], we can leverage the same PTM with diverse controls. The major drawback is that prompt-based methods usually have coarser control granularity or smaller control power, thus incapable of handling hard constraint tasks like copying a span of text. (2) *Distribution modification methods.* By incorporating different constraints in the decoding stage of the language model [78], the generated text from the same PTM can be steered from different directions. Its limitation is that distribution modification methods may hinder the fluency of generation [59]. Hence, how to combine the two approaches or develop novel approaches for unification are still open questions.

Compositionality of Controls In addition to the diversity, controllable generation is also expected to be *multidimensional* and *multi-grained* to allow more intricate combinations of controls. As discussed in Chap. 3, compositionality, which studies how to use low-level linguistic units to form high-level semantics, is a topic of considerable interest in text representations [86] and natural language understanding. It is less explored in the context of controllable generation due to the dispersal of control approaches. To this end, the advocates of a unified framework for generation can contribute to compositionality. To steer the generation toward multiple control requirements simultaneously, combining prompts with individual functionalities can be explored to form more comprehensive capabilities [92]. Nonetheless, the exploration is still primitive, with the simple concatenation of prompts as the composition method. As yet, we do not have an understanding of the internal mechanism of controllable generation for PTMs, making it difficult to develop advanced compositional control methods. Of course, we also look forward to other novel approaches that can achieve compositionality of controls.

Well-Recognized Evaluation Benchmark As ImageNet [28] in computer vision and GLUE Benchmark [120] in natural language understanding have demonstrated, a recognized benchmark can foster benign competition among researchers and identify promising approaches. However, such a benchmark is absent for generation tasks, especially controllable generation. The problem is further compounded by the fact that researchers may use different assessment methods and different data when focusing on the same aspects of controllability [64, 78]. We highlight three aspects of the difficulty of establishing a benchmark for controlled generation and the potential improvements. (1) Firstly, human language is rich in expressions, and the same meaning can take on many nuances. So any golden answer is not sufficient. A possible solution is to create semantic matches between utterances. This requires a powerful semantic understanding model that can provide reliable matching scores from diverse angles. The previous works, e.g., BERT-Score [146], are still insufficient in this regard. Whether the large PTMs like GPT-3 could be used to provide powerful semantic matching is still an open problem. (2) Secondly, control requirements are intractable and diverse. For example, topic satisfaction or emotional tendencies are difficult to measure quantitatively. Considering the diversity issue, how to integrate the criteria into a unified implementation that can be used across the community is a complicated but urgent task. (3) Thirdly, evaluation should take into account potential degraded factors such as quality and efficiency. Some works [78] point out that there is an inevitable trade-off between the control's satisfaction rate and text quality. Additionally, either increasing the length of input via prompts or applying complex decoding strategies will sacrifice generation efficiency, which should also be taken into consideration for a well-rounded evaluation. Due to the aforementioned challenges, few attempts have been made to unify the evaluation, and a universally recognized benchmark is still urgently needed.

Controllable generation is important in all areas of AI. The approaches to controllable generation are not unified across tasks, and this in turn leads to difficulties in compositionality of various control approaches. Further, the challenge of controllable generation is exacerbated by the lack of a well-recognized evaluation benchmark. Advances in the above three directions will greatly contribute to the controllability of generation and thus make generation techniques better serve practical needs.

14.2.6 *P6: Safe and Ethical Big Models*

With the exciting progress made in recent years, big models are deemed as cornerstones of modern NLP as well as AI. However, responsible AI research calls for clear recognition of both benefits and risks. While the benefits of big models are under extensive exploration, we should also be concerned about the underlying negative impacts and harms to individuals and society before deploying big models in the real world. In Chap. 8, we have discussed the robustness requirements for NLP models, and most topics are related to model safety or ethics. Although considerable efforts have been devoted, there still remain major challenges to solve and possible future directions to explore. In this section, we discuss open problems toward safe and ethical big models from the perspective of evaluation, governance, and construction.

Evaluating Safety and Ethical Levels The very first challenge in building safe and ethical big models is *how to conduct rigorous and comprehensive evaluations*. For model safety, we have introduced several essential threats against NLP models in Chap. 8, including backdoor attacks, adversarial attacks, and distribution shifts. However, a golden standard of model safety has not been reached, which means we still have no comprehensive safety evaluations. As the deployed models are continually exposed to complex external environments, there are emerging risks, and we wonder if the models are robust to such risks. Tramer et al. [117] figure out that the majority of adversarial defense methods fail to work when attackers adapt their attack strategies accordingly. This suggests that safety over known threats is not enough, and the underlying unknown threats should also be taken into consideration.

Measuring the ethical level of models is even more complicated. It is observed that big models could generate stereotypical or hateful comments about certain groups of people [131], disseminate false or misleading information [144], and leak private information from training data [14]. Obviously, these behaviors violate human values and thus are undesirable. However, it is easy to find individual cases, but rather difficult to conduct rigorous measurements since the human values are hard to specify. Given the social and regional diversity, there does not exist a static and universal rule to assess ethical levels. Worse still, values about politics, religions, and ethnicity are always conflicted across groups, making it even harder to evaluate. Under such conditions, datasets and benchmarks in this research field need to be carefully checked for valid measurement. We also suggest researchers cooperate with sociologists to gain theoretical insights.

Governing Big Models Given the potential safety and ethical risks of big models, how to cooperate correctly with big models is an essential problem for the AI community, which is referred to as model governance. However, big model governance is challenging both technically and non-technically. On the technical side, big models are capable of completing various downstream tasks via simple adaptation, which also include harmful ones such as generating offensive speech or fake news. Due to the black-box architecture of big models, finding and disabling these harmful functionalities can be difficult. Although practitioners adopt some effective approaches like keyword filters, they cannot guarantee the models are fully governed [119], leaving this problem open for future research. On the non-technical side, model governance is not only about the research community but also about achieving principles and laws across model providers and users, which requires multi-party cooperation. We are glad to see that some responsible organizations are contributing in this area [24] and appeal to more researchers to help advance this important direction.

Building Inherently Safe Models Another fundamental question about model safety is *how can we learn inherently safe models?* In Chap. 8, we introduce approaches to solve robustness issues, but most methods we mentioned are targeted at specific problems except pre-training. However, while it has been widely acknowledged that bigger models may make fewer mistakes, we still argue that scaling models and data sizes is not the elixir to eliminate safety problems because an inherently safe model does not equal a model making no mistakes. Instead, to achieve human-level robustness, the models should (1) know what they know and do not know (i.e., calibrated) and (2) learn from mistakes and correct themselves [69, 83]. In this regard, current big models are still far from inherently safe, and we hope to see more efforts devoted to this fundamental problem. Toward inherently safe models, we figure out two possible directions. (1) Incorporating knowledge. In Chap. 9, we see the remarkable success made by injecting knowledge into PTMs. On model safety, incorporating knowledge can help as well. For example, models won't be fooled by "U r stupid!" if they possess phonetic knowledge. Hence, we recognize building knowledgeable big models as a reliable approach for model safety. (2) Cognitive learning. Nowadays learning paradigm for big models is still data-driven, which cannot fully reflect the underlying risks in the real world. Different from models, we human beings can actively interact with the world and consistently gain knowledge. Moreover, we also largely benefit from the "trial and error" process and learn how to avoid mistakes. Therefore, we address the importance of learning from cognition and interaction for building safe models [65], and we further elaborate on this topic in Sect. 14.2.8.

Safety and ethics are two long-standing topics in AI, which are even extensively discussed in literature and artworks (e.g., Isaac Asimov's "Three Laws of Robotics" [5]). In the worry of runaway powerful machines, we present several key challenges and future directions for this open problem. We stress that, in the context of nowadays AI hype, we researchers especially need careful consideration before we take every single step and take responsibility for the healthy development of big models.

14.2.7 P7: Cross-Modal Computation

Building intelligent agents that can think and behave like humans is a long-standing goal of AI. An important and appealing characteristic of human intelligence is the impressive capability of perceiving and handling information from different modalities. Recently PTMs have greatly pushed forward the development of intelligent agents in single modalities (such as text [29], image [44], and audio [31]) and also led to breakthroughs in cross-modal computation. By exploiting self-supervised signals in large-scale cross-modal data, generic representations connecting different modalities can be effectively pre-trained and transferred to facilitate various downstream tasks. Cross-modal PTMs based on the *pre-training-fine-tuning* paradigm seem to constitute a promising foundation to realize such cross-modal intelligence. To this end, we discuss several promising directions for advancing cross-modal PTMs in this subsection, including big cross-modal models with efficient pre-training and adaptation, more unified representation with more modalities, and embodied cross-modal reasoning and cognition.

Big Cross-Modal Models with Efficient Pre-training and Adaptation Existing works show that impressive capabilities can emerge in pre-trained language models when the model capacity (e.g., number of parameters) substantially scales up. For example, the 175B GPT-3 is able to perform in-context few-shot learning and chain-of-thought prompting for complex tasks. However, although cross-modal pre-training on deep Transformers has pushed forward the state of the art of various tasks, compared with language models, cross-modal models are typically limited in parameter sizes. This hinders the exploration of cross-modal PTMs to more advanced capabilities and tasks. An important reason is that compared with big language models, it can be even more expensive to pre-train and adapt big models that deal with multiple modalities. Some works have explored more efficient pre-training by reusing unimodal models that have been well pre-trained and focusing on connecting PTMs from different modalities [3]. Some works have investigated the efficient adaptation of vision-language models in terms of both data [3, 126, 139] and parameters [150]. In the future, more efforts can be devoted to efficient pre-training and adaptation of big cross-modal representation learning models.

More Unified Representation with More Modalities Traditional cross-modal works typically design highly specialized model architectures to maximally exploit the inductive bias of modalities and tasks. For example, RNNs are designed to model the sequential dependency of text, and CNNs are developed to model the shift and scale invariance of images. The learning signals usually come from the human annotation of specific tasks. However, designing specific model architectures and learning signals for different modalities and tasks requires extensive expert knowledge, and it can be problematic to maintain a model for each of the large number of tasks. With the development of deep cross-modal pre-training, cross-modal representation learning models are becoming more unified in terms of model architectures and learning mechanisms [74, 138]. Most recently, some works have shown promising results in using unified model architectures, parameters, and learning mechanisms for unimodal, cross-modal, and embodied tasks [97, 123, 124]. Some works have explored pre-training with more modalities, including text, image, and audio [79]. In the future, building a unified representation learning model that can simultaneously deal with various modalities and tasks will be a promising foundation and path to realizing general intelligent systems.

Embodied Cross-Modal Reasoning and Cognition Semantic recognition capability has been extensively investigated in different modalities, e.g., named entity detection from text and object detection from images. For more complex reasoning and cognition capabilities, obstacles have been encountered in different ways: (1) For modalities with low information density, such as images and audios, semantic recognition can already be a challenging task [98], let alone more complex reasoning [143]. (2) For text which has high information density, it can be more natural to perform complex reasoning based on the abstract symbolic tokens, and recently big language models have shown promising results in commonsense and mathematical reasoning [130]. However, many AI researchers believe that true recognition capability cannot arise from learning only from text [8]. Research in cognitive science also shows that the human mind is highly shaped by embodied learning [133]. Therefore, a more promising direction will be an embodied cross-modal reasoning model. The concrete signals from other modalities can be effectively aggregated into a text-based central unit for high-level semantic reasoning. Some attempts have been made [10], and we believe that the direction is worth more exploration.

In summary, as an important interdisciplinary area that connects information in different modalities, cross-modal computation is essential and beneficial to various real-world AI applications and is also one of the key problems to more general intelligent systems. With their recent rapid development, cross-modal PTMs have become a new foundation in advancing toward this goal. We believe that developing an efficient big cross-modal PTM that can deal with various complex embodied reasoning tasks in a unified fashion will be a promising direction.

14.2.8   P8: Cognitive Learning

An essential measurement of general AI is whether neural models can correctly perceive, understand, and interact with the world, i.e., the cognitive ability. A prototype of general intelligence can be viewed as the capability of manipulating existing tools (e.g., search engines, databases, web-side mail systems, etc.), conducting cognitive planning with complex reasoning, and interacting with the real world to acquire and organize information/knowledge.

Serving as the foundation for AI, PTMs have pushed state-of-the-art performance in a variety of downstream tasks. The rich language knowledge, world knowledge, and commonsense knowledge stored in PTMs determine their unique advantages in cognitive modeling. Efficiently utilizing such knowledge conduces to stimulating the cognitive ability of PTMs, based on which PTMs could effectively interact with the real world in complex scenes. Despite the great success, current PTMs still cannot handle advanced cognitive tasks. To bring PTMs human-level cognitive intelligence, we identify three core challenges for achieving general cognitive intelligence:

Understanding Human Instructions and Interacting with Tools How could PTMs better understand the user's instructions and interact with existing tools to complete a specific task? Fulfilling this goal requires precisely (1) mapping the natural language instructions in the semantic space to the cognitive space of the model and (2) mapping the cognitive ability of the model to the action space of the tool, so as to correctly perform the operation and use the tool. The realization of this goal has profound practical significance:[1] (1) for one thing, an ideal next generation of human-computer interaction (HCI) will be based on natural language rather than a graphical user interface (GUI). The user only needs to inform the model of the goals that need to be achieved, and the model can perform a series of operations in response; (2) for another, the bar of utilizing complex tools will be greatly lowered. In this sense, any beginner can quickly get started with a new software or tool with the help of the model, making it more convenient to fulfill an intended complex task. However, PTMs trained on general domains are not designed for instruction understanding or tool manipulation by nature. To this end, a potential solution is continual pre-training, which adapts the PTM from the original pre-training domain to the human instruction domain, so as to better grasp the semantics of human

---

[1] https://www.adept.ai/act.

instruction. In addition, it is also promising to design knowledge-enhanced tuning methods to improve the PTMs' semantic understanding of specific domains under the guidance of structured human knowledge.

Cognitive Planning and Reasoning for Complex Tasks Based on the proper understanding of human instructions, PTMs could form implicit solution chains, i.e., thoughts for complex tasks. This process requires the ability of reasoning and planning for complex tasks. Such an ability has a variety of applications, including theorem proving [68], tool manipulation [137], etc. The recent emergence of chain-of-thought (COT) prompting techniques [130] can be leveraged to further enhance PTMs' reasoning ability. Through a sequence of intermediate natural language reasoning steps, COT prompting helps PTMs decompose a complex task into relatively simple atomic tasks and solve them one by one. Ultimately, the correct decision-making path can be found to achieve the goal of the user. Another potential solution for complex reasoning is to "learn from experiences." That is, generalizing the reasoning process of a specific task to form its "thoughts" of planning for other tasks. To achieve this goal, we need to train models to understand how different tasks are intrinsically related, so as to break the barriers between different tasks. In this way, models can learn various tools by analogy. Such a capability is related to the concept in cognitive psychology, that is, human beings generalize a property from one stimulus to another stimulus if both of them are similar in an appropriate psychological space [103].

Integrating Information from the Real World By interacting with the real world, we may finally gather a series of fragmented information separately. It is of great importance for PTMs to integrate information returned by existing tools into a self-contained and well-organized one. Rendering such organized information to humans completes a closed loop for a cognitive task. Integrating information for PTMs is challenging because newly retrieved information may inherently contradict the original knowledge/belief of PTMs themselves, and it is under-explored how to combine the implicit knowledge of PTMs and the retrieved knowledge from the real world. In fact, recent efforts have been paid to address this challenge. For instance, in open-domain QA, WebGPT [89] and GopherCite [85] are proposed to leverage externally retrieved knowledge to increase the reliability, faithfulness, factuality, and interpretability of the outputs produced by PTMs. Specifically, researchers teach PTMs to learn to interact with reliable IR systems like Microsoft Bing and Google Search, so that the system can retrieve more faithful and relevant documents. After that, PTMs are trained to organize supporting facts into a coherent and self-contained answer. Although many efforts have been spent on integrating textual information from the real world, less is studied about the exploration of other types of information (e.g., graphical information, tabular information, etc.).

To sum up, the ultimate goal of cognitive learning is to move toward the next generation of machine intelligence. Cognitive intelligence will enable PTMs to play a more involved role in all walks of life and interact with the real world on behalf of humans, posing a huge impact on both academia and industry.

14.2.9 P9: Innovative Applications of Big Models

AI is a discipline that emphasizes practical applications and is widely expected to play a role in a broad range of downstream fields and task scenarios. Among these applications, many of them express both immense value and challenges, such as autonomous driving [110], medical assistance [35], etc. Traditional solutions for AI applications can be divided into two main ideas. The first one is to implement symbolic systems driven by human knowledge (like expert systems in the 1980s), while it is difficult to cover all the scenarios encountered in practical applications based on manual rules. The other idea is to conduct data-driven deep learning systems, which still face obstacles due to high labeling costs in various fields that lack sufficient high-quality training data.

The emergence of big models has brought new possibilities for achieving innovative applications. Big models are equipped with a substantial amount of human knowledge, which is scattered in the large-scale unlabelled corpus and can be gained in an unsupervised manner to avoid the high annotation cost. Representative instances for big model applications can be classified into two types: new breakthroughs and new scenarios.

New Breakthroughs This type refers to the big model systems that achieve surprisingly good performances in already existed application problems. For example, the Critical Assessment of protein Structure Prediction (CASP) challenge has been held for over 20 years, and machine learning systems made just slow progress on this task until the appearance of AlphaFold [57], as we have introduced in Chap. 12). Image generation is also a classical task, while DALLE-2 [94] historically achieves a high-resolution generation that can precisely express the meanings of the given text, providing realistic results that humans can hardly tell whether they are real. Further, DALLE-2 can even imitate paintings of a particular style or even create something that is never seen in the real world.[2] This greatly inspires and expands the boundaries of artistic creation and has gained a new wave of AI-generated contents (AIGC).

New Scenarios This refers to the problems that are newly proposed or solved firstly by AI methods. For instance, the characteristic of COVID-19 is a new and significant research topic in recent years. Big models are applied in precision diagnostics, drug repurposing, spread forecasting for Epidemiology, and other problems [105]. Ancient writing research, on the contrary, is an old topic, while AI never played a central role until DeepMind proposes Ithaca [6], which is designed for ancient Greek inscriptions. In this case, the big model can achieve textual restoration, geographical attribution, and chronological attribution. And it helps historians improve their accuracy from 25 to 72% and provide evidence for history and civilization research.

---

[2] Due to the copyright of the generated images, please go to the official website to enjoy them: https://openai.com/dall-e-2/.

In the above examples, the improvement of parameter scales allows greater knowledge capability and generalization toward various domains, which brings a leap in performance. By observing these success cases, we propose the following two prerequisites that an application scenario can turn to big model systems for help: plenty of domain data and documented domain knowledge.

Plenty of Domain Data Big models need more data for training (e.g., 650M training images for DALLE-2). Luckily, the requirement for the data form has been quite lower, and the unlabeled/heterogeneous data can be well utilized by big models. Most of the models follow the basic paradigm of pre-training-fine-tuning and can use large-scale unlabeled data to learn the general comprehension ability of basic elements (e.g., words for a language, pixels for an image) by themselves. From there, it is relatively easy to transfer to any specific downstream domain and solve the tasks with as little supervision as possible. For instance, recent works have explored the necessity and advantage of adopting models pre-trained on natural images for medical image processing, especially when the scale of the downstream dataset is small [111]. Besides, researchers also explore large-scale PTMs for domains with versatile formats of data materials, such as the collaborative processing of chemical and natural language as we have introduced in Chap. 12. Nevertheless, after creating a new scenario, there still must be corresponding domain data to unleash the potential of big models.

Documented Domain Knowledge For fields that humans already have a basic understanding of, the architecture and training strategy of big models can be sophisticatedly designed based on corresponding prior knowledge, and documented knowledge also provides the basic conditions for big models to access and utilize knowledge. In the previous chapters (e.g., Chaps. 9 and 11), we have explained how to conduct knowledge-guided representation learning, such as architecture reformulation and input augmentation methods. In addition, big models have been shown to have behavioral imitation capabilities to access knowledge as human beings. A typical example is WebGPT [89] which can automatically search commonsense and facts to generate more reasonable answers, as we have introduced in cognitive learning. From these examples, we can see that there are more sufficient conditions to realize innovative applications in scenarios with existing domain knowledge bases or ontologies.

Spread the wings of imagination, and we can realize that there are so many fields that big models can dabble in, from sophisticated scientific predictions (such as weather data) to smart home services in our daily life. More innovative applications are waiting for us to explore.

14.2.10 P10: Big Model Systems Accessible to Users

Due to the generalizability of pre-trained models in terms of architecture and capability, big models are expected to become a foundational infrastructure for many information services supported by NLP and AI [9], e.g., search engines, personalized recommendation, and virtual assistants, and domain-specific information organization, e.g., financial, medical, legal, and academic domains.

In particular, recent findings on parameter-efficient delta tuning [30] show that, by keeping a central big model fixed, we can simply design task-specific delta objects to adapt the central model to handle multiple downstream tasks. These breakthroughs indicate a new technique paradigm in NLP, from training a task-specific model for each task separately to stimulating task-specific knowledge scattered in a unified and versatile big model for downstream tasks. Intuitively, with pre-trained big models, our focus is no longer limited to how to learn model parameters for specific tasks but how to stimulate the knowledge of big models to handle specific tasks.

Although the development trend of building unified big models for multiple tasks is clear, it is still not easy for most institutions and individuals to enjoy the power of big models due to the computation and expertise barriers as when have discussed in Chap. 13. We argue that, like the historical successful cases that database management systems (DBMS) are proposed to manage massive data and big data analytics systems (BDS) are proposed to big data mining, it is time for us to build unified management systems of big models, i.e., big model systems (BMS). Similar to DBMS and BDS that store and analyze data in a unified view, we should also design BMS to build and organize big models in a unified view. BMS is expected to provide easy and standardized interfaces for the deployment and application of big models. We should consider the following principles to design BMS accessible to general institutions and individuals.

Data Form and Operation Abstraction of Big Models Both data form abstraction and operation abstraction enable DBMS and BDS to serve as a standard infrastructure in most companies and organizations. Examples of the data form abstraction are tables in relational DBMS (RDBMS) supported by the relational model of data [23] and resilient distributed datasets (RDD) in the Spark BDS [141]. Examples of the operation abstraction are structured query language (SQL) in RDBMS and the *map* and *reduce* functions in the MapReduce BDS [27]. Intuitively, these abstract methods can isolate users and developers of DBMS and BDS. Take DBMS for example: users only consider how to use DBMS to manage data through a series of unified interfaces, without learning how the underlying modules of DBMS that perform data management; developers, by ensuring that the interfaces provided to users remain unchanged, can have more freedom to develop and optimize the underlying modules of DBMS.

We believe big models will also serve as an infrastructure beyond DBMS and BDS for information services. The general-purpose BMS is expected to enable more persons with basic programming skills to use big models. Hence, we should have data form abstraction and operation abstraction specifically designed for big models. BMS relies on data form abstraction to support learning big models from various types of data and provide a unified operation scheme for model manipulation. With the help of prompt learning as a natural language interface between humans and big models [55, 99], we can design high-level and unified programming languages for BMS to manipulate big models and protect big model users from directly interacting with big models by sophisticated deep learning programming.

Efficient Computation and Management of Big Models BMS should support comprehensive management of big models based on many techniques in above-mentioned topics, such as high-performance computing mentioned in P3, parameter-efficient delta tuning mentioned in P4, and safety mentioned in P6. Since the techniques of big models are still developing rapidly, BMS will actively evolve internally in physical implementation by taking advantage of these advances while keeping user interface stable.

We further argue that, with the novel adaptation technique of delta tuning, BMS should manage and schedule central big models as well as massive task-specific delta objects to support the high concurrency of user requests. Hence, we need to design efficient model scheduling manager (MSM) responsible for storing or distributing big models and delta objects in computing devices. There are many real-world scenarios that should be addressed by MSM, such as continual learning and adaptation of big models, efficient scheduling of multiple big models of various sizes and purposes, fault tolerance that can recover from hardware or network failures, and supporting heterogeneous device architectures such as cloud-edge-terminal cooperation.

In summary, we have shown the broad prospects of big models in the above-mentioned nine key problems, and we need big model systems to turn these prospects into reality accessible to general institutions and individuals. The OpenBMB introduced in Chap. 13 can be regarded as our preliminary attempt at building big model systems. As discussed in this key problem, BMS actually brings many open problems with the deployment of big models in the real world, which requires the collaboration of researchers and practitioners from deep learning and AI, high performance computing, software engineering, networking, and edge/cloud computing. We believe an efficient and effective big model system will play an essential role in making the growing capabilities of AI accessible to everyone.

14.3 Summary

In this chapter, we outlook the future of representation learning standing on the new giants of big models in 2023, as the final chapter of the book. We list ten key problems of big models, including theoretical foundation, next-generation architecture, high-performance computing, parameter-efficient delta tuning, controllable generation, safety and ethics, cross-modality, cognitive learning, innovative applications, and big model systems.

Although the summarized problems may be biased by our research experiences, we still hope they can help readers of the book find your interests. Any suggestions and comments are welcome from our community. Let's work together on these exciting topics to contribute novel techniques and applications of AI in the future.

Acknowledgments The contributions of all authors are as follows: Zhiyuan Liu, Yankai Lin, and Maosong Sun designed the overall architecture of this chapter; Zhiyuan Liu initiated the discussions about ten key problems of pre-trained models with all authors and drafted the introduction; Ning Ding drafted P1 and P4, Weize Chen drafted P2, Zhengyan Zhang drafted P3, Shengding Hu drafted P5, Ganqu Cui drafted P6, Yuan Yao drafted P7, Yujia Qin drafted P8, Zheni Zeng drafted P9, and Xu Han drafted P10; Ning Ding, Yankai Lin, and Zhiyuan Liu revised and proofread the chapter.

This is the newly complemented chapter about the outlook for representation learning of the second edition of the book *Representation Learning for Natural Language Processing*. The first edition of the book was published in 2020 [81].

References

1. Samira Abnar, Mostafa Dehghani, Behnam Neyshabur, and Hanie Sedghi. Exploring the limits of large scale pre-training. In *Proceedings of ICLR*, 2021.
2. David H Ackley, Geoffrey E Hinton, and Terrence J Sejnowski. A learning algorithm for boltzmann machines. *Cognitive science*, 9(1):147–169, 1985.
3. Jean-Baptiste Alayrac, Jeff Donahue, Pauline Luc, Antoine Miech, Iain Barr, Yana Hasson, Karel Lenc, Arthur Mensch, Katherine Millican, Malcolm Reynolds, Roman Ring, Eliza Rutherford, Serkan Cabi, Tengda Han, Zhitao Gong, Sina Samangooei, Marianne Monteiro, Jacob Menick, Sebastian Borgeaud, Andrew Brock, Aida Nematzadeh, Sahand Sharifzadeh, Mikolaj Binkowski, Ricardo Barreira, Oriol Vinyals, Andrew Zisserman, and Karen Simonyan. Flamingo: a visual language model for few-shot learning. In *Proceedings of NeurIPS*, 2022.
4. Sanjeev Arora, Hrishikesh Khandeparkar, Mikhail Khodak, Orestis Plevrakis, and Nikunj Saunshi. A theoretical analysis of contrastive unsupervised representation learning. *arXiv preprint arXiv:1902.09229*, 2019.
5. Isaac Asimov. *I, robot*. 1950.
6. Yannis Assael, Thea Sommerschield, Brendan Shillingford, Mahyar Bordbar, John Pavlopoulos, Marita Chatzipanagiotou, Ion Androutsopoulos, Jonathan Prag, and Nando de Freitas. Restoring and attributing ancient texts using deep neural networks. *Nature*, 603(7900):280–283, 2022.
7. Bowen Baker, Ilge Akkaya, Peter Zhokhov, Joost Huizinga, Jie Tang, Adrien Ecoffet, Brandon Houghton, Raul Sampedro, and Jeff Clune. Video pretraining (vpt): Learning to act by watching unlabeled online videos. *arXiv preprint arXiv:2206.11795*, 2022.
8. Emily M Bender and Alexander Koller. Climbing towards NLU: On meaning, form, and understanding in the age of data. In *Proceedings of ACL*, 2020.
9. Rishi Bommasani, Drew A Hudson, Ehsan Adeli, Russ Altman, Simran Arora, Sydney von Arx, Michael S Bernstein, Jeannette Bohg, Antoine Bosselut, Emma Brunskill, et al. On the opportunities and risks of foundation models. *arXiv preprint arXiv:2108.07258*, 2021.
10. Anthony Brohan, Yevgen Chebotar, Chelsea Finn, Karol Hausman, Alexander Herzog, Daniel Ho, Julian Ibarz, Alex Irpan, Eric Jang, Ryan Julian, et al. Do as I can, not as I say: Grounding language in robotic affordances. In *Proceedings of CoRL*, 2022.
11. Michael M Bronstein, Joan Bruna, Taco Cohen, and Petar Veličković. Geometric deep learning: Grids, groups, graphs, geodesics, and gauges. *arXiv preprint arXiv:2104.13478*, 2021.

12. Michael M Bronstein, Joan Bruna, Yann LeCun, Arthur Szlam, and Pierre Vandergheynst. Geometric deep learning: going beyond euclidean data. *IEEE Signal Processing Magazine*, 34(4):18–42, 2017.
13. Tom Brown, Benjamin Mann, Nick Ryder, Melanie Subbiah, Jared D Kaplan, Prafulla Dhariwal, Arvind Neelakantan, Pranav Shyam, Girish Sastry, Amanda Askell, et al. Language models are few-shot learners. In *Proceedings of NeurIPS*, 2020.
14. Nicholas Carlini, Florian Tramer, Eric Wallace, Matthew Jagielski, Ariel Herbert-Voss, Katherine Lee, Adam Roberts, Tom Brown, Dawn Song, Ulfar Erlingsson, et al. Extracting training data from large language models. In *Proceedings of USENIX Security*, 2021.
15. Jean Cea. Conception optimale ou identification de formes, calcul rapide de la dérivée directionnelle de la fonction coût. *ESAIM: Mathematical Modelling and Numerical Analysis-Modélisation Mathématique et Analyse Numérique*, 20(3):371–402, 1986.
16. Ines Chami, Adva Wolf, Da-Cheng Juan, Frederic Sala, Sujith Ravi, and Christopher Ré. Low-dimensional hyperbolic knowledge graph embeddings. In *Proceedings of ACL*, 2020.
17. Zhengping Che, Sanjay Purushotham, Kyunghyun Cho, David Sontag, and Yan Liu. Recurrent neural networks for multivariate time series with missing values. *Scientific reports*, 8(1):1–12, 2018.
18. Jie Chen, Tengfei Ma, and Cao Xiao. FastGCN: Fast learning with graph convolutional networks via importance sampling. In *Proceedings of ICLR*, 2018.
19. Mark Chen, Jerry Tworek, Heewoo Jun, Qiming Yuan, Henrique Ponde de Oliveira Pinto, Jared Kaplan, Harri Edwards, Yuri Burda, Nicholas Joseph, Greg Brockman, et al. Evaluating large language models trained on code. *arXiv preprint arXiv:2107.03374*, 2021.
20. Ricky TQ Chen, Yulia Rubanova, Jesse Bettencourt, and David K Duvenaud. Neural ordinary differential equations. In *Proceedings of NeurIPS*, 2018.
21. Tianqi Chen, Bing Xu, Chiyuan Zhang, and Carlos Guestrin. Training deep nets with sublinear memory cost. *arXiv preprint arXiv:1604.06174*, 2016.
22. Weize Chen, Xu Han, Yankai Lin, Hexu Zhao, Zhiyuan Liu, Peng Li, Maosong Sun, and Jie Zhou. Fully hyperbolic neural networks. In *Proceedings of ACL*, 2022.
23. Edgar F Codd. A relational model of data for large shared data banks. *Communications of the ACM*, 13(6):377–387, 1970.
24. Allan Dafoe. Ai governance: a research agenda. *Governance of AI Program, Future of Humanity Institute, University of Oxford: Oxford, UK*, 1442:1443, 2018.
25. Kahneman Daniel. Thinking, fast and slow, 2017.
26. Mike Davies, Narayan Srinivasa, Tsung-Han Lin, Gautham Chinya, Yongqiang Cao, Sri Harsha Choday, Georgios Dimou, Prasad Joshi, Nabil Imam, Shweta Jain, et al. Loihi: A neuromorphic manycore processor with on-chip learning. *Ieee Micro*, 38(1):82–99, 2018.
27. Jeffrey Dean and Sanjay Ghemawat. Mapreduce: simplified data processing on large clusters. *Communications of the ACM*, 51(1):107–113, 2008.
28. Jia Deng, Wei Dong, Richard Socher, Li-Jia Li, Kai Li, and Li Fei-Fei. ImageNet: A large-scale hierarchical image database. In *Proceedings of CVPR*, 2009.
29. Jacob Devlin, Ming-Wei Chang, Kenton Lee, and Kristina Toutanova. BERT: pre-training of deep bidirectional transformers for language understanding. In *Proceedings of NAACL-HLT*, 2019.
30. Ning Ding, Yujia Qin, Guang Yang, Fuchao Wei, Zonghan Yang, Yusheng Su, Shengding Hu, Yulin Chen, Chi-Min Chan, Weize Chen, et al. Delta tuning: A comprehensive study of parameter efficient methods for pre-trained language models. *arXiv preprint arXiv:2203.06904*, 2022.
31. Linhao Dong, Shuang Xu, and Bo Xu. Speech-Transformer: a no-recurrence sequence-to-sequence model for speech recognition. In *Proceedings of ICASSP*, 2018.
32. Alexey Dosovitskiy, Lucas Beyer, Alexander Kolesnikov, Dirk Weissenborn, Xiaohua Zhai, Thomas Unterthiner, Mostafa Dehghani, Matthias Minderer, Georg Heigold, Sylvain Gelly, et al. An image is worth 16x16 words: Transformers for image recognition at scale. In *Proceedings of ICLR*, 2021.

33. David Eigen, Marc'Aurelio Ranzato, and Ilya Sutskever. Learning factored representations in a deep mixture of experts. *arXiv preprint arXiv:1312.4314*, 2013.
34. Nelson Elhage, Neel Nanda, Catherine Olsson, Tom Henighan, Nicholas Joseph, Ben Mann, Amanda Askell, Yuntao Bai, Anna Chen, Tom Conerly, Nova DasSarma, Dawn Drain, Deep Ganguli, Zac Hatfield-Dodds, Danny Hernandez, Andy Jones, Jackson Kernion, Liane Lovitt, Kamal Ndousse, Dario Amodei, Tom Brown, Jack Clark, Jared Kaplan, Sam McCandlish, and Chris Olah. A mathematical framework for transformer circuits, 2021.
35. Andre Esteva, Alexandre Robicquet, Bharath Ramsundar, Volodymyr Kuleshov, Mark DePristo, Katherine Chou, Claire Cui, Greg Corrado, Sebastian Thrun, and Jeff Dean. A guide to deep learning in healthcare. *Nature medicine*, 25(1):24–29, 2019.
36. Angela Fan, Mike Lewis, and Yann Dauphin. Hierarchical neural story generation. In *Proceedings of ACL*, 2018.
37. Chris Finlay, Jörn-Henrik Jacobsen, Levon Nurbekyan, and Adam Oberman. How to train your neural ODE: the world of jacobian and kinetic regularization. In *Proceedings of ICML*, 2020.
38. Wulfram Gerstner and Werner M Kistler. *Spiking neuron models: Single neurons, populations, plasticity*. Cambridge university press, 2002.
39. Walter Goodwin, Sagar Vaze, Ioannis Havoutis, and Ingmar Posner. Semantically grounded object matching for robust robotic scene rearrangement. In *Proceedings of ICRA*, 2022.
40. Kelvin Guu, Kenton Lee, Zora Tung, Panupong Pasupat, and Ming-Wei Chang. Realm: Retrieval-augmented language model pre-training. In *Proceedings of ICML*, 2020.
41. Song Han, Huizi Mao, and William J. Dally. Deep compression: Compressing deep neural network with pruning, trained quantization and huffman coding. In *Proceedings of ICLR*, 2016.
42. Xu Han, Zhengyan Zhang, Ning Ding, Yuxian Gu, Xiao Liu, Yuqi Huo, Jiezhong Qiu, Yuan Yao, Ao Zhang, Liang Zhang, et al. Pre-trained models: Past, present and future. *AI Open*, 2021.
43. Jeff Z HaoChen, Colin Wei, Adrien Gaidon, and Tengyu Ma. Provable guarantees for self-supervised deep learning with spectral contrastive loss. In *Proceedings of NeurIPS*, 2021.
44. Kaiming He, Xinlei Chen, Saining Xie, Yanghao Li, Piotr Dollár, and Ross Girshick. Masked autoencoders are scalable vision learners. In *Proceedings of CVPR*, 2022.
45. Kaiming He, Xiangyu Zhang, Shaoqing Ren, and Jian Sun. Deep residual learning for image recognition. In *Proceedings of CVPR*, 2016.
46. Geoffrey Hinton, Oriol Vinyals, Jeff Dean, et al. Distilling the knowledge in a neural network. *arXiv preprint arXiv:1503.02531*, 2(7), 2015.
47. Sepp Hochreiter and Jürgen Schmidhuber. Long short-term memory. *Neural Computation*, 1997.
48. John J Hopfield. Neural networks and physical systems with emergent collective computational abilities. *Proceedings of the national academy of sciences*, 79(8):2554–2558, 1982.
49. Neil Houlsby, Andrei Giurgiu, Stanislaw Jastrzebski, Bruna Morrone, Quentin De Laroussilhe, Andrea Gesmundo, Mona Attariyan, and Sylvain Gelly. Parameter-efficient transfer learning for nlp. In *Proceedings of ICML*, 2019.
50. Edward J Hu, Phillip Wallis, Zeyuan Allen-Zhu, Yuanzhi Li, Shean Wang, Lu Wang, Weizhu Chen, et al. LoRA: Low-rank adaptation of large language models. In *Proceedings of ICLR*, 2021.
51. Zhiting Hu, Zichao Yang, Xiaodan Liang, Ruslan Salakhutdinov, and Eric P. Xing. Toward controlled generation of text. In *Proceedings of ICML*, 2017.
52. Yanping Huang, Youlong Cheng, Ankur Bapna, Orhan Firat, Dehao Chen, Mia Chen, HyoukJoong Lee, Jiquan Ngiam, Quoc V Le, Yonghui Wu, et al. GPipe: Efficient training of giant neural networks using pipeline parallelism. In *Proceedings of NeurIPS*, 2019.
53. David H Hubel and Torsten N Wiesel. Receptive fields of single neurones in the cat's striate cortex. *The Journal of physiology*, 148(3):574, 1959.
54. RA Jacobs, MI Jordan, SJ Nowlan, and GE Hinton. Adaptive mixtures of local experts. *Neural Computation*, 3(1):79–87, 1991.

55. Ellen Jiang, Kristen Olson, Edwin Toh, Alejandra Molina, Aaron Michael Donsbach, Michael Terry, and Carrie Jun Cai. Prompt-based prototyping with large language models. 2022.
56. Yimin Jiang, Yibo Zhu, Chang Lan, Bairen Yi, Yong Cui, and Chuanxiong Guo. A unified architecture for accelerating distributed DNN training in heterogeneous GPU/CPU clusters. In *Proceedings of OSDI*, 2020.
57. John Jumper, Richard Evans, Alexander Pritzel, Tim Green, Michael Figurnov, Olaf Ronneberger, Kathryn Tunyasuvunakool, Russ Bates, Augustin Žídek, Anna Potapenko, et al. Highly accurate protein structure prediction with AlphaFold. *Nature*, 596(7873):583–589, 2021.
58. Nitish Shirish Keskar, Bryan McCann, Lav R Varshney, Caiming Xiong, and Richard Socher. Ctrl: A conditional transformer language model for controllable generation. *arXiv preprint arXiv:1909.05858*, 2019.
59. Muhammad Khalifa, Hady Elsahar, and Marc Dymetman. A distributional approach to controlled text generation. In *Proceedings of ICLR*, 2021.
60. Saeed Reza Kheradpisheh, Mohammad Ganjtabesh, Simon J Thorpe, and Timothée Masquelier. Stdp-based spiking deep convolutional neural networks for object recognition. *Neural Networks*, 99:56–67, 2018.
61. Patrick Kidger, James Foster, Xuechen Li, and Terry J Lyons. Neural SDEs as infinite-dimensional gans. In *Proceedings of ICML*, 2021.
62. Marisa Kirisame, Steven Lyubomirsky, Altan Haan, Jennifer Brennan, Mike He, Jared Roesch, Tianqi Chen, and Zachary Tatlock. Dynamic tensor rematerialization. In *Proceedings of ICLR*, 2020.
63. Teuvo Kohonen. Self-organized formation of topologically correct feature maps. *Biological cybernetics*, 43(1):59–69, 1982.
64. Ben Krause, Akhilesh Deepak Gotmare, Bryan McCann, Nitish Shirish Keskar, Shafiq Joty, Richard Socher, and Nazneen Fatema Rajani. Gedi: Generative discriminator guided sequence generation. In *Findings of EMNLP*, 2021.
65. Ranjay Krishna, Donsuk Lee, Li Fei-Fei, and Michael S Bernstein. Socially situated artificial intelligence enables learning from human interaction. *The National Academy of Sciences*, 119(39):e2115730119, 2022.
66. Alex Krizhevsky et al. Learning multiple layers of features from tiny images. 2009.
67. Alex Krizhevsky, Ilya Sutskever, and Geoffrey E Hinton. ImageNet classification with deep convolutional neural networks. In *Proceedings of NeurIPS*, 2012.
68. Guillaume Lample, Marie-Anne Lachaux, Thibaut Lavril, Xavier Martinet, Amaury Hayat, Gabriel Ebner, Aurélien Rodriguez, and Timothée Lacroix. Hypertree proof search for neural theorem proving. *arXiv preprint arXiv:2205.11491*, 2022.
69. Yann LeCun. A path towards autonomous machine intelligence. *Openreview*, 2022.
70. Yann LeCun, Léon Bottou, Yoshua Bengio, and Patrick Haffner. Gradient-based learning applied to document recognition. In *Proceedings of the IEEE*, 1998.
71. Brian Lester, Rami Al-Rfou, and Noah Constant. The power of scale for parameter-efficient prompt tuning. In *Proceedings of EMNLP*, 2021.
72. Patrick Lewis, Ethan Perez, Aleksandra Piktus, Fabio Petroni, Vladimir Karpukhin, Naman Goyal, Heinrich Küttler, Mike Lewis, Wen-tau Yih, Tim Rocktäschel, et al. Retrieval-augmented generation for knowledge-intensive nlp tasks. In *Proceedings of NeurIPS*, 2020.
73. Hao Li, Zheng Xu, Gavin Taylor, Christoph Studer, and Tom Goldstein. Visualizing the loss landscape of neural nets. In *Proceedings of NeurIPS*, 2018.
74. Junnan Li, Ramprasaath Selvaraju, Akhilesh Gotmare, Shafiq Joty, Caiming Xiong, and Steven Chu Hong Hoi. Align before fuse: Vision and language representation learning with momentum distillation. In *Proceedings of NeurIPS*, 2021.
75. Xiang Lisa Li and Percy Liang. Prefix-tuning: Optimizing continuous prompts for generation. In *Proceedings of ACL-IJCNLP*, 2021.
76. Xuechen Li, Ting-Kam Leonard Wong, Ricky TQ Chen, and David K Duvenaud. Scalable gradients and variational inference for stochastic differential equations. In *Proceedings of AABI*, 2020.

77. Zheng Li, Zijian Wang, Ming Tan, Ramesh Nallapati, Parminder Bhatia, Andrew Arnold, Bing Xiang, and Dan Roth. DQ-BART: Efficient sequence-to-sequence model via joint distillation and quantization. In *Proceedings of ACL*, 2022.

78. Alisa Liu, Maarten Sap, Ximing Lu, Swabha Swayamdipta, Chandra Bhagavatula, Noah A. Smith, and Yejin Choi. DExperts: Decoding-time controlled text generation with experts and anti-experts. In *Proceedings of ACL*, 2021.

79. Jing Liu, Xinxin Zhu, Fei Liu, Longteng Guo, Zijia Zhao, Mingzhen Sun, Weining Wang, Hanqing Lu, Shiyu Zhou, Jiajun Zhang, et al. OPT: omni-perception pre-trainer for cross-modal understanding and generation. *arXiv preprint arXiv:2107.00249*, 2021.

80. Ze Liu, Yutong Lin, Yue Cao, Han Hu, Yixuan Wei, Zheng Zhang, Stephen Lin, and Baining Guo. Swin transformer: Hierarchical vision transformer using shifted windows. In *Proceedings of ICCV*, 2021.

81. Zhiyuan Liu, Yankai Lin, and Maosong Sun. *Representation Learning for Natural Language Processing*. Springer, 2020.

82. Yiping Lu, Aoxiao Zhong, Quanzheng Li, and Bin Dong. Beyond finite layer neural networks: Bridging deep architectures and numerical differential equations. In *Proceedings of ICML*, 2018.

83. Yi Ma, Doris Tsao, and Heung-Yeung Shum. On the principles of parsimony and self-consistency for the emergence of intelligence. *Frontiers of Information Technology & Electronic Engineering*, pages 1–26, 2022.

84. Emile Mathieu and Maximilian Nickel. Riemannian continuous normalizing flows. In *Proceedings of NeurIPS*, 2020.

85. Jacob Menick, Maja Trebacz, Vladimir Mikulik, John Aslanides, Francis Song, Martin Chadwick, Mia Glaese, Susannah Young, Lucy Campbell-Gillingham, Geoffrey Irving, et al. Teaching language models to support answers with verified quotes. *arXiv preprint arXiv:2203.11147*, 2022.

86. Jeff Mitchell and Mirella Lapata. Composition in distributional models of semantics. *Cognitive science*, 34(8):1388–1429, 2010.

87. Milad Mozafari, Mohammad Ganjtabesh, Abbas Nowzari-Dalini, Simon J Thorpe, and Timothée Masquelier. Bio-inspired digit recognition using reward-modulated spike-timing-dependent plasticity in deep convolutional networks. *Pattern recognition*, 94:87–95, 2019.

88. Basil Mustafa, Carlos Riquelme, Joan Puigcerver, Rodolphe Jenatton, and Neil Houlsby. Multimodal contrastive learning with limoe: the language-image mixture of experts. *arXiv preprint arXiv:2206.02770*, 2022.

89. Reiichiro Nakano, Jacob Hilton, Suchir Balaji, Jeff Wu, Long Ouyang, Christina Kim, Christopher Hesse, Shantanu Jain, Vineet Kosaraju, William Saunders, et al. WebGPT: Browser-assisted question-answering with human feedback. *arXiv preprint arXiv:2112.09332*, 2021.

90. Deepak Narayanan, Aaron Harlap, Amar Phanishayee, Vivek Seshadri, Nikhil R Devanur, Gregory R Ganger, Phillip B Gibbons, and Matei Zaharia. PipeDream: generalized pipeline parallelism for DNN training. In *Proceedings of SOSP*, 2019.

91. Maximillian Nickel and Douwe Kiela. Poincare embeddings for learning hierarchical representations. In *Proceedings of NeurIPS*, 2017.

92. Jing Qian, Li Dong, Yelong Shen, Furu Wei, and Weizhu Chen. Controllable natural language generation with contrastive prefixes. In *Findings of ACL*, 2022.

93. Colin Raffel, Noam Shazeer, Adam Roberts, Katherine Lee, Sharan Narang, Michael Matena, Yanqi Zhou, Wei Li, and Peter J Liu. Exploring the limits of transfer learning with a unified text-to-text transformer. *Journal of Machine Learning Research*, 21:1–67, 2020.

94. Aditya Ramesh, Prafulla Dhariwal, Alex Nichol, Casey Chu, and Mark Chen. Hierarchical text-conditional image generation with CLIP latents. *arXiv preprint arXiv:2204.06125*, 2022.

95. Jeff Rasley, Samyam Rajbhandari, Olatunji Ruwase, and Yuxiong He. DeepSpeed: System optimizations enable training deep learning models with over 100 billion parameters. In *Proceedings of KDD*, 2020.

96. Tiernan Ray. Meta's ai guru lecun: Most of today's ai approaches will never lead to true intelligence, 2022.
97. Scott Reed, Konrad Zolna, Emilio Parisotto, Sergio Gomez Colmenarejo, Alexander Novikov, Gabriel Barth-Maron, Mai Gimenez, Yury Sulsky, Jackie Kay, Jost Tobias Springenberg, et al. A generalist agent. *arXiv preprint arXiv:2205.06175*, 2022.
98. Shaoqing Ren, Kaiming He, Ross Girshick, and Jian Sun. Faster R-CNN: Towards real-time object detection with region proposal networks. In *Proceedings of NeurIPS*, 2015.
99. Laria Reynolds and Kyle McDonell. Prompt programming for large language models: Beyond the few-shot paradigm. In *Proceedings of CHI*, 2021.
100. Minsoo Rhu, Natalia Gimelshein, Jason Clemons, Arslan Zulfiqar, and Stephen W Keckler. vDNN: Virtualized deep neural networks for scalable, memory-efficient neural network design. In *Proceedings of MICRO*, 2016.
101. Yulia Rubanova, Ricky TQ Chen, and David K Duvenaud. Latent ordinary differential equations for irregularly-sampled time series. In *Proceedings of NeurIPS*, 2019.
102. Frederic Sala, Chris De Sa, Albert Gu, and Christopher Ré. Representation tradeoffs for hyperbolic embeddings. In *Proceedings of ICML*, 2018.
103. Roger N Shepard. Toward a universal law of generalization for psychological science. *Science*, 237(4820):1317–1323, 1987.
104. Mohammad Shoeybi, Mostofa Patwary, Raul Puri, Patrick LeGresley, Jared Casper, and Bryan Catanzaro. Megatron-LM: Training multi-billion parameter language models using model parallelism. *arXiv preprint arXiv:1909.08053*, 2019.
105. Connor Shorten, Taghi M Khoshgoftaar, and Borko Furht. Deep learning applications for covid-19. *Journal of Big Data*, 8(1):1–54, 2021.
106. Mohit Shridhar, Lucas Manuelli, and Dieter Fox. Cliport: What and where pathways for robotic manipulation. In *Proceedings of CoRL*, 2022.
107. SB Shriram, Anshuj Garg, and Purushottam Kulkarni. Dynamic memory management for GPU-based training of deep neural networks. In *Proceedings of IPDPS*, 2019.
108. Uriel Singer, Adam Polyak, Thomas Hayes, Xi Yin, Jie An, Songyang Zhang, Qiyuan Hu, Harry Yang, Oron Ashual, Oran Gafni, et al. Make-a-video: Text-to-video generation without text-video data. *arXiv preprint arXiv:2209.14792*, 2022.
109. Linfeng Song, Ante Wang, Jinsong Su, Yue Zhang, Kun Xu, Yubin Ge, and Dong Yu. Structural information preserving for graph-to-text generation. In *Proceedings of ACL*, 2020.
110. Qingyuan Song, Weiping Fu, Wen Wang, Yuan Sun, Denggui Wang, and Jincao Zhou. Quantum decision making in automatic driving. *Scientific reports*, 12(1):1–15, 2022.
111. Nima Tajbakhsh, Jae Y Shin, Suryakanth R Gurudu, R Todd Hurst, Christopher B Kendall, Michael B Gotway, and Jianming Liang. Convolutional neural networks for medical image analysis: Full training or fine tuning? *IEEE transactions on medical imaging*, 35(5):1299–1312, 2016.
112. Guangzhi Tang, Neelesh Kumar, Raymond Yoo, and Konstantinos Michmizos. Deep reinforcement learning with population-coded spiking neural network for continuous control. In *Proceedings of CoRL*, 2021.
113. David Tarditi, Sidd Puri, and Jose Oglesby. Accelerator: using data parallelism to program GPUs for general-purpose uses. *ACM SIGPLAN Notices*, 41(11):325–335, 2006.
114. Michel Thiebaut de Schotten and Stephanie J Forkel. The emergent properties of the connected brain. *Science*, 378(6619):505–510, 2022.
115. Richard F Thompson and Jeansok J Kim. Memory systems in the brain and localization of a memory. *PNAS*, 93(24):13438–13444, 1996.
116. Christopher Tosh, Akshay Krishnamurthy, and Daniel Hsu. Contrastive learning, multi-view redundancy, and linear models. In *Proceedings of ALT*, 2021.
117. Florian Tramer, Nicholas Carlini, Wieland Brendel, and Aleksander Madry. On adaptive attacks to adversarial example defenses. In *Proceedings of NeurIPS*, 2020.
118. Ashish Vaswani, Noam Shazeer, Niki Parmar, Llion Jones, Jakob Uszkoreit, Aidan N Gomez, and Lukasz Kaiser. Attention is all you need. In *Proceedings of NeurIPS*, 2017.

119. Eric Wallace, Shi Feng, Nikhil Kandpal, Matt Gardner, and Sameer Singh. Universal adversarial triggers for attacking and analyzing nlp. In *Proceedings of EMNLP-IJCNLP*, 2019.
120. Alex Wang, Amanpreet Singh, Julian Michael, Felix Hill, Omer Levy, and Samuel Bowman. GLUE: A multi-task benchmark and analysis platform for natural language understanding. In *Proceedings of EMNLP Workshop*, 2018.
121. Hanrui Wang, Zhanghao Wu, Zhijian Liu, Han Cai, Ligeng Zhu, Chuang Gan, and Song Han. Hat: Hardware-aware transformers for efficient natural language processing. In *Proceedings of ACL*, 2020.
122. Haojie Wang, Jidong Zhai, Mingyu Gao, Zixuan Ma, Shizhi Tang, Liyan Zheng, Yuanzhi Li, Kaiyuan Rong, Yuanyong Chen, and Zhihao Jia. PET: Optimizing tensor programs with partially equivalent transformations and automated corrections. In *Proceedings of OSDI*, 2021.
123. Peng Wang, An Yang, Rui Men, Junyang Lin, Shuai Bai, Zhikang Li, Jianxin Ma, Chang Zhou, Jingren Zhou, and Hongxia Yang. OFA: Unifying architectures, tasks, and modalities through a simple sequence-to-sequence learning framework. In *Proceedings of ICML*, 2022.
124. Wenhui Wang, Hangbo Bao, Li Dong, Johan Bjorck, Zhiliang Peng, Qiang Liu, Kriti Aggarwal, Owais Khan Mohammed, Saksham Singhal, Subhojit Som, et al. Image as a foreign language: BEiT pretraining for all vision and vision-language tasks. *arXiv preprint arXiv:2208.10442*, 2022.
125. Zhehui Wang, Tao Luo, Rick Siow Mong Goh, and Joey Tianyi Zhou. Edcompress: Energy-aware model compression for dataflows. *IEEE Transactions on Neural Networks and Learning Systems*, 2022.
126. Zirui Wang, Jiahui Yu, Adams Wei Yu, Zihang Dai, Yulia Tsvetkov, and Yuan Cao. SimVLM: Simple visual language model pretraining with weak supervision. In *Proceedings of ICLR*, 2021.
127. Colin Wei, Kendrick Shen, Yining Chen, and Tengyu Ma. Theoretical analysis of self-training with deep networks on unlabeled data. *arXiv preprint arXiv:2010.03622*, 2020.
128. Colin Wei, Sang Michael Xie, and Tengyu Ma. Why do pretrained language models help in downstream tasks? an analysis of head and prompt tuning. In *Proceedings of NeurIPS*, 2021.
129. Jason Wei, Yi Tay, Rishi Bommasani, Colin Raffel, Barret Zoph, Sebastian Borgeaud, Dani Yogatama, Maarten Bosma, Denny Zhou, Donald Metzler, et al. Emergent abilities of large language models. *arXiv preprint arXiv:2206.07682*, 2022.
130. Jason Wei, Xuezhi Wang, Dale Schuurmans, Maarten Bosma, Ed Chi, Quoc Le, and Denny Zhou. Chain of thought prompting elicits reasoning in large language models. *arXiv preprint arXiv:2201.11903*, 2022.
131. Laura Weidinger, John Mellor, Maribeth Rauh, Conor Griffin, Jonathan Uesato, Po-Sen Huang, Myra Cheng, Mia Glaese, Borja Balle, Atoosa Kasirzadeh, et al. Ethical and social risks of harm from language models. *arXiv preprint arXiv:2112.04359*, 2021.
132. E Weinan. A proposal on machine learning via dynamical systems. *Communications in Mathematics and Statistics*, 1(5):1–11, 2017.
133. Margaret Wilson. Six views of embodied cognition. *Psychonomic bulletin & review*, 9(4):625–636, 2002.
134. Sang Michael Xie, Aditi Raghunathan, Percy Liang, and Tengyu Ma. An explanation of in-context learning as implicit bayesian inference. *arXiv preprint arXiv:2111.02080*, 2021.
135. Ji Xin, Raphael Tang, Jaejun Lee, Yaoliang Yu, and Jimmy Lin. Deebert: Dynamic early exiting for accelerating bert inference. In *Proceedings of ACL*, 2020.
136. Ziqing Yang, Yiming Cui, and Zhigang Chen. TextPruner: A model pruning toolkit for pretrained language models. In *Proceedings of ACL Demonstrations*, 2022.
137. Shunyu Yao, Howard Chen, John Yang, and Karthik Narasimhan. Webshop: Towards scalable real-world web interaction with grounded language agents. *arXiv preprint arXiv:2207.01206*, 2022.

138. Yuan Yao, Qianyu Chen, Ao Zhang, Wei Ji, Zhiyuan Liu, Tat-Seng Chua, and Maosong Sun. PEVL: Position-enhanced pre-training and prompt tuning for vision-language models. In *Proceedings of EMNLP*, 2022.
139. Yuan Yao, Ao Zhang, Zhengyan Zhang, Zhiyuan Liu, Tat-Seng Chua, and Maosong Sun. CPT: Colorful prompt tuning for pre-trained vision-language models. *arXiv preprint arXiv:2109.11797*, 2021.
140. Anthony Zador, Blake Richards, Bence Ölveczky, Sean Escola, Yoshua Bengio, Kwabena Boahen, Matthew Botvinick, Dmitri Chklovskii, Anne Churchland, Claudia Clopath, et al. Toward next-generation artificial intelligence: Catalyzing the neuroai revolution. *arXiv preprint arXiv:2210.08340*, 2022.
141. Matei Zaharia, Reynold S Xin, Patrick Wendell, Tathagata Das, Michael Armbrust, Ankur Dave, Xiangrui Meng, Josh Rosen, Shivaram Venkataraman, Michael J Franklin, et al. Apache spark: a unified engine for big data processing. *Communications of the ACM*, 59(11):56–65, 2016.
142. Elad Ben Zaken, Shauli Ravfogel, and Yoav Goldberg. Bitfit: Simple parameter-efficient fine-tuning for transformer-based masked language-models. In *Proceedings of ACL*, 2021.
143. Rowan Zellers, Yonatan Bisk, Ali Farhadi, and Yejin Choi. From recognition to cognition: Visual commonsense reasoning. In *Proceedings of CVPR*, 2019.
144. Rowan Zellers, Ari Holtzman, Hannah Rashkin, Yonatan Bisk, Ali Farhadi, Franziska Roesner, and Yejin Choi. Defending against neural fake news. In *Proceedings of NeurIPS*, 2019.
145. Saizheng Zhang, Emily Dinan, Jack Urbanek, Arthur Szlam, Douwe Kiela, and Jason Weston. Personalizing dialogue agents: I have a dog, do you have pets too? In *Proceedings of ACL*, 2018.
146. Tianyi Zhang*, Varsha Kishore*, Felix Wu*, Kilian Q. Weinberger, and Yoav Artzi. Bertscore: Evaluating text generation with bert. In *Proceedings of ICLR*, 2020.
147. Yizhe Zhang, Guoyin Wang, Chunyuan Li, Zhe Gan, Chris Brockett, and Bill Dolan. POINTER: Constrained progressive text generation via insertion-based generative pre-training. In *Proceedings of EMNLP*, 2020.
148. Zhengyan Zhang, Baitao Gong, Yingfa Chen, Xu Han, Guoyang Zeng, Weilin Zhao, Yanxu Chen, Zhiyuan Liu, and Maosong Sun. BMCook: A task-agnostic compression toolkit for big models. In *Proceedings of EMNLP: System Demonstrations*, 2022.
149. Ningxin Zheng, Bin Lin, Quanlu Zhang, Lingxiao Ma, Yuqing Yang, Fan Yang, Yang Wang, Mao Yang, and Lidong Zhou. SparTA: Deep-learning model sparsity via tensor-with-sparsity-attribute. In *Proceedings of OSDI*, 2022.
150. Kaiyang Zhou, Jingkang Yang, Chen Change Loy, and Ziwei Liu. Learning to prompt for vision-language models. *International Journal of Computer Vision*, 130(9):2337–2348, 2022.

Printed in the United States
by Baker & Taylor Publisher Services